Comparative vertebrate endocrinology

Comparative vertebrate endocrinology

P. J. BENTLEY

Mount Sinai School of Medicine of
The City University of New York

SECOND EDITION

CAMBRIDGE UNIVERSITY PRESS

CAMBRIDGE

LONDON   NEW YORK   NEW ROCHELLE

MELBOURNE   SYDNEY

QP187

B46

1982

Published by the Press Syndicate of the University of Cambridge
The Pitt Building, Trumpington Street, Cambridge CB2 1RP
32 East 57th Street, New York, NY 10022, USA
296 Beaconsfield Parade, Middle Park, Melbourne 3206, Australia

First published 1976
Reprinted 1980
Second edition 1982

Printed in the United States of America

Library of Congress Cataloging in Publication Data
Bentley, P. J.
Comparative vertebrate endocrinology.
Bibliography: p.
Includes index.
1. Endocrinology, Comparative.
2. Vertebrates – Physiology. I. Title.
[DNLM: 1. Endocrine glands – Physiology.
2. Physiology, Comparative. 3. Vertebrates –
Physiology. QP187.B477c]
QP187.B46 1982 596'.0142   82-1205
ISBN 0 521 24653 9 hard covers   AACR2
ISBN 0 521 28878 9 paperback

For Hans Heller and Harry Waring
who introduced me to
comparative endocrinology

Contents

Preface to
the second edition

It is over 6 years since the first edition of this book went to press. Interest in comparative endocrinology has not waned in that time, as shown by the steady stream of papers and the organization of meetings and symposia on this subject. Several new hormones have been identified and described in the interim. Information about the synthesis of proteins that act as prohormones has provided enlightenment about the existence of more "hormone families" with consequent speculation about their evolution. An increased utilization of radioimmunoassays and immunohistochemistry has promoted many of these advances. There has also been an increased appreciation of commonalities of the endocrine and nervous systems, as described in the discipline of neuroendocrinology. However, because the basic information about the endocrine system has not really changed, it has been unnecessary to alter significantly the conclusions at the end of each chapter.

In view of the great expansion of the literature, the preparation of this edition has been especially challenging. I have generally refrained from substituting new references for old ones, a practice that would ignore the seniority of discoveries and distort the historical perspective of the subject. There are thus many more references in the text. I hope that this does not distract the students for whom this book is primarily intended. They should "read around" the references and use them as a source if necessary. More senior readers may find the expanded bibliography more useful. Finally, I would like to apologize to the many endocrinologists whom it has not been possible to quote but without whose discoveries our knowledge of this subject would be much poorer.

P. J. Bentley

New York
February 1982

Preface to
the first edition

This book has been written primarily for use as a textbook by undergraduate, as well as graduate, students. It is hoped that it may serve as a basis for course work in comparative endocrinology and also as an auxiliary text to aid in the teaching of comparative animal physiology. In order to gain the most from this book, the reader should have a basic knowledge of zoology and animal physiology. I have nevertheless attempted to put the endocrinology that is described into a broader biological framework by relating it to the animal's physiology, ecology, and evolutionary background. This is one of the reasons why I have departed from the more usual format of previous textbooks in this area, which generally deal with each endocrine gland in succession, chapter by chapter. Instead, I have attempted to describe certain broad and basic biological processes, the functioning of which is often coordinated by the secretion from several endocrine glands.

No attempt has been made to describe invertebrate endocrinology, as the rapid growth of this area really justifies a separate textbook. The book by K. G. Highnam and L. Hill (*Comparative Endocrinology of the Invertebrates*, Elsevier: Amsterdam, 1970) deals admirably with this subject.

It has not been possible in a book of this nature to give a complete list of original references. There are far too many of these, and many of the earlier observations are already a part of the "classical literature." Instead, I have attempted to refer the reader to more recent papers and reviews that contain references to the material described and can act as useful "starting points" for the students who wish to study the subject further. In order to keep abreast of developments in the various subject areas described, the current literature should be consulted. The principal journals where papers on these subjects are published are *General and Comparative Endocrinology, Journal of Endocrinology, Endocrinology*, and *Comparative Biochemistry and Physiology*. In addition, many papers appear in the standard physiological journals, especially *Journal of Physiology* and *American Journal of Physiology*.

P.J.B.

Mount Sinai School of Medicine of
The City University of New York
September 1974

Some commonly used abbreviations in endocrinology

ACTH	adrenocorticotropic hormone
ADH	antidiuretic hormone
AMP	adenosine 3′,5′-monophosphate
AVP	arginine–vasopressin
CBG	cortisol-binding globulin
COMT	catechol-O-methyltransferase
CRH	corticotropin-releasing hormone (= CRF)
CT	calcitonin
FSH	follicle-stimulating hormone
GH	growth hormone
Gn-RH	gonadotropin-releasing hormone (= LHRH)
hCG	human chorionic gonadotropin
hCS	human chorionic somatomammotropin (= hPL)
HIOMT	hydroxyindole-O-methyltransferase
hPL	human placental lactogen
HTF	heterothyrotropic factor
-IF	- inhibiting-factor
-IH	- inhibiting-hormone
LH	luteinizing hormone (=ICSH)
LVP	lysine–vasopressin
MAO	monoamine oxidase
MI	melanophore index
MSH	melanocyte- (or melanophore-)stimulating hormone
MRH	melanocyte-stimulating hormone-releasing hormone
PNMT	phenylethanolamine-N-methyltransferase
P-	prolactin (prefix as in P-RH)
PTH	parathyroid hormone
-RF	- releasing factor
-RH	- releasing hormone
-R-IH	- release inhibiting hormone

SHBG	sex hormone-binding globulin
T_3	triiodothyronine
T_4	tetraiodothyronine ($=$ thyroxine)
TBG	thyroid hormone-binding globulin
TRH	thyrotropin-releasing hormone
TSH	thyroid-stimulating hormone ($=$ thyrotropin)

1. Introduction

This book describes a method of transferring information within vertebrates. Such communication is necessary in order to coordinate physiological processes with each other and to the happenings in the external environment. Even unicellular organisms synchronize their various internal life processes. In such small creatures, however, local accumulations of metabolites may exert a direct control on biochemical reactions, whereas external stimuli have relatively widespread effects so that specialized pathways for communication may not be as necessary. Thus, when the distances involved are short, physical processes such as conduction, convection, and diffusion may be adequate for the integration of the physiological processes. Nevertheless, even unicellular organisms possess specific coordinating systems such as that seen in the protozoan *Tetrahymena* (Blum, 1967), which possesses epinephrine. This hormone has similar metabolic actions in this protozoan to those that it has in vertebrates.

The problems of communication and coordination are greater in multicellular than in unicellular organisms. There are several reasons for this, especially their larger size. As the linear distances between the different parts of an animal increase, simple physical communications become relatively slower and less precise, and so not as effective. In multicellular organisms, the cells are usually specialized and perform different functions that, in combination, are essential for the animal's life. Thus, some tissues may be concerned with the formation of reproductive germ cells, several others with the preparation of suitable nutritive materials, and yet others with building morphological structures. The ultimate successful completion of these processes will be determined by the effectiveness of the communication between the tissues themselves and the external environment.

The transfer of information in animals

There are three principal ways by which cells in multicellular organisms can communicate with each other. First, when they are in close juxtaposition and are only separated by narrow fluid-filled spaces, direct electrical and chemical interactions can occur. Cells also maintain some structural connections with each other and secrete special excitants by which they may also communicate.

[1]

Second, contact between more remote cells can be maintained along tracts of nerve cells that are merely tissues specialized for such exchanges of information. Third, chemicals may be released, for example, from the endocrine glands, into the blood, which carries them to special sites that are physico-chemically programmed to react and respond to them.

The endocrine glands are tissues that, unlike exocrine glands, have no ducts but release their secretions, called "hormones," directly into the blood passing through them. It is with the diversity of such hormonally controlled processes that we will be principally concerned in this book. It should always be re-called, however, that the endocrine gland represents only a single facet of the animal's communication network and that nerves are also important. Endo-crinologists and neurophysiologists have often only concentrated on their own special fields of study to the relative exclusion of the rest of the animal's physiology. This is unfortunate as the complete animal is an academically and aesthetically pleasing thing to see and contemplate, and any single facet taken from the whole becomes less interesting and is physiologically nonsensical. The relations of nerves and endocrines, however, can also be considered from a more direct standpoint as it is apparent that the functions of each are related to each other. They are often mutually interdependent and may even act together to control a single process. Nerve cells thus can respond to hormones in a manner that influences behavior, and endocrine glands often receive information and directions from the brain. Both hormones and nerves can act together to control the melanophores in certain fishes. Many hormones, in-cluding epinephrine, vasopressin, oxytocin, and the hypophysiotrophic hor-mones are even made by nerve cells.

Neural versus humoral coordination

It is uncertain which came first, nerves or hormones. Why do animals have both? It may help us to understand hormones if we compare their respective properties and roles in the body.

The neural transfer of information occurs along distinct morphological pathways made up of chains of nerve cells with their long axons. Transmis-sion along these avenues is fast (up to about 100 m/sec) and is directed precisely to specific sites in the body. Neural transmission involves a series of electrical events interrupted at intervals by a local release of chemicals (trans-mitters) and is concluded by the release of these, principally acetylcholine or a catecholamine (such as norepinephrine), but also peptides, close to the effec-tor tissue. Such a transmitter is then rapidly destroyed near its site of action. Further stimulation will be dependent on subsequent neural transmission. The effect is thus rapid in onset, short in duration, and can be localized with considerable accuracy.

The hormones, on the other hand, are released into the blood, which carries them toward their effector site(s). In most instances this is outside the cardio-vascular system so that the hormone must also cross capillaries and diffuse through the intercellular spaces to the site of its action. Not surprisingly, hormonal responses are slower than those mediated by nerves. Hormones are dispersed very widely in the body and so come into contact with a great variety of cells with which an interaction, in most instances, would not be fruitful. The problem of ensuring that hormones only act at specific sites is largely solved by the multiplicity in their chemical structures. (There are over 40 different known hormones in a mammal.) Complementing such variations are parallel differences in the chemical structures of the sites where they interact ("receptors") with their target (or "effector") cells.

A hormone can exert widespread effects by interacting with different effector tissues (for instance, estrogens act on the uterus, mammary glands, liver, brain, etc.). The characteristics of the receptors in each may differ just as the response will vary. A hormone thus may act very specifically at each of several sites in the body and yet, at the same time, exert many different actions.

Perhaps the most physiologically significant difference between neural and humoral communication is in the duration of the actions of the transmitters involved. Because their transmitters are rapidly destroyed, nerves must be repetitively stimulated if their effects are to be prolonged. Although hormones also have a finite period of survival, the duration of their effects varies from less than a minute to several days. Some hormones, once released into the circulation, survive many hours. When some reach their receptor sites, the initiated response may be of a persistent nature that is not readily terminated. Thus, if an endocrine gland is removed, it may be several days or even months before physiological signs of its absence become apparent. Hormones are thus sometimes described as exerting their effects slowly but persistently, in contrast to the more rapid and transient actions of nerves. There are, however, exceptions to such a generalization.

What is comparative endocrinology?

Comparative endocrinology concerns the study of the endocrine glands in different species of animals, both vertebrates and invertebrates. Its aims are analogous to the older and more classical disciplines of comparative anatomy and comparative physiology. The prime academic objective is to reconstruct evolutionary pathways by the study of extant species. Figure 1.1 shows the phylogenetic relationships of the vertebrates and this emphasizes the extant groups that may be particularly interesting in such studies. The mere examination of the endocrine system of some bizarre and exotic vertebrate does not

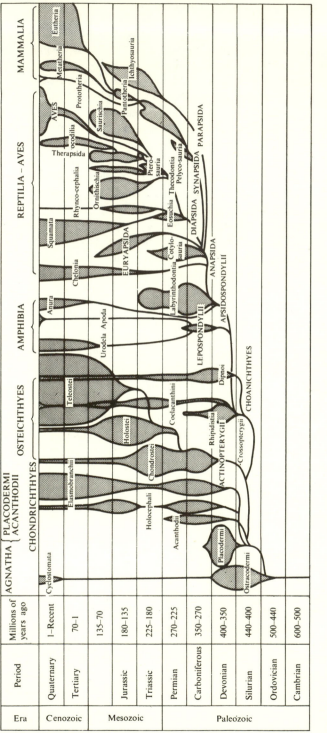
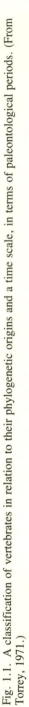

Era	Period	Millions of years ago
Cenozoic	Quaternary	1–Recent
	Tertiary	70–1
Mesozoic	Jurassic	135–70
		180–135
	Triassic	225–180
Paleozoic	Permian	270–225
	Carboniferous	350–270
	Devonian	400–350
	Silurian	440–400
	Ordovician	500–440
	Cambrian	600–500

Fig. 1.1. A classification of vertebrates in relation to their phylogenetic origins and a time scale, in terms of paleontological periods. (From Torrey, 1971.)

alone constitute "Comparative Endocrinology" (it may be "Animal Endocrinology") unless the data can be considered in relation to that in other, phyletically related species. Such information can be used to help confirm, complete, and even extend our knowledge of the phylogenetic relationships between vertebrates and to follow the evolution of endocrine mechanisms. The lungfishes (Dipnoi) may afford us an example. These fishes have long been considered, on the basis of morphological information, to be close to the original line of evolution connecting the bony fishes (Osteichthyes) and the Amphibia. As we shall see later, homologous vertebrate hormones often exhibit considerable differences in their chemical structure. Many such differences are apparent between the hormones in fishes and tetrapods. The structure of several hormones present in lungfishes, however, shows a greater similarity to those in tetrapods than those in other fishes. For instance, a neurohypophysial hormone called mesotocin is present in amphibians, reptiles, and birds, but in bony fishes the homologous hormone is isotocin (which differs from mesotocin by a single amino acid substitution), with the exception of the lungfishes, which have mesotocin. It has also been found that the growth hormone and prolactin present in lungfishes are more like those in tetrapods than in other fishes.

Apart from contributing to the overall phyletic study of vertebrates, the comparative endocrinologist aspires to reconstruct the lines of evolution within the endocrine system itself. This can be done by examining and comparing in different species, the morphology of the endocrine tissues, the structures and activities, both immunological and pharmacological, of their secreted hormones and their different physiological roles.

Occurrence of vertebrate hormones in other organisms

Although this book is concerned with the occurrence and physiological roles of hormones in vertebrates, these chemical messenger molecules are not necessarily confined to this phylum. All living creatures possess mechanisms to aid their coordination and, as we have seen, protozoans even may utilize a molecule, epinephrine, which is also present and used for a similar purpose in vertebrates. Molecules that are identical or close chemical relatives of several vertebrate hormones have been identified in a variety of invertebrates and even in plants.

In the plant kingdom such substances have been sufficient to cause diseases in grazing animals. Serious reproductive disorders, including abortion, have been observed in sheep that eat red clover, and this disease was found to be due to a substance with activity like that of estrogens, called genistein (Wong and Flux, 1962). Such substances, or phytoestrogens, are not uncommon in plants and may contribute in a desirable manner to the growth and nutrition of herbivorous animals, especially in the period of the "spring flush" of vegetation. In other instances, such chemicals apparently may serve a biological use, as seen in desert quail in California, where they limit breeding in dry seasons (Leopold et al., 1976). They can thus act as natural contraceptives. Such plant estrogens are usually different chemically from the animal hormones, but they share active chemical groups that endow them with their special activity. Cattle in parts of Argentina and the United States (Florida) that graze on certain types of shrubs have been observed to sicken and die from a disease associated with a calcification of tissues, especially the heart. The toxic material was found to be $1\alpha,25$-dihydroxycholecalciferol, which is the active hormonal form of vitamin D_3 (Wasserman et al., 1976). A substance that cross-reacts immunologically with thyrotropin-releasing hormone (TRH) has been found in alfalfa plants (Jackson, 1981). It is unknown whether it has any action in the animals that consume it. The role of such substances in the plants that contain them is unknown.

The presence of molecules related to vertebrate hormones among the invertebrates is not surprising when one recalls that the principal neurotransmitters acetylcholine, catecholamines, 5-hydroxytryptamine, and γ-aminobutyric acid (GABA) are all shared on many occasions by both of these major groups of animals. Information about the occurrence of hormones in both is less common, however, and that which is available is largely dependent on immunological evidence, especially immunohistochemical procedures. Insulin-like molecules have been identified in the gut tissues of some molluscs and a honeybee (Plisetskaya et al., 1978; Moreau, Raoelison, and Sutter, 1981), calcitonin and somatostatin in nerve tissue of a sea squirt (a tunicate, Protochordata) (Falkmer et al., 1977; Girgis et al., 1980), and cholecystokinin (CCK) in the brain of a blowfly (Duve and Thorpe, 1981). The occurrence of iodinated tyrosine molecules with activity like that of thyroid hormones in Protochordates is also well known. A molecule that cross-reacts with antibodies to mammalian vasopressin (antidiuretic hormone, ADH) has been found in the brain and other tissues of a locust (Proux and Rougon-Rapuzzi, 1980). The physiological roles of such substances in these invertebrates are uncertain, but it has been suggested that they may act as neurotransmitters or hormones and, in the instance of insulin and epinephrine, they may even have comparable metabolic effects to the ones they have in vertebrates.

It is tempting to suggest that the presence of vertebrate types of hormones in invertebrates may reflect some evolutionary relationship. In the instance of the protochordates and even the molluscs this may be so. Their presence in insects, however, makes such common origins appear rather tenuous. It seems likely that nature has been somewhat frugal or has a limited inventiveness in its provision of molecules suitable for roles in chemical coordination mechanisms. They may then have been invented by nature on several separate occasions. There are even instances, such as among the neurohypophysial-type hormones (e.g., vasotocin and mesotocin) where chemists have synthesized such reactive molecules before they were known to exist as natural hormones.

The uses of comparative endocrinology

The classic, or academic, aims of comparative endocrinology have been described. The provision of such intellectual satisfaction is not, however, sufficient justification for all! There are, indeed, a number of other contributions that such studies can make to biology, and some examples of these are given in this section.

The process of reproduction in vertebrates is dependent on the endocrine secretions, and an understanding of this relationship can provide information that may be usefully applied when, for aesthetic or economic reasons, we may wish to increase, or decrease, the fecundity of a species. This type of study thus constitutes a contribution to the field of "biological control" (Bern, 1972).

Knowledge of the endocrine system in man has largely been made possible by experiments on other animals. This has principally involved mammals like rats, rabbits, and monkeys but also some more exotic and bizarre creatures. Quantitative measurements of gonadotropins and melanocyte-stimulating hormone (MSH) were originally made (and sometimes are still) using the responses of the clawed toad (*Xenopus laevis*), and prolactin levels can be measured by its effects on the pigeon's crop-sac or on the behavior of a newt. Oxytocin is assayed by utilizing its ability to decrease the blood pressure of chickens, and the rate of water movement across the toad's urinary bladder can be used to distinguish between two, chemically different mammalian antidiuretic hormones (ADHs).

The responsiveness of a toad's urinary bladder to ADH and aldosterone is used to study the "mechanism of action" of these hormones on membrane permeability. Such preparations provide useful "models" of hormonal effects on the mammalian kidney.

The relationship of the structure of a molecule, to its biological activity, is a field of considerable interest to biologists. The diversity, or polymorphism, in the structure of vertebrate hormones, together with their disparate effects on different tissues and in various species, offers a natural "laboratory" for such studies. Nature has had a long time and wide opportunities to experiment with the effects of changes in molecular structures on the activities of such excitants. At present, this is most clearly seen among the neurohypophysial hormones, of which there are at least 10 known chemical variants among the vertebrates. These hormones are peptides containing eight amino acids and often only differ from one another by a substitution at a single chemical locus. They are very reactive molecules and can exert actions at many different sites ranging from the uterus and mammary gland to blood vessels, the kidney, and the amphibian skin and urinary bladder. Analogous effector tissues in different phyletic groups exhibit different abilities to respond to each such hormone, be it a natural one or a variant made in the chemist's laboratory. There are available, and in use, more than 20 different effector preparations that can be used to study the effects of changes in chemical structure among these hormones on its biological effectiveness. Natural variants of hormones, in which the biological activity has been altered in some way, may be of potential use to man. For instance, calcitonin (a hormone concerned with the regulation of calcium in the body) from the salmon ultimobranchial bodies is far more potent in man than the natural hormone. Salmon calcitonin is used to treat a bone disease (Paget's disease) in man.

The diversity of vertebrates as a background for endocrine variation

There are some 42,000 extant species of vertebrate animals. The vertebrates originated some 400 million years ago as creatures who apparently lived in the sea or, possibly, in fresh water. They subsequently evolved and occupied almost every conceivable habitat in the oceans, in freshwater rivers and lakes, and on the land. Their abodes range from the cold polar regions to hot equatorial ones, from deserts to swamps, from high mountains to the ocean deeps. The considerable morphological and physiological diversity of vertebrates mirrors their success in this multitude of environmental conditions. It is thus not surprising to find that the endocrine system exhibits interspecific differences that reflect adaptations to such different environments. Nevertheless, it is also somewhat unexpected to find that considerable similarities are still apparent in the endocrine systems of species as distantly related as the hagfish (Cyclostomata) and man.

The endocrine glands of vertebrates have special roles to play in the regulation of many types of physiological processes that include reproduction, osmoregulation, intermediary and mineral metabolism, and growth and development (Table 1.1). The nature of the responses to hormones differs considerably but can be classified into several major groups including their actions on membrane permeability, muscular contraction, the transformation of substrates involved in intermediary metabolism and growth, and a controlling (or tropic) action on other endocrine glands (Fig. 1.2).

Many, though not all, of the endocrine glands are essential for life and the reproduction and survival of the species. In other instances, however, their immediate importance for survival is not clear. Animals cannot reproduce if the endocrine function of their gonads is compromised, and death soon fol-

Table 1.1. *The secretions of the endocrine glands*

Gland	Hormones	Target tissues
Pituitary		
Adenohypophysis		
Pars distalis	Follicle-stimulating hormone, FSH	Ovary and testis
	Luteinizing hormone, LH (also called interstitial cell-stimulating hormone, ICSH)	Ovary and testis
	Thyrotropic hormone, TSH	Thyroid
	Corticotropic hormone, ACTH (adrenocorticotropic hormone)	Adrenocortical tissue
	Growth hormone, GH (somatotropic hormone)	Liver forms somatomedins, which alter tissue metabolism (liver, muscle, adipose tissue)
	Prolactin	Mammary glands, fish gills, tadpole metamorphosis, corpus luteum, kidney, skin, etc.
	Endorphins	Nerve and endocrine cells
Pars intermedia	Melanocyte-stimulating hormone, MSH	Melanocytes, pigmentation and color change
Neurohypophysis		
Pars nervosa	Vasopressin, ADH, vasotocin	Kidney, amphibian skin, and urinary bladder
	Oxytocin	Mammary gland, uterus
Hypothalamus	Pituitropins; LHRH, CRH, TRH,[a] somatostatin, etc.	Release of hormones by the adenohypophysis
Thyroid gland	Thyroxine (T_4), triiodothyronine (T_3)	Tissue metabolism and differentiation; calorigenic (mammals), morphogenetic (amphibians)

Table 1.1 (*cont.*)

Gland	Hormones	Target tissues
Parathyroid glands	Parathyroid hormone, PTH	Bone and kidney
Ultimobranchial bodies ("C"-cells in mammalian thyroid)	Calcitonin, CT (also called thyrocalcitonin)	Bone and kidney
Adrenal glands		
Cortex (interrenals in sharks and rays)	Cortisol, corticosterone, cortisone, 1α-hydroxycorticosterone	Tissue metabolism (liver, muscle), proteins to amino acids, gluconeogenesis
	Aldosterone	Na and K in kidney, sweat and salivary glands, gut, amphibian skin, and bladder Intestine (teleosts)
Medulla (chromaffin tissue)	Norepinephrine (noradrenaline) Epinephrine (adrenaline)	Tissue metabolism (liver, muscle, adipose tissue), glycogenolysis, mobilization fatty acids, calorigenic, constriction and relaxation of smooth muscle
Islets of Langerhans		
Alpha-cells	Glucagon	Liver (glycogenolysis), adipose tissue (fatty acid release)
Beta-cells	Insulin	Liver, muscle and adipose tissue (amino acids to protein, glucose to fat and glycogen)
Delta-cells	Somatostatin	α- and β-cells
Gonads		
Ovary		
Graafian follicle	Estrogens (estradiol)	Female sex organs and characters, mammary glands, brain
Corpus luteum and interstitial tissue	Progestins (progesterone)	Uterus and mammary glands
Testis		
Interstitial tissue (Leydig cells)	Androgens (testosterone)	Male sex organs and characters, sperm maturation, brain

Source	Hormone	Function
Sertoli cells (?)	Androgens	Sperm maturation
Placenta (pregnant eutherian mammals)	Estrogen (estriol), progesterone	Uterus, mammary glands, fetus
	Chorionic gonadotropin, HCG	Corpus luteum
	Placental lactogen, HPL (somatomammotropin, HCS)	Mammary glands
Gut		
Stomach (pyloric mucosa)	Gastrin	Stimulates secretion of gastric juice
Intestine (mucosa)	Gastric inhibitory peptide (GIP)	Inhibits secretion of gastric juices
	Secretin	Stimulates secretion of pancreatic juices from exocrine pancreas and hormones from endocrine pancreas
	Cholecystokinin–pancreozymin (CCK)	Enzyme secretion from pancreas and hormones from endocrine pancreas
	Enteroglucagon	As for glucagon
Kidney		
Tubular cells	$1\alpha,25$-dihydroxycholecalciferol $(1,25\text{-}(OH)_2\text{-vitamin }D_3)$	Intestine, bone
Juxtaglomerular cells	Renin	Plasma α2-globulin-angiotensinogen → angiotensin (targets: adrenal cortex, vascular smooth muscle, thirst center)
Putative endocrine glands		
Pineal gland	Melatonin	Hypothalamus (inhibits release MSH? and gonadotropins), melanocytes (larval anurans, cyclostomes?)
Corpuscles of Stannius (some bony fishes)	Vasotocin? "Hypocalcin"	Calcium metabolism
Urophysis (some fishes)		Osmoregulation, smooth muscle contractions (?)
Thymus	Thymic hormone(s)	Immunological maturation via induction of immunological competence of lymphocytes, etc.

[11]

a For explanation see list of abbreviations.

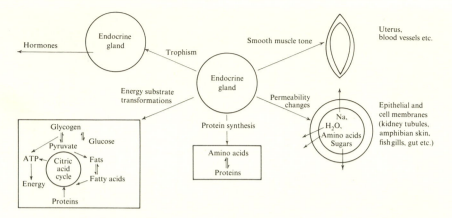

Fig. 1.2. A diagrammatic summary of the basic types of actions of hormones in vertebrates.

lows complete destruction of the adrenal cortex. Life may be shortened if the islets of Langerhans fail to product sufficient insulin, and normal growth, development, and maturation of the young will not occur if the secretion of pituitary growth hormone or thyroid hormone is inadequate. On the other hand, antidiuretic hormone from the neurohypophysis is not essential for life though in its absence very large volumes of urine are secreted by the kidney. In man this is an annoying condition as prolonged sleep is not possible and even during the waking hours it can lead to social difficulties; but it is not fatal. If drinking water were in limited supply, however, dehydration could be a potential problem and absence of this hormone might then affect survival. It should also be remembered that whereas too little of a hormone can constitute a problem, too much may also result in physiological difficulties. Hormone imbalances can result from genetic abnormalities, the presence of tumors, and accidental disruption of the events controlling secretion of the hormone. A few examples of such endocrine dysfunction and their effects are summarized in Table 1.2.

Endocrine glands, or tissues, have been identified among all of the vertebrates. Those common to the major groups (from the Cyclostomata to the Mammalia) are the pituitary, thyroid, endocrine pancreas, adrenal chromaffin and cortical tissues, and gonads. The parathyroid glands have been found in the tetrapods but not in the fishes. The ultimobranchial bodies or their homologues, the thyroid "C"-cells, have been identified in all groups except the cyclostomes. Such tissues secret more than 40 different hormones in a mammal. If we include all the naturally occurring analogues of these hormones that occur among the vertebrates, we can account for at least twice this number of hormones and there are undoubtedly many more.

Table 1.2. *Some effects of endocrine dysfunction in man*

Gland	Secretory activity[a]	Abnormality	Principal effects
Pituitary			
Adenohypophysis (growth hormone)	↑	Giantism, acromegaly	Excessive growth
	↓	Dwarfism	Retarded growth
Neurohypophysis	↓	Diabetes insipidus, ADH	Excessive loss of water in urine
	↑	Schwartz–Bartter syndrome, ADH	Low plasma Na
Thyroid gland	↑	Graves's disease	High metabolic rate and nerve cell activity
	↓	Myxedema	Low metabolic rate
	↓	Cretinism	Inadequate development and growth
Parathyroids	↑	Hyperparathyroidism	Hypercalcemia, polyuria, reduced bone calcium
	↓	Hypoparathyroidism	Muscle tetany
Islets of Langerhans			
Beta-cells	↓	Diabetes mellitus	Hypoglycemia, muscle wasting
Adrenal cortex	↓	Addison's disease	Renal Na loss and K retention (low plasma Na, high plasma K), low blood pressure, muscle weakness
	↑	Cushing's syndrome	High blood pressure, obesity, retarded growth
Adrenal medulla	↑	Pheochromocytoma	Hyperglycemia, high blood pressure
Ovaries	↓		Sterility, failure to develop or maintain secondary sex characters
Testis	↓		As above

[a] ↑ Increased activity; ↓ decreased activity.

Apart from these glands, other tissues that may have an endocrine function (putative endocrine glands) have also been identified in a number of vertebrates. A tissue is considered to have an endocrine function if it releases a product into the circulation that has an excitatory or inhibitory effect on some distant effector gland, organ, or tissue. The precise status of the pineal body as an endocrine gland is still equivocal though many consider that it has such a function especially as one of its products, melatonin, has been identified in the blood. Two other tissues that are present in some fishes, the urophysis and corpuscles of Stannius, also have a putative endocrine status.

It is conceivable, indeed likely, that other endocrine glands exist among the vertebrates. Even within the last 20 years a new endocrine tissue, the ultimobranchial bodies (the thyroid C-cells in mammals), has been identified. This tissue is concerned with the regulation of calcium levels in some vertebrates. Despite the busy, even frantic, activity of endocrinologists the hormonal role of this gland previously had not been confirmed.

Despite their anatomical and embryological homologies, the endocrine glands of vertebrates display considerable diversity in their morphological arrangements, the chemical nature of their secretions, and even their physiological role in the body. It is principally about these differences that we will be concerned in the succeeding chapters.

Conclusions

Physiological processes are coordinated with the aid of both nerves and hormones. Each of these mechanisms has special characteristics that may be suited to the needs of the particular process involved and they often operate together. During the course of geological time vertebrates have evolved and acquired morphological features and physiological processes that have permitted them to adapt to changing environments and to occupy a variety of ecological habitats. Such biologically important changes are accompanied by the neural and humoral processes necessary for their coordination. Contemporary species of vertebrates exhibit considerable structural and functional diversity that can be related to the nature of the life they lead and to their ancestry. They are classified into systematic groups that are also thought to reflect their evolution. Thus, a comparison of the endocrine function of contemporary species of vertebrates is of importance not only in fully understanding how they live today, in a particular environmental situation, but also may tell us how such hormonally mediated processes evolved.

2. Comparative morphology of the endocrine tissues

Endocrine glands and tissues display a diversity in their gross morphological and histological patterns. This is particularly apparent when comparing species from phyletically distant groups. In some instances the physiological significance of these differences has been recognized but in most this is not so and may be related to the initial pattern of embryonic growth. If, however, one intuitively suspects a close relationship between structure and function, then the lack of a known correlation may merely reflect our ignorance.

The endocrines may display several different types of morphological variation. Their position in the body may not be the same. This variation can be of a minor nature, such as is seen with the ultimobranchial bodies, which can be situated near the heart or the thyroid gland. In some fish, however, thyroid tissue may vary in position from the branchial region to the kidney. Endocrine cells may show varying degrees of association and be scattered as individual cells, in small segments, or "islets," or be closely associated as a compact gland enclosed in a capsule. Such aggregation of an endocrine tissue is commonly seen as one ascends the evolutionary (or the phyletic) scale. In addition, different endocrine tissues may display diverse associations with each other, as for instance the conglomeration of chromaffin and interrenal (or adrenocortical) tissue in the adrenal gland. Their relationship to the neural and vascular tissues can be very important. Pituitary tissues thus cannot function properly if they are transplanted to other parts of the body (ectopic transplant) or if the small blood vessels between the gland and the brain are cut. The major blood vessels not only carry hormones away from endocrine tissues but also supply them with nutrients and controlling stimuli. The pattern of the vasculature within the gland can also be important for its correct functioning.

The types of cells that make up an endocrine gland are, not surprisingly, similar in homologous glands among the vertebrates. Such similarities as reflected by their microscopic anatomy (size, shape, the presence of inclusions, granules, etc.) and their reactions with dyes (tinctorial relationships) serve to aid in their identification. More recently, antibodies to specific hormones have been used to identify the cells where they are formed. These antibodies may be labelled with radioactive materials or fluorescent dyes so that the precise locus where they react can be seen. The histological appearance of endocrine cells may change somewhat at different times depending on

[15]

their secretory state. This characteristic can be used to predict their activity and physiological role. Inactive thyroid cells thus have a flattened, rather than columnar appearance that is typical of their active state, whereas neurohypophysial tissue that is depleted of its hormone has little stainable (with Gomori chrome–alum hematoxylin) neurosecretory material.

The pituitary gland

The pituitary is a conglomerate of tissues and cells that reflect the 10 or so major hormones it secretes. These hormones help regulate the activities of the thyroid, adrenal cortex, and gonads and contribute to the control of various other physiological activities, including water and salt metabolism, growth, lactation, parturition, and the pigmentation of the skin. A comparative account of the anatomy of this gland has been provided by Holmes and Ball (1974). Embryologically, the pituitary arises as a result of a downgrowth of tissue (the infundibulum) from the brain and an upgrowth (the hypophysis) from the roof of the mouth. Enclosed within these tissues is a piece of mesoderm that forms a net of blood vessels sometimes called the "mantle plexus." The pituitary lies in close apposition to the hypothalamus at the base of the brain. In mammals it is usually enclosed in a small, bony chamber, the sella turcica, and it is connected by a stalk of nervous tissue to the brain, just behind the optic chiasma (Fig. 2.1). The hypophysis partly differentiates into the adenohypophysis that secretes seven or eight hormones, which, so the histologists tell us, are formed by a number of distinctive types of cells. These are most descriptively labelled by the name of the hormone they secrete followed by the suffix *trope*. We thus have thyrotropes, gonadotropes, somatotropes, and so on. An alternative terminology utilizes the Greek alphabet: α-cells = somatotropes, β-cells = thyrotropes, and so on. The adenohypophysis can be divided on a gross morphological basis into three or four sections: the pars tuberalis, the pars distalis (sometimes with a rostral and caudal section), and the pars intermedia. The latter gives rise to the melanocyte-stimulating hormone (MSH) while the rest of the hormones come from the pars distalis. The pars tuberalis lies between the pars distalis and the brain (in the region of the median eminence) and is associated with the blood vessels that connect the two.

The neural, or infundibular, tissue forms the neurohypophysis, which basically lies caudally to the adenohypophysis, hence the terms anterior and posterior lobes of the pituitary. The neurohypophysis is connected to the brain by the infundibular stalk. The two hormones (ADH and oxytocin in mammals) it secretes are formed in nerve cells (by a process called neurosecretion) that originate in the supraoptic and paraventricular nuclei in the brain of amniotes or the preoptic nucleus of amphibians and fishes. The axonal tract

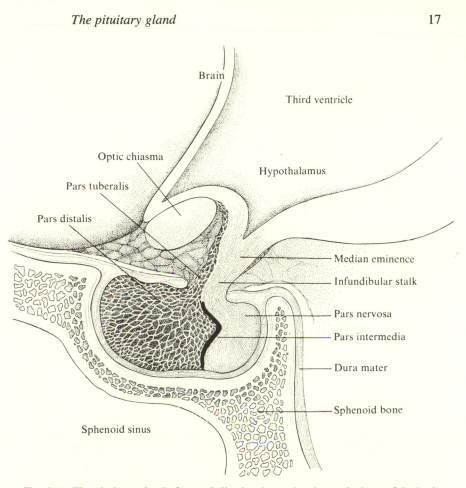

Fig. 2.1. The pituitary gland of man. It lies in a bony chamber at the base of the brain to which it is connected by a stalk. The pars intermedia is quite small in man but may be much larger in other species. (From R. Guillemin and R. Burgus, *The Hormones of the Hypothalamus*. Copyright © 1972 by Scientific American, Inc. All rights reserved.)

running from the bodies of these nerve cells, in the nuclei, to the periphery, where they are stored and released, is called the supraopticohypophysial tract. The neurohypophysis thus consists of the distal parts of nerve cells interspersed with glial cells and pituicytes. The function of the latter is unknown.

Three regions of the neurohypophysis can be distinguished. The rostral *median eminence* is part of the wall of the hypothalamus and lies in close conjunction with the adenohypophysis to which it is usually connected by a system of portal blood vessels that originate from the mantle plexus. The median eminence is contiguous with the *infundibular stalk*, which connects it to the most prominent part of the neurohypophysis, the *pars nervosa* (or *neural lobe*). The latter is much more highly developed in terrestrial tetrapods than in the fishes. The phyletic development of the amniote neurohypophysis is shown in Fig. 2.2.

Comparative morphology of the pituitary

The diverse morphology of the vertebrate pituitary provides us with some information (albeit equivocal) about the nature of the evolutionary changes that may have taken place in this gland. Attempts have been made to choose or construct the pituitary that is considered most typical of each major phyletic group. Considerable differences from a "median gland" may nevertheless exist among various species within each systematic group.

The structures of the pituitaries of fishes, from the Cyclostomata to the Dipnoi, are shown in Fig. 2.3. The cyclostomes have a simple type of pituitary in which the different regions are only loosely associated with each other. The parts of the adenohypophysis in these phyletic prototypes are often termed the pro-, meso-, and metaadenohypophysis. They are thought to correspond respectively to the cephalic part of the pars distalis (or possibly the pars tuberalis), the caudal pars distalis, and the pars intermedia of other vertebrates. The close proximity of the adenohypophysis to the brain may not be functionally essential in these lowly fishes. Considerable intraspecific variation occurs, and ectopic transplants of the adenohypophysis to other parts of the body do not appear to compromise its function, at least in hagfishes (Myxinoidea) (Fernholm, 1972).

The actinopterygian fishes possess a pituitary in which there is a close association between the various component tissues. The homologies of these tissues to those in tetrapods have on occasion been difficult to recognize but they undoubtedly exist. The neurohypophysis is not a very discrete tissue in fishes (there is no distinct neural lobe) and shows considerable admixture with the pars intermedia into which it sends finger-like projections and with which it shares a common blood supply (see Fig. 2.3). Portal blood vessels connecting the median eminence and adenohypophysis have been described in all groups of actinopterygians except the teleosts. Considerable variation has been observed among the latter, in which the blood supply to the adenohypophysis passes initially through the neurohypophysis. No clear portal system, as seen in other actinopterygians, is apparent in teleosts. At least five distinct types of cells have been identified in the fish pituitary as shown in Fig. 2.4, which is that of a teleost, the eel *Anguilla anguilla*. These cells are present in separate zones in contrast to the tetrapods and lungfishes (Dipnoi) where they are intermingled with each other.

The chondrichthyean fish (sharks and rays) have a pituitary that on superficial examination looks rather different from that of other fishes. It displays,

Primitive type as in some reptiles

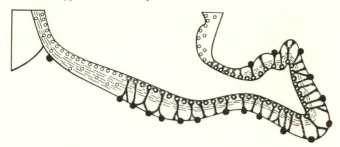

Birds and reptiles

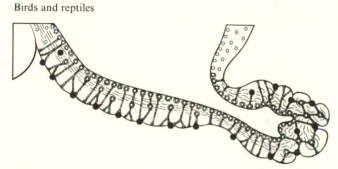

Mammals

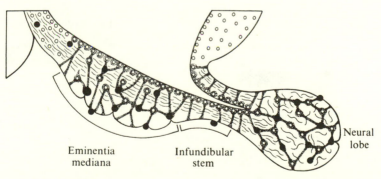

Eminentia
mediana

Infundibular
stem

Neural
lobe

Fig. 2.2. Histological differentiation of the amniote neurohypophysis. The "primitive form" is seen in reptiles such as the Rhynchocephalia, Chelonia, and some Lacertilia. Solid black lines are the blood vessels; nerve fibers are thinner lines. (From Wingstrand, 1951.)

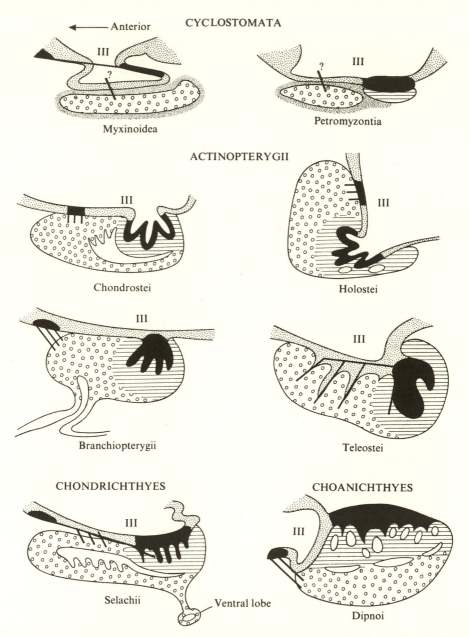

Fig. 2.3. The pituitary glands of fishes. Diagrammatic representation from midsagittal section. Small dots = nervous tissue; black = neurohypophysial tissue; large open dots = pars distalis; horizontal lines = pars intermedia. Thick black lines = blood vessels that carry neurosecretory products to the adenohypophysis, or in Myxinoidea to the neurohypophysis. III = third ventricle. (From Ball and Baker, 1969, modified slightly according to Holmes and Ball, 1974.)

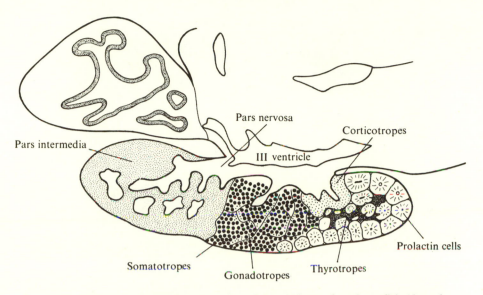

Fig. 2.4. Diagram from midsagittal section of the pituitary of a teleost fish (the eel *Anguilla anguilla*). The cells (tropes) that produce the hormones in the pars distalis can be seen to lie generally in distinct zones. (From Olivereau, 1967.)

however, a similar basic structure, though gross differences, such as a rather large pars intermedia, are often apparent. A characteristic and distinct lobe of the adenohypophysis lies below the pars distalis; this is called the "ventral lobe," which has been shown to be the site of formation of a gonadotropin. Like most actinopterygians, chondrichthyeans have a distinct portal blood system connecting the hypothalamus and adenohypophysis, but this special blood supply does not extend to the ventral lobe.

The pituitary of lungfishes shows more similarities to that of tetrapods than to those of other fishes. This is especially interesting in view of the special phyletic relationship that is usually considered to exist between lungfishes and tetrapods. The gross similarities between the amphibian and lungfish pituitaries can be seen in Fig. 2.5. The different types of cells in the adenohypophysis are intermingled in lungfishes (not separated as in other fishes), just as in the tetrapods. The neurohypophysis of the lungfishes also displays the beginnings of the differentiation of a neural lobe.

The neural lobe, which is a characteristic of the tetrapods, is formed by the enlargement, in a posterior direction, of the neurohypophysis. Wingstrand (1966) has suggested that this change may be related to a terrestrial manner of life in which the secreted hormones had a special significance. This theory is consistent with measurements showing a much greater amount of stored hormonal material in the neurohypophysis of tetrapods than in that of fishes (Follett, 1963).

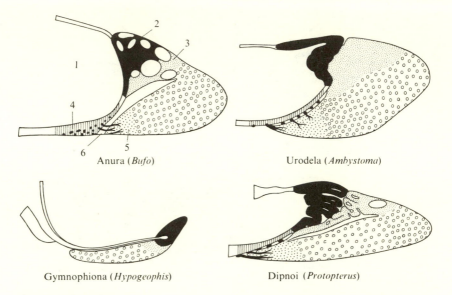

Fig. 2.5. The pituitaries of the three main groups of the Amphibia compared to that of a dipnoan (lungfish). 1 = Saccus infundibuli; 2 = neural lobe; 3 = pars intermedia; 4 = median eminence; 5 = "zona tuberalis"; 6 = portal blood vessels. (From Wingstrand, 1966. Originally published by the University of California Press; reprinted by permission of The Regents of the University of California.)

The basic morphological pattern of the tetrapod pituitary is well exemplified in the reptiles (Fig. 2.6). It can be seen, however, that even here differences between the major systematic groups occur. Such variations usually reflect the relative degree of development of the neurohypophysis and the presence, reduction, or absence of the pars tuberalis. The reptilian adenohypophysis has distinct cephalic and caudal zones.

The pituitaries from more than 100 species of birds have been examined and, as in the reptiles, the pars distalis has two distinct regions. It is interesting that a pars intermedia has not been identified among the birds. The hormone typically secreted by this tissue, MSH, has nevertheless been identified in the pituitary of the domestic chicken (Shapiro et al., 1972). Whether this MSH is a normally secreted hormone in birds is unknown, however. This absence of a pars intermedia is not unique to birds; it is not present in elephants or whales, either.

Among the mammals, considerable morphological differences exist in the intimate arrangement of the tissues within the pituitary (see Hanstrom, 1966). The detailed embryonic development of the pars distalis differs from that of other amniotes. The mammalian pars distalis arises mainly from the aboral lobe of Rathkes pouch, which in birds and reptiles forms the caudal section of

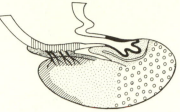
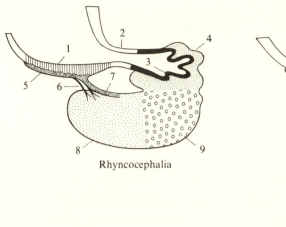

Rhyncocephalia

Chelonia

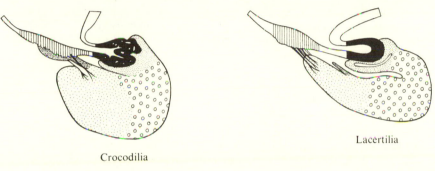

Crocodilia

Lacertilia

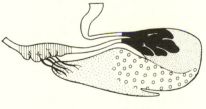

Ophidia

Fig. 2.6. The pituitaries of the five main groups of the Reptilia. 1 = Median eminence; 2 = infundibular stem; 3 = neural lobe (pars nervosa); 4 = pars intermedia; 5 = pars tuberalis; 6 = portal blood vessels; 7 = pars tuberalis interna; 8 = cephalic lobe of pars distalis; 9 = caudal lobe of pars distalis. (From Wingstrand, 1966. Originally published by the University of California Press; reprinted by permission of The Regents of the University of California.)

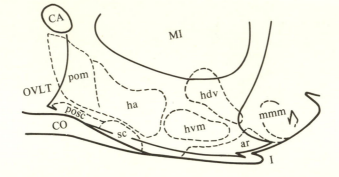

Fig. 2.7. Parasagittal section through the hypothalamic region of the rat, showing the distribution of nuclei of nerve cells. OVLT = organus vasculosum lamina terminalis; I = median eminence; ar = arcuate nucleus; ha = nucleus anterior hypothalamus; hdv = nucleus dorsomedialis hypothalamus; hvm = nucleus ventromedialis hypothalamus; pom = nucleus preopticus medialis; posc = nucleus peopticus, pars suprachiasmatic; mmm = nucleus mammilaris medialis, pars medialis; sc = nucleus suprachiasmaticus; CA = commissura anterior; CO = chiasma opticum; MI = massa intermedia.

the pars distalis. The oral lobe, which forms the cephalic part of the pars distalis in the last two groups, fails to develop in mammals (Wingstrand, 1966). In addition to being absent in whales and elephants, the pars intermedia is very much reduced in adult primates, including man. The simplest type of pituitary is seen in the echidna, *Tachyglossus aculeatus* (an egg-laying monotreme) and in some rodents and insectivores. It is interesting that the echidna shows a pattern that is considered to be like that of a "primitive" mammal. The echidna's pituitary, however, is typically mammalian in its embryonic origins. It nevertheless has some features, including a prominent portotuberal tract between the median eminence and pars distalis, which are seen more often in birds and some reptiles.

The hypothalamus

The activities of the pituitary gland are mainly controlled from the hypothalamus, both as a result of nerve impulses and the secretion of hormones. The hypothalamus is that region of the brain that lies on its ventral surface just below the third ventricle. It is situated just behind the optic chiasma and in man it is about 2.5 cm long and weighs approximately 4 g. It consists mainly of nerve cells, some of which originate in other parts of the brain and the autonomic nervous system, but it also contains its own complete neural network. The latter nerve cells originate in "nuclei" of tissue. These aggregations of cells have been carefully mapped (Fig. 2.7) and each appears to have special functions. The axons of the nerves, whether they originate inside or even outside the hypothalamus, tend to converge in the ventral peripheral region of the tissue and make up the median eminence.

The hypothalamus contains two main types of nerve cells: those that transmit nerve impulses in the usual way and others that have a neurosecretory function. Various neurotransmitters have been identified in the hypothalamus, including acetylcholine, dopamine, norepinephrine, 5-hydroxytryptamine, histamine, and GABA. The presence of these substances is thought to mainly reflect the normal neural activities of the tissue. The hypothalamus also has been shown to contain a variety of peptides formed by the process of neurosecretion. Some of the neurons that make them originate in the supraoptic and paraventricular nuclei outside the hypothalamus. These are called the *magnocellular neurons* and, in mammals, form antidiuretic hormone and oxytocin. Within the hypothalamus are smaller, *parvicellular neurons* that form such peptides as somatostatin, thyrotropin-releasing hormone, and luteinizing hormone-releasing hormone. These products of neurosecretion can be released from their nerve terminals in the median eminence and pass into the blood vessels of the hypophysial–portal system, in which they flow to the pars distalis.

The release of such peptides is regulated within the hypothalamus as a result of the activities of its nerve network and also in response to the presence of hormones and metabolites in its blood supply. In the latter instance, these may arrive via the internal carotid artery or, as has been recently shown (see Bergland and Page, 1979), they may pass in a retrograde direction from the pituitary gland up certain of the hypophysial–portal vessels. The neurons in the hypothalamus may synapse with each other or with those nerves that arise from areas outside of the hypothalamus, and either of these nerves may also synapse with the neurosecretory cells. The hypothalamus is clearly a primary site where the activities of the nervous and endocrine systems intermix.

The endocrine glands of the pharynx: thyroid, parathyroids, and ultimobranchial bodies

Apart from the adenohypophysis, which has its origins in the roof of the mouth, three (or four, if one includes the thymus) other endocrine glands arise from the pharyngeal tissues: the thyroid gland from the floor of the pharynx, the parathyroids from the II, III, and IV gill pouches, and the ultimobranchial bodies from the last, VI, pair of these.

The thyroid

Thyroid tissue is present in all vertebrates though its gross morphological arrangement varies somewhat. Its hormones have ubiquitous effects on tissue metabolism, differentiation, and maturation in tetrapods, but its role in fishes is not clear. The "thyroid unit" is a follicle in which a group of epithelial cells surrounds a central cavity that is filled with a glycoprotein secretion called thyroglobulin (Fig. 2.8). The encompassing cells have a columnar appearance when they are most active and a flattened one when they are least active. Thyroid follicles have a remarkable ability to trap inorganic iodide, which can be stored and incorporated into hormones that are, in turn, stored in the follicle cavity. It is probably the only endocrine gland that stores its products outside the cells.

In man, the thyroid gland is situated in the region of the neck and it has a generally comparable position in other vertebrates. In cyclostomes and most teleost fishes the thyroid follicles lie scattered along the blood vessels under the pharynx. Occasionally they may be found further afield (heterotopic), even in the kidneys. In chondrichthyean fish (sharks and rays), some teleosts, like the Bermuda parrot fish and tuna, the lungfishes, and the coelacanth (Sarcopterygii) (Chavin, 1976), the follicles are aggregated into a distinct glandular mass. This pattern persists in higher vertebrates; there are two such aggregates in amphibians, birds, and many reptiles. In lizards these two lobes are joined, which is also usually characteristic of mammals.

The thyroid appears to have the longest phylogenetic history of any endocrine gland (see Table 2.1). It is present not only in vertebrates; tissues that may be homologous, though not having the characteristic follicular units, also have been identified in protochordates, including *Amphioxus* (Cephalochordata) and various ascidians (sea squirts, Urochordata). The development of the thyroid in lampreys can be followed during the metamorphosis of its ammocoete larva. This beast collects small particles of food by filtering water that passes, with the help of ciliary action, through its pharynx. This process is aided by a ventral outgrowth from the floor of the mouth called the endostyle or subpharyngeal gland. An analogous tissue also exists in *Amphioxus* and ascidians. It secretes a sticky mucus that traps the food particles before they can pass out across the gills. This action has been likened to that of "moving flypaper." Embryologically, the endostyle of the lamprey ammocoete larva differentiates to form the adult thyroid. This has given rise to speculation as to whether the endostyle in the ammocoete and in protochordates has some thyroid function.

The endostyle does not contain thyroid-like follicles. It has, however, been shown (see Barrington, 1962), like the thyroid, to be able to accumulate selectively and concentrate radioactive iodide. This has been demonstrated not only in the lamprey ammocoete but also in *Amphioxus* and several ascidians. The iodine formed is bound in organic form with tyrosine and organoiodine compounds (*iodothyronines*), including, possibly, small amounts of thyroxine (for a summary see Table 2.1).

(a)

(b)

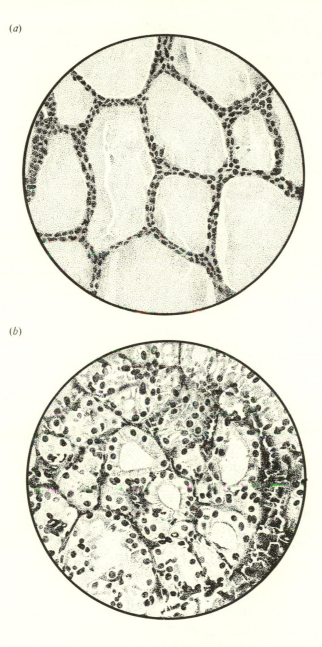

Fig. 2.8. The thyroid gland of the laboratory rat showing the follicles surrounded by epithelial cells. (a) The inactive condition where the cells are flattened and the follicles are distended with "colloid," which contains the thyroglobulin. (b) The active condition where the epithelial cells are columnar and little colloid is present.

Table 2.1. *The thyroid in the phylum Chordata*

Subphylum	Class	Species[a]	Thyroid gland	Thyroid-like[b] activity
Hemichordata[c]		*Glossobalanus minutus*	—	—
Protochordata				
Urochordata	Ascideacea	*Ciona intestinalis*	—	+
(tunicate)	(sea squirt)	*Clavelina lepadiformis*	—	+
	Larvacea		—	+
	Thaliacea	*Salpa maxima*	—	+
Cephalochordata	Amphioxi	*Branchiostoma lanceolatum* (amphioxus)	—	+
Vertebrata	Agnatha	(Lamprey)		
	(cyclostoma)	ammocoete larva	—	+
		adult	+	+
		(Hagfish)	+	+
	Chondrichthyes	(Shark)	+	+
	(elasmobranch)			
		(Skate)	+	+
	Osteichthyes		+	+
	(teleost)		+	+
	Amphibia		+	+
	Reptilia		+	+
	Aves		+	+
	Mammalia		+	+

[a] Not a complete list. Common names are given in parentheses.
[b] That is, synthesis of iodothyronines.
[c] Not usually classified as Chordata at present time, but included for reference in the light of earlier discussions.
Source: Rall, Robbins, and Lewallen, 1964.

Apart from iodoproteins, the endostyle of ascidians has been shown to contain two components that are concerned with the synthesis of thyroid hormones in vertebrates: a molecule that behaves immunologically like thyroglobulin and a peroxidase enzyme (Thorndyke, 1978; Dunn, 1980). Thyroid-like hormones have also been identified by radioimmunoassay in the serum of *Ciona intestinalis* (Urochordata) though the amounts were very small and their physiological significance is in doubt. It would appear, however, that urochordates may have the potential to produce thyroid hormone-like substances though thyroid follicles have not been observed in nonvertebrates.

Iodine readily reacts with proteins containing the amino acid tyrosine. Indeed, in one extensive investigation in which cows were being fed experimental diets containing thyroid compounds to improve their milk yields, these were made by incubating proteins, such as casein, with iodine at an appropriate pH and temperature. Thus, the spontaneous formation of organoiodine compounds in nature would not be surprising. Indeed, among the ascidians, the outer tunic or coat contains scleroproteins that combine with iodine, possibly even more readily than the tissues associated with the endostyle. Iodinated tyrosines also have been isolated in many other nonvertebrates, including coelenterates. Barrington (1962) has conjectured about the possibility that the spontaneous occurrence and availability of such compounds in nature may have led to their use as hormones. Subsequently, their formation may have become more localized in special tissues.

Parathyroid glands and ultimobranchial bodies

These glands secret hormones that contribute to the control of calcium in the body fluids. Embryologically, there are initially three pairs of parathyroids in tetrapods but those from the II pair of branchial pouches usually disappear; two or sometimes only one pair persist (see Roth and Schiller, 1976). When one pair persists they are usually derived from the III gill pouch. Two pairs of parathyroids are usually present in amphibians. They are absent in larval and neotenic urodeles. Among reptiles the number varies. One pair is present in the Crocodilia, two pairs in the Chelonia and Ophidia, and one to three pairs in the Lacertilia. Birds and mammals have one or two pairs. The number found does not seem to follow any phyletic pattern. The parathyroid gland contains two main types of cells: chief cells, which contain granules and secrete parathyroid hormone, and oxyphil cells, whose role is unknown.

A pair of ultimobranchial bodies is present in all the vertebrates from the birds to the chondrichthyean fish. They are apparently absent in cyclostomes. In mammals this tissue is embedded in the thyroid gland, where it makes up the parafollicular or C- (for calcitonin) cells. They may be widely dispersed in the thyroid gland, as seen in the dog, or localized in certain regions, as in rabbits and man. The structure, origin, and phyletic distribution of the ultimobranchial bodies and C-cells have been reviewed by Copp (1976) and Pearse (1976).

The association of the thyroid, parathyroids, and ultimobranchial tissues may be somewhat complex. The last two tissues are histologically different from the thyroid and they lack the typical follicular structure of that gland. Their positions in the domestic fowl are shown in Fig. 2.9. The morphological distribution of these three glands has often made it difficult to dissociate the effects of the latter two, which both elaborate secretions having opposite effects on blood calcium concentrations. The parathyroids, especially in mammals, are usually closely associated with the thyroid though not to the same extent as the ultimobranchial tissues. Removal of the mammalian thyroid, including that in man, is often associated with low blood calcium levels and an associated muscle tetany. This is the result of removal of or damage to the parathyroid gland, an observation that furnished an important clue as to its

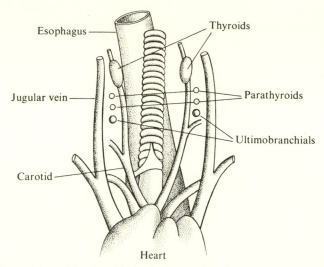

Fig. 2.9. The position of the thyroid gland, the parathyroids, and the ultimobranchial bodies in the domestic fowl, *Gallus domesticus*. (From Copp, Cockcroft, and Keuk, 1967*b*. Reproduced by permission of the National Research Council of Canada.)

possible endocrine significance. The concomitant absence of the C-cells was not initially apparent and only became so after examination of the effects of thyroid extracts on plasma calcium levels.

In order to elucidate the respective roles of the C-cells and the parathyroids in mammals, morphological variations between species have been usefully exploited. In dogs, which are common experimental animals, there are two pairs of parathyroids, one embedded deeply in the thyroid. Rats, however, only have a single pair of parathyroids, which are at the surface of the thyroid and so can be destroyed with a cautery. In neither species is it possible to isolate the blood supply of the parathyroids from that of the thyroid, which contains the C-cells, so that many crucial endocrine experiments cannot be performed on them. Sheep and goats have two pairs of parathyroids and one of these is situated near the thymus, where it has a separate blood supply from that of the thyroid. These animals have played an important role in elucidating the respective roles of the endocrines in calcium metabolism (Hirsch and Munson, 1969). In addition, the pig has also proved to be useful and in this species the thyroid has no attached parathyroid tissue so that one can deal with the C-cells in relative isolation from the former.

In nonmammals the ultimobranchial bodies are usually separated from the thyroid. Nevertheless, in birds, for example, the domestic fowl, they may contain parathyroid tissue (Copp, 1972). In pigeons, calcitonin is found not only in the ultimobranchials but also in the parathyroids and thyroid.

The admixture of these three distinct endocrines may have some fundamental significance but this is unknown. It has been suggested that the differentiation of the C-cells is aided by their association with the thyroid tissue. It is also possible that, as has been observed in the adrenals, some functional symbiosis may occur. Much of the variation, however, would appear to be the result of embryological complications. Although this has certainly helped to hide their effective roles from endocrinologists, it has nevertheless provided some fascinating intellectual exercises during which much of the interspecific variation that initially served to deceive has been productively utilized.

The adrenals

The adrenal glands are so named from their position adjacent to the kidneys. In mammals, they are a composite gland made up of two distinct tissues arranged in two zones, an outer cortex surrounding an inner medulla. The cortex is mesodermal in its origins and secretes several steroid hormones involved in the regulation of intermediary and mineral metabolism. It is called the interrenal or adrenocortical tissue. The medulla, on the other hand, is neural tissue homologous to that of the sympathetic ganglia. Because it stains dark brown with chromic acid it is called chromaffin tissue. It secretes the catecholamine hormones epinephrine (from A-cells) and norepinephrine (from NA-cells). These hormones are also called, respectively, adrenaline and noradrenaline. They have several roles, including the mobilization of fats and carbohydrates, as well as exerting an influence on the tone of the muscle surrounding many blood vessels and the heart. Although the chromaffin and adrenocortical tissues are closely associated in mammals and other tetrapods, they are often quite separate in fishes. In cyclostome fishes the chromaffin tissues are widely dispersed in small islets along certain blood vessels. Putative adrenocortical tissue ("yellow bodies") has been identified embedded in the posterior cardinal veins of the pro- and mesonephroi (Idler and Burton, 1976). This tissue, however, has not been conclusively shown to secret steroid hormones, but it appears to be a target organ for corticotropin, which usually stimulates the activity of such tissue. In bony fishes, including teleosts, holosteans, and the coelacanth, adrenocortical tissues lie along the posterior cardinal veins in the anterior part of the kidney (the head kidney) (see Lagios and Stasko-Concannon, 1979). In the lungfishes it is dispersed more widely. In sharks and rays (Chondrichthyes), the adrenocortical tissue forms a more compact glandular mass lying between the kidneys; hence the name interrenals. In the dogfish (Fig. 2.10*a*) islets of chromaffin tissue lie along the inner borders of the kidneys whereas the interrenal forms a fairly complete mass between them. The adrenocortical tissue of the skate, on the other hand, forms several lobules (Fig. 2.10*b*).

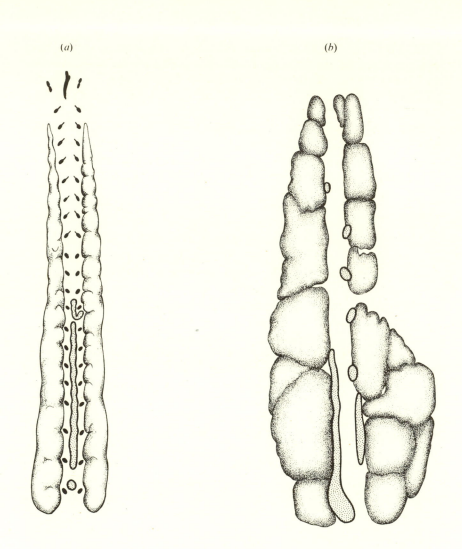

Fig. 2.10. Adrenal tissues in the Chondrichthyes. (*a*) The smooth dogfish (*Mustelus canis*). Double row of black dots = chromaffin tissue lying between the two kidneys. Interrenal (adrenocortical) tissue = stippled. This lies in several pieces between the kidneys. (*b*) The skate (*Raja laevis*). The broken U-shaped interrenal is shown lying between the kidneys. (From Hartman and Brownell, 1949.)

Table 2.2. *Weights of the adrenal medulla and cortex in various species*

Animal	Weight (g)	Medullae (g)	Cortexes (g)	Proportion
Fowl	2000	0.1	0.1	1:1
Dog	15,000	0.25	1.25	1:5
Cat	3000	0.02	0.35	1:17.5
Rat	200	0.0002	0.04	1:20
Rabbit	3000	0.01	0.4	1:40
Guinea pig	500	0.008	0.5	1:62.5

Source: Hartman and Brownell, 1949.

In the Amphibia, chromaffin and adrenocortical tissues are usually associated with each other, lying in islets on the ventral surface of the kidney (Fig. 2.11). Considerable differences can be seen between various species. In urodeles, they are in scattered groups; in *Siren* (Fig. 2.11a) they lie in rows between the kidney, and in *Necturus* and *Amphiuma* (Fig. 2.11b, c), they are on its surface. Anurans, like the leopard frog (*Rana pipiens*), have contiguous strips of adrenal tissue (Fig. 2.11d). It is interesting that in the African lungfish (*Protopterus*), adrenocortical tissues lie in islets along the postcardinal veins and on the ventral surface of the kidney, a pattern similar to that seen in urodeles (Janssens et al., 1965).

In anamniotes, the adrenocortical tissues and the mesonephric kidney have a common embryological origin so that their close association is not unexpected. Amniotes, however, have a metanephric kidney (the mesonephros is not seen in adults) so that the kidneys and adrenals are less intimately connected. The adrenals, more predictably, form separate compact masses of tissue lying near the kidneys. Considerable variations nevertheless still exist, as seen among the different major groups of the reptiles (Fig. 2.12). The chromaffin tissues of reptiles are more closely intermingled with the adrenocortical tissues than they are in amphibians. This admixture of the two tissues is even more apparent in birds (Fig. 2.13) where the adrenals may be fused to form a single gland. Mammals have paired adrenal glands (Fig. 2.14) with a distinct cortex and medulla. It is interesting that this is not as well defined in the echidna *Tachyglossus aculeatus* (an egg-laying monotreme), whose adrenals are considered to be similar to those of reptiles (Wright, Chester Jones, and Phillips, 1957).

The relative amounts of adrenocortical and chromaffin tissues vary in different species. As shown in Table 2.2, there are similar amounts of both in the domestic fowl; in the dog, adrenocortical tissue is five times more predominant, and in the guinea pig there is more than 60 times as much adrenocortical as chromaffin tissue. Adrenocortical tissues may also show considerable variability depending on the season, diet, and physiological condition of the

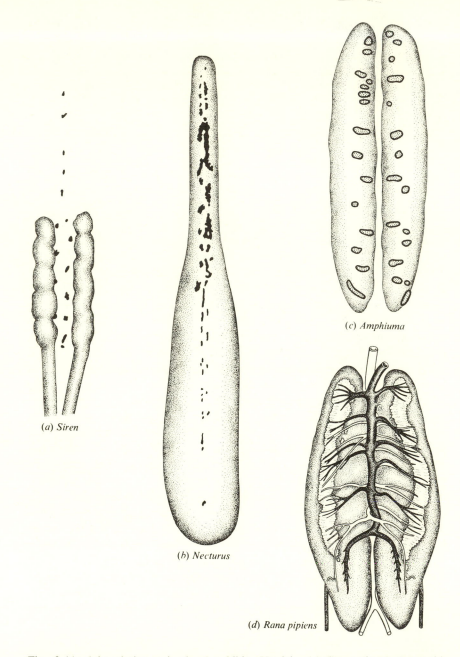

(a) Siren

(b) Necturus

(c) Amphiuma

(d) Rana pipiens

Fig. 2.11. Adrenal tissues in the Amphibia. Urodela: (*a*) *Siren*; (*b*) *Necturus*; (*c*) *Amphiuma*. The adrenal tissue is shown as the dark area lying on the ventral surface or between the kidneys. Anura: (*d*) *Rana pipiens*. The adrenal tissues lie in two strips (light color) along the outer ventral border of each kidney. (From Hartman and Brownell, 1949.)

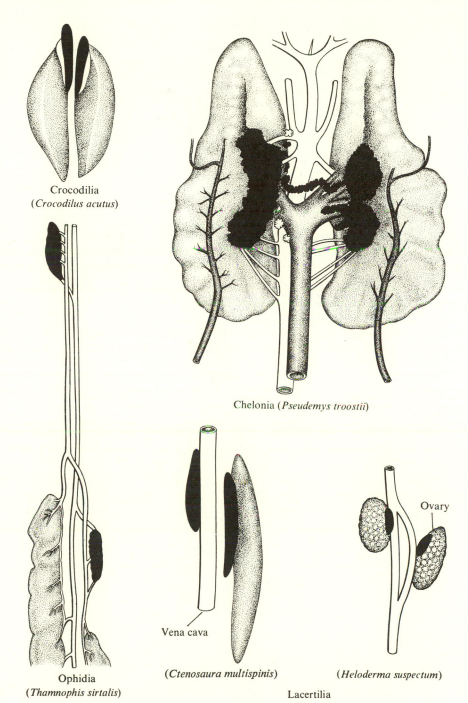

Crocodilia
(*Crocodilus acutus*)

Chelonia (*Pseudemys troostii*)

Ovary

Vena cava

Ophidia
(*Thamnophis sirtalis*)

(*Ctenosaura multispinis*)

(*Heloderma suspectum*)

Lacertilia

Fig. 2.12. The adrenal tissues in the Reptilia. The adrenals are shown in black in relationship to the kidney(s) (shaded). (From Hartman and Brownell, 1949.)

[35]

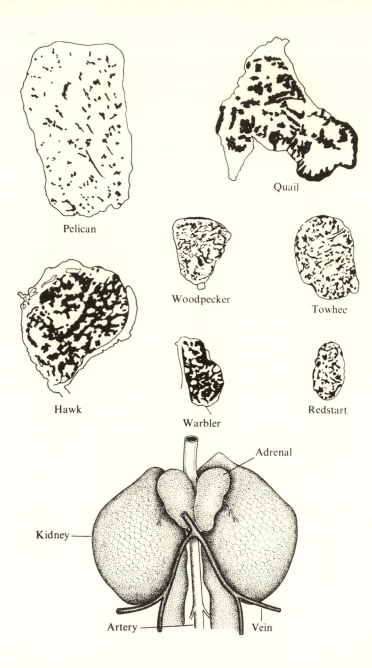

Fig. 2.13. The adrenals in birds. *Top*. Cross section of the adrenal glands from various species showing the distribution of the chromaffin tissue (black) and the adrenocortical tissue (white). *Bottom*. The adrenals of the herring gull (*Larus argentatus*). (From Hartman and Brownell, 1949.)

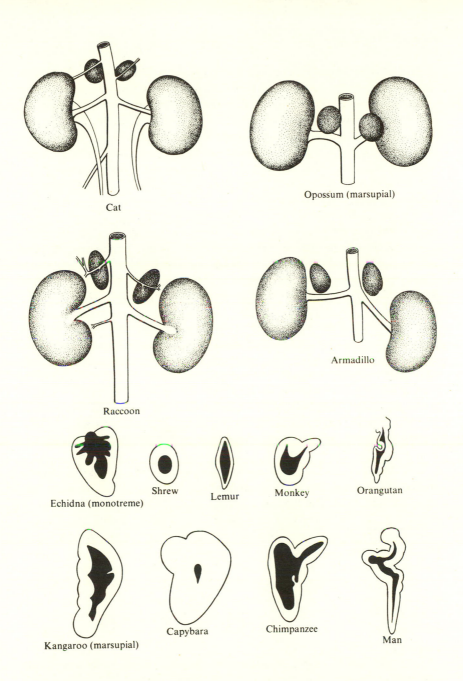

Fig. 2.14. The adrenals in mammals. *Top*. The position of the adrenals in relation to the kidneys in a variety of species. *Bottom*. Cross section of the adrenals from various species showing the chromaffin tissue (black) and adrenocortical tissue (white). (From Hartman and Brownell, 1949.)

animal. It is well known that in many reptiles and amphibians the adrenocortical tissue regresses in winter and proliferates in summer. This can also be related to breeding and has been observed in teleost fish (see, for instance, Robertson and Wexler, 1959; Chan and Phillips, 1971; Lofts, Phillips, and Tam, 1971). Birds from marine habitats, where a lot of salt is present in the diet, have larger adrenals than those species where fresh water is freely available. Glaucous-winged gulls (*Larus glaucescens*) reared with only salt solutions to drink have much larger adrenals than those given fresh water (Holmes, Butler, and Phillips, 1961).

The mammalian adrenal cortex is histologically composed of three types of cells situated in three layers or zones. These are the outer *zona glomerulosa* (round cells, rich in mitochondria and poor in lipids), an intermediate *zona fasciculata* (columnar cells, rich in lipids), and a smaller inner *zona reticularis* (flattened cells poor in lipids) (see Fig. 2.15). The zona glomerulosa is not apparent in all mammals, such as some mice, lemurs, and monkeys. It appears that the three zones are each principally (though possibly not exclusively) involved in the formation of different hormones; the zona glomerulosa forms aldosterone and corticosterone; the zona fasciculata and zona glomerulosa, cortisol and corticosterone. The zona reticularis may secrete androgenic steroids; this zone hypertrophies in certain conditions associated with an excess production of these hormones in man. Aldosterone assists regulation of sodium metabolism in mammals. It is therefore not unexpected to find that the zona glomerulosa can undergo considerable hypertrophy in mammals, such as rabbits and kangaroos, that live in areas where the salt content of the diet is low (Fig. 2.16) (Blair-West et al., 1968).

It should be emphasized that there is a crossover in the abilities of each zone of the adrenal cortex to produce particular steroids. It seems likely (see Tait and Tait, 1979) that the basic difference in the cells is in their ability to respond to the different types of stimuli rather than an innate biochemical distinction. Thus, the zona fasciculata strongly responds to corticotropin while the zona glomerulosa increases its steroid secretion only slightly in the presence of this pituitary hormone. In contrast, the zona glomerulosa secretes lots of aldosterone in response to angiotensin II. Possibly, the cells in each zone possess different types of hormone receptors. Despite a lack of discrete tissue zonation it seems likely that the production of aldosterone by nonmammalian tetrapods also occurs in distinct types of cells, as shown in the bullfrog *Rana catesbeiana* (Varma, 1977).

Some mammals possess an additional type of adrenocortical tissue that usually lies between the medulla and definitive cortex. The fetuses of a number of mammals, especially the primates, have an enlarged region of tissue in this position, which regresses after birth. As the adrenal cortex of the fetus is an important source of steroids, which may even play a role in

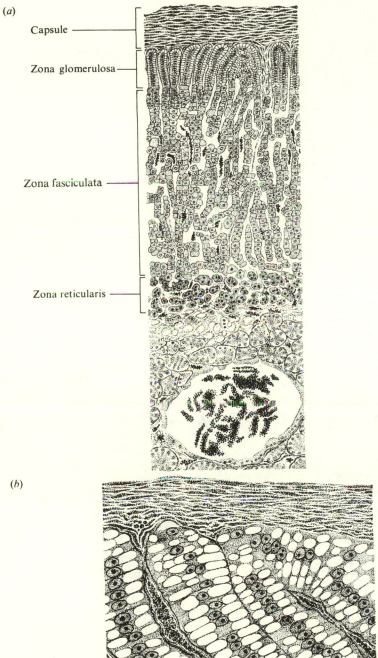

Fig. 2.15. (*a*) Histological section of the adrenal of a mammal, the raccoon, showing the zonation of the adrenal cortex. (*b*) Enlargement of the capsular glomerular zone. (From Hartman and Brownell, 1949.)

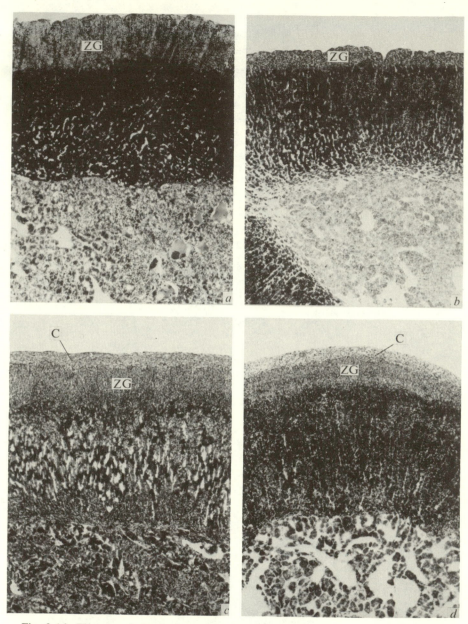

Fig. 2.16. The adrenal glands, in section, from two marsupials from sodium-deficient and sodium-replete areas. *Macropus giganteus* (kangaroo): (*a*) Sodium deficient; (*b*) sodium replete. *Vombatus hirsutus* (wombat): (*c*) Sodium deficient; (*d*) sodium replete. Note that the zona glomerulosa is wider in sodium-deficient animals. In the wombat a thicker capsule lies at the outer border of this cell layer. C = capsule, ZG = zona glomerulosa. (From Blair-West et al., 1968, and J. R. Blair-West, personal communication.)

initiating birth, it has been speculated that this zone may be involved in this event (Vinson and Kenyon, 1979). Several rodents, especially mice, have also been shown to possess a special layer of adrenocortical tissue at this site that has been called the "X-zone" or "transient" zone (see Idelman, 1979). This tissue regresses in the males at puberty and in the females during pregnancy. Its physiological role is unknown. The adrenal cortex of some female marsupials, including the brush-tailed possum, *Trichosurus vulpecula*, wombats, and some kangaroos, is about twice as large, on a unit body weight basis, as that of the male. This difference, especially in the possum, largely reflects the presence of an additional juxtamedullary layer of cells that were originally likened to the X-zone in mice. However, as it exhibits different properties it is now most often called the "special" zone. This tissue, in contrast to the X-zone in mice, hypertrophies during pregnancy and lactation. The collection and analysis of venous blood and *in vitro* incubation experiments using pieces of this tissue zone show that it can produce androgenic steroids. It appears to be under the control of pituitary gonadotropins rather than corticotropin.

Intraspecific differences in the size of the adrenals, the volumes of the cortex and medulla, and the cells of the cortex may be determined genetically. Shire (1970) has observed such differences in several strains of mice. In one strain, "CBA," both the cortex and medulla are larger than in another, "strain A." The volume of the medulla in CBA mice is $0.35 \text{ mm}^3/25$ g body weight, whereas it is only 0.18 mm^3 in strain A. Similarly, the cortex has a volume of 1.5 mm^3 in the CBA strain and 0.82 in the other mice. The differences have been shown to reflect genetic variation at one or two gene loci for the cortex and at at least two such loci in the case of the medulla. The CBA mice have a well-developed zona glomerulosa, but attempts to correlate this with more production of aldosterone (as compared to a less well-endowed strain) have not been successful (Stewart et al., 1972).

Among the vertebrates, there appears to be an evolutionary trend toward a more intimate association of the adrenocortical and the chromaffin tissue. This tendency may partially reflect their embryogenesis, such as the tissue aggregation that follows the loss of the mesonephros. The relationship of the two endocrine tissues has functional significance. The adrenocortical tissue can certainly function in the absence of the chromaffin tissue as seen *in vitro* in the laboratory. However, the chromaffin cell's ability to convert norepinephrine to epinephrine depends largely on the presence of corticosteroids. The NA-cells (which secrete norepinephrine) can even be converted to A-cells during neonatal life if they are exposed to corticosteroids (Coupland, 1968). The adrenal medulla contains an enzyme, phenylethanolamine-*N*-methyltransferase (PNMT), that, by methylating norepinephrine, converts it to epinephrine. The formation of this enzyme is induced by steroid hormones from the adrenal cortex (Pohorecky and Wurtman, 1971). The concentrations of the steroids must be high, far higher than normally present in the systemic circulation. This is achieved by the direct transfer of the steroids to the medulla through a local portal blood system. PNMT activity has also been identified in nonmammals, and the association of chromaffin and adrenocortical tissue in other vertebrates may have an important, even determining, role in their abilities to form epinephrine.

The endocrine hormones of the gut

The gut is the largest and most diffuse endocrine organ. Several hormones are formed and released from the posterior part of the foregut and its derivative glands (gastroenterohepatic endocrine system, GEP). The most notable of these are insulin, glucagon, and somatostatin from the pancreas; gastrin from the stomach; and secretin, cholecystokinin–pancreozymin, enteroglucagon, and gastric inhibitory peptide from the duodenum and upper parts of the jejunum. These hormones integrate and control the processes that result from feeding, which include the secretion of acid, digestive enzymes, and the concentrations of the absorbed nutrients in the blood.

The gastrointestinal tract provided the first hormone to be scientifically identified. This was secretin and it is formed by the S-cells, which lie in the crypts and villi of the duodenum and jejunum. In 1938, F. Feyrter described the numerous but scattered types of cells called "clear cells." He proposed that these cells, which did not stain by the methods he used, had a paracrine function and helped to regulate the activities of their neighbors. Many such types of cells have now been identified (Table 2.3) along the entire length of the gut, where they each have a distinct but extended and overlapping distribution (Fig. 2.17). Many of these cells have been shown to send tufts of microvilli into the lumen of the gut ("open-type"), by which means it is thought that they receive stimuli and so can sense or "taste" acidity and the presence of products of digestion. Mainly with the aid of immunocytochemical techniques (see Larsson, 1980), the endocrine and paracrine products of many such cells have been identified. For instance, secretin was found to be produced by the S-cells, gastrin by G-cells, cholecystokinin (CCK) by I-cells, and so on (see Table 2.3). The official terminology for such types of cells has frequently been changed, and the cells are now often called by the name of their hormonal product: For instance, I-cells are also called CCK-cells. Most information about such gut cells is related to mammals but the phyletic distribution of a few has been followed in other vertebrates. Thus, the somatostatin-containing D-cells in the gut have also been identified in birds and reptiles but were not observed in teleost or cyclostome fishes (Seino, Porte, and Smith, 1979*a*). The gut of the sea squirt *Ciona intestinalis* (Urochordata) has been shown to contain cells that are similar histologically to those that secrete

Table 2.3. *The Lausanne 1977 classification of gastroenteropancreatic endocrine cells*

Pancreas	Stomach		Small intestine		Large intestine	Function proposed or ascertained
	Oxyntic	Pyloric	Upper	Lower	intestine	
[P]	P	P	P			Bombesin-like?
(EC)	EC	EC	EC	EC	EC	5-HT (substance P, EC_1) (motilin, EC_2) (others? EC_n)
D_1	D_1	D_1	D_1	D_1	D_1	(VIP-like) Disputed
PP	(PP)	(PP)	(PP)	(PP)	(PP)	Pancreatic polypeptide
D	D	D	D			Somatostatin
B						Insulin
A	[(A)]		?			Glucagon
	X	(X)				Unknown
	ECL					Unknown (H or 5-HT)
		G	G			Gastrin
			S	S		Secretin
[(G)]		I	I			Cholecystokinin
			K	K		GIP
				N		Neurotensin
			L	L	L	GLI

Note: [ ], Fetus or newborn, not or only exceptional in adult beings; ( ), animals, not in man.
Source: Solcia et al., 1978.

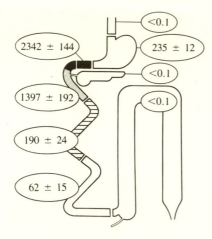

Gastrin

Somatostatin

Secretin

Cholecystokinin

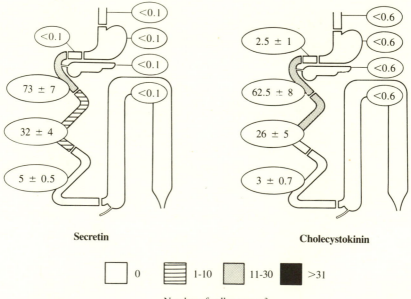

Number of cells per mm²

Fig. 2.17. The distribution of some gut hormones in the different regions of the gastrointestinal tract of man. The number refers to the mean concentrations (in pmole/g) in the whole thickness of gut sections. The shading indicates the density of each cell type, as observed by quantitative immunocytochemistry. (From Bloom and Polak, 1978.)

polypeptides in vertebrates (Fritsch and Sprang, 1977; Falkmer et al., 1978). The hormone-secreting cells of the gastroenterohepatic endocrine system thus may have had very ancient forebears.

Insulin, glucagon, and somatostatin are found in special tissues situated in the pancreas called (after their discoverer) *islets of Langerhans* (see Falkmer and Patent, 1972; Falkmer and Östberg, 1977). Several types of cells have been identified in these tissues. The hormone-secreting cells contain granules that exhibit characteristic differences in each cell type. They are named in several ways: *A-cells* (also called α- or α_2-), which form glucagon, *B-cells* (β-), forming insulin, and *D-cells* (δ- or α_1-), which contain somatostatin. B-Cells are present in all vertebrates whereas A-cells are absent in cyclostome fishes and possibly some urodele amphibians. The D-cells have been identified as islet-like tissues in cyclostome fishes, chondrichthyeans, and the Brockmann bodies of teleosts (Falkmer et al., 1977; Falkmer, Östberg, and van Noorden, 1978). There are other types of cells present in the islets of Langerhans but the role of each of these is not clear. One of them, however, has been shown to secrete a peptide called *pancreatic polypeptide*, and the cells have thus been called *PP-cells*. Insulin-like hormones have also been identified in a number of invertebrates, and cells that react with insulin antibodies have been identified in the gut of two molluscs (Plisetskaya et al., 1978). These cells had a structure that was reminiscent of the B-cells in cyclostomes.

The morphological disposition of hormone-secreting cells of the pancreas differs considerably among vertebrates (for summary, see Table 2.4). They are most often associated with the exocrine pancreas. This gland is formed from one or more diverticula of the gut and contains exocrine acinar cells and their ducts. The endocrine secretory cells are formed from the ducts, and in several fishes they lie around them. The cyclostome fishes lack such a discrete pancreas, and the B-cells and D-cells are found in the submucosal part of the anterior region of the intestine or around the bile duct.

In mammals the endocrine cells lie in small groups (or islets) that are about 100–200 μm in diameter. They are quite distinct from the exocrine tissue. A man has about 2 million of these islets and the guinea pig has 15–40 thousand. These islets may be an admixture of the three or more types (A-, B-, D-, and PP-cells) of cells or principally composed of a single type, as seen in rabbits and rats. B-cells are about five times more common than A-cells in mammals. Cyclostome fishes only have B- and D-type cells. In the hagfishes (Myxinoidea), they are situated in a distinct gland, around the bile duct near its entrance to the intestine (Fig. 2.18a). Lampreys (Petromyzontia) also have a well-defined region of B-cells that extends along the submucosa of the duodenum. The B-cells form follicles (*follicles of Langerhans*, Barrington, 1942) enclosing a cavity containing material that may represent the storage site for the insulin.

Table 2.4. *Evolution of the endocrine pancreas in vertebrates*

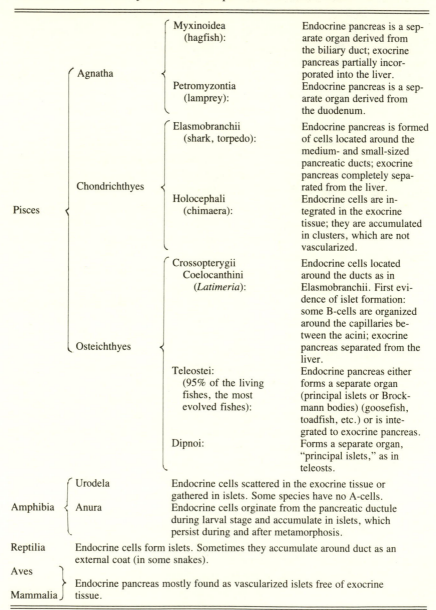

Pisces	Agnatha	Myxinoidea (hagfish):	Endocrine pancreas is a separate organ derived from the biliary duct; exocrine pancreas partially incorporated into the liver.
		Petromyzontia (lamprey):	Endocrine pancreas is a separate organ derived from the duodenum.
	Chondrichthyes	Elasmobranchii (shark, torpedo):	Endocrine pancreas is formed of cells located around the medium- and small-sized pancreatic ducts; exocrine pancreas completely separated from the liver.
		Holocephali (chimaera):	Endocrine cells are integrated in the exocrine tissue; they are accumulated in clusters, which are not vascularized.
	Osteichthyes	Crossopterygii Coelocanthini (*Latimeria*):	Endocrine cells located around the ducts as in Elasmobranchii. First evidence of islet formation: some B-cells are organized around the capillaries between the acini; exocrine pancreas separated from the liver.
		Teleostei: (95% of the living fishes, the most evolved fishes):	Endocrine pancreas either forms a separate organ (principal islets or Brockmann bodies) (goosefish, toadfish, etc.) or is integrated to exocrine pancreas.
		Dipnoi:	Forms a separate organ, "principal islets," as in teleosts.
Amphibia	Urodela		Endocrine cells scattered in the exocrine tissue or gathered in islets. Some species have no A-cells.
	Anura		Endocrine cells orginate from the pancreatic ductule during larval stage and accumulate in islets, which persist during and after metamorphosis.
Reptilia			Endocrine cells form islets. Sometimes they accumulate around duct as an external coat (in some snakes).
Aves Mammalia			Endocrine pancreas mostly found as vascularized islets free of exocrine tissue.

Source: Pictet and Rutter, 1972.

The Chondrichthyes also have different distribution patterns of A-, B-, and D-cells. In the Selachii (sharks and rays), the cells are usually situated near or around the ducts of the exocrine pancreatic tissue. In the Holocephali (chimaeroid fishes), this tissue extends more among the exocrine cells though it retains its association with the ducts. (see Fig. 2.18*b*).

There is also considerable diversity in the arrangement of the A-, B-, and D-cells among bony fishes (Osteichthyes). Some teleosts (toadfish and goose-fish) show an aggregation of these cells into two pea-sized glands called *principal islets* or *Brockmann bodies* (Fig. 2.18*c*). Others, such as the eel (Fig. 2.18*c*), have islets of tissue scattered among the exocrine tissue as in mammals. In lungfishes (Dipnoi), the tissue is congregated into teleostean-like principal islets; the coelacanth (Crossopterygii) has a Selachii-type arrangement where the tissue is associated with the ducts of the acinar cells.

The islets of Langerhans in the Amphibia are usually quite small and contain mainly B-cells; some urodeles may even lack A-cells completely. The reptiles and birds have a much higher proportion of A-cells, which probably accounts for the large amount of glucagon that can be extracted from their pancreases. In birds three types of islets are present: "dark" islets containing mainly A-cells, "light" islets with B-cells, and mixed islets with both types. As in mammals, the islets in birds contain D-cells at their periphery, but the relative distribution in the three types of islets does not appear to have been described.

What is the significance of all this morphological variation in the distribution of A- and B-cells? "Why are the islets of Langerhans?" (Henderson, 1969). Why, in so many different vertebrates, is the endocrine tissue dispersed so widely among the exocrine tissue? Although in some fishes the endocrine tissues may tend to retain the site of their embryonic origins from the acinar duct epithelium, there seems to be an evolutionary tendency toward aggregation of the cells. This aggregation sometimes results in relatively large pieces of tissue, the principal islets of the teleosts and dipnoans. Among tetrapods, the small islet arrangement is universal. Henderson (1969) has conjectured that this may reflect a dependence of the exocrine acinar cells on the endocrine tissues that secrete insulin, glucagon, and somatostatin. The small islets increase the surface area of the endocrine tissues and promote their contact with the exocrine tissue so that high local levels of hormones can be maintained. It has indeed been noticed that, in diabetes mellitus (a lack of insulin) in man, the exocrine pancreas tends to atrophy and become invaded with excess fat and fibrous tissue. It has also been observed that the level of the pancreatic digestive enzyme amylase declines in rats with experimental diabetes mellitus, but this defect can be corrected by injected insulin (Korc et al., 1981). The admixture of the A-cells and the B-cells also may have functional significance, and there is evidence to suggest that glucagon may have an insulinogenic effect on the A-cells.

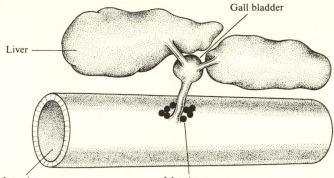

(a)   CYCLOSTOME (*Myxine*)

Gall bladder

Liver

Intestine

Islet tissue

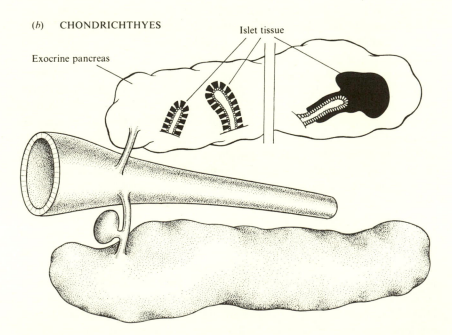

(b)   CHONDRICHTHYES

Islet tissue

Exocrine pancreas

Fig. 2.18. The various types of pancreas in fishes. (*a*) Cyclostome-type (*Myxine*): ring-like arrangement around the bile duct. (*b*) Chondrichthyean-types. Left = many elasmobranches; right = Holocephali.

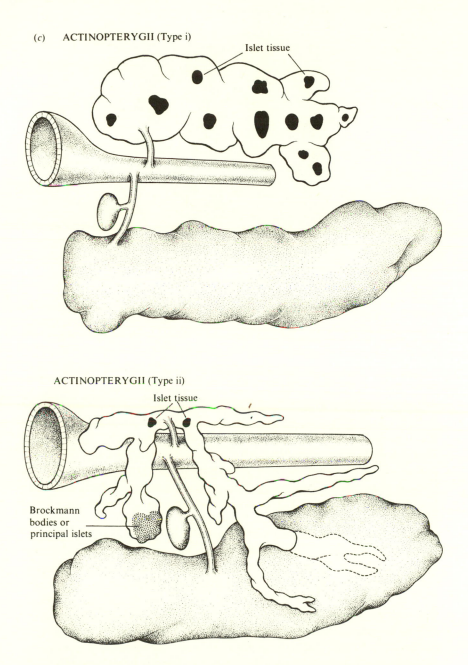

(c) ACTINOPTERYGII (Type i)

Islet tissue

ACTINOPTERYGII (Type ii)

Islet tissue

Brockmann
bodies or
principal islets

(c) Actinopterygian-types. (i) *Anguilla*; this is present in a few teleosts and is similar to that in tetrapods. (ii) The more general teleost with Brockmann bodies or "principal islets." (From Epple, 1969.)

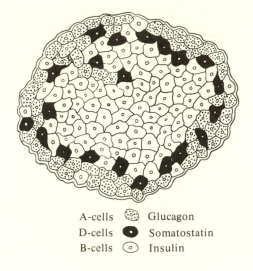

A-cells Glucagon
D-cells Somatostatin
B-cells Insulin

Fig. 2.19. Schematic representation of an islet of Langerhans in man showing the distribution of glucagon, somatostatin, and insulin-containing cells. Islet cell types for which a function has not yet been positively established are omitted. (From Orci and Unger, 1975.)

An idealized reconstruction of a human islet of Langerhans is shown in Fig. 2.19. There is a core of B-cells surrounded on the peripheral surface by A-cells. Underlying the latter and also in contact with the B-cells, are D-cells. It has been suggested that the inner B-cells supply the basic insulin needs and that finer control of blood glucose concentration is exerted by the outermost B-cells and glucagon released from the A-cells. This latter process is inhibited by the release of somatostatin from the D-cells whereas both hormones can increase or decrease, respectively, the release of insulin from their adjoining B-cells. Morphological associations of B- and D-cells have been observed in the GEP system of other vertebrates (Falkmer et al., 1977).

As has been described, insulin-like immunoreactivity has been identified in cells that are present in the gut of several invertebrates, including the mollusc *Mytilus edulis* and the protochordate *Amphioxus*. These observations are consistent with reports of insulin-like activity in tissue extracts from nonvertebrates. Such cells, however, are not aggregated as islets, and it has been suggested that such an arrangement may reflect their primeval condition as secretory cells in the gut. Islets of B-cells first occur in the cyclostomes and as they are not associated with pancreatic acinar cells they may represent a transitional evolutionary situation between invertebrates and gnathostome vertebrates.

The APUD concept – a hypothesis regarding the origins of peptide-secreting endocrine cells

The cells of endocrine tissues that secrete peptide hormones have a number of common ultrastructural and cytochemical characteristics. In many instances these properties are shared with nerve cells. Endocrine cells associated with regions such as the gut, pharynx, and brain may synthesize similar peptides. Thus, somatostatin is present in the hypothalamus as well as the D-cells of the islets of Langerhans and gut. Gastrin is a peptide that is secreted by the G-cells of the stomach, but gastrin-like immunoreactivity has also been identified in the brain. In 1968, A. G. E. Pearse proposed that many types of endocrine cells that secrete peptide hormones may share a common ancestry and originate embryologically in nervous tissue. A large part of the endocrine system may thus essentially be a neuroendocrine extension of the nervous system, which, as its cells often do not aggregate in distinct masses, has been called the "diffuse neuroendocrine system" (Pearse and Takor, 1979). This system has two divisions: a central one, which includes the pars distalis, hypothalamus, and pineal body, and a peripheral part consisting of the gut, islets of Langerhans, parathyroids, thyroid C-cells, and the adrenal medulla.

These peptide-secreting cells, apart from sharing structural features, also, like many nerve cells, may contain amines or have an ability to metabolize such compounds. This may include the process of the uptake of dopamine and 5-hydroxytryptamine, or that of their precursors, dopa and 5-hydroxytryptophan. These amino acids may be converted to the amines by decarboxylation. These types of cells were then appropriately called APUD cells, which is an acronym for amine content and/or precursor uptake and decarboxylation. It has been proposed that cells exhibiting such properties have a common embryological origin from neuroectoderm in the neural crest. During development they may migrate from this site and, following differentiation, assume their final special functions. The adrenal medullary and thyroid C-cells have been clearly shown to be formed in this manner, but there remain considerable doubts as to whether it applies to all such cells, including those of the gut (see Fontaine, Le Lièvre, and Le Douarin, 1977; Pearse and Takor, 1979; Dockray and Gregory, 1980).

The gonads

Although the ovaries and testes are not essential for the survival of the individual, they are for the propagation of the species. They have a dual but related function: the production of ova and sperm as well as several hormones that are concerned with the development of the germ cells and the fertilized egg.

The gonads are formed from the dorsal celomic epithelium. The adrenocortical (or interrenal) tissue has a similar origin, and both secrete related steroid hormones. The primordial gonadal tissue consists of a cortex and a medulla. The latter differentiates into the testis, the former the ovary. The steroidogenic cells that are present in the gonads (as well as the adrenals) have a distinctive structure. They contain a very well developed endoplasmic reticulum, mitochondria that have tubular cristae, and usually lipids that histochemically behave like cholesterol. A notable characteristic that helps histological identification of steroidogenic cells in the gonads is the presence of the enzyme 3β-HSDH (Δ^5-3β-hydroxysteroid dehydrogenase). It is responsible for the conversion of certain precursors of the steroid hormones into progesterone and androstenedione, which may subsequently be converted to other hormones.

The gonads are usually paired structures lying in the body cavity near the kidneys. The testes of most mammals (except a few such as whales, the elephant, and guinea pig), however, are suspended outside the abdominal cavity in the scrotal sac. Some species have only a single gonad, which usually reflects the degeneration of the other or possibly their fusion. The cyclostome fishes only have a single testis and ovary in the median line, whereas in nearly all birds only the left ovary reaches full development. A single ovary is also sometimes present in teleost and chondrichthyean fishes.

Considerable diversity exists in the relationship of the gonads to their excretory ducts. These are lacking altogether in cyclostomes where the eggs and sperm are released into the body cavity and thence through pores to the exterior. Some fishes, especially teleosts, have gonaducts that are merely extensions of the ovaries. In most vertebrates, including many bony fishes, sharks and rays, and tetrapods, the germ cells pass through a homologous series of ducts derived from the Wolffian duct in the male and the Müllerian duct in the female.

Testis

The testis shows a considerable degree of uniformity among the vertebrates (see Dodd, 1960; Lofts, 1968; Lofts and Bern, 1972). The formation of sperm occurs in two principal ways. In amniotes, it differentiates from the germinal epithelium situated in the wall of the seminiferous tubules. Anamniotes, on the other hand, do not have such continuous tubules but those they have may be branched and divided into lobules. The sperm in these instances is differentiated into groups contained within small envelopes called "cysts" (cystic spermatogenesis). Two principal types of cells are concerned with the secretion of the male sex hormones. (1) In tetrapods (except for urodele amphibians) groups of cells lying between the seminiferous tubules and called *interstitial* or *Leydig cells* (Fig. 2.20a) are the site of formation of testosterone. In

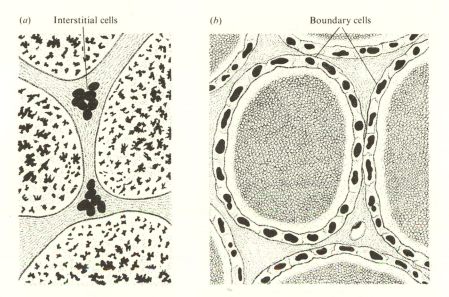

Fig. 2.20. The testis. (*a*) Sections from a reptile, the viper, showing the lipid-filled interstitial (Leydig) cells. (*b*) A teleost fish, the pike, showing the boundary cells in the wall of the testicular lobule. These are the homologues of the interstitial cells that are absent in most of the fishes. (From Gorbman and Bern, 1962, based on data from B. Lofts and A. J. Marshall.)

urodele amphibians and fishes, the walls of the testicular lobules contain cells called "boundary cells" that apparently have a similar role (Fig. 2.20*b*). Embryologically, the interstitial and boundary cells are thought to be homologous. Some fishes, including lampreys, possess tetrapod-like interstitial cells. (2) Associated with the basement membrane of the seminiferous tubules are the *Sertoli cells*. Their function has for a long time been controversial and it was suggested that they were concerned with the nourishment of the sperm (hence they are also called "sustentacular cells"). It now seems clear that they are the site of production of sex hormones that are probably involved in the growth and maturation of the sperm. These cells have been identified in nearly all vertebrates that have been examined. In those species that have a cystic spermatogenesis, they lie juxtaposed to the heads of the sperm in the spermatic cysts.

The testes undergo considerable structural changes associated with the periodic, or cyclical, breeding behavior. These changes are reflected in the size and lipid content of the interstitial and Sertoli cells.

The ovaries

The ovaries contain several distinct types of cells and tissues (see Dodd, 1960; Lofts and Bern, 1972) (Fig. 2.21). (1) The *germinal epithelium* that envelops

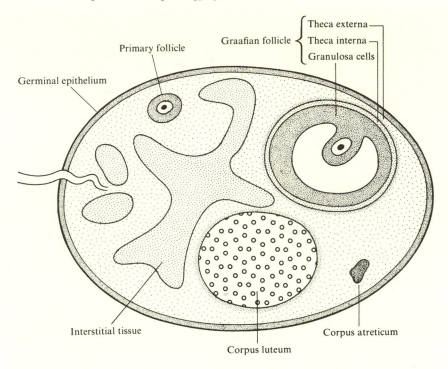

Fig. 2.21. The ovary. Diagrammatic representation of the various tissues present. (From Lofts and Bern, 1972.)

the ovaries and from which the developing eggs and their surrounding sheet of granulosa cells are formed. (2) The latter is called the *theca interna*, which, in turn, is surrounded by a fibrous *theca externa*. The complete structure, ovum and its surrounding membranes, is called the *Graafian follicle*. (3) Following the maturation and expulsion of the ova the granulosa cells may form the *lutein cells* that compose the *corpus luteum*. Also present in the ovary are structures formed as the result of the degeneration (atresia) of unovulated follicles, the *corpora atretica*. (4) Lying in between all these structures is the *interstitial tissue*, which anatomically, and probably functionally, is analogous to the tissue of the same name in the testis.

Estrogenic hormones are formed by the cells of the theca interna and the granulosa cells of the developing follicles. Progesterone arises from the lutein cells of the corpus luteum and, at least in the rat, the interstitial tissue. The latter also produces androgenic hormones.

The basic structure and endocrine role of the ovaries persist throughout the vertebrates. There are some points worthy of comment. The corpus luteum is not present in all species, especially oviparous ones. It is absent in birds, which nevertheless have progesterone in their ovaries and blood. Corpora lutea appear sporadically among other vertebrate groups. They are present in ovoviviparous and viviparous amphibians, chondrichthyeans, and some teleosts. A corpus luteum also appears following ovulation in *Myxine* (hagfish) whereas all reptiles (including oviparous ones) develop a corpus luteum after ovulating. Indirect, histological, and more direct chemical evidence indicates that it produces progesterone in all groups. It is possible that the corpora atretica also produce hormones, but many consider this to be doubtful.

The juxtaglomerular apparatus

The kidney is the site of formation of a protein called *renin*, considered by some to be an enzyme and by others a hormone, which can be released into the circulation. This secretion initiates the formation of a peptide, called *angiotensin*, in the plasma, which contributes to the regulation of sodium retention in the body and possibly also increases blood pressure. Renin is formed by cells situated near the renal glomerulus at a site called the juxtaglomerular apparatus. In mammals (see Barajas, 1979) it consists of *juxtaglomerular cells* (on the afferent glomerular arteriole) and the *macula densa*, which is a thickening of the distal renal tubule in the region where it abuts onto a glomerular area between the glomerular arterioles called the *polkissen* (or extraglomerular mesangium) (Fig. 2.22*a*). In nonmammals the situation is less complex (Fig. 2.22*b*) as the macula densa and polkissen are absent (the former may be present in birds). Many species, however, still have the juxtaglomerular cells (see Fig. 2.23). They contain "granules" that can be stained histologically in a distinctive way (Bowie's method). Such juxtaglomerular granules occur in arterioles often distant from glomeruli in teleost, dipnoan, and coelacanth fishes and even in aglomerular fishes but in most vertebrates they occur in the afferent glomerular arteriole. The latter type of cells is present in birds, reptiles, amphibians, and teleost fishes but neither these cells nor Bowie's granules have been identified in the elasmobranchs or cyclostomes (see Ogawa et al., 1972). These observations parallel the identification of renin in the kidneys of these groups. An inconsistency exists in that juxtaglomerular cells with granules stainable by the Bowie method have not been found in many of the Chondrostei or Holostei even though there is evidence to suggest the presence of renin in such fish. Renin has also been tentatively identified in the corpuscles of Stannius in teleost fishes (Chester Jones et al., 1966; Sokabe et al., 1970), but it is not clear if Bowie's granules are also present in these tissues (compare for instance Krishnamurthy and Bern, 1969; Sokabe et al., 1970).

An additional putative component of the juxtaglomerular complex has been identified recently (Ryan, Coghlan, and Scoggins, 1979). Granulated epithelial-

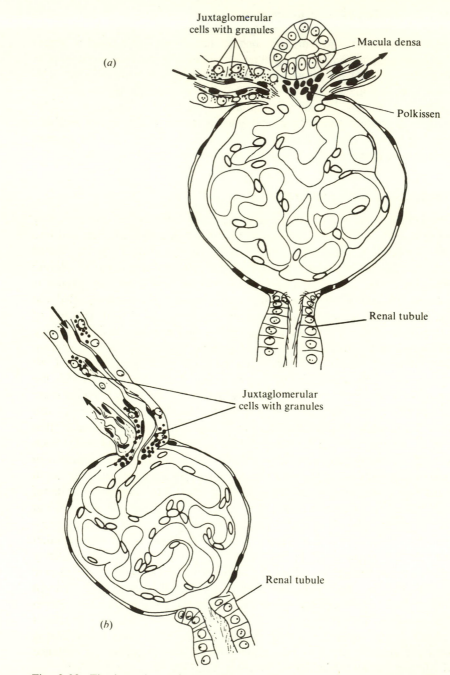

Fig. 2.22. The juxtaglomerular apparatus. (*a*) The laboratory rat. (*b*) The bullfrog (*Rana catesbeiana*) showing the absence of a macula densa and polkissen. (From Sokabe et al. 1969.)

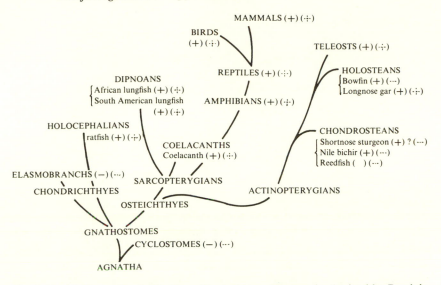

Fig. 2.23. Phylogenetic distribution of juxtaglomerular granules (stained by Bowie's method) and renin activity in the kidneys of vertebrates: (+) or (−), presence or absence of renin; (∹) or (⋯), presence or absence of granules. It can be seen that while the two are always associated with each other in tetrapods and most fishes this is not so in all holosteans or chondrosteans. (From Nishimura, Ogawa, and Sawyer, 1973, modified by H. Nishimura.)

type cells encircling the polar region of the glomerular tuft have been observed in several mammals, including sheep and man, though they are much smaller and are difficult to identify in rats and mice. They have been called granulated peripolar epithelial cells. It has been suggested that these cells could be secreting a product into the glomerular filtrate that may influence the absorption of fluid from the proximal tubule.

The macula densa may be concerned with the regulation of the release of renin in mammals (and birds?). Renin and angiotensin contribute to the regulation of sodium levels in the bodies of mammals. It has been suggested that local changes in the sodium and/or chloride permeability of the macula densa may regulate the release of renin from the juxtaglomerular cells (Peach, 1977). Sodium depletion increases the release of renin and this response has been shown to be dependent on a renal "vascular receptor" (Gotshall et al., 1973). In this respect it is noteworthy that a reduction in renal blood flow, such as results from hemorrhage or stimulation of the renal nerves, also promotes renin release in mammals. This response can be blocked by a β-adrenergic blocking drug (propranolol) that acts distally to the renal vascular muscle (Coote et al., 1972). Thus, several steps appear to be involved in the release of renin but more precise evidence as to the role (if any) of the macula

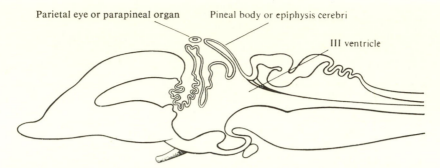

Fig. 2.24. The pineal body in relation to the brain (in section) of a lower vertebrate. (From Bargmann, 1943.)

densa is lacking. It is interesting that nonmammals may also utilize the renin–angiotensin system to aid sodium regulation in the body and as these animals lack a macula densa it presumably does not have an essential role in this process in all vertebrates. It remains possible, however, that it could mediate some effect on the release of renin that is special to the mammals.

Putative endocrine glands

The pineal body

Some argue that, at least in mammals and probably also in birds, the pineal body deserves the title of gland and should not be classified as a putative endocrine gland. It forms melatonin, which exerts an antigonadotropic action in some vertebrates and pales the skin in larval amphibians and cyclostomes. A peptide resembling neurohypophysial vasotocin has also been identified in the pineal that may have a similar effect on the gonads.

The pineal (see Ariëns Kappers, 1965, 1970; Oksche, 1965; Wurtman, Axelrod, and Kelly, 1968) originates as a sac-like evagination from the dorsal part of the brain (the diencephalon) (Figs. 2.24 and 2.25). It lies beneath the cranium in the midline position and is also called, especially in mammals, the *epiphysis cerebri*. Embryologically, its hollow stem maintains contact with the III ventricle (into which secretion may occur) and this pattern persists in many nonmammals.

The pineal body is present in most vertebrates. It is absent, however, in hagfish (myxinoid cyclostomes), crocodiles, at least two species of chondrichthyeans (*Torpedo ocellata* and *T. marmorata*), and several mammals, including whales. The pineal is very small in the elephant and rhinoceros.

In man, the pineal is a relatively simple knob of tissue but in other vertebrates it is more complex. It appears to have undergone considerable changes in its structure (and no doubt function) during its evolution. Noteworthy is the differentiation of an associated tissue that comes to lie nearer to the roof of the skull. This is variously called the *parapineal* (in fishes), *frontal-* (in amphibians), and *parietal-* (in reptiles) *organ*. In many species, especially lizards, it penetrates through the brain case and lies just beneath the skin. The parapineal is connected to the roof of the brain, or the pineal body proper by nerve cells. In fishes, amphibians, and certain reptiles (not the snakes), the parapineal and even the pineal body contain photoreceptor cells. In lizards and the tuatara (Rhyncocephalia), the pineal takes the form of a well-differentiated "third eye" that even contains structures analogous to the cornea and lens of lateral eyes. In such vertebrates the pineal has a sensory function.

The pineal of mammals lacks functioning photosensory cells but these have been modified to form secretory cells (*pinealocytes*) that seem to have an endocrine function. The presence of photosensory cells in the bird pineal is equivocal but pinealocytes are present. Whether or not the pineal of other vertebrates secretes hormone-like products is uncertain, but seems likely. In the simplest interpretation of pineal evolution, it may be viewed as evolving from a sensory structure in fish to an endocrine gland in birds and mammals. In intermediate forms it may possess both types of activity, and the photosensory cells have thus also been called secretory rudimentary photoreceptor, or SRP, cells.

In addition to the pinealocytes the pineal also contains other types of cells, including neurons and an ependymal-type called supporting (or supportive) cells. These three types of cells usually have a tubulofollicular arrangement, the supporting cells sending microvilli into luminal space. Granules have been identified in all these cells that may correspond to storage sites for amines and peptides. The pinealocytes are usually considered to be the principal site of melatonin and, possibly, vasotocin secretion, but the contributions of the other cells have not been satisfactorily delineated.

Embryologically, the pineal body is a hollow sac-like structure and this persists in most vertebrates. In mammals, it is filled with parenchymous cells (the pinealocytes). Some birds (many Passeriformes) exhibit the primitive hollow condition whereas others, like the domestic fowl, have a solid pineal containing pinealocytes. In mammals and birds, the pineal organ has an autonomic sympathetic innervation arising from the superior cervical ganglion, the destruction of which in rats has profound effects. In other vertebrates, the pineal seems to maintain direct neural contact with the brain, but an autonomic innervation is also evident in fish, amphibians, and reptiles. The pineal has been shown to form several substances that have potent biological effects. These include norepinephrine (in nerve cells) and serotonin, melatonin, and, possibly, vasotocin (in the pinealocytes). Melatonin has been identified in the circulation and may represent the pineal's endocrine product.

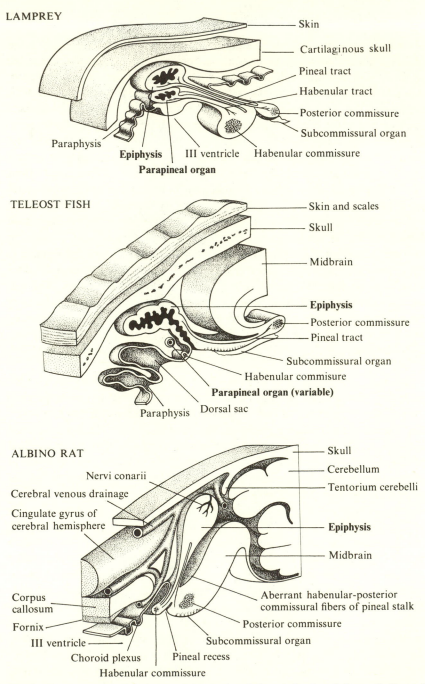

LAMPREY

Skin

Cartilaginous skull

Pineal tract

Habenular tract

Posterior commissure

Subcommissural organ

Paraphysis

Epiphysis   III ventricle   Habenular commissure

Parapineal organ

TELEOST FISH

Skin and scales

Skull

Midbrain

Epiphysis

Posterior commissure

Pineal tract

Subcommissural organ

Habenular commisure

Parapineal organ (variable)

Paraphysis   Dorsal sac

ALBINO RAT

Skull

Cerebellum

Tentorium cerebelli

Nervi conarii

Cerebral venous drainage

Cingulate gyrus of
cerebral hemisphere

Epiphysis

Midbrain

Corpus
callosum

Aberrant habenular-posterior
commissural fibers of pineal stalk

Fornix

Posterior commissure

III ventricle

Subcommissural organ

Choroid plexus   Pineal recess

Habenular commissure

Fig. 2.25. The pineal of various vertebrates in relation to the dorsal diencephalic roof region. (From Wurtman, Axelrod, and Kelly, 1968.)

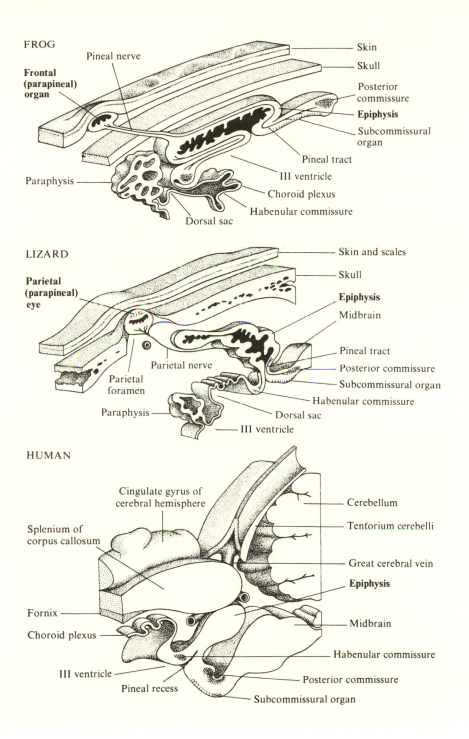

FROG

Pineal nerve

Frontal
(parapineal)
organ

Skin

Skull

Posterior
commissure

Epiphysis

Subcommissural
organ

Pineal tract

III ventricle

Choroid plexus

Habenular commissure

Paraphysis

Dorsal sac

LIZARD

Parietal
(parapineal)
eye

Skin and scales

Skull

Epiphysis

Midbrain

Parietal nerve

Pineal tract

Posterior commissure

Subcommissural organ

Parietal
foramen

Habenular commissure

Paraphysis

Dorsal sac

III ventricle

HUMAN

Cingulate gyrus of
cerebral hemisphere

Cerebellum

Tentorium cerebelli

Splenium of
corpus callosum

Great cerebral vein

Epiphysis

Fornix

Midbrain

Choroid plexus

Habenular commissure

III ventricle

Posterior commissure

Pineal recess

Subcommissural organ

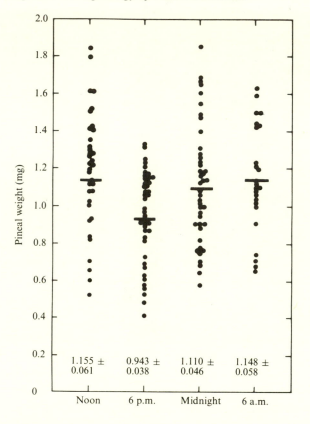

Fig. 2.26. Rhythmical changes in the weight of the rat's pineal throughout the day. These changes are associated with changes in various biochemical products that are present in the gland. (From Axelrod, Wurtman, and Snyder, 1965.)

Melatonin and serotonin have also been identified in the bird and amphibian pineal. One of the enzymes responsible for its formation, HIOMT (hydroxyindole-*O*-methyltransferase), has been found in the pineal of fish and all the tetrapod groups as well as in the lateral eyes of fish, amphibians, reptiles, and birds (but not in mammals). The retina of the trout has been shown to synthesize melatonin (Gern and Ralph, 1979).

The pineal of rats undergoes a number of changes associated with the exposure of the animals to light. It increases in weight in rats kept in the dark and decreases on exposure to light. Weight changes can be seen to follow a circadian rhythm over a normal 24-hour period (Fig. 2.26). As we shall see later, this change in weight may reflect its role (?) as a "neuroendocrine transducer" in the control of reproductive cycles.

The pineal body has often been described as a "vestigial" organ. Despite the implied slight about its usefulness, it has shown considerable phylogenetic persistence during which it appears to have evolved from a mainly sensory organ to an endocrine one.

The corpuscles of Stannius

These tissues are so named after their discoverer (Stannius, 1839). They appear to form a hormone that decreases blood calcium levels and may also have osmotic effects in fishes. They are distinct well-vascularized masses of tissue ("white bodies") that are present near the kidney of some, but not all, fishes. Their number varies considerably; 40–50 are present in the bowfin, *Amia calva* (Holostei), whereas in the Teleostei the number varies from two to six. They are absent in some teleosts, including the Salmonidae, as well as the Chondrostei (sturgeons) (Krishnamurthy and Bern, 1969). In the corpuscles, one, or sometimes two, types of cells can be distinguished that are usually arranged in rows to form lobules around a central core. These cells have the appearance of peptide-secreting cells. They contain granules, the number and appearance of which can be seen to vary conspicuously with changes in the breeding cycle, the life cycle, and the osmotic environment. This has led to the suggestion that they may have an endocrine role. Indeed, as will be related, extracts of these tissues have been shown to contain materials that in some instances may alter sodium and calcium metabolism as well as exhibit a vasopressor effect similar to that of the kidney hormone renin.

The corpuscles of Stannius are usually derived from the pronephric duct (in some instances they are formed from the mesonephric duct). The presence of the renin-like material in the tissue extracts may be a reflection of this origin.

The Holostei have the largest number of corpuscles of Stannius and it has been suggested that the reduction among the Teleostei, culminating in their absence in the Salmonidae, may reflect an evolutionary trend. In one study (Krishnamurthy and Bern, 1969), 28 species of teleosts from 16 different families were examined, but no conclusions as to any evolutionary relationships in their morphology could be drawn.

The urophysis

Tucked away beneath the vertebrae in the tail of teleost fishes is a lump of tissue that has been called the urophysis and which may influence their osmoregulation and assist smooth muscle contraction in the urinogenital tract. This "gland" was first described in 1813 and has since been identified in some 400 different species of teleost fishes. Despite its widespread distribution and systematic persistence, its physiological role is still not understood but has captured considerable attention from several endocrinologists (see Fridberg and Bern, 1968).

The urophysis, like the neurohypophysis, is composed of neural tissue, whose cell bodies are situated in the posterior part of the spinal cord. Axons of these cells pass outside the spinal column ventrally where they make contact with blood vessels (that lead through the kidneys) to form a neurohemal junction. Such an arrangement is admirably suited to the discharge of endocrine secretions into the circulation. Whether this occurs is equivocal, however. These nerve cells contain granules and appear to be typical neurosecretory cells such as those seen in the neurohypophysis, though their tinctorial characteristics differ. Extracts of this tissue show several biological activities; they can alter the permeability of some membranes to water and sodium, contract certain smooth muscle preparations, increase the blood pressure of eels, and lower the blood pressure of rats.

A distinct neurohemal urophysis has only been identified in teleost fish. The chondrichthyeans, however, also possess neurosecretory-type cells in the caudal part of the vertebral column. These cells are giant neurons, about 20 times the size of an ordinary motor neuron and, in *Raja batis*, extend along the last 55 vertebrae. They are called *Dahlgren cells* after their discoverer and send their axons out of the ventral part of the spinal cord to make contact with blood vessels there. The tissue is thus more widespread along the spinal cord. Comparable tissues appear to be present in holostean fishes but not in cyclostomes.

The types of such structures present in chondrichthyeans and teleosts are shown in Fig. 2.27. The diffuse distribution of tissue seen in the sharks and rays may represent a primitive pattern that has subsequently evolved among teleosts to form a discrete aggregation. The widespread distribution and systematic persistence make one suspect that the urophysis serves a physiological role. Its histological and cytological appearance, similarity to the neurohypophysis, and its neural and vascular connections suggest an endocrine gland. Conclusive evidence is not yet available, however.

The thymus

The thymus, like the parathyroids and ultimobranchial bodies, is derived, embryologically, from the gill pouches and so is situated in the region of the neck or upper thorax. In fishes all the gill pouches may be involved but in amniotes usually only numbers III and IV. The thymus is present in all groups of vertebrates, from cyclostomes to mammals, but is more prominent in young and larval forms than adults where, in some species, it may even disappear. The thymus tissue may exist in varying degrees of agglomeration, from a single bilobed organ, as in mammals, to paired structures, as in frogs, or in several dispersed nodes, as in the domestic fowl. The

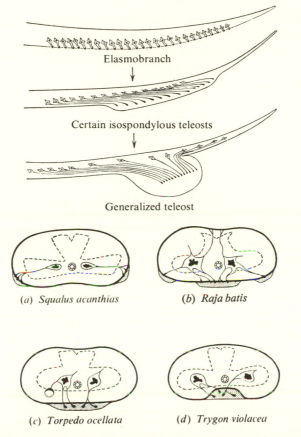

Elasmobranch

Certain isospondylous teleosts

Generalized teleost

(*a*) *Squalus acanthias*   (*b*) *Raja batis*

(*c*) *Torpedo ocellata*   (*d*) *Trygon violacea*

Fig. 2.27. The urophysis. *Top.* Proposed evolution of the teleost urophysis from elasmobranchs, which have neurosecretory Dahlgren cells. Longitudinal section through the tail. *Bottom. (a)* to *(d).* Different configurations of the Dahlgren cells among the elasmobranchs. Transverse sections through the spinal cord. The vascular beds are shaded and the menix is represented by a heavy line. (From Fridberg and Bern, 1968.)

ultimobranchial bodies and even parathyroids may on occasions be associated with the thymus.

The endocrine function of the thymus is not an established one. A number of biologically active materials have been extracted from it, however, and these (several proteins and a steroid) have been called "thymic hormones" (see Luckey, 1973; Bach, 1977). During early life the thymus produces lymphocytes that are involved in the development of immunity. It has been proposed that the so-called thymic hormone(s) may initiate development of immune competence (ability of cells to react to the presence of foreign substances or antigens) in the T-lymphocytes and bone marrow.

Such "hormones" thus could be considered to be involved in embryonic differentiation.

One of these "hormones" (see Bach et al., 1972; Goldstein et al., 1974) has been isolated from the thymus and is a polypeptide with a molecular weight of about 12,000. It has been called thymosin and has been identified in the blood of the mouse and man. The circulating concentrations decline with age and after thymectomy. In man the thymosin concentration in the blood starts to decrease at 30–40 years of age and it disappears after about 50 years. There is considerable medical interest in these discoveries as the blood levels have also been shown to be low in diseases where immune responses are deficient, and it fails to decline with age in people suffering from myasthenia gravis. This disease (a failure of skeletal muscles to be able to contract adequately) can often be ameliorated by thymectomy. Another intriguing possibility is that the decline in the activity of the thymus contributes to the process of aging and an increased susceptibility to autoimmune and infectious diseases. It seems likely that this interesting tissue will gain the status of a true endocrine gland.

Conclusions

It can be seen that although the endocrine glands display considerable inter-specific differences in their morphology many of these variations can be placed into categories that correspond to major systematic groups of vertebrates. It would thus appear that the endocrines have evolved in a relatively orderly manner that may be influenced by broad structural considerations, such as the animal's shape, size, and pattern of embryonic differentiation, as well as its particular hormonal requirements. Such evolutionary changes are not, however, confined to the glands' morphology for, as we shall see in the next chapter, considerable variation also occurs in the chemical structures of the hormones themselves.

3. The chemical structure, polymorphism, and evolution of hormones

In a mammal the endocrine glands secrete more than 40 distinct hormones. In addition, different species may form many hormones that, although structurally analogous, nevertheless display chemical differences. Such natural variants are usually characteristic of a single species and represent a polymorphism of the excitant's molecular structure. This change has a genetic basis. For example, it may only be the substitution of a single amino acid in the molecule of a peptide hormone or it may be much more extensive. The biological effects of such differences can be considerable or negligible.

Vertebrate hormones belong to two principal classes of chemical compounds. Some are made from cholesterol. These are the steroid hormones from the adrenal cortex and the gonads. The others are made up of amino acids and range in complexity from those, like epinephrine, that are derived from a single tyrosine molecule, to others like the pituitary growth hormone that contain about 190 such units. The molecular weights can vary from about 200 to 30,000.

What properties do these molecules have that make them suitable to be hormones? What characteristics may be important for their utilization as such? Armed with considerable hindsight about endocrine physiology some answers can be offered. The basic requirements will not be the same for all hormones but will depend on what they do. The steroid hormones are poorly soluble in water but readily soluble in lipids. This will facilitate their penetration into the cell and fixation at intracellular sites. Such lipid solubility will also be important if a hormone is to penetrate the blood–brain barrier. Transport in the blood is essential for a hormone to fulfill its physiological role so that, if they are hydrophobic molecules, they must either be effective at very low concentrations or be attachable to protein components that carry them to their sites of action. This binding is especially prominent among the steroid and thyroid hormones. An ability to interact with other biological molecules is also important for "triggering" the excitant effects of hormones. They must be capable of interacting with a receptor molecule in, or on, the effector tissue. Such an interaction must not be of a strong covalent nature but must involve chemical forces whereby an equilibrium of a reversible nature occurs. Above all, a hormone must have a high degree of specificity toward its target receptor site. Necessarily this is a property of both structures. The manner by which it is accomplished is still largely conjectural. Hormones have complex three-dimensional structures that contain various components that may be electrically charged, hydrophilic or hydrophobic, acidic or basic, and so on. Such properties together may constitute a "key" to which the receptor acts as a complementary "lock." Obviously, large hormone molecules offer the possibilities of more complex "keys" and greater specificity. They are also more liable to genetically mediated structural changes. The latter are very common events in nature. The contribution of size to specificity of effect is, however, not at all clear. Indeed, it is difficult to comprehend why such gigantic molecules as the pituitary adenohypophysial hormones are necessary to mediate their effects. There is no evidence to indicatethat evolution toward smaller, more compact hormone molecules has occurred, such as would perhaps be expected if they initially contained much superfluous material. Indeed, some hormone molecules, for instance, parathyroid hormone, contain sections that are not needed for their biological activity.

In order to function optimally, a hormone molecule needs to possess some other properties consonant with its physiological role. For adequate control, hormonal responses often need to be rapidly terminated. The excitant can either be readily excretable in the urine or bile, or, by virtue of the presence of chemical groups that can be changed by metabolic processes, be converted to an inactive form. The synthesis of hormones is not always rapid enough to meet the immediate demands for their release, so that their accumulation and storage in glandular tissues may be necessary. In this instance the molecule should possess a considerable measure of innate stability and be able to interact with cellular (or even extracellular, as for thyroxine) binding proteins that facilitate this storage. Related to such a process is an ability to undergo rapid mobilization from such storage sites so that the hormone can be released into the blood.

Clearly, hormones are highly specialized molecular structures incorporating (or programmed for) several important interrelated properties. They are not just keys that fit various locks. In order to function as a hormone a molecule must also exhibit a variety of other physical and chemical properties consistent with the hormone's synthesis, storage, release, transport, and removal from the body.

Structural differences between hormones are tentatively assumed on the basis of differences in their biological actions. They are confirmed by the demonstration of variations in their chemical behavior and ultimately by the determination of their molecular structure. It is usually a comparatively simple procedure to show that two hormones differ from each other. Tests for biological activity, for instance, changes in blood glucose levels, an ability to alter blood pressure, decrease urine flow, and so on, are reasonably straightforward laboratory procedures. Broad chemical differences in even very impure preparations can often be seen when, for instance, one compares their stabilities at different temperatures and pH's, solubilities in different solvents, relative rates of destruction when incubated with various enzymes, chromatographic mobilities, and so on. Such biological and chemical characterization can be used to identify and measure the relative quantities of the hormonal material present in an extract. Determination of chemical composition and structure is a more complex procedure and, before this can be done, highly purified preparations must be made.

Differences in the structure of related homologous hormones in distinct (or even on occasion the same) species are more difficult to detect. The molecules may only exhibit quite small quantitative, in contrast to qualitative, differences in a common biological activity.

Several procedures are available to help us make such distinctions.

1. Biological activities can be compared. A simple cross-test between two or more species can be informative. For instance, a comparison between the action of extracts of the neurohypophyses of frogs with those from mammals indicated the presence of different but analogous hormones. Heller (1941) found that extracts from the neurohypophyses of European frogs (*Rana temporaria*) were about 20 times as active in eliciting water retention in frogs than a comparable amount of a hormone extract from the mammalian pituitary. The two extracts (frog and mammal) were each standardized by their ability to contract the guinea pig uterus and to increase the blood pressure of cats. Equal amounts of activity, as measured in these ways, showed that the homologous frog hormone was much more active in frogs than the mammalian one. As we shall see, this change reflects a single amino acid substitution in the octapeptide molecule of mammalian antidiuretic hormone. Similar comparisons have been made between many species and usually the more distantly related they are the greater their effects differ. Even two closely related species may show distinct differences; thus pituitary growth hormone from animals, including other primates, is completely ineffective in man even though that from man is active in many other species. Hormones often exhibit a variety of biological actions and, by comparing them in several assay systems, a pharmacological profile or "fingerprint" can be made. Such bioassays can be used to characterize some hormones with a considerable degree of accuracy.

2. The chemical behavior of hormones can be shown to differ. In closely related molecules this may be difficult to detect. It may depend on differences in electrophoretic or chromatographic mobilities in different solvent systems. A correlation (isopolarity), however, does not necessarily confirm the chemical identity of two molecules. Steroid molecules are often chemically altered, for instance, by methylation, and their chromatographic behavior is again

compared. If it still corresponds, the evidence of identity is much stronger. The absorption spectra of extracts can also be compared. This measurement may involve the use of ultraviolet and infrared light when the hormones are dissolved in different solvents. Mass spectrometry and nuclear magnetic resonance spectrometry are effective but expensive methods for identifying many hormonal materials. Using such procedures a chemical "fingerprint" can be obtained for comparison with standard preparations of known structure. Such chemical methods are particularly useful in aiding the identification of steroid hormones (see Brooks et al., 1970; Sandor and Idler, 1972).

3. Immunological responses are used to predict the differences in the structure of protein hormones. Antibodies (or antisera) to purified hormones can be made following their injection into another species, often a rabbit. The degree of interaction of protein hormones with such antisera can be followed in a precipitin test or by comparing their abilities to displace known radioiodinated hormones from binding with the antibodies. Apart from indicating similarities and differences between hormones, such immunological procedures are widely used to measure the quantities of hormones by radioimmunoassay. This procedure is not confined to protein hormones, as antibodies to other hormones, including steroids and thyroid hormones, can be made by conjugating them to a protein molecule and using this to elicit formation of antibodies.

4. The chemical structure of a hormone may be determined. This is ultimately desirable and allows one to relate the structure of the molecule to its biological actions and even predict the genetic basis of its formation and possible evolution. As some hormones are very complex structures containing as many as 200 amino acids, knowledge of their precise chemical structure has been somewhat slow in coming. A special relationship with immunological procedures also exists, as these can be used to confirm the chemical proposals. Initial chemical analyses of human growth hormone and prolactin indicated that they had identical chemical structures, but this was not confirmed immunologically. Small, but then undetectable, differences existed in these hormones.

Although the chemical structure of many hormones is known, this knowledge is mainly confined to the mammals, especially with respect to the larger protein hormones. In addition, although the disposition of chemical groups and the sequence of amino acids may be known, less information is available as to their three-dimensional (tertiary) arrangement. Such data will ultimately be required if we are to understand properly how the hormones work.

Steroid hormones

Steroids are chemical compounds derived from cholesterol. They consist of a series of carbon rings, the basic unit being the cycloperhydrophenanthrene

nucleus. Such compounds occur widely in nature and are not con
animal kingdom. Plants contain many steroids and some of the
exhibit activities reminiscent of those of the mammalian horm
ample, catkins of the pussywillow plant contain steroids that have the
of female sex hormones (estrogenic activity). Steroids obtained from plants
are indeed often used as starting materials for the preparation of vertebrate-
like hormones in the laboratory.

Several different types of steroids function as hormones in vertebrates.
These and the parent cholesterol molecule are shown in Fig. 3.1*a, b*. They are
often classified in the following manner: (1) Those based on pregnane and
containing 21 carbon atoms (C_{21}). These include the adrenocortical steroids
and progesterone, which, apart from being a metabolic intermediate in the
formation of most steroid hormones, also acts as a sex hormone, especially
during pregnancy. (2) Androstane compounds with 19 carbons (C_{19}) and
which include the androgens that have the actions of male sex hormones. (3)
Estrane (C_{18}) compounds that have actions of female sex hormones (estrogenic). (4) Vitamin D (Fig. 3.1*c*), a group of sterols the precursors of which
are commonly obtained in the diet and which can be converted (see Chapter 6)
into hormones that influence calcium metabolism.

The hormones from the gonads and adrenal cortex are all derived from
cholesterol compounds. In the instance of C_{18}, C_{19}, and C_{21} steroids, various
metabolic pathways, usually involving several hydrolase enzymes, lead to the
formation of the ultimate hormone product (Fig. 3.2). These include the
female sex hormones, called *estrogens* (C_{18}), estradiol-17β, estrone, and es-
triol; the *androgens* (C_{19}), mainly testosterone but also its metabolic precursor
androstenedione and more active metabolite 5α-dihydrotestosterone; *progestins*
(C_{21}), progesterone and the *adrenocorticosteroids* (C_{21}) cortisol, corticosterone,
aldosterone, and 1α-hydroxycorticosterone. Some other adrenocorticosteroids,
such as cortisone, are also sometimes found in the blood.

Many other steroids are found in the steroidogenic tissues where they
constitute intermediates of the hormones; others may represent products of
steroid catabolism. The chemical structures of these hormones are shown in
Fig. 3.2.

The C_{18}, C_{19}, and C_{21} steroids have been identified in tissues and also often
in the blood of all the main groups of vertebrates. The compounds present are,
however, not identical in all of these while the evidence of their precise
identity in some (for instance, cyclostomes) has been noted as "tentative" or
"only suggestive" (see Idler, 1972). Steroids of a hormonal nature, nevertheless, undoubtedly have a wide phyletic distribution among vertebrates.

The sex hormones show a remarkable uniformity; testosterone, progesterone, and estradiol-17β are common throughout the vertebrates. This possibly
reflects the "conservative" nature of the sexual process and the early evolution

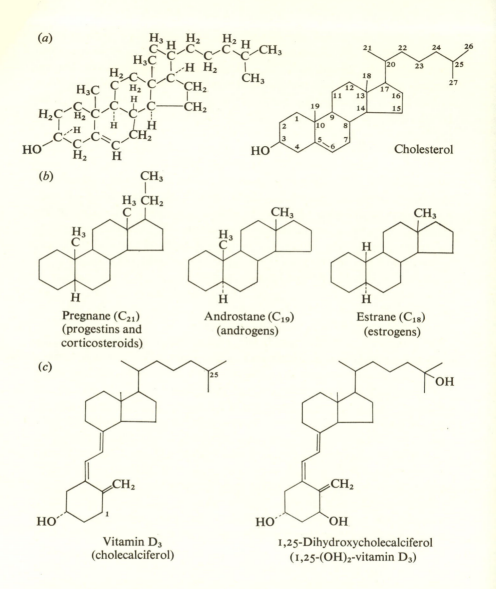

Fig. 3.1. (a) The chemical structure of cholesterol and the conventional manner of numbering the carbon atoms. (b) The parent steroid compounds for the progestins and corticosteroids (C_{21}), androgens (C_{19}), and estrogens (C_{18}). (c) Vitamin D_3 and its active metabolite $1\alpha,25$-dihydroxycholecalciferol.

of a mechanism of such efficiency that little subsequent endocrine modification of the hormonal excitants could be advantageous. It can be seen in Fig. 3.2 that, when the structures of the steroid hormones are compared, the chemical differences appear surprisingly minor. Nevertheless, each molecule exerts distinct effects. A high degree of specificity based on such simple structural differences probably allows little room for subsequent successful evolutionary "experiments."

Among the adrenocorticosteroids different molecules have emerged and these often have a distinct systematic distribution (Fig. 3.3). Such corticosteroid hormones exist in the cyclostome fishes. The steroid 1α-hydroxycorticosterone is widespread in the adrenal tissues and blood of the Chondrichthyes. This hormone is, however, only present in the Selachii (sharks and rays) and not the Holocephali (chimaeroids), which instead have cortisol (see Idler and Truscott, 1972). Among the Actinopterygii (including the Holostei, Chondrostei, and Teleostei), cortisol is the predominant corticosteroid in the blood, but corticosterone is also present. Cortisone, aldosterone, and corticosterone have also been identified in teleosts but the quantities appear to be small compared to cortisol. The criteria for such identifications are, however, sometimes in doubt. The Dipnoi (lungfishes) possess cortisol, like other bony fishes, and a fascinating discovery has been the additional identification of aldosterone in the South American lungfish *Lepidosiren paradoxa* (Idler, Sangalang, and Truscott, 1972) though it is not present in other members of this order. Aldosterone has been identified in the blood of representatives of all the tetrapod groups so that its presence in the Dipnoi, but not apparently in most other fish (though it has been found in a few species) is consistent with their suggested phylogenetic relationships to tetrapods. A second major corticosteroid, corticosterone, is also present in amphibians, reptiles, and birds. This hormone is also the major corticosteroid in some mammals whereas in most others cortisol is predominant.

The ratio of cortisol to corticosterone varies among the mammals. Rats, rabbits, and mice secrete little or no cortisol from their adrenal cortexes; corticosterone (aldosterone is also present) predominates. Other mammals secrete a mixture of cortisol and corticosterone, usually with the former predominant. It was at one time suggested that the ratio cortisol:corticosterone may be a characteristic of a species and therefore determined genetically. It has, however, been found that this ratio can vary considerably, even in a single animal, depending on the physiological conditions. Nevertheless, the inability of the rat to form cortisol reflects the inactivity of an enzyme, 17α-hydroxylase, and it seems likely that this, at least, is genetic. Most mammals, including placentals and marsupials, secrete more cortisol than corticosterone. An interesting exception is the echidna *Tachyglossus aculeatus*, a monotreme in which corticosterone predominates (Weiss and McDonald, 1965). This pattern is more like that in reptiles and birds than that in most other

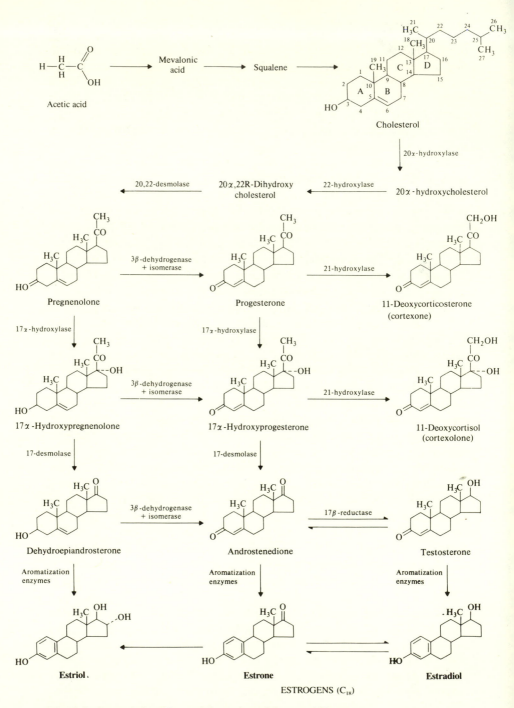

Fig. 3.2. Interrelationships and formation of the steroid hormones.

[74]

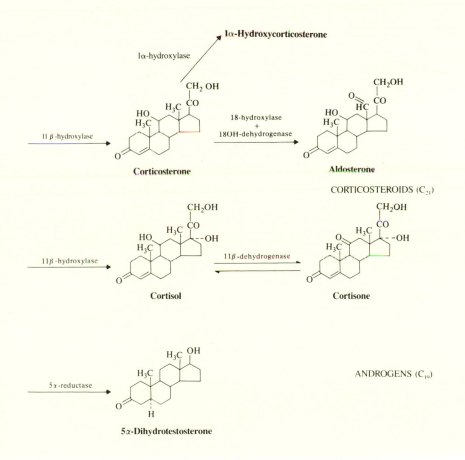

1α-Hydroxycorticosterone

1α-hydroxylase

11 β-hydroxylase

Corticosterone

18-hydroxylase
+
18OH-dehydrogenase

Aldosterone

CORTICOSTEROIDS (C$_{21}$)

11β -hydroxylase

Cortisol

11β-dehydrogenase

Cortisone

5 α-reductase

5α-Dihydrotestosterone

ANDROGENS (C$_{19}$)

(a)

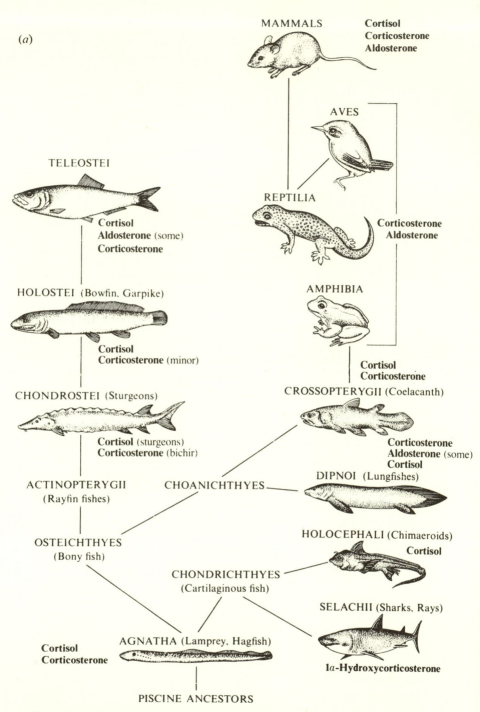

Fig. 3.3. (a) Phyletic distribution of the adrenocorticosteroids in the vertebrates.

(b)

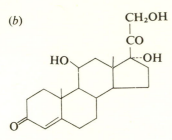

Cortisol

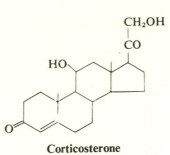

Corticosterone

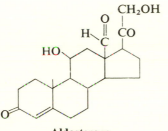

Aldosterone

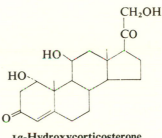

1α-Hydroxycorticosterone

Fig. 3.3 (b) The principal structures of the corticosteroid hormones.

mammals, but it is not seen in another monotreme, the platypus (Weiss, 1980).

The corticosteroids, in contrast to the sex steroids, display different chemical structures that probably reflect evolutionary changes. Sex is a relatively uniform process, but the roles of corticosteroids show some variation. This may be reflected in the different structures of these steroids. The role of 1α-hydroxycorticosterone in the Selachii is uncertain. In other vertebrates, cortisol and corticosterone influence intermediary metabolism, which is a basic function in all vertebrates. Aldosterone and, to a lesser extent, corticosterone exert a prominent effect on sodium and potassium metabolism in tetrapods. These animals have special osmotic problems not faced by their piscine ancestors, so that it is conceivable that the solutions to them were accompanied not only by the evolution of special effector mechanisms but also hormones to fit them.

Hormones made from tyrosine

Catecholamines

The adrenal medulla and other chromaffin tissues secrete two hormones, epinephrine (adrenaline) and norepinephrine (noradrenaline) (Fig. 3.4). These

Table 3.1. *Norepinephrine as percent of total catecholamines in the adrenals of various species of vertebrates*

	Norepinephrine as % total catecholamines
Whale	83
Domestic fowl	80
Dogfish	68
Turtle	60
Pigeon	55
Frog	55
Toad	55
Pig	49
Sheep	33
Ox	26
Man	17
Rat	9
Rabbit	2
Guinea pig	2

Source: Based on West, 1955.

are amine derivatives of catechol; hence their name catecholamines. Such compounds are found in all vertebrates where they also act as neurotransmitters in the sympathetic nervous system and brain. They are also present in many invertebrates and even in the ciliated protozoan *Tetrahymena* where they influence metabolism in a manner reminiscent of that in more sophisticated metazoan animals (Blum, 1967).

Using phenylalanine, then tyrosine, as substrates, norepinephrine (Fig. 3.4*a*) is formed in chromaffin tissue. This may, under the influence of a methyltransferase enzyme PNMT (phenylethanolamine-*N*-methyltransferase), have a methyl group added to become epinephrine. These two hormones have differing actions though a crossover of their effects occurs. Their actions are separately classified as α-adrenergic and β-adrenergic that can each separately be prevented by specific drugs. β-Adrenergic actions include stimulation of the heart (β_1), relaxation of the bronchi (β_2), dilatation of certain blood vessels (β_2), and increase in blood glucose levels. These are principally seen in response to epinephrine rather than norepinephrine, which does, however, affect the heart. The constriction of peripheral blood vessels and the sphincter muscles in the gut and bladder are α-adrenergic effects. Both norepinephrine and epinephrine are effective at such sites, but epinephrine is more versatile and also has β-adrenergic effects.

The ratio of epinephrine to norepinephrine in the adrenal chromaffin tissue varies considerably among vertebrates. This is illustrated in Table 3.1 where it can be seen that norepinephrine makes up about 80% of the total catecholamines in whales and the domestic fowl (an ill-assorted phyletic pair) and as little as 2% in rabbits and guinea pigs. No phyletic pattern in the distribution can be seen. Young and fetal animals possess a predominance of norepinephrine because of the relative lack of the methylating enzyme (see West, 1955).

Thyroid hormones

These are unique hormones as they contain, as part of their structure, the halogen iodine. The formation of thyroxine (tetraiodothyronine or T_4) and triiodothyronine (T_3) is shown in Fig. 3.4b. The thyroid gland also contains, probably mainly as metabolic intermediates, mono- and diiodotyrosine. Thyroxine and triiodothyronine differ quantitatively in their effects; those of the former are much slower in onset but longer in duration than those of the latter. In order to act, T_4 must be converted to T_3. T_4 can be bound more strongly to plasma proteins in a complex, which may contribute to the difference in the time course of its effects from T_3. In mammals, T_4 is probably secreted at about five times the rate of T_3.

Thyroid hormones have been identified in the blood of representatives of several groups of lower vertebrates, including cyclostomes, teleosts, amphibians, and reptiles (see Higgs and Eales, 1973; Chiu et al., 1975; Packard, Packard, and Gorbman, 1976). Biologically active iodothyronine compounds occur throughout the vertebrates and have also been identified in a number of protochordates (see Chapter 2). Iodine readily combines with proteins containing tyrosine (*in vitro*) so that the natural occurrence of such compounds, especially in iodine-rich solutions like seawater, is perhaps not surprising. Their transformation to iodothyronine compounds, however, seems to depend on specialized metabolic pathways and conditions such as those that occur in the thyroid gland, which has a unique ability to trap and oxidize iodide. Some of these mechanisms have been demonstrated in ascidians (Urochordata) (see Chapter 2). This process can be imitated *in vitro* provided the appropriate amounts of iodine and tyrosine-containing proteins are incubated together.

The spontaneous occurrence of thyroxine compounds in nature, even before the origin of the thyroid gland, is not inconceivable. Whether such compounds did arise and acquire a usefulness as hormonal excitants is sheer conjecture. If this did occur, subsequent specializations may have led to the hormones' more efficient formation in the thyroid gland.

Finally, we may consider the question of why iodine is present in the thyroid hormone molecules. It has been suggested that it plays a vital part in the initiation of the hormonal effect, the hormone acting as a "carrier" moving iodine into the cell. Thyroxine-like molecules that contain no iodine have been made, however (Taylor, Tu, and Barker, 1967). These contain bromine and *iso*propyl groups instead of iodine and yet they still exhibit considerable

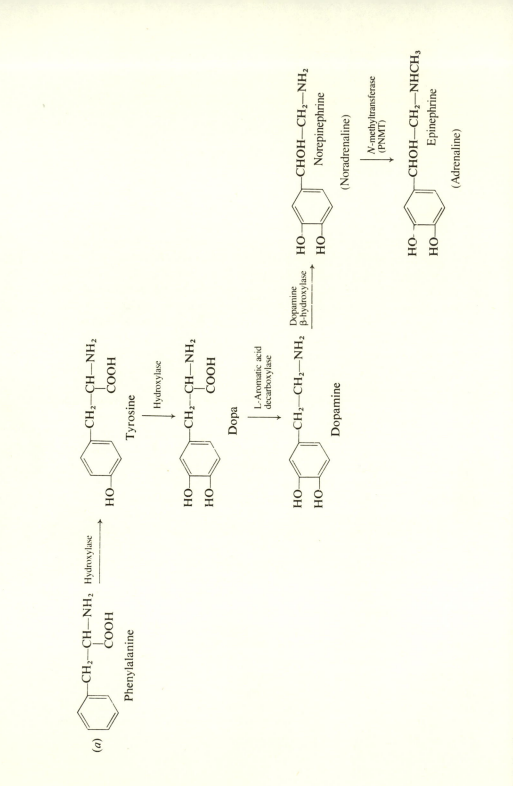

(a)

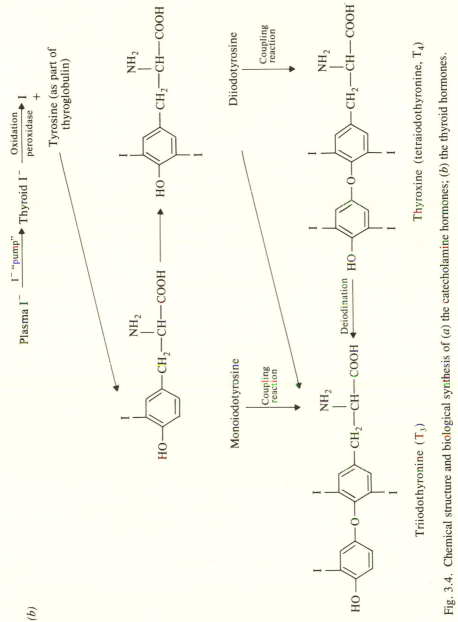

Fig. 3.4. Chemical structure and biological synthesis of (*a*) the catecholamine hormones; (*b*) the thyroid hormones.

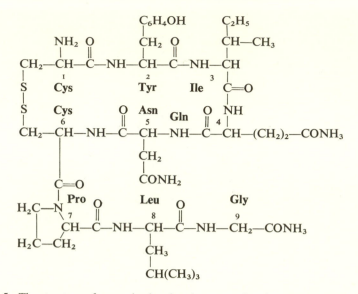

Fig. 3.5. The structure of oxytocin showing the conventional numbering of the amino acids.

biological activity. Iodine thus does not appear to have a unique role in the action of thyroid hormones. Presumably, however, it was available in the environment and this, along with its ready chemical reactivity with tyrosine, resulted in its utilization.

The thyroid and catecholamine hormones are clearly very "conservative" with respect to evolutionary changes in their chemical structure. This would seem to be, at least partly, because of their small size, which limits the possibility for change in their molecules, and the fact that they are made from simple precursors that are abundant in nature. The catecholamines and thyroid hormones (and to a slightly lesser extent the steroid hormones) provide us with an illustration of the dictum that, "it is not the hormones that have evolved but the uses to which they have been put." As will become particularly apparent in the succeeding sections, this is not always true.

The peptide hormones of the neurohypophysis

Two chemically related hormones are usually secreted by the neurohypophysis. These are peptides containing eight amino acids. They are arranged in a five-membered ring, joined by a disulfide bridge (contributed by two half-cystine residues) and a side chain with three amino acids (Fig. 3.5 and Table 3.2). In most mammals the two hormones are arginine–vasopressin (AVP, also called antidiuretic hormone or ADH) and oxytocin. These differ by two

Table 3.2. *Amino acid sequences of known neurohypophysial hormones*

Common structure [Variations in positions 2, 3, 4, and 8 indicated by (X)]	1 Cys	2 (X)	3 (X)	4 (X)	5 Asn	6 Cys	7 Pro	8 (X)	9 Gly(NH$_2$)
			Amino acids in position						
		2	3	4				8	
Basic peptides									
Arginine–vasopressin (AVP)		Tyr	Phe	Gln				Arg	
Lysine–vasopressin (LVP)		Tyr	Phe	Gln				Lys	
Phenypressin[a]		Phe	Phe	Gln				Arg	
Arginine–vasotocin (AVT)		Try	Ile	Gln				Arg	
Neutral (= oxytocin-like) peptides									
Oxytocin			Ile	Gln				Leu	
Mesotocin			Ile	Gln				Ile	
Isotocin (= ichthyotocin)			Ile	Ser				Ile	
Glumitocin			Ile	Ser				Gln	
Valitocin			Ile	Gln				Val	
Aspargtocin			Ile	Asn				Leu	

[a]See Chauvet et al., 1980.
Source: Modified from Heller, 1974.

amino acid substitutions; vasopressin has phenylalanine and arginine at positions 3 and 8 in the molecule, where oxytocin has *iso*leucine and leucine. This change confers considerable differences in biological activity; vasopressin enhances water reabsorption across the renal tubule, and so reduces urine flow, while oxytocin can contract the uterus and initiate "milk letdown" from the mammary glands. There is little crossover in their actions.

Homologous hormones have been identified in the neurohypophyses of representatives of all the systematic groups of vertebrates. Considerable differences in chemical structure exist, however, so that, so far, 10 such peptides have been identified in nature. Amino acid substitutions occur at the 2, 3, 4, and 8 positions in the molecule (Table 3.2). The occurrence of these natural analogues has a well-defined systematic distribution (Fig. 3.6). For example, arginine–vasopressin is confined to mammals whereas arginine–vasotocin (a combination of the ring of oxytocin and the side chain of vasopressin) is present in all other vertebrates. The second, oxytocin-like (or neutral) peptide in nonmammals exists in five variant forms; mesotocin (*iso*leucine instead of leucine at position 8) is present in birds, reptiles, amphibians, and lungfishes; isotocin (*iso*leucine at 8, serine instead of glutamine at 4) is found in all the myriad of bony fishes except lungfishes. The chondrichthyeans exhibit more variability, vasotocin and oxytocin being present in the Holocephali, and vasotocin as well as glumitocin, valitocin, and aspargtocin are distributed

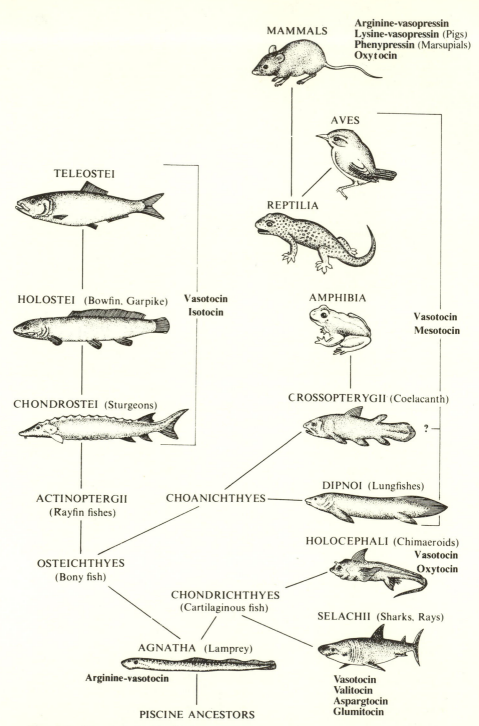

MAMMALS **Arginine-vasopressin**
Lysine-vasopressin (Pigs)
Phenypressin (Marsupials)
Oxytocin

AVES

REPTILIA

TELEOSTEI

HOLOSTEI (Bowfin, Garpike)   **Vasotocin**
Isotocin

AMPHIBIA

Vasotocin
Mesotocin

CHONDROSTEI (Sturgeons)

CROSSOPTERYGII (Coelacanth)

?

DIPNOI (Lungfishes)

ACTINOPTERGII   CHOANICHTHYES
(Rayfin fishes)

HOLOCEPHALI (Chimaeroids)
Vasotocin
Oxytocin

OSTEICHTHYES
(Bony fish)

CHONDRICHTHYES
(Cartilaginous fish)

SELACHII (Sharks, Rays)

AGNATHA (Lamprey)

Arginine-vasotocin

Vasotocin
Valitocin
Aspargtocin
Glumitocin

PISCINE ANCESTORS

Fig. 3.6. The phyletic distribution of the neurohypophysial hormones among the vertebrates.

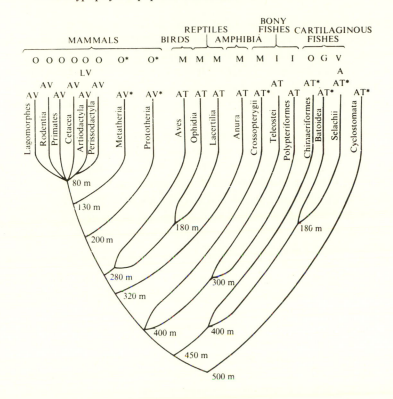

Fig. 3.7. Evolution of the neurohypophysial hormones. Letters represent hormones that have been identified in extant species from each group. O = oxytocin; AV = arginine–vasopressin; LV = lysine–vasopressin; A = aspargtocin; V = valitocin; G = glumitocin; AT = arginine–vasotocin; M = mesotocin. 500 m indicates 500 million years since divergence, and so on. *, Identification is pharmacological, not chemical. (From Acher, Chauvet, and Chauvet, 1972.) Lysine–vasopressin and phenypressin (Phe^2-Arg^2-vasopressin) have been identified in the Metatheria (Chauvet et al., 1980).

among the Selachii (see Fig. 3.7). The physiological roles of vasopressin and oxytocin in mammals, and vasotocin in tetrapods, are reasonably well understood, but the functions of the other peptides remain unknown, particularly in fish. They are nevertheless present and, from our knowledge of extant species, apparently have persisted for about 500 million years since the first cyclostomes evolved (see Fig. 3.7).

Such polymorphism of the hormones is genetically determined. It can be examined more closely among mammals where a variant of arginine–vasopressin occurs. Many pig-like mammals (Suiformes, including the true pigs, Suidae, the peccaries, Tayassuidae, and the hippopotamus, Hippopotamidae) possess a vasopressin with lysine instead of arginine present in the 8 position. This probably arose as a result of a single-step mutation from arginine–vasopressin. Its present distribution suggests that this transformation occurred in an ancestor of the Suiformes before the hippopotami broke away from the pig–peccary stock (Ferguson and Heller, 1965). The change occurred in the Eocene, about 60 million years ago. The neurohypophyses of domestic pigs contain only lysine–vasopressin (and oxytocin) but among other Suiformes, such as peccaries, warthogs, and hippopotami, both arginine– and lysine–vasopressin may, or may not, be present in the same individual. It is possible that the homozygotes contain one such peptide, the heterozygotes both. The evolutionary persistence of lysine–vasopressin seems to reflect the fact that its biological potency is only a little less than that of its arginine-containing relative so that it is not appreciably disadvantageous. In addition, an adaptive increase in sensitivity of the kidney to lysine–vasopressin may occur (Stewart, 1973). It is, of course, possible that the presence of this hormone also confers adaptive advantages that we do not know about.

The marsupials (or Metatheria) have been separated from Eutheria for about 130 million years. It has recently been shown (Chauvet et al., 1980) that they may possess two vasopressin-like hormones: lysine–vasopressin, which is the predominant one, and phenypressin (Phe^2-Arg^8-vasopressin). It was suggested that this duality of vasopressins in a single animal may have resulted from duplication of the vasopressin gene with subsequent single-step mutations. These two peptides appear to be quite widespread among Australian marsupials and have even been observed in a South American species (an opossum). Both of these peptides have an antidiuretic activity that is only a little less than that of arginine–vasopressin so that their acquisition may not be expected to be disadvantageous in this respect.

Mutations that result in such changes in the amino acid composition of the neurophypophysial peptides may occur periodically. A strain of mice descended from some wild Peruvian specimens has been shown to contain lysine–vasopressin, which is not usual in this species (Stewart, 1968). Also, a strain of laboratory rats (Brattleboro) has been found to lack vasopressin altogether and so cannot regulate its urine flow normally (Valtin, Sawyer, and Sokol, 1965). When these rats are bred with normal rats the pituitary glands of the heterozygotes contain reduced amounts of vasopressin. Such genetic changes may result in the complete absence of such a peptide or a mutational "mutilation" of the molecule so that it has no biological action. The physiological result to the animal would be the same.

A number of biologists have played genetic games with the neurohypophysial peptides. Forearmed with the structures of the natural analogues and the genetic code it is possible to construct feasible lines for their evolution. These are usually calculated on the basis of the minimum number of mutations needed to produce a change from one amino acid at a certain position in a molecule, to another. Four such schemes (that do not include the diverse chondrichthyean hormones or those of the marsupials) are shown in Fig. 3.8. Most consider vasotocin to be the "parent," or original, ancestral molecule.

Chemists have made more than 600 analogues of the neurohypophysial hormones that have not been identified in nature (see Berde and Boissonnas, 1968). By looking at the biological activity of these we can speculate about why the 10 identified natural analogues have arisen. Mutations have been perpetuated at only four positions, 2, 3, 4, and 8, in these molecules. With regard to the 8 position, basic amino acids like arginine and lysine endow it with the most activity on the mammalian kidney and nonmammalian effectors such as the frog and toad skin and urinary bladder (osmotic water transfer across these preparations is increased by such peptides). Less basic amino acids such as ornithine and histidine are much less effective. Similarly, leucine at the 8 position in oxytocin results in a hormone that is most potent in its ability to contract the mammalian uterus and stimulate milk letdown from the mammary gland. Radical substituents at positions other than 2, 3, 4, or 8 usually result in drastic reductions in biological activity. It seems that of all the thousands of possibilities available, nature has, in the course of time, provided the hormones with a structure optimal to that of the receptors. There is, however, an interesting exception. If threonine is substituted for glutamine in the 4 position of oxytocin a peptide is formed that is about four times as effective on the mammalian uterus as oxytocin itself (Manning and Sawyer, 1970). Why then is this not found in nature? The transition to this molecule from oxytocin would require two successive mutations, the first being the substitution of lysine or proline in the 4 position. Such analogues have a very low activity and so may not survive in nature. The succeeding mutation to threonine may thus not have been possible.

The quantities of the hormones stored in the neurohypophysis as well as the ratios of their concentrations are also determined genetically. Storage of these peptides in the neurohypophysis of fishes is more than 10 times less than that normally seen in tetrapods. Five inbred strains of mice have been shown to exhibit a two-fold range of variation in the stores of vasopressin and oxytocin in their neurohypophyses (Stewart, 1972). The molecular ratio of vasopressin to oxytocin (V/O ratio), however, remained a steady 1.5. As seen in Table 3.3, systematic variations in this V/O ratio occur among the mammals. It is interesting that two geographically separated members of the Tylopoda, the camel (from Asia) and the llama (from South America) both have a V/O ratio of about 3, which is higher than that observed in other placentals. The nature of such hereditarily determined differences could be in the ratio of neurosecretory fibres, if one type made oxytocin and another vasopressin. Alternatively, if the two hormones are made together, in the same neuron, this synthesis may be regulated by a control system of genes.

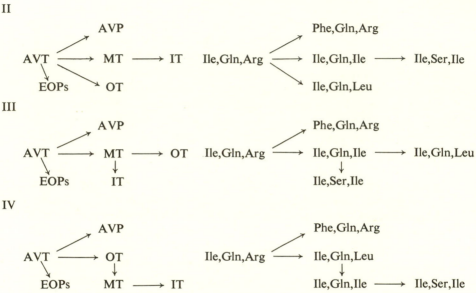

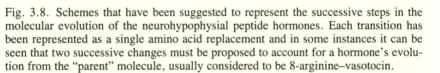

Fig. 3.8. Schemes that have been suggested to represent the successive steps in the molecular evolution of the neurohypophysial peptide hormones. Each transition has been represented as a single amino acid replacement and in some instances it can be seen that two successive changes must be proposed to account for a hormone's evolution from the "parent" molecule, usually considered to be 8-arginine–vasotocin.

Such changes in amino acid composition can be described according to codon base changes in the genetic code. These transformations are often consistent with a single base change (one-step mutation), but in other instances, two base replacements would be needed, such as in the transition from isotocin to mesotocin. In this instance the intermediate could be 4-proline, 8-isoleucine–oxytocin, but this peptide has not been identified in nature. At the time that the proposed schemes were advanced the structures of the chondrichthyean neurohypophysial peptides (EOPs; glumitocin, aspargtocin, and valitocin; see Table 3.2) were unknown but it was suggested that they also arose from vasotocin. This is still thought to be likely but the transitions may involve more than a single unknown intermediate "hormone." 8-Lysine–vasopressin (present in the Suina and marsupials) probably evolved from arginine–vasopressin by a single base replacement. AVT = arginine–vasotocin; AVP = arginine–vasopressin; IT = isotocin; MT = mesotocin; OT = oxytocin; EOPs = the chondrichthyean neurohypophysial hormones; and X = a "parent" molecule of unknown composition. (From Geschwind, 1969.)

Table 3.3. *Examples of the distribution of mole ratios of neurohypophysial hormones in mammalian taxonomic groups. The ratio of vasopressin to oxytocin is shown*

Order or suborder	Species	V/O
Order: Marsupialia[a,b]	American opossum	2.9
	Australian opossum	6.2
	Wallaby	3.8
	Red kangaroo	4.8
Order: Perissodactyla[b]	Horse	0.93
	Zebra	0.44
	Tapir	0.72
	Black rhinoceros	0.77
Order: Artiodactyla[b]		
Suborder: Ruminantia	African buffalo	0.9
	Kongoni	1.5
	Topi	1.4
	Blue wildebeest	1.3
	Kob	1.4
	Bushbuck	1.4
Suborder: Tylopoda[b]	Llama	3.1
	Camel	3.6

[a] Lysine–vasopressin.
[b] Arginine–vasopressin.
Source: Heller, 1966.

The neurohypophysial hormones are stored in the granules of the neurohypophysis in association with proteins that have been called *neurophysins*. Two forms may be present in a single gland and it seems likely that one may be associated with oxytocin and the other with vasopressin (see Acher, 1978). These proteins and the attached peptide hormones are probably products of a common biosynthetic process, being derived from a larger molecule. The two neurophysins contain 93–95 amino acids and they principally differ from each other by amino acid substitutions at the 2, 3, 6, and 7 positions (Fig. 3.9). Depending on the particular amino acids present they have been called MSEL-neurophysin and VLVD-neurophysin (M = methionine, S = serine, and so on). Differences in amino acid composition may occur in other parts of the molecules depending on the species. It is thought that each main type represents a phylogenetic line that originally arose from a common ancestral neurophysin following gene duplication. The nature of the ancestral form is unknown but this may become apparent when the structures of neurophysins in nonmammalian vertebrates are investigated.

Arginine–vasotocin is present in the neurohypophyses of nonmammals but

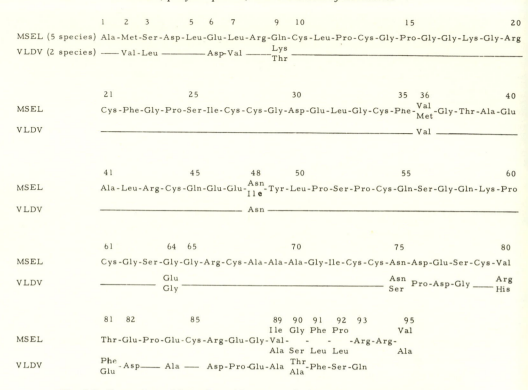

Fig. 3.9. Comparison of the amino acid sequences in the MSEL- and VLDV-neurophysins. The upper line shows the sequence in the MSEL-neurophysins from ox, sheep, pig, horse, and whale, including the substitutions that have been observed among these species. The lower line shows the structures of the VLDV-neurophysins from the pig and ox, together with observed amino acid substitutions. The solid line indicates regions of the molecules where the structures are identical (From Chauvet et al., 1979.)

not in those of adult mammals; it was fascinating to find that vasotocin occurs in the fetuses of sheep and seals (Vizsolyi and Perks, 1969).

The neurohypophysial peptide hormones apparently are not confined to the neurohypophysis, as vasotocin has also been identified in the pineal gland (see Pavel et al., 1973). Vasopressin has also been identified in some cells from human tumors from nonendocrine tissues.

The neurohormones of the hypothalamus

The hypothalamus at the base of the brain contains a host of biologically active materials some of which act as neurotransmitters whereas others are released into the portal blood vessels that supply the adenohypophysis. These

Table 3.4. *Biologically active peptides and putative neurotransmitters in the median eminence of the rat*

Substance	Content pmole/mg protein
LHRH	19.0
TRH	110.0
Somatostatin	189.0
Vasopressin	717.0
Oxytocin	416.0
Norepinephrine	118.0
Dopamine	523.0
Epinephrine	3.4
5-Hydroxytryptamine	87.0
Histamine	160.0

Source: Based on Brownstein, 1977.

compounds are usually either amines or peptides, and they tend to accumulate in the median eminence, which is adjacent to the portal vessels. The concentrations of some of these substances in this region of the hypothalamus are shown in Table 3.4. There is physiological, pharmacological, and even chemical evidence that possibly nine or more such substances can be secreted and influence the release of hormones from the adenohypophysis (Table 3.5). Most of these hypophysiotrophic hormones (or factors) appear to be peptides (Fig. 3.10) (see Vale, Rivier, and Brown, 1977) that are formed by special neurons in the hypothalamus. They may either increase (and thus have the suffix -*releasing hormone*, -RH) or decrease (-*release-inhibiting hormone*, -R-IH) the release of an adenohypophysial hormone. Two such hormones with opposite actions may regulate the secretion of a single pituitary hormone.

The amino acid sequences of three such hypophysiotrophic hormones are known: TRH, which is a tripeptide; LHRH (or LH/FSH-RH), a decapeptide; and somatostatin (or GH-R-IH), a tetradecapeptide. The physiological significance of the observed effect of exogenous MSH-releasing hormone and MSH-R-IH is not clear as they have not been unequivocally identified in the hypothalamus. TRH can also increase prolactin release in mammals, and it may be identical to the prolactin-releasing hormone. Dopamine appears in the hypophysial portal vessels and can inhibit the release of prolactin acting as a P-R-IH (Macleod and Lehmeyer, 1974). The exact chemical structure of the corticotropin-releasing factor has been extremely elusive but it appears to be a peptide. Vasopressin can exert a similar action and it has been identified in high concentrations in the hypophysial portal vessels (Zimmerman et al., 1973). It is possible that CRH is actually a complex of more than one sub-

Table 3.5. *Hypothalamic hormones believed to control the release of pituitary hormones*

Hypothalamic hormone (or factor)	Abbreviation
Corticotropin (ACTH)-releasing hormone	CRH or CRF
Thyrotropin (TSH)-releasing hormone[a]	TSH-RH or TRH or TRF
Luteinizing hormone (LH)-releasing hormone[a]	LH-RH or LH-RF
Follicle-stimulating hormone (FSH)-releasing[a] hormone	FSH-RH or FSH-RF
Growth hormone (GH)-releasing hormone[a]	GH-RH or GH-RF
Growth hormone (GH) release-inhibiting hormone somatostatin	Gh-R-IH or GIF
Prolactin release-inhibiting hormone	P-R-IH or PIF[b]
Prolactin-releasing hormone	PRH or PRF[c]
Melanocyte-stimulating hormone (MSH) release-inhibiting hormone	MSH-R-IH or MRIH or MIF
Melanocyte-stimulating hormone (MSH)-releasing hormone	MRH or MRF

Note: The evidence for the presence of some of these hormones is equivocal. LH-RH and FSH-RH may be identical.

[a] Or regulating hormone.

[b] ? Dopamine.

[c] ? TRH.

Source: Based on Schally et al., 1973. Copyright © 1973 by the American Association for the Advancement of Science.

Thyrotropic-stimulating hormone-releasing hormone, TRH or TSH-RH

(Pyro)   Glu—His—Pro—(NH$_2$)

Luteinizing hormone/follicle-stimulating hormone-releasing hormone, LHRH, LH/FSH-RH, Gn-RH

(Pyro)   Glu—His—Trp—Ser—Tyr—Gly—Leu—Arg—Pro—Gly—NH$_2$

Growth hormone-release-inhibiting hormone (somatostatin)

H—Ala—Gly—Cys—Lys—Asn—Phe—Phe—Trp—Lys—Thr—Phe—Thr—Ser—Cys—OH

Melanocyte-stimulating hormone-release-inhibiting hormone, MSH-R-IH or MRIH

Pro—Leu—Gly—NH$_2$

Melanocyte-stimulating hormone-releasing hormone, MRH

H—Cys—Tyr—Ile—Gln—Asn—OH

Fig. 3.10. The amino acid sequences of some of the hormones from the median eminence that influence the release of adenohypophysial hormones. The structure MSH-RH is tentative; it exhibits such activity but has not been positively identified in the median eminence. Some doubt has also been expressed about MSH-R-IH.

stance consisting principally of vasopressin, which is modulated by other substances (Gillies and Lowry, 1979). However, a larger peptide has recently been identified in extracts of the ovine hypothalamus that can initiate the release of corticotropin in rats (Vale et al., 1981). This putative CRH contains 41 amino acid residues. Its structure shows some remarkable similarities to urotensin I, which is present in the teleost urophysis (see later) and a peptide called *sauvagine*, which has been extracted from the skin of a South American frog.

Using an immunoassay, TRH has been identified in the hypothalamus of a variety of nonmammals, including the domestic fowl, a reptile, an amphibian, a teleost fish, and even from the brain of a larval cyclostome and the head region of a protochordate *Amphioxus* (Jackson and Reichlin, 1974). It is also interesting that TRH has been identified in other parts of the brain, apart from the hypothalamus, in both mammals and nonmammals, suggesting that it may have a more widespread physiological role. It has even been identified in frog skin, where it is present in very high concentrations (Jackson and Reichlin, 1977). Rather remarkably, TRH has also been found in alfalfa plants (Jackson, 1981*b*).

Porcine and ovine TRH are identical. As this hormone is a tripeptide there is little opportunity for change and substituted synthetic analogues have little activity. It is noteworthy, however, that although mammalian TSH can stimulate the thyroid gland in the African lungfish, mammalian TRH is ineffective in these fish (Gorbman and Hyder, 1973). This observation suggests that, as in frogs, a different mechanism or molecular variant of TRH may be present in nonmammals.

LHRH has been identified in representatives of most of the phyletic groups of vertebrates (see King and Millar, 1980*a*). Mammalian LHRH has also been shown to exert similar qualitative effects on the release of gonadotropins in lower vertebrates. However, quantitative differences in the biological responses resulted in the suspicion that the structure of LHRH may not be identical in all species. Immunological studies that measured the binding of different preparations of LHRH to antibodies to the mammalian hormone have also indicated that the structure of this hormone differs in the various phyletic groups. Thus, mammalian and amphibian LHRH preparations behaved similarly; the curves showing displacement of LHRH binding to the antibodies were parallel (Fig. 3.11). However, although LHRH from birds, reptiles, and teleost fish also displaced bound mammalian LHRH, the curves were not parallel to the mammalian and amphibian ones, though they were parallel to each other. It was therefore suggested that LHRH from the latter two groups may be identical, or rather similar, and that from the other groups of vertebrates differs from these. It is possible that a single amino acid substitution may be involved.

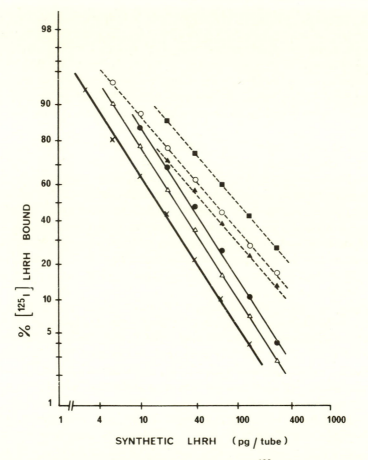

Fig. 3.11. A comparison of the relative displacement of [^{125}I]LHRH from binding to an antiserum by synthetic LHRH and hypothalamic extracts from various species. X, Synthetic LHRH; ●, rat; ○, chicken; ▲, tortoise; △, frog; and ■, teleost. Elasmobranch hypothalamic extracts had no effect in this assay. Pigeon and toad hypothalamic extracts (not shown) gave curves that were parallel, respectively, to those of the chicken and frog. (From King and Millar, 1980*a*.)

Somatostatin has also been identified in the brain of representatives of all vertebrates, including cyclostome fish (King and Millar, 1979). The binding of somatostatin to its antibodies displayed similar characteristics to hormone preparations from all the vertebrate species. It was suggested that the ancestral molecule has persisted unchanged through 400 million years of evolution. However, a putative hormone formed in the urophysis of many osteichthyean fishes, urotensin II, has been shown to exhibit a number of structural similari-

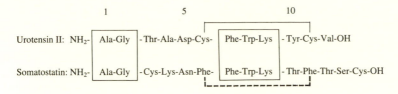

Fig. 3.12. The amino acid sequence of urotensin II extracted from the caudal neurosecretory system of teleost fish compared to somatostatin. The homologous parts of the molecules are enclosed. The disulfide bridges are shown by the continuous line and phenylalanine interactions by the broken line. (From Pearson et al., 1980).

ties to somatostatin (Fig. 3.12). Whether this peptide may have a common ancestry to somatostatin is unknown.

The peptides identified in the hypothalamus, which regulate the release of pituitary hormones, have also often been found to occur in other tissues where they may exhibit other types of effects. It seems likely that such peptides originally played a more general role in neural transmission and that during evolution they were co-opted for regulating the function of the adenohypophysis (Jackson, 1981*b*).

The renin–angiotensin system

It has been known since the turn of the century that saline extracts of the mammalian kidney, when injected into mammals, produce a large increase in the blood pressure. This effect is due to the interaction of an enzyme present in the kidney called renin, which, as described in the last chapter, is formed by the juxtaglomerular cells. Renin interacts with an α-2 globulin in the blood plasma to form angiotensin I, which is converted by a "converting" enzyme to angiotensin II, which is the hormone that actively constricts the peripheral blood vessels. Angiotensinogen is the substrate.

Renin has a wide phyletic distribution in bony fishes and tetrapods. It is, however, absent in cyclostomes and chondrichthyean fishes (Nishimura et al., 1970; Taylor, 1977). Renin is a glycoprotein with a molecular weight of about 50,000 and appears to exist in several forms (isorenins) (see Skeggs et al., 1977; Peach, 1977). "High" and "low" molecular weight renins have been isolated from the kidney. Renin may be stored and secreted as a less active prorenin (Sealey, Atlas, and Laragh, 1980). Renins, sometimes called pseudorenins, have been isolated from such tissues as brain, uterus, the submaxillary gland of mice, and the corpuscles of Stannius in bony fishes (see Chapter 2).

Angiotensin I is a decapeptide (Fig. 3.13) with little biological activity, but it can be converted enzymically to angiotensin II, which is the most potent

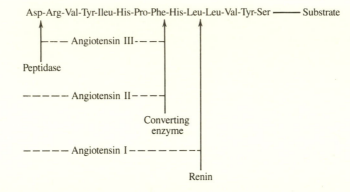

Fig. 3.13. The structure and synthesis of the angiotensins, shown in relation to that of their precursor angiotensinogen. A decapeptide (angiotensin I) is cleaved from the latter under the influence of renin, and this peptide is subsequently further broken down (or "activated") under the influence of "converting enzyme," to form an octapeptide (angiotensin II) and then a peptidase to form the heptapeptide angiotensin III. (From Peart, 1977.)

vasoconstrictor substance known. This change takes place as a result of the splitting off of the two amino acids at the C-terminus, under the influence of "converting enzyme," which is present mainly in the lungs. *Angiotensin II* has several other actions, including an ability to stimulate aldosterone secretion (Blair-West et al., 1971) from the zona glomerulosa cells of the adrenal cortex, and it also induces thirst and hence drinking. Angiotensin II undergoes a series of degradative changes one of which involves the removal of aspartic acid from the N-terminus. This heptapeptide has little vasoconstrictor activity, but it retains most of its action on the adrenal cortex. It may also function, therefore, as a hormone and has been called *angiotensin III*.

Species differences in the structure of angiotensinogen and angiotensin exist. Thus, cattle angiotensin has valine at position 5, whereas in man, horses and pigs, isoleucine is present at this site. There is also evidence that marsupial angiotensinogen (from kangaroos) differs from a placental (sheep) substrate (Simpson and Blair-West, 1972; Blair-West and Gibson, 1980). The structures of angiotensin have not been reported in nonmammals, but differences in their chemical and pharmacological behavior indicate that structural variation is widespread (Sokabe and Nakajima, 1972).

Considerable differences have also been observed in the interactions between renins and angiotensinogens from the various major phyletic groups (Fig. 3.14). Fish (teleost) renin interacts with angiotensinogens (to form angiotensin) from all the tetrapods except mammals. Mammalian renin will not interact with the plasma substrate from nonmammals. Bird renin is also very specific but that of other vertebrates is less so. Amphibian renin interacts with

Renins

	Fish	Amphi-bian	Reptile	Aves	Mammal
Mammal	—	—	—	—	+
Aves	—	—	—	+	—
Reptile	—	−/+	+	+	—
Amphi-bian	—	+	+	+	—
Fish	+	+	+	+	—

Angiotensinogens

Fig. 3.14. The interactions of renin and angiotensinogen, from different vertebrates, to form angiotensin. On the ordinate are the renins from the different vertebrates and on the abscissa, the angiotensinogens. + , Angiotensin (or angiotensin-like) formed; − , no interaction. (From Nolly and Fasciola, 1973. Reprinted with permission of Pergamon Press.)

the angiotensinogens from birds and reptiles and reptilian renin reacts with those of birds and amphibians.

Parathyroid hormone and calcitonin

Parathyroid hormone and calcitonin are, respectively, the peptide hormones originating in the parathyroids and ultimobranchial bodies (or in mammals the thyroid C-cells). Parathyroid hormone increases calcium levels in the plasma and calcitonin decreases them. Parathyroid hormone contains 84 amino acids and is present in all tetrapods but not fishes. Variations in its structure undoubtedly exist but have not been chemically elucidated in nonmammals. Human parathyroid hormone differs (Fig. 3.15) from both the porcine and bovine hormones by substitutions of amino acids at 11 positions, whereas the latter two species differ from each other at 7 positions. All the changes, except at position 43, may be accounted for genetically with a single base change. When tested in the rat (*in vivo*, blood calcium levels) these two

Parathyroid Hormone

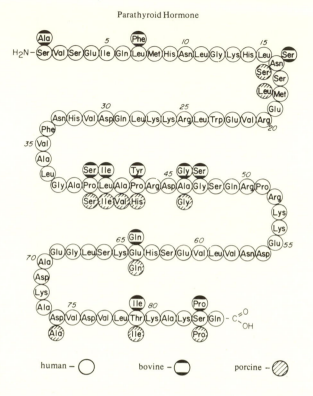

Fig. 3.15. Amino acid sequences of human, bovine, and porcine parathyroid hormone. (Reprinted with permission from H. T. Keutmann et al. Copyright © 1978 by the American Chemical Society.)

animal parathyroid hormones do not exhibit different biological activities, though human parathyroid hormone is only about one-third as active. When bovine and porcine parathyroid hormone are compared *in vitro* (by their ability to activate renal adenylate cyclase) the porcine hormone is less effective. This observation reflects differences in their rates of inactivation by the kidney tissue *in vitro* (Aurbach et al., 1972). It is likely that the active hormone at the effector site represents only a portion of the whole polypeptide molecule and that the hormone, once released, is converted at some peripheral site into an active fragment. The complete molecule is not essential for the exertion of a biological effect as it has been shown that a portion, the amino acids 1–34 at the amino-terminal of the bovine hormone, has a similar activity to the complete molecule with its 84 amino acids (Tregear et al., 1973). This observation suggests that considerable polymorphism of the parathyroid hormone molecule is possible.

Table 3.6. *Calcitonin concentration in glands from various vertebrates*

Class and species	Units (MRC)/g fresh gland wt			
	Thryoid	Ultimo-branchial	Internal parathyroid	Unit/kg body wt
Mammalia				
Man, *Homo sapiens*				
Normal thyroid	0.4	—	0.1–0.5	0.16
Medullary cell carcinoma of thyroid	17	—	—	—
Rat, *Rattus rattus*	5–15	—	—	0.2–0.6
Hog, *Sus scrofa*	2–5	—	—	0.4–0.8
Dog, *Canis familiaris*	1–4	—	1.5–3.3	0.25–0.50
Rabbit, *Oryctolagus cuniculus*				
Lower pole	1.5–2	—	2.1–2.5	—
Upper pole	a	—	—	—
Aves				
Domestic fowl, *Gallus domesticus*	a	30–120	—	0.5–0.8
Turkey, *Meleagris gallopavo*	a	60–100	—	0.5–0.9
Reptilia				
Turtle, *Pseudemys concinna suwaniensis*	a	3–9	—	0.002–0.006
Amphibia				
Bullfrog, *Rana catesbeiana*	—	0.5–0.8	—	0.001–0.002
Teleosti				
Chum salmon, *Oncorhynchus keta*	—	25–40	—	0.4–0.6
Gray cod, *Gadus macrocephalus*	—	10–20	—	0.2–0.4
Elasmobranchii				
Dogfish shark, *Squalus suckleyi*	a	25–35	—	0.25–0.40

[a] No detectable hypocalcemic activity.
Source: Copp, 1969.

Parathyroid hormone is formed from a precursor (pre-proparathyroid hormone) that contains 115 amino acids. During synthesis of the hormone the "pre-" segment, containing 25 amino acids, is removed from the N-terminus followed by the six residues in the "pro-" part of the precursor. Internal structural homologies within the pre-proparathyroid molecule suggest that it may have arisen following the duplication and fusion of an ancestral gene (Cohn, Smardo, and Morrissey, 1979). This hypothesis is given credence by the observation that segments of parathyroid hormone that contain as few as 27 amino acids are still biologically active. Such a small molecule may reflect the general nature of the ancestral hormone.

Calcitonin activity has been measured in all vertebrates except the cyclostomes (Table 3.6). The hormones contain 32 amino acids. Chemical analysis

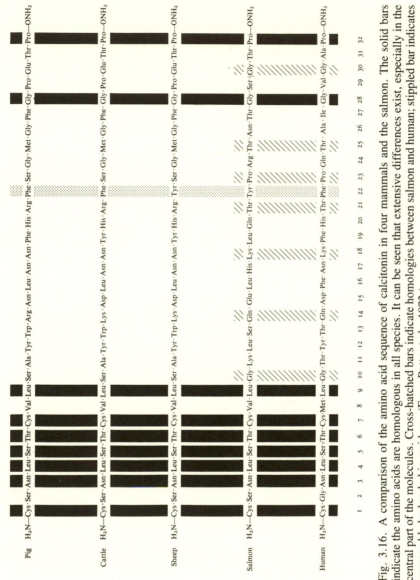

Fig. 3.16. A comparison of the amino acid sequence of calcitonin in four mammals and the salmon. The solid bars indicate the amino acids are homologous in all species. It can be seen that extensive differences exist, especially in the central part of the molecules. Cross-hatched bars indicate homologies between salmon and human; stippled bar indicates comparable hydrophobic residues. (From Potts et al., 1972.)

Pig   H₂N—Cys·Ser·Asn·Leu·Ser·Thr·Cys·Val·Leu·Ser·Ala·Tyr·Trp·Arg·Asn·Leu·Asn·Asn·Phe·His·Arg·Phe·Ser·Gly·Met·Gly·Phe·Gly·Pro·Glu·Thr·Pro—ONH₂

Cattle   H₂N—Cys·Ser·Asn·Leu·Ser·Thr·Cys·Val·Leu·Ser·Ala·Tyr·Trp·Lys·Asp·Leu·Asn·Asn·Tyr·His·Arg·Phe·Ser·Gly·Met·Gly·Phe·Gly·Pro·Glu·Thr·Pro—ONH₂

Sheep   H₂N—Cys·Ser·Asn·Leu·Ser·Thr·Cys·Val·Leu·Ser·Ala·Tyr·Trp·Lys·Asp·Leu·Asn·Asn·Tyr·His·Arg·Tyr·Ser·Gly·Met·Gly·Phe·Gly·Pro·Glu·Thr·Pro—ONH₂

Salmon   H₂N—Cys·Ser·Asn·Leu·Ser·Thr·Cys·Val·Leu·Gly·Lys·Leu·Ser·Gln·Glu·Leu·His·Lys·Leu·Gln·Thr·Tyr·Pro·Arg·Thr·Asn·Thr·Gly·Ser·Gly·Thr·Pro—ONH₂

Human   H₂N—Cys·Gly·Asn·Leu·Ser·Thr·Cys·Met·Leu·Gly·Thr·Tyr·Thr·Gln·Asp·Phe·Asn·Lys·Phe·His·Thr·Phe·Pro·Gln·Thr·Ala·Ile·Gly·Val·Gly·Ala·Pro—ONH₂

1   2   3   4   5   6   7   8   9   10   11   12   13   14   15   16   17   18   19   20   21   22   23   24   25   26   27   28   29   30   31   32

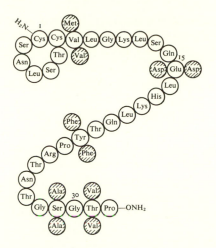

Fig. 3.17. The three variant forms of salmon calcitonin. The main amino acid sequence is that of calcitonin I. Calcitonin II differs by only four residues; valine is still present at position 8. Calcitonin III has five substitutions. Among the four species of salmon examined all form calcitonin I; the chum, pink, and sockeye salmon also form calcitonin II, while the coho salmon has calcitonin III. (From Potts et al., 1972.)

indicates that considerable differences in their sequence occur that result in quantitative differences of biological activity. The amino acid sequence of the calcitonins in four mammals and a teleost fish (salmon) is shown in Fig. 3.16. Only nine amino acid positions are commonly shared by all five species. The differences, however, can nearly all be accounted for by single-base changes in the genetic code (Potts et al., 1972).

The salmon calcitonins are of special interest. Three variants have been identified among four different species of salmon with amino acid substitutions at four or five positions (Fig. 3.17). All species have a common hormone, calcitonin I, but others, calcitonin II or III, may also be present. Salmon calcitonin is much more active when tested in mammals (20–100-fold) than is the natural (homologous) hormone. This is probably a unique situation and is due to two factors: a slow rate of destruction of the piscine hormone as well as a greater affinity for the receptor in the kidney and in bone (Marx, Woodward, and Aurbach, 1972). This is a very interesting situation theoretically and shows that a species need not necessarily have evolved a hormone with a structure that has the maximal possible biological activity. It is nevertheless conceivable that other factors may be involved in the "choice" of such a molecule, and sheer persistency in the circulation may even be a disadvantage. Calcitonin could even have other roles in mammals that are not reflected in its ability to increase calcium concentrations in the bioassay system.

A calcitonin-like molecule has been identified, using immunohistochemical methods, in the nervous tissues and brain of birds, reptiles, and cyclostome fishes (see Girgis et al., 1980). It has also been found in the pituitary gland of mammals and teleost fishes (Deftos et al., 1980). Even more remarkably, it is present in the nervous system of several protochordates, including amphioxus and the ascidian *Ciona intestinalis*. Calcitonin may thus be a very ancient molecule whose structure has been highly conserved during evolution. Originally it may have had a role in regulating the activity of the nervous system. Calcitonin receptors have been identified in the brain of rats, which is consistent with this possibility (Rizzo and Goltzman, 1981). Its functioning as a calcemic hormone may have been a subsequent development, possibly associated with the migration of the neuroectodermal calcitonin-forming cells ("C"-cells; see Chapter 2) to the region of the thyroid gland.

The hormones of the islets of Langerhans and the gastrointestinal tract

Insulin

This hormone has an important role in controlling several processes in intermediary metabolism that involve levels of glucose, fatty acids, and proteins. It has been identified in all vertebrate groups and, in addition, in extracts of the gut, and its associated tissues, of some invertebrates. Some coelenterates, crustaceans, molluscs, and protochordates have been shown to possess insulin-like substances. This activity has been demonstrated in biological assays in mammals and also by the interaction with antibodies to bovine insulin (see Falkmer and Patent, 1972). The chemical nature of the prevertebrate "insulin" is uncertain and it may be a large molecule, a proinsulin, that can react with vertebrate antibodies and from which biologically active fragments may be split off. It has been suggested (Steiner et al., 1972) that insulin may have originated from a large proteolytic digestive enzyme, a proto-proinsulin, in such a manner that when it was absorbed into the blood it became associated with certain metabolic processes in the body.

Several of the vertebrate insulins have been described chemically. The molecule consists of two main parts, an A-chain with 21 amino acids and a B-chain, usually with 31. These are joined by two disulfide bridges contributed by four cysteine residues (Table 3.7). Among the species so far examined (but excluding the hagfish) amino acid substitutions have been recorded at 29 of the 51 positions in the insulin molecules. The A-chain is identical in the insulin present in man, pigs, rabbits, dogs, and sperm whales, and the B-chain is the same in pig, horse, ox, dog, sheep, goat, sperm whale, and sei whale. The intact insulin in the pig, dog, sperm whale, and fin whale are identical. Most of the differences that occur in mammals are localized at

Table 3.7. *Amino acid sequence in vertebrate insulins*

(a) Amino acid sequences of insulin A chains

Type of insulin	1	2	3	4	5	6	7	8	9	10	11	12	13	14	15	16	17	18	19	20	21
Man[a]	Gly	Ile	Val	Glu	Gln	Cys	Cys	Thr	Ser	Ile	Cys	Ser	Leu	Tyr	Gln	Leu	Glu	Asn	Tyr	Cys	Asn
Sei whale	Gly	Ile	Val	Glu	Gln	Cys	Cys	*Ala*	Ser	*Thr*	Cys	Ser	Leu	Tyr	Gln	Leu	Glu	Asn	Tyr	Cys	Asn
Horse	Gly	Ile	Val	Glu	Gln	Cys	Cys	Thr	*Gly*	Ile	Cys	Ser	Leu	Tyr	Gln	Leu	Glu	Asn	Tyr	Cys	Asn
Beef	Gly	Ile	Val	Glu	Gln	Cys	Cys	*Ala*	Ser	*Val*	Cys	Ser	Leu	Tyr	Gln	Leu	Glu	Asn	Tyr	Cys	Asn
Sheep, goat	Gly	Ile	Val	Glu	Gln	Cys	Cys	*Ala*	*Gly*	*Val*	Cys	Ser	Leu	Tyr	Gln	Leu	Glu	Asn	Tyr	Cys	Asn
Elephant	Gly	Ile	Val	Glu	Gln	Cys	Cys	Thr	*Gly*	*Val*	Cys	Ser	Leu	Tyr	Gln	Leu	Glu	Asn	Tyr	Cys	Asn
Rat, mouse (I and II)	Gly	Ile	Val	*Asp*	Gln	Cys	Cys	Thr	Ser	Ile	Cys	Ser	Leu	Tyr	Gln	Leu	Glu	Asn	Tyr	Cys	Asn
Guinea pig	Gly	Ile	Val	*Asp*	Gln	Cys	Cys	Thr	*Gly*	*Thr*	Cys	*Thr*	*Arg*	*His*	Gln	Leu	Glu	*Ser*	Tyr	Cys	Asn
Chicken, turkey	Gly	Ile	Val	Glu	Gln	Cys	Cys	*His*	*Asn*	*Thr*	Cys	Ser	Leu	Tyr	Gln	Leu	Glu	Asn	Tyr	Cys	Asn
Cod	Gly	Ile	Val	*Asp*	Gln	Cys	Cys	*His*	*Arg*	*Pro*	Cys	*Asp*	*Ile*	*Phe*	*Asp*	Leu	*Gln*	Asn	Tyr	Cys	Asn
Tuna (II)	Gly	Ile	Val	Glu	Gln	Cys	Cys	*His*	*Lys*	*Pro*	Cys	*Asn*	*Ile*	*Phe*	*Asp*	Leu	*Gln*	Asn	Tyr	Cys	Asn
Angler fish	Gly	Ile	Val	Glu	Gln	Cys	Cys	*His*	*Arg*	*Pro*	Cys	*Asn*	*Ile*	*Phe*	*Asp*	Leu	*Gln*	Asn	Tyr	Cys	Asn
Toadfish (I)	Gly	Ile	Val	Glu	Gln	Cys	Cys	*His*	*Arg*	*Pro*	Cys	*Asp*	*Ile*	*Phe*	*Asp*	Leu	*Gln*	*Ser*	Tyr	Cys	Asn
Toadfish (II)	Gly	Ile	Val	Glu	Gln	Cys	Cys	*His*	*Arg*	*Pro*	Cys	*Asp*	*Lys*	*Phe*	*Asp*	Leu	*Gln*	*Ser*	Tyr	Cys	Asn
Hagfish	Gly	Ile	Val	Glu	Gln	Cys	Cys	*His*	*Lys*	*Arg*	Cys	Ser	*Ile*	Tyr	*Asp*	Leu	*Gln*	Asn	Tyr	Cys	Asn

(b) Amino acid sequences of insulin B chains

Type of insulin	−1	1	2	3	4	5	6	7	8	9
Pig**		Phe	Val	Asn	Gln	His	Leu	Cys	Gly	Ser
Man, elephant		Phe	Val	Asn	Gln	His	Leu	Cys	Gly	Ser
Rabbit		Phe	Val	Asn	Gln	His	Leu	Cys	Gly	Ser
Rat, mouse (I)		Phe	Val	*Lys*	Gln	His	Leu	Cys	Gly	*Pro*
Rat, mouse (II)		Phe	Val	*Lys*	Gln	His	Leu	Cys	Gly	Ser
Guinea pig		Phe	Val	*Ser*	*Arg*	His	Leu	Cys	Gly	Ser
Chicken		*Ala*	*Ala*	Asn	Gln	His	Leu	Cys	Gly	Ser
Cod	*Met*	*Ala*	*Pro*	*Pro*	Gln	His	Leu	Cys	Gly	Ser
Tuna (II)	*Val*	*Ala*	*Pro*	*Pro*	Gln	His	Leu	Cys	Gly	Ser
Angler fish	*Val*	*Ala*	*Pro*	*Ala*	Gln	His	Leu	Cys	Gly	Ser
Toadfish (I)	*Met*	*Ala*	*Pro*	*Pro*	Gln	His	Leu	Cys	Gly	Ser
Toadfish (II)	*Met*	*Ala*	*Pro*	*Pro*	Gln	His	Leu	Cys	Gly	Ser
Hagfish		*Arg*	*Thr*	*Thr*	*Gly*	His	Leu	Cys	Gly	*Lys*

Table 3.7 (*cont.*)

(b) Amino acid sequences of insulin B chains (cont.)												
10	11	12	13	14	15	16	17	18	19	20	21	22
His	Leu	Val	Glu	Ala	Leu	Tyr	Leu	Val	Cys	Gly	Glu	Arg
His	Leu	Val	Glu	Ala	Leu	Tyr	Leu	Val	Cys	Gly	Glu	Arg
His	Leu	Val	Glu	Ala	Leu	Tyr	Leu	Val	Cys	Gly	Glu	Arg
His	Leu	Val	Glu	Ala	Leu	Tyr	Leu	Val	Cys	Gly	Glu	Arg
His	Leu	Val	Glu	Ala	Leu	Tyr	Leu	Val	Cys	Gly	Glu	Arg
Asn	Leu	Val	Glu	*Thr*	Leu	Tyr	*Ser*	Val	Cys	*(Gln*	*Asp*	*Asp)*
His	Leu	Val	Glu	Ala	Leu	Tyr	Leu	Val	Cys	Gly	Glu	Arg
His	Leu	Val	*Asp*	Ala	Leu	Tyr	Leu	Val	Cys	Gly	*Asp*	Arg
His	Leu	Val	*Asp*	Ala	Leu	Tyr	Leu	Val	Cys	Gly	*Asp*	Arg
His	Leu	Val	*Asp*	Ala	Leu	Tyr	Leu	Val	Cys	Gly	*Asp*	Arg
His	Leu	Val	*Asp*	Ala	Leu	Tyr	Leu	Val	Cys	Gly	*Asp*	Arg
Asp	Leu	Val	Asn	Ala	Leu	Tyr	*Ile*	*Ala*	Cys	Gly	*Val*	*Arg*

23	24	25	26	27	28	29	30	
Gly	Phe	Phe	Tyr	Thr	Pro	Lys	Ala	
Gly	Phe	Phe	Tyr	Thr	Pro	Lys	*Thr*	
Gly	Phe	Phe	Tyr	Thr	Pro	Lys	*Ser*	
Gly	Phe	Phe	Tyr	Thr	Pro	Lys	*Ser*	
Gly	Phe	Phe	Tyr	Thr	Pro	*Met*	*Ser*	
Gly	Phe	Phe	Tyr	*Ile*	Pro	Lys	*Asp*	
Gly	Phe	Phe	Tyr	*Ser*	Pro	Lys	Ala	
Gly	Phe	Phe	Tyr	*Asn*	Pro	Lys		
Gly	Phe	Phe	Tyr	*Asn*	Pro	Lys		
Gly	Phe	Phe	Tyr	*Asn*	Pro	Lys		
Gly	Phe	Phe	Tyr	*Asn*	Pro	Lys		
Gly	Phe	Phe	Tyr	*Asn*	*Ser*			
Gly	Phe	Phe	Tyr	*Asp*	Pro	*Thr*	*Lys*	*Met*

Note: The italicized amino acids indicate the principal differences.
[a] Sequence is identical in man, rabbit, dog, pig, and sperm whale.
[b] Sequence is identical in pig, horse, ox, dog, sheep, sperm whale, and sei whale.
Source: Humbel, Bosshard, and Zahn, 1972. The hagfish sequence is from Peterson et al., 1975.

three positions (8, 9, 10) in the A-chain and in one position (30) in the B-chain.

Hagfish insulin differs from pig insulin by 19 amino acid substitutions and the B-chain is elongated by the addition of methionine at its C-terminus (Table 3.7) (Peterson et al., 1975). The structure of this insulin is especially interesting as it is that of a species from the bottom of the phyletic scale and so it may provide information about the evolution of the hormone. The particular changes that occur at 16 individual positions in hagfish insulin have never been observed in any other vertebrate. However, it has been observed previously that 24 positions in different insulin molecules appear never to change, and this situation persists for 23 of these positions in hagfish insulin. These invariant amino acid residues are assumed to be vital to the basic physiological properties of the hormone and it appears that they have been retained over about 450 million years of evolution. The chemical properties and biological activities of hagfish insulin, however, do display some differences from mammalian insulin (Emdin, Gammeltoft, and Gliemann, 1977). When tested for its ability to increase lipogenesis in fat cells from rats, hagfish insulin was only about 5% as potent as pig insulin. However, provided the concentration of the hormone was increased enough, it could still elicit a similar response but about five times as many receptors needed to be occupied. Hagfish insulin can form dimers but, in contrast to mammalian insulin, cannot aggregate in a hexameric form.

Such large differences in the structure and properties of vertebrate insulin are not confined to species so distant on the phyletic scale. It can be seen (see Table 3.7) that guinea pig insulin differs from pig insulin by 17 amino acid substitutions. A large amount of such variability is common in this subgroup (Hystricomorpha) of the rodents (Horuk et al., 1979). Thus, the insulin from the casiragua (*Proechimys quairae*) differs from pig insulin by 18 amino acid substitutions. The porcupine *Hystrix cristata* only differs by six such amino acids, but this results in some notable changes in its properties (Horuk et al., 1980). For instance, it cannot aggregate and only exists in a single monomeric form. It also has a low potency and a relatively poor ability to combine with insulin receptors. It was suggested that these properties are characteristic of insulin from hystricomorphs, though surprisingly no common structural features of the molecule that can account for the generally low potency have been observed within this group of rodents. Possibly the hystricomorph insulins all evolved in a similar direction, reflecting a generally reduced importance of the hormone in the control of their carbohydrate metabolism.

The laboratory rat and mouse and some fishes each have two insulins (see Smith, 1966, and Fig. 3.18). The two rodents have insulins which differ by two amino acids (in the B-chain) and both hormones are present in the same individuals. This is thought to be a homozygous condition that nevertheless could be the result of a gene duplication so that two genes are present, each controlling the form of one insulin. It is unknown whether the intraspecific polymorphism among the fish insulins is similar to that in the rodents or whether each hormone is present in separate, individual fish.

Despite the chemical differences in the structure of the vertebrate insulins, there is surprisingly little demonstrable variation in their specific biological activities when they are tested on mammalian preparations.

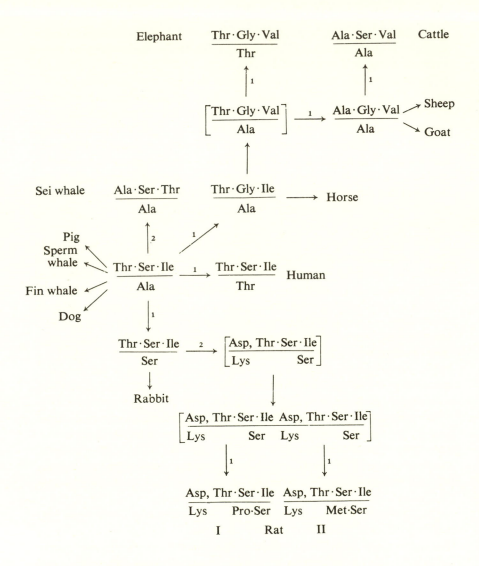

Fig. 3.18. A scheme for the evolution of the mammalian insulins. The sequences of the A-chain are given above the line and the B-chain below it. Numbers alongside the arrows are the minimum number of base changes (mutations) required for the amino acid substitution. The sequences given in brackets are postulated intermediates that have not been identified in nature. The amino acids in the A-chain are those at positions 8, 9, and 10 or, in the rat, 4, 8, 9, and 10. In the B-chain they refer to the 30 position or, in the rat, 3, 29, and 30. The rat produces two insulins (I and II), a process that may be due to a gene duplication and a mutation that occurred subsequently in one of the gene pairs. (From L. F. Smith, 1966, and personal communication.)

Attempts to relate the similarities and differences in the amino acid composition of the insulins to the closeness of the relationship and systematic position of the species have not been very successful. Guinea pig insulin thus differs from that in man by 17 amino acid substitutions (16 in a close relative, the elephant). That of the chicken and man (and the elephant) have only six such disparities. Nevertheless, an attempt has been made to trace the evolution of the mammalian insulin. The "parent," or prototype, may be the most common form, which is found in the pig and others. The successive mutations that would be required to produce the hormones present in other mammals can then be traced (Fig. 3.18). As with the neurohypophysial hormones this must be considered a "game" that is fun and may even be partly correct.

Glucagon

This hormone is a smaller molecule than insulin, consisting of a single chain of 29 amino acids. It has been identified in teleost fishes and all the main tetrapod groups. Glucagon has not been identified in the Chondrichthyes or Cyclostomata though immunological evidence suggests its possible presence in invertebrates including some molluscs and protochordates (see Falkmer and Patent, 1972). The teleost's "glucagon," although biologically effective (it has a hyperglycemic action) in fish, has no effect in the rabbit. This fact suggests structural differences from the homologous mammalian hormone(s). Failure to detect glucagon in other fishes could reflect even greater structural disparities.

Biological and immunological glucagon-like activity has been identified in extracts of the stomach and intestine (Samols et al., 1966; Heding, 1971; Holst, 1978). This material includes a peptide with a molecular weight of 3500 that appears to be identical to pancreatic glucagon but, in deference to its origin, is called *enteroglucagon*. The major portion of the glucagon-like immunoreactivity (GLI), however, exhibits little glucagon-like effect. It has been isolated and found to have a molecular weight of about 11,400 (Jacobsen et al., 1977; Tager and Markese, 1979). This protein is called *glicentin* and contains the same amino acid sequence as pancreatic proglucagon (Patzelt et al., 1979), which is the precursor of the hormone. Glucagon-like immunoreactivity has been identified in the gut of many vertebrates, including cyclostomes (Zelnik, Hornsey, and Hardisty, 1977).

Pancreatic polypeptide

In 1968, Kimmel, Pollock, and Hazelwood (see Kimmel, Pollock, and Hayden, 1978) extracted a peptide from the pancreas of chickens that exhibited a number of biological activities, including an ability to inhibit gastric acid secretion and decrease liver glycogen. This substance was purified and found to contain 36 amino acids. Radioimmunoassay showed that it is secreted into the circulation. A similar immunological activity was identified in 10 more species of birds as well as mammals, amphibians, and reptiles (Langslow, Kimmel, and Pollock, 1973). It originates from cells (PP-cells) in the pancreas, some of which are associated with the islets of Langerhans. This peptide was originally referred to as avian pancreatic polypeptide (APP), but as it is now known to have a wider phyletic distribution it is called pancreatic polypeptide (PP). In domestic chickens ^{125}I-APP has been observed to bind to specific sites in several tissues, including the duodenum, pancreas, spleen, and bone marrow (Kimmel and Pollock, 1981).

Mammalian pancreatic polypeptide has been isolated from bovine pancreas (see Lin et al., 1977), but although this molecule also contains 36 amino acids, 20 of these have been substituted, as compared with the avian molecule. Pancreatic polypeptide may have a function as a hormone, or have some other endocrine role, but at present this is not clear. When injected into dogs it can suppress the rate of secretion of the pancreatic juices (Lonovics et al., 1981).

Insulin-like hormones

Insulin in blood plasma can be detected *in vitro* by its ability to increase the incorporation of glucose into fats. However, in 1963, Froesch and his collaborators found that this response of human plasma could only be partly blocked by insulin antibodies. An activity remained which, although reminiscent of that of insulin, was clearly not due to normal insulin. This material was called *nonsuppressible insulin-like activity*, or *NSILA*. It can, like insulin, promote the growth of tissues but it is far more active than the hormone in this respect. It is less potent, however, in its ability to increase the incorporation of glucose into fats. NSILA can also combine with insulin receptors but not as readily as insulin. NSILA belongs to a family of growth-promoting factors that have been broadly classified as *somatomedins* (growth mediating). The first such substance to be identified was present in rat plasma; it increased the incorporation of sulfate into cartilage (Salmon and Daughaday, 1957). It was initially called "sulfation factor" and later somatomedin (Daughaday et al., 1972). There appears to be a family of such substances. Some of these are synthesized under the influence of growth hormone and androgens.

Two types of NSILA have been isolated from human plasma. They are linear peptides with molecular weights of about 7500. NSILA I contains 70 amino acids, the sequence of which resembles that in the A- and B-chains of insulin (Fig. 3.19). The similarities between NSILA I (also called insulin-like growth factor I) and insulin are quite remarkable: 25 residues are identical to

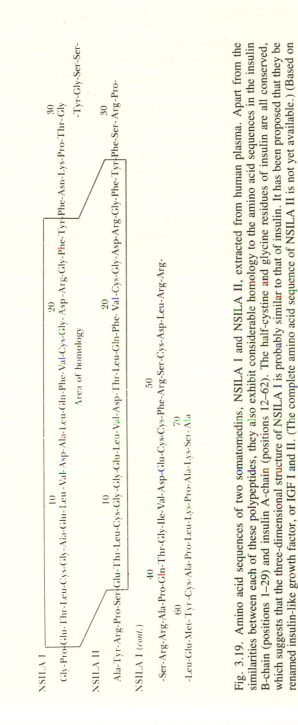

NSILA I

Gly-Pro-Glu-Thr-Leu-Cys-Gly-Ala-Glu-Leu-Val-Asp-Ala-Leu-Gln-Phe-Val-Cys-Gly-Asp-Arg-Gly-Phe-Tyr-Phe-Asn-Lys-Pro-Thr-Gly

Area of homology

-Tyr-Gly-Ser-Ser-

NSILA II

Ala-Tyr-Arg-Pro-Ser-Glu-Thr-Leu-Cys-Gly-Gly-Glu-Leu-Val-Asp-Thr-Leu-Gln-Phe-Val-Cys-Gly-Asp-Arg-Gly-Phe-Tyr-Phe-Ser-Arg-Pro-

NSILA I (cont.)

-Ser-Arg-Arg-Ala-Pro-Gln-Thr-Gly-Ile-Val-Asp-Glu-Cys-Cys-Phe-Arg-Ser-Cys-Asp-Leu-Arg-Arg-

-Leu-Glu-Met-Tyr-Cys-Ala-Pro-Leu-Lys-Pro-Ala-Lys-Ser-Ala

Fig. 3.19. Amino acid sequences of two somatomedins, NSILA I and NSILA II, extracted from human plasma. Apart from the similarities between each of these polypeptides, they also exhibit considerable homology to the amino acid sequences in the insulin B-chain (positions 1–29) and insulin A-chain (positions 12–62). The half-cystine and glycine residues of insulin are all conserved, which suggests that the three-dimensional structure of NSILA I is probably similar to that of insulin. It has been proposed that they be renamed insulin-like growth factor, or IGF I and II. (The complete amino acid sequence of NSILA II is not yet available.) (Based on Rinderknecht and Humbel, 1976*a*, *b*, 1978.)

those in human insulin and these include most of the invariant amino acids. The molecules indeed have many similarities to the precursor of insulin, proinsulin (see Chapter 4). It has been suggested (Rinderknecht and Humbel, 1978) that insulin and NSILA may have arisen from a common ancestral gene following its duplication, possibly in an invertebrate ancestor.

Somatomedins have been identified in the plasma of many species of vertebrates. NSILA was found in representatives of all major classes except the Chondrichthyes (Poffenbarger, Burns, and Bennett-Novak, 1976). Sulfation factor was also identified in the latter group of fishes (Shapiro and Pimstone, 1977). Cyclostomes were not investigated. Neither activity was observed in invertebrates.

Relaxin is a putative hormone that was first identified in the blood of pregnant guinea pigs by Hisaw in 1926. It is a peptide that can relax the pubic ligament and cervix and so may be involved in the process of birth. It has been identified in the blood of a number of mammals including man. Relaxin appears to be formed in the ovary, probably the corpus luteum during pregnancy (Weiss, O'Byrne, and Steinetz, 1976). Like insulin (Fig. 3.20), it consists of two chains of amino acids linked by two disulfide bridges (Kwok and Bryant-Greenwood, 1977; Schwabe and McDonald, 1977). Many of the amino acid residues are homologous, and the two molecules also have a similar tertiary structure (Bedarkar et al., 1977). Relaxin, NSILA I, and insulin may each have arisen from a common ancestral peptide resembling proinsulin.

Relaxins from rats and pigs have been compared and, surprisingly, they only showed a 40% homology in their amino acid sequences compared with 92% in their insulins (John et al., 1981). This observation suggests that the evolution of mammalian relaxins has been much less constrained than that of the insulins. Relaxin has also been isolated from the ovaries of the sand tiger shark (Reinig et al., 1981). It relaxes the pubic symphysis ligaments of guinea pigs but not of mice. The chemical structure of this shark relaxin was more similar to porcine insulin than porcine relaxin, suggesting that its evolution has been more constrained than that of mammalian relaxins. The presence of relaxin in a chondrichthyean suggests that duplication of the proinsulin gene, which gave rise to the relaxin gene, occurred quite early in evolution.

The secretin family

Secretin is a hormone that is secreted by the S-cells of the upper part of the intestinal tract. It has a number of actions, the most notable being an ability to stimulate the secretion of digestive juices by the acinar cells of the pancreas. Secretin is a peptide containing 27 amino acid residues, many of which it has in common with other intestinal peptides and glucagon. For instance, porcine

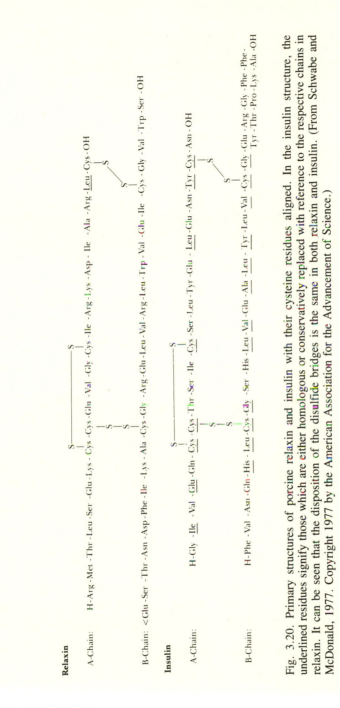

Fig. 3.20. Primary structures of porcine relaxin and insulin with their cysteine residues aligned. In the insulin structure, the underlined residues signify those which are either homologous or conservatively replaced with reference to the respective chains in relaxin. It can be seen that the disposition of the disulfide bridges is the same in both relaxin and insulin. (From Schwabe and McDonald, 1977. Copyright 1977 by the American Association for the Advancement of Science.)

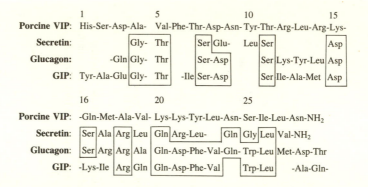

Fig. 3.21. The secretin family of hormones: the amino acid sequences of porcine vasoactive intestinal peptide (VIP), secretin, glucagon, and gastrointestinal polypeptide (GIP). (Only the first 29 of the 43 residues in the GIP are shown.) Amino acids that are identical to those in VIP are shown by the blank spaces while those common to secretin, glucagon, and GIP are in boxes. (From Dockray, 1978.)

secretin and glucagon have 15 amino acids at identical positions (Fig. 3.21). Two other peptides have been identified in the gut that also have common structural features to secretin and glucagon. *Vasoactive intestinal peptide* (*VIP*) relaxes vascular and other types of smooth muscle and, like secretin, can also stimulate the secretion of pancreatic juices (see Rayford, Miller, and Thompson, 1976). It contains 28 amino acids and it shares 9 of these with secretin (Fig. 3.21). *Gastric inhibitory peptide* (*GIP*) (see Rayford, Miller, and Thompson, 1976) was first found in extracts of pig duodenum; it can inhibit the release of gastric acid and pepsin, increase intestinal secretion, and release glucagon and insulin. GIP is released into the circulation in response to the presence of fat in the intestine and is usually considered to be a true hormone. It contains 43 amino acids and among the first 27 of these at the N-terminal, 10 are identical to those in secretin. Thus, secretin, VIP, GIP, and glucagon share many structural features, which in relation to a crossover in their biological activities suggests that they have evolved from a common ancestral molecule. Secretin-like biological activity has been identified in the gut of a number of vertebrate classes including birds, teleosts, and cyclostomes (Dockray, 1975, 1978). VIP has, apart from mammals, also been found in the gut of elasmobranch and teleost fish (Fouchereau-Peron et al., 1980). This phyletic distribution of these peptides suggests that they may have appeared quite early in vertebrate evolution (see Weinstein, 1968).

The gastrin–cholecystokinin family

Gastrin is a hormone that is formed by the G-cells in the pyloric region of the stomach and the duodenum. Its principal action is to increase the secretion of

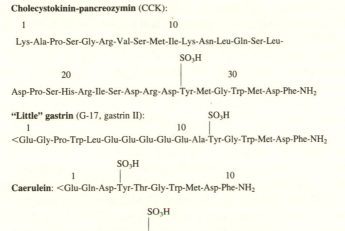

Cholecystokinin-pancreozymin (CCK):

1   10

Lys-Ala-Pro-Ser-Gly-Arg-Val-Ser-Met-Ile-Lys-Asn-Leu-Gln-Ser-Leu-

                            SO$_3$H

20 | 30

Asp-Pro-Ser-His-Arg-Ile-Ser-Asp-Arg-Asp-Tyr-Met-Gly-Trp-Met-Asp-Phe-NH$_2$

"Little" gastrin (G-17, gastrin II):   SO$_3$H

1   10 |

<Glu-Gly-Pro-Trp-Leu-Glu-Glu-Glu-Glu-Glu-Ala-Tyr-Gly-Trp-Met-Asp-Phe-NH$_2$

             SO$_3$H

1 | 10

Caerulein: <Glu-Gln-Asp-Tyr-Thr-Gly-Trp-Met-Asp-Phe-NH$_2$

             SO$_3$H

             |

Active terminal octapeptide: Asp-Tyr-Met-Gly-Trp-Met-Asp-Phe-NH$_2$

Fig. 3.22. The gastrin family of hormones: the amino acid sequences of human cholecystokinin–pancreozymin and gastrin II ("little" gastrin; gastrin I contains 34 amino acids). Caerulein is a peptide that was first isolated from the skin of a frog (*Hyla caerulea*) and which has a similar structure to the active C-terminus of the mammalian hormones. It has been suggested that the ancestral parent hormone may have been similar to this peptide.

gastric acid but it displays a number of other effects on glandular secretion and smooth muscle contractility in the gut. Cholecystokinin (CCK) is formed by the I- or CCK-cells in the upper parts of the intestine. It has two prominent actions: contraction of the gall bladder and increased secretion of an enzyme-rich fluid from the pancreas. The latter effect initially led to its being named pancreozymin, which was once thought to be a separate hormone from CCK (see Jorpes and Mutt, 1966).

Both gastrin and CCK are linear peptides that have a common pentapeptide sequence at their C-terminals (see Rehfeld, 1981). This segment is vital to the biological activities of both hormones, the configuration of the rest of the molecules conferring their abilities to each act at special target sites.

Gastrin exists in two main forms in the body: "big" gastrin (or gastrin I), which contains 34 amino acids, and "little" gastrin (gastrin II), which has 17 such residues (Fig. 3.22). Both forms appear to be secreted by the G-cells in mammals. Little gastrin is more potent than big gastrin, but it is destroyed much more rapidly in the circulation. In mammals amino acid substitutions have been described at three loci in gastrin I, but the C-terminal pentapeptide appears to be invariant. When tested for the ability to increase gastric acid secretion in cats, the ratio of the activity ovine:porcine:human:canine gastrin was 1.8:1.0:0.5:0.3 (see Bromer, 1972). Gastrin-like activity has also been identified in the stomach of amphibians and teleost fish (Gibson et al., 1976; Vigna, 1979).

Cholecystokinin contains 33 amino acid residues (Fig. 3.22). Its physiological actions are usually distinct from those of gastrin though in pharmacological doses it can also increase gastric acid secretion. This effect appears to reflect the chemical similarities of the two hormones. The octapeptide sequence, with the sulfated tyrosine at position 7 from the C-terminal, instead of the 6 position as in gastrin, appears to be important for the differences in activity. Apart from mammals, CCK-like activity has been identified in the intestines of cyclostomes, chondrichthyeans, teleosts, amphibians, reptiles, and birds (see Barrington and Dockray, 1970, 1972; Nilsson, 1970; Dockray, 1979). CCK-like activity has also been identified in the stomach of dogfish (Chondrichthyes), but they lack gastrin (Vigna, 1979). The gut of lampreys also lacks gastrin activity but CCK is present (Holmquist et al., 1979). Immunological evidence shows that gastrin and CCK cannot be identified in different types of cells in amphibians and teleosts; they react with antisera to either hormone (Larsson and Rehfeld, 1977). CCK-like molecules have been identified in nerve cells of a coelenterate. It appears that gastrin and CCK may have evolved from a common ancestral molecule with CCK-like activity, and this event probably occurred in an amphibian ancestor of the reptiles.

Caerulein is a decapeptide that was first identified in the skin of frogs. With the exception of a single amino acid substitution (Fig. 3.22), it is identical in structure to the C-terminal decapeptide sequence of CCK and it also shares the active pentapeptide sequence with gastrin. In frogs caerulein is much more potent than gastrin in stimulating gastric acid secretion. Caerulein-like molecules have also been identified immunohistochemically in the stomach of amphibians and teleosts (Larsson and Rehfeld, 1977, 1978). It is possible that the molecular ancestor of gastrin and CCK was similar to caerulein.

Putative hormones of the gut

A number of other peptides that exhibit a variety of biological activities have been identified in the gut (see Rayford, Miller, and Thompson, 1976). Some, notably *somatostatin*, clearly have a paracrine function, acting to control the activity of adjacent cells. They may also, however, be released into the general circulation and so act as hormones. These peptides include *motilin*, *enterooxytin*, and *chymodenin*. Immunohistochemical evidence indicates that cells secreting such peptides may also be present in a number of nonmammals (see Chapter 2). Somatostatin has been identified in gastroenteropancreatic tissues of cyclostome, chondrichthyean, and teleost fishes, as well as amphibians and reptiles (Falkmer et al., 1977). It has even been found in a sea squirt

(Tunicata) and amphioxus (Cephalochordata), suggesting that it is an "old" hormone. As described earlier (see Fig. 3.10), mammalian somatostatin contains 14 amino acids. Cloning experiments using DNA isolated from the pancreas of anglerfish indicate that two such peptides are present in these bony fish. One of these (somatostatin I) has the same structure as the mammalian hormone, but the other (somatostatin II) differs from it by two amino acid substitutions (Hobart et al., 1980). It has tyrosine instead of phenylalanine at position 7 and glycine instead of threonine at position 10. It is thus possible that somatostatins may also be polymorphous in other species, and this could play a role in determining an appropriate specificity for the diverse actions this peptide is known to have. Larger molecular weight forms of somatostatin, including somatostatin-28, have been identified. It is uncertain whether these represent precursors of the hormone or, as they also have biological activity, if they are distinct hormones.

The corticotropin (ACTH)–lipotropin family

The pars distalis and pars intermedia secrete a family of peptides that exhibit many similarities in the sequences of the amino acids they contain. These relationships reflect the syntheses of these hormones from a common precursor. This molecule is a glycoprotein with a molecular weight of about 31,000 (see Eipper and Mains, 1980). It is formed in the corticotrope and melanotrope cells of the pars distalis and pars intermedia, respectively, and is processed by a series ("cascade") of enzymically controlled cleavage reactions to produce several hormones, as well as a number of peptides of uncertain physiological significance. It is interesting that many of these peptides, from the pituitary gland, have also been identified at other sites in the body, especially the brain (Guillemin, 1978; Krieger and Liotta, 1979). Their presence in the brain is not influenced, however, by the removal of the pituitary gland so that they appear to have separate origins at each site. However, it is possible that in some instances such peptides may be transported from the pituitary to the brain (see Bergland et al., 1980).

The established hormones in the corticotropin–lipotropin family are as follows:

i. *Corticotropin* (adrenocorticotropin, ACTH) is a linear peptide containing 39 amino acids. This hormone acts on the zona fasciculata cells of the adrenal cortex to stimulate the secretion of cortisol and corticosterone. Fragments of ACTH, when injected, have been shown to influence memory and learning (see de Wied, 1977), but the physiological significance of such effects is not clear; they do not appear to reflect the endocrine function of the molecule.

```
              5                 10                15                20
H-Glu-Leu-Thr-Gly-Gln-Arg-Leu-Arg-Gln-Gly-Asp-Gly-Pro-Asn-Ala-Gly-Ala-Asn-Asp-Gly-
```

```
             25                30                35  ┌───────────    40
Glu-Gly-Pro-Asn-Ala-Leu-Glu-His-Ser-Leu-Leu-Ala-Asp-Leu-Val-Ala-Ala-Glu-Sys-Lys-
```

```
──────── β-MSH ───────────────────────────────────────────────────────────┐
             45                50                55    │            60
Asp-Glu-Gly-Pro-Tyr-Arg-Met-Glu-His-Phe-Arg-Trp-Gly-Ser-Pro-Pro-Lys-Asp-Lys-Arg-
```

```
┌───────────── α-Endorphin ─────────────────────────────────────┐
│Methionine│
│Enkephalin      65 ┐           70               75    │           80
Tyr-Gly-Gly-Phe-Met-Thr-Ser-Glu-Lys-Ser-Gln-Thr-Pro-Leu-Val-Thr-Leu-Phe-Lys-Asn-
└── β-Endorphin or "C"-fragment of LPH ────────────────────────────────────
```

```
┌─────────────────────────────────────┐
           85                90        │
Ala-Ile-Ile-Lys-Asn-Ala-Tyr-Lys-Lys-Gly-Glu-OH
```

Fig. 3.23. Amino acid sequence of human β-lipotropin. The sections of the molecule that correspond to some of the known active fragments have been superimposed. (From Li and Chung, 1976.)

ii. β-*Lipotropin* (β-LPH) is a peptide containing 91 amino acids (Fig. 3.23) that was originally isolated from sheep pituitary glands by Li and his collaborators in 1965. It was characterized at that time by its ability to metabolize lipids in an *in vitro* tissue preparation. β-LPH has since been identified in other mammals including cattle, pigs, and man. It has also been measured in the blood (Desranlau, Gilardeau, and Chrétien, 1972). β-LPH has no known direct role as a hormone but it may represent a precursor of other pituitary gland hormones (see the section on the pituitary glycoprotein hormones).

iii. *The endorphins* include a series of peptides that can be formed by enzymic cleavage of β-lipotropin or its parent molecule. β-Endorphin (= C-fragment of β-lipotropin) thus corresponds in its amino acid sequence to β-LPH-(61–91) (see Fig. 3.23) and α-endorphin to β-LPH-(61–76). The β-LPH-(1–58) segment is called γ-lipotropin.

In 1975, two peptides were isolated from the brain (Hughes et al., 1975) that were able to mimic the effects of the opiate drug morphine in inhibiting contractions (*in vitro*) of the mouse vas deferens and guinea pig ileum. They were also found to be able to bind to proteins in the brain that act as receptors for morphine and which are the primary sites for its pain-relieving (analgesic) effects (see Goldstein, 1976; Lord et al., 1977; Snyder, 1980). These peptides were called *enkephalins* and it appears that they may act as neurotransmitters in the brain, possibly by influencing activity that is associated with the sensation of pain. They are also associated with neurons in the hypophysis where they could be contributing to the release of oxytocin and vasopressin (Martin and Voigt, 1981). The enkephalins belong to a family of peptides that have been called *endorphins* (from endogenous and morphine). The enkephalins are pentapeptides and it was of special interest to observe that one of them, Met-enkephalin (the other is Leu-enkephalin) has the same amino acid sequence as that of β-LPH-(61–65). β-Lipotropin has no detectable opioid analgesic action but isolated sections of its molecule, especially β-endorphin are even more effective than the enkephalins (β-endorphin contains the amino acid sequence of Met-enkephalin). Whether endorphins secreted by the pituitary gland have a physiological role is uncertain but it is possible that they may aid the alleviation of pain in stressful situations (see Lord et al., 1977).

iv. The *melanocyte-stimulating hormones* (MSH, melanotropin) are peptides that in lower vertebrates may induce a darkening of the skin by promoting the dispersal of melanin in the melanophores (see Chapter 7). In mammals they can increase the formation of melanin in melanocytes and so increase the pigmentation of the skin. Two general types of MSH have been described: α-MSH, which contains a linear sequence of 13 amino acids, and β-MSH, which, depending on the species, may contain 16–22 amino acids (see Fig. 3.24). In mammals both of these types of MSH contain a common heptapeptide sequence, Met-Glu-His-Phe-Arg-Trp-Gly-, which may be important in their actions. This sequence corresponds to that of mammalian ACTH-(4–10) and may account for the small amount of MSH-like activity exhibited by corticotropin. The pentapeptide His-Phe-Arg-Trp-Gly has MSH-like activity, and it has been called the "active core" of the molecule. The MSH hormones appear to be secreted only by animals with a distinct pars intermedia. In man, where this lobe is not readily apparent, the previous identification of MSH is now generally thought to reflect the presence of an artifact of the chemical extraction procedures used.

The corticotropin–lipotropin family of hormones all originate in corticotrope and melanotrope cells in the pars distalis and pars intermedia, respectively. These cells synthesize a large precursor molecule ("31 k"), the complete amino acid sequence of which has been described by Nakanishi et al. (1979), whose results are summarized in Fig. 3.25. This molecule contains amino acid sequences that correspond to those of various hormones and peptides that can be secreted by the pituitary gland. Lowry and Scott (1975) have suggested that the processing of such a precursor molecule is different in the corticotrope and melanotrope cells (Fig. 3.26). Thus, in the latter, the products are α-MSH and β-MSH whereas in the former ACTH and β-LPH are formed. The site(s) of the release of endorphins is equivocal; both types of cells have been sug-

Corticotropin (ACTH)

<pre>
                          Mammalian α-MSH   ┊
               ┊   Active core of   ┊   ┊
               ┊   α and β-MSH   ┊   ┊
    Man   ┊   ┊   ┊
    H-Ser-Tyr-Ser-Met-Glu-His-Phe-Arg-Trp-Gly-Lys-Pro-Val-Gly-Lys-Lys-Arg-Arg-Pro-Val
        1   2   3   4   5   6   7   8   9 10 11 12 13 14 15 16 17 18 19 20
    Pig

    Ox

    Sheep

    Dogfish   Met   Arg   Ile
</pre>

β-MSH (in mammals β-lipotropin 41–58)

Pig, sheep	Asp	Glu-Gly-Pro-	Tyr-Lys	Met-Glu	His-Phe-Arg-Trp	Gly-Ser	Pro	Pro-Lys-Asp
Dogfish						
(*Squalus acanthias*)	Asp	Gly-Asp-Asp-	Tyr-Lys	Phe-Gly	His-Phe-Arg-Trp	Ser-Val	Pro	Leu
(*Scyliorhinus canicula*)	Asp	Gly-Ile-Asp-	Tyr-Lys	Met-Gly	His-Phe-Arg-Trp	Gly-Ala	Pro	Met-Asp-Lys

Fig. 3.24. The amino acid sequence of ACTH from various mammals and the dog-fish. In the lower section are shown the structures of β-MSH from pigs and sheep compared to that in two species of dogfish. The boxed sections indicate the presence of identical amino acids. (From Lowry and Scott, 1975.)

gested. The original precursor molecule has been appropriately called proopiocortin (Rubinstein, Stein, and Udenfriend, 1978).

Homologous hormones from the corticotropin–lipotropin family have been identified throughout the vertebrates, though there is a paucity of information about their precise structures in nonmammals. Differences in their biological and immunological behavior, however, suggest that there are numerous poly-morphic forms of these hormones.

Several variants of corticotropin have been observed in mammals (see Fig. 3.24). The sequences in ACTH-(1–24) appear to be invariant in mammals, and chemical analogues with this structure have full biological activity. Variations in amino acid sequence are confined to the ACTH-(24–39) region (the "immunogenic tail"), specifically between amino acids 25 and 33, the last six residues being invariant. These differences in structure are sufficient to induce the formation of antibodies when the hormones are administered to heterolo-gous species. Nonmammalian corticotropin, however, may exhibit more vari-ation than observed in mammals. Thus, dogfish (*Squalus acanthias*) corticotropin differs from human corticotropin by 10 amino acid substitutions, 8 of which are at invariant sites in mammals, 3 in the ACTH-(1–24) region and 5 in the "tail" region. There are even changes in the heptapeptide "core" [ACTH-(4–10)].

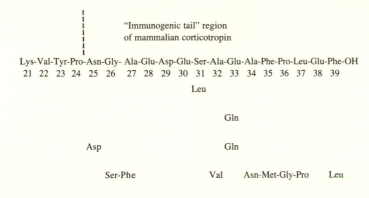

Lipotropin has been identified in a number of mammals. The structure of β-endorphin is the same in sheep and the camel while in man it differs from these by two amino acid substitutions (see Li, Chung, and Doneen, 1976). The structure of this peptide thus appears to be quite conservative in mammals, which would be consistent with its having a discrete physiological role. Enkephalins have been identified in the brains of representatives of all groups of vertebrates (Simantov et al., 1976; King and Millar, 1980*b*). However, their presence in invertebrates is inconsistent though Met-enkephalin has been found in nerve tissue of a snail. The concentrations of enkephalins in the

Fig. 3.25. A diagrammatic representation of the bovine proopiocortin molecule that is the precursor of pituitary ACTH and β-lipotropin. The sequences of amino acids can also be seen to contain several other peptides, including α- and β-MSH and β-endorphin. Met-enkephalin (not shown) is also present at the 61–65 position of β-lipotropin. The active hormone fragments are progressively cleaved, in a cascade, from the precursor. (Based on Nakanishi et al., 1979. From Krieger and Liotta, 1979. Copyright © 1979 by the American Association for the Advancement of Science.)

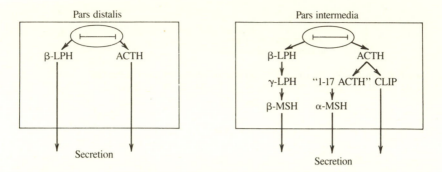

Fig. 3.26. A diagram summarizing the hypothesis that α- and β-MSH are formed from the same precursor molecule as β-lipotropin and ACTH but only in cells in the pars intermedia. In the latter, enzymes are present that can cleave the parent molecule (now known to be proopiocortin) at certain positions (see Fig. 3.25) that result in the separation of the melanocyte-stimulating hormones. (From Lowry and Scott, 1975.)

brains of nonmammals are generally much higher than in mammals. α-Endorphin has been identified in the pituitary gland of trout (Follénius and Dubois, 1978). Opiate activity has also been found in the pars distalis and pars inter-media of eels and trout (Hunter and Baker, 1979; Carter and Baker, 1980). A physiological role for endorphins appears to be likely in nonmammals but remains to be confirmed.

Comparison of the chromatographic, immunological, and biological behavior, as well as the chemical structures, of MSH prepared from different species indicates that considerable polymorphism occurs in its structure (see Burgers, 1963; Shapiro et al., 1972; Lowry and Scott, 1975; Dickhoff and Nicoll, 1979). α-MSH has the same structure in many species, and even that in dogfish (Chondrichthyes) only differs from the mammalian hormone by a single amino acid substituent (see Fig. 3.24). β-MSH is much more variable however. Substantial differences have even been observed between two species of dogfish (see Fig. 3.24); *Scyliorhinus canicula* contains 18 amino acids like most mammals, but *Squalus acanthias* only has 16 such residues. They also differ by five amino acid substitutions. Changes have been observed in the pentapeptide "core" in these fish. Such variability suggests that either β-MSH may not have unique physiological importance as a hormone in many species or that its effector tissues can have quite catholic requirements.

A considerable number of changes in the amino acid sequences of the peptides in the corticotropin–lipotropin family have been described. It is tempting to try to ascribe to each of these an optimal physiological functioning of the particular hormones or the mechanisms of their biosynthetic processing from the parent precursor. However, it should also be recalled that two or more hormones or peptides may share sequences in the precursor. Thus, α-MSH in mammals has an identical sequence to ACTH-(1–13) and β-MSH to γ-LPH-(41–58). There is little variation in α-MSH and the active region (1–24) of ACTH. Thus, the requirements of one hormone may place restraints on evolutionary changes in another with which it shares a common sequence in the precursor. The greater variability in the structure of β-MSH may partly reflect a lack of evolutionary restraint on γ-LPH, which has no known physiological role. The evolution of the hormones in the corticotropin–lipotropin family may be expected to exhibit a "togetherness." The evolution of such family hormones may reflect changes not only in the gene coding for the precursor but also in the mechanisms of the post-translational processing. Such an event probably led to the perpetuation of the separate corticotrope and melanotrope cells, each of which produces distinctive products from the same precursor.

The pituitary glycoprotein hormones: luteinizing hormone (LH), follicle-stimulating hormone (FSH), and thyroid-stimulating hormone (TSH)

The anterior lobe of the pituitary gland is the site of synthesis and release of hormones that control the activities of the testes and ovaries, as well as the thyroid gland. These hormones (see Papkoff, 1972; Licht et al., 1977) are proteins with a molecular weight of about 30,000 and they contain about 12%–20% carbohydrate. They exhibit strong analogies in their general tertiary structure and amino acid sequences. All of these hormones consist of two nonidentical subunits (α- and β-), which are not covalently bound. In mammals they have been named according to their particular actions on the gonads (see Chapter 9): *follicle-stimulating hormone* (FSH) and *luteinizing hormone* (LH). As the latter stimulates hormone production from interstitial cells in the testis it has also been called interstitial cell-stimulating hormone (ICSH), but this name has now been generally discarded. These two hormones are included in the broad category of gonadotropic hormones. Other such hormones with a similar structure and lineage have been identified in members of all classes of the vertebrates where they seem to perform similar functions. Two structurally related variants of the pituitary gonadotropins are found in mammals, where they are formed during pregnancy in the placenta. They are human *chorionic gonadotropin* (hCG) (other primates also appear to possess an analogous hormone) and *pregnant mare's serum gonadotropin* (PMSG), which has recently been renamed *equine chorionic gonadotropin* (eCG). A molecule that cannot be distinguished from hCG has recently been isolated from the pituitary gland (Matsuura et al., 1980). The pituitary gland also secretes a *thyroid-stimulating hormone* (TSH, thyrotropin), which has remarkable similarities to the gonadotropins. Indeed, hCG has endogenous TSH-like activity (Azukizawa et al., 1977) and it is not clear if some lower vertebrates, especially fishes, actually secrete a TSH that is distinct from the gonadotropin(s) they possess.

hLH-α:

Val-Gln-Asp-Cys-Pro-Glu-Cys-Thr-Leu-Gln-Glu-Asn-Pro-Phe-Phe-Ser-Gln-Pro-Gly-

hCG-α:

Ala-Pro-Asp-Val-Gln-Asp-Cys-Pro-Glu-Thr-Leu-Gln-Glu-Asp-Pro-Phe-Phe-Ser-Gln-Pro-Gly-

Alignment
position: 1   10

Ala-Pro-Ile-Leu-Gln-Cys-Met-Gly-Cys-Cys-Phe-Ser-Arg-Ala-Tyr-Pro-

Ala-Pro-Ile-Leu-Gln-Cys-Met-Gly-Cys-Cys-Phe-Ser-Arg-Ala-Tyr-Pro-

20   30

Thr-Pro-Leu-Arg-Ser-Lys-Lys-Thr-Met-Leu-Val-Gln-Lys-Asn-Val-
                                        CHO

Thr-Pro-Leu-Arg-Ser-Lys-Lys-Thr-Met-Leu-Val-Gln-Lys-Asn-Val-

40   50

Thr-Ser-Glx-Ser-Thr-Cys-Cys-Val-Ala-Lys-Ser-Tyr-Asn-Arg-Val-Thr-Val-
Thr-Ser-Glu-Ser-Thr-Cys-Cys-Val-Ala-Lys-Ser-Tyr-Asn-Arg-Val-Thr-Val-

60

Met-Gly-Gly-Phe-Lys-Val-Glx-Asn-His-Thr-Ala-Cys-His-Ser-Cys-Thr-Cys-Tyr-Tyr-His-Lys-Ser
                    CHO

Met-Gly-Gly-Phe-Lys-Val-Glu-Asn-His-Thr-Ala-Cys-His-Cys-Ser-Thr-Cys-Tyr-Tyr-His-Lys-Ser-

70   80   90

Fig. 3.27. The amino acid sequences of the α-subunits of human luteinizing hormone (hLH) and human chorionic gonadotropin (hCG). The α-subunit of human thyroid-stimulating hormone (TSH-α) has the same structure as human LH-α. (Based on Sairam, Papkoff, and Li, 1972; Morgan, Birken, and Canfield, 1975.)

The pituitary and placental gonadotropins and thyrotropin each contain an α-subunit and a β-subunit, which have different amino acid sequences. The particular subunits of each hormone are named LH-α, LH-β, TSH-α, TSH-β, and so on. These subunits can be chemically separated and later recombined, either with each other or with subunits from other hormones. Each subunit by itself lacks biological activity; however, if TSH-α is combined with LH-β the product regains activity but this corresponds to LH. Conversely, if LH-α is combined with TSH-β the molecule has TSH activity. Thus, the β-subunit determines which type of action the molecule has, but the α-subunit is necessary for its expression. Such recombinations of subunits from different sources also suggest that it is the β-subunit that determines the species specificity of a particular hormone. The ability of the hybrid turtle -LH-α ovine LH-β to increase secretion of testosterone from turtle testis was only 12% of that seen by the homologous combination (turtle -LH-α turtle LH-β) whereas that of ovine -LH-α turtle LH-β was 64% of that elicited by the native hormone (Licht, Farmer, and Papkoff, 1978).

The precise sequences of the amino acids and carbohydrates have been determined in the subunits of the mammalian glycoprotein hormones. In man (Fig. 3.27) it can be seen that LH-α, TSH-α, and hCG-α are almost identical in their structures. The β-subunits show variation but they are still in many respects similar (Fig. 3.28). Thus, the homologies in the sequences of FSH-β and hCG-β are about 40%, and LH-β and hCG-β are over 80%. There are also differences in the chain length: FSH-β contains 118 amino acids; LH-β, 115; TSH-β, 112; and hCG-β, 145. The similarities between hCG and both LH and FSH may be reflected in the observation that hCG has both LH-like and FSH-like activity (Siris et al., 1978), though the former predominates.

Each subunit of the glycoprotein hormones is coded by a separate gene (Daniels-McQueen et al., 1978; Kourides and Weintraub, 1979; Godine, Chin and Habener, 1980; Fiddes and Goodman, 1980). The similarities in the structures of the products of all the genes have led to the suggestion (Fontaine and Burzawa-Gerard, 1977) that they may have evolved from an ancestor to form an α-subunit and a family of β-subunit genes coding for the specific structure in each type of hormone (Fig. 3.29).

The pituitary glands of nonmammals have all been shown to possess a biological activity that indicates the presence of a hormone or hormones similar to the mammalian gonadotropins. They also appear to be glycoproteins though information about their precise structure is limited. Members of all the classes of tetrapod vertebrates have been shown to possess two such hormones (see Licht et al., 1977) though in some species only one has been identified. A survey of several families of snakes (Ophidia, Reptilia), for instance, has failed to uncover a second gonadotropin (Licht et al., 1979). The one that is present has neither a distinct FSH nor LH activity. Most investigations on teleost fish have also revealed only a single gonadotropin (Burzawa-Gerard and Fontaine, 1972), and the same situation seems to exist in a chondrostean, the sturgeon (Burzawa-Gerard, Goncharov, and Fontaine, 1975) and an elasmobranch, the dogfish (Sumpter et al., 1978). However, in one laboratory two types of gonadotropins have been isolated from the pituitary glands of several teleosts (Idler and Ng, 1979). Nonmammalian gonadotropins all conform to the general pattern in possessing two subunits (see Burzawa-Gerard, 1974; Licht et al., 1977) Even the single gonadotropin isolated from the pituitary gland of the carp showed close structural homologies to the α- and β-subunits of the mammalian hormones (Jollès et al., 1977). Immunologically, the carp gonadotropin-β behaved more like mammalian LH-β than FSH-β (Burzawa-Gerard, Dufour, and Fontaine, 1980).

The phyletic distribution of gonadotropins suggests that in ancestral forms, possibly like in present-day fishes, there may have been a single gonadotropic hormone that fulfilled all the required functions. A contemporary gonadotropin that can clearly exhibit such a dual activity is pregnant mare's serum gonadotropin. It has been suggested that this ability may reside in a special C-terminal region of its β-subunit (Moore and Ward, 1980). Subsequently, another such hormone may have evolved in the tetrapods that facilitated a more specialized control over gonadal function. It is nevertheless disturbing hFSH-β: Asn- Ser | Cys | Glu-Leu-Thr-Asn(CHO)-Ileu

hTSH-β: Phe | Cys | Ileu-Pro-Thr-Glx   -Tyr

hLH-β: | Ser Arg-Glu- | Pro-Leu-Arg-Pro -Trp | Cys | His-Pro-Ileu-Asn   -Ala

hCG-β: | Ser Lys-Gln- | Pro-Leu-Arg-Pro -Arg | Cys | Arg-Pro-Ileu-Asn(CHO)-Ala

Alignment position: 5   10

hFSH-β: -Thr-Ileu-Ala- Ileu-Glu-Lys-Glu -Glu | Cys | Arg-Phe | Cys | Leu | Thr | Ileu

hTSH-β: -(Met,Thr,His, Val,Glx)-Arg-Arg -Glx | Cys | Ala-Tyr | Cys | Leu | Thr | Ileu

hLH-β: -Ileu-Leu-Ala- Val-Gln-Lys-Glu -Gly | Cys | Pro-Val | Cys | Ileu | Thr | Val

hCG-β: -Thr-Leu-Ala- Val-Glu-Lys-Glu -Gly | Cys | Pro-Val | Cys | Ileu | Thr | Val

             15   20   25

hFSH-β: | Asn(CHO)-Thr-Thr | - Trp | Cys-Ala-Gly-Tyr-Cys | -------- 112 Tyr-OH

hTSH-β: | Asn(CHO)-Thr-Thr | - Ileu | Cys-Ala-Gly-Tyr-Cys | -------- 118 Tyr-OH

hLH-β: | Asn(CHO)-Thr-Thr | - Ileu | Cys-Ala-Gly-Tyr-Cys | -------- 115 Gly-OH

hCG-β: | Asn(CHO)-Thr-Thr | - Ileu | Cys-Ala-Gly-Tyr-Cys | -------- 145 Gln-OH

             30   35

Fig. 3.28. Comparison of the N-terminal amino acid sequences of hFSH-β, hLH-β, hTSH-β, and hCG-β. The boxed-in sections indicate regions of the molecules where the amino acid sequences are identical. The remaining parts also show considerable homologies. (Based on Saxena and Rathnam, 1976.)

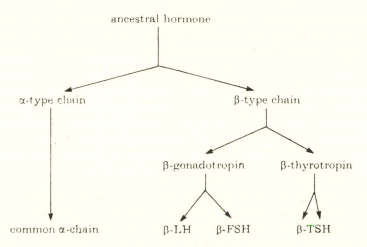

Fig. 3.29. A scheme showing the possible pathway of the evolution of the pituitary glycoprotein hormones; luteinizing hormone (LH), follicle-stimulating hormone (FSH), and thyrotropin (TSH). Each of these hormones consists of two subunits, α and β. The α-subunit is remarkably similar in all these hormones but differences exist in the β-subunits, which confer on the hormone its particular selective, physiological effect and species specificity. (The α-subunit, however, is vital for the response to occur.) It is suggested that each subunit originally arose as a result of gene duplication. The structure of the α-subunit was then largely conserved while that of the β-subunit underwent further duplications and point mutations to provide the materials needed to assemble the three hormones. (From Acher, 1980).

to observe a single such hormone in snakes though this could have been a secondary reversion. However, it cannot be excluded that such an event has not also occurred in contemporary fishes. It is unknown whether cyclostome fishes possess one or two gonadotropins, and the answer to this question may be especially enlightening.

The principal evidence that vertebrate gonadotropin(s) and TSH exist in various polymorphic forms comes from comparative measurements of their activities using bioassay preparations derived from different species. The ratio of the biological activity of two glandular extracts can be initially compared in one type of assay system and then repeated in other, different types of preparations. If the ratio differs in the two or more systems used for the measurements it suggests that the activities result from hormones that differ somewhat in their chemical structures. For the measurement of gonadotropic activity such assay preparations include stimulation of the uptake of radioactive phosphorus (^{32}P) by the testes of day-old chicks or eels, spermiation in amphibians and teleosts, maintenance and development of the gonads and secondary sex characters in lizards, and ovulation in amphibians and mammals. Comparison of TSH activity depends on measurements of ^{131}I uptake by the thyroid gland, the release of thyroxine, and the histological appearance and size of the cells in the thyroid. Phyletically diverse species have been used for these comparative assays ranging from mammals to teleosts and chondrichthyean fishes.

"There are no clear cut, well documented cases of species specificity of gonadotrophic hormones" (Nalbandov, 1969). Different species invariably show *some* response to heterologous gonadotropins. There are, however, considerable variations in the biological potency of such hormones, indicating that polymorphic variations exist. Mammalian gonadotropins exhibit some activity in all vertebrates. In teleost fish, mammalian LH (and human chorionic gonadotropin, hCG) are sometimes effective but FSH is inactive (Burzawa-Gerard and Fontaine, 1972). Teleost gonadotropin, on the other hand, while being very active in teleosts, has little effect in mammals. Amphibians, reptiles, and birds show considerable responsiveness to gonadotropins from teleost, chondrichthyean, and dipnoan fishes as well as those from mammals. The avian hormones are more effective in lizards than in mammals (Licht and Stockell Hartree, 1971; Burzawa-Gerard and Fontaine, 1972; Donaldson et al., 1972; Scanes et al., 1972). While a reptilian gonadotropin preparation, from the snapping turtle, is ineffective *in vivo* on mammalian test preparations it stimulates the ovarian granulosa cells of a monkey *in vitro* and is also active in birds, amphibians, and other reptiles (Channing et al., 1974; Licht and Papkoff, 1974a). No gonadotropin preparations thus would appear to be completely species specific, but they exhibit considerable differences in their activity when tested on preparations from other phyletic groups. Such differences are assumed to reflect variations in their molecular structures.

Immunological cross-reactions are also indicative of chemical relationships between hormones. Antibodies to highly purified preparations of LH from the domestic fowl have been tested on a variety of vertebrates (Scanes, Follett, and Goos, 1972). Such antiserum reacted with pituitary gland extracts and plasma from 10 other species of birds as well as three species of reptiles and a dogfish. Nonparallel dose–response reactions were observed with preparations from three amphibians, a lungfish, and one teleost (the goldfish), but no reaction could be seen in another teleost (the carp). A similar type of survey has been made comparing the abilities of FSH from various tetrapods to interfere with the binding of human FSH to antibodies to ovine FSH (Fig. 3.30). FSH from the various eutherian mammals was usually most effective in preventing the binding of human FSH; birds and reptiles showed less cross-reactivity. The amphibians showed a very poor reactivity and the curves were also not parallel to those of FSH from all the other species. The snakes were an interesting exception as they showed no activity in this system. One marsupial, the kangaroo *Macropus eugenii*, showed less cross-reactivity than all of the eutherians, falling among the birds and reptiles. One cannot trace the phylogenetic history of hormones in this way but it emphasizes the variations

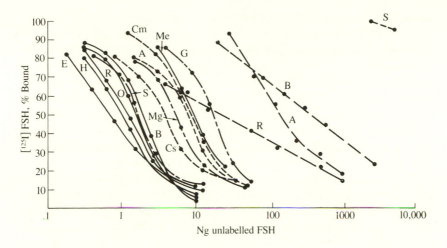

Fig. 3.30. Competitive inhibition of the binding of ^{125}I-labelled human FSH to anti-ovine FSH serum by various tetrapod FSH preparations. Mammalia (solid curves): *Bos* (B), *Equus* (E), *Homo* (H), *Ovis* (O), *Rattus* (R), *Macropus* (Me); Reptilia (short dash curves): *Alligator* (A), *Chelonia* (Cm), *Chelydra* (Cs), "Snakes" (S); Aves (line–dash curves): *Gallus* (G), *Meleagris* (Mg), *Struthio* (S). Amphibia (long dash curves): *Ambystoma* (A), *Bufo* (B), *Rana* (R). (From Licht and Bona Gallo, 1978.)

between them. It should be emphasized that similarities and differences in biological and immunological responses of hormones need not parallel each other, as the associated changes in the molecules may have evolved independently for each type of activity.

In 1940, Gorbman found that the goldfish thyroid tissue was stimulated by pituitary extracts from a teleost fish, two amphibians, a bird, and a mammal. This suggested that TSH had a wide phyletic distribution. Subsequent measurements using a greater variety of species to compare the activity of such glandular extracts (in addition to the goldfish, a salamander, lizard, and guinea pig were used) indicated the hormones present in the various species were not identical, though they exerted the same general biological effects. Although mammalian TSH preparations are active in teleosts, teleost TSH has little activity in mammals (Fontaine, 1969*a*, *b*). As the phyletic scale is ascended it is found that TSH preparations from a lungfish (*Protopterus*), amphibians, reptiles, and birds can exert well-defined effects on the thyroid of both a mammal (mouse) and teleost fish (trout). The thyroid of chondrichthyean fish (the stingray, *Dasyatis sabina*) although responding to its own, homologous TSH shows no response to the mammalian or even teleost hormones (Jackson and Sage, 1973). Chondrichthyean TSH, on the other hand, stimulates the mammalian thyroid (Dodd and Dodd, 1969). Mammalian TSH increases thyroidal activity in the Pacific hagfish, *Eptatretus stouti* (Cyclostomata), though TSH activity (when tested in mammals) has not been demonstrated in the pituitary of the Atlantic hagfish, *Myxine glutinosa* (Dodd and Dodd, 1969; Kerkof, Boschwitz, and Gorbman, 1973). One obviously cannot construct an ordered story from these observations, but they serve to show that TSH, like other hormones, has suffered, during evolution, changes in its structure.

The mammalian gonadotropins have a thyrotropic effect when they are injected into teleost fishes, an action that was initially attributed to the presence of a distinct heterothyrotropic factor, or HTF, in the mammalian pituitary (Fontaine, 1969*a*). As described earlier hCG is also thyrotropic in man. This crossover in the actions of gonadotropins and TSH in mammals and fishes presumably reflects similarities in the chemical structure of teleost TSH and mammalian FSH and LH. A molecule with TSH activity has been isolated from the bullfrog (*Rana catesbeiana*) pituitary gland (MacKenzie, Licht, and Papkoff, 1978). This substance was thyrotropic in the bullfrogs but not in the reptiles. It was quite distinct from bullfrog LH and FSH, which had little effect on the activity of their thyroid glands. It was surprising, however, to observe that bullfrog LH, in contrast to bullfrog TSH, had a pronounced thyrotropic action in the reptiles. TSH would appear to have undergone substantial evolutionary change even in these neighboring classes of vertebrates. It is notable that the potentially dual activities of TSH and the gonadotropins are also apparent in this instance.

It has been proposed (Fontaine, 1969*a*) that the gonadotropins and TSH in extant species evolved from a common ancestral molecule, probably by a process of gene duplication and subsequent genetic change. It is interesting that there is no clear evidence for the presence of a distinct TSH in cyclostome fish where hypophysectomy (in the lamprey) has no effect on thyroidal activity (see Sage, 1973). It has also been observed that reproductive rhythms in fish are associated with parallel changes in the activity of the thyroid tissue. It has thus been suggested that the ancestral, or parent, molecule had a gonadotropic role that was extended to a thyrotropic one when the thyroid gland assumed a role in reproduction in fishes.

Growth hormone, prolactin, and placental lactogen

These three hormones are proteins containing about 190 amino acids and they show many structural and functional homologies to one another. Growth hormone and prolactin are formed in the adenohypophysis, the former being concerned with the regulation of growth and the latter with diverse processes ranging from lactation in mammals to osmoregulation in some fish. Placental lactogen (hPL = human placental lactogen, also called chorionic somatomammotropin) has been isolated from human placenta and that of several other primates, as well as from members of the Artiodactyla, Lagomorpha, and Rodentia – 15 genera in all (Talamantes, 1975; Talamantes et al., 1980). It has not been identified, however, in all placental mammals that have been investigated.

The amino acid sequence of the hormones has been described. That of human growth hormone is shown in Fig. 3.31. Human placental lactogen hormone and ovine prolactin have similar structures to this hormone. Human prolactin has not been isolated but sensitive bioassay and immunological techniques indicate that it exists as a distinct entity, apart from growth hormone, although this was once in doubt. The amino acid sequence of the three hormones has been compared (Fig. 3.32) and considerable homologies exist. There are thus 160 (out of 190) identical residues in human growth hormone and placental lactogen although, of the remainder, only seven positions are occupied by what are considered "nonhomologous" amino acids (Li, 1972). Such similarities have led to the suggestion (Bewley and Li, 1970; Niall et al., 1971) that the three hormones may have arisen from a common ancestral molecule. Various segments of each hormone molecule also bear considerable similarities to each other (internal homologies). The ancestral molecule may have been a smaller peptide of 25–50 amino acids that, by a process of genetic reduplication in a "tandem" manner, led to an increase in the chain length of the hormones.

The chemical similarities in the molecules are reflected in their biological activities. Apart from the dual effects of placental lactogen on growth and lactation, prolactin exhibits considerable growth hormone-like activity whereas growth hormone has (though more limited) prolactin-like actions.

Growth hormone and prolactin are present throughout the vertebrates (with the possible exception of the cyclostomes).

Growth hormone can be measured by its ability to increase growth of the tibia of young, hypophysectomized rats. Pituitary gland extracts from all the groups of tetrapods exhibit this effect as well as teleostean, chondrichthyean, and chondrostean fish (see Geschwind, 1967; Farmer et al., 1976, 1981; Hayashida and Lewis, 1978). Other bioassay test systems are responsive to a phyletic range of growth hormone preparations. Such tests include the incorporation of sulfate into the cartilage of embryonic chicks as well as growth in toads (Zipser, Licht, and Bern, 1969; Meier and Solursh, 1972), lizards (Licht and Hoyer, 1968), and turtles (Nichols, 1973). These tests also show that not only the hormone, but also a similar biological response to it occurs in many vertebrates.

Growth hormones from all species do not always exhibit an effect when injected into an heterologous species. This is seen very clearly in man who is unresponsive to all animal growth hormones, including those from other primates. As indicated earlier, this is an observation of some practical signifi-

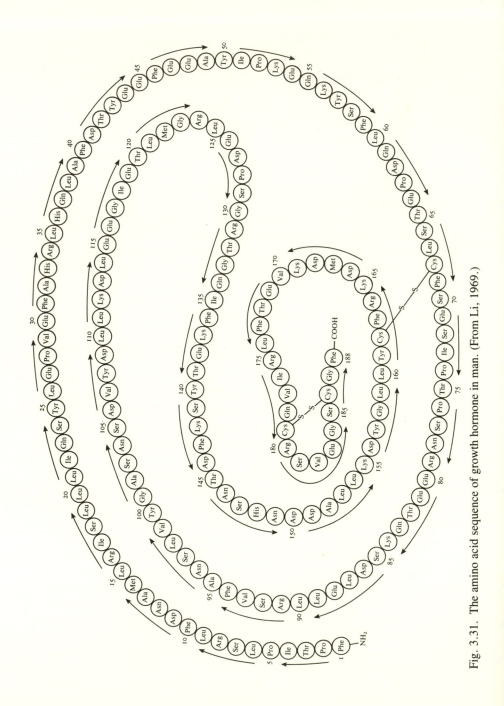

Fig. 3.31. The amino acid sequence of growth hormone in man. (From Li, 1969.)

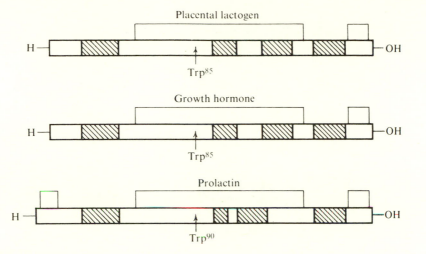

Fig. 3.32. Diagrammatic representation of the structures of placental lactogen (chorionic somatomammotropin) and growth hormone from man, and prolactin from sheep. The cross-hatched areas represent regions of internal homology in the sequence of the amino acids. Other similarities can be seen in the presence of disulfide bridges (narrow lines) and the tryptophan residues at position 85 in placental lactogen and growth hormone and 90 in prolactin. (From Niall et al., 1971.)

cance as the supply of growth hormone for administration to man has to be obtained from human cadavers and is thus limited. Its production using cloning techniques in microorganisms is, however, being developed.

Immunological evidence also emphasizes the differences, and similarities, among growth hormones from different vertebrates. Growth hormones show varying activities as antigens, depending on the species of the donor and the recipient. Rat growth hormone is not antigenic in rabbits but is very effective when injected into monkeys. Primate growth hormone is, however, antigenic in rabbits. Rabbit antiserum to human growth hormone has been shown to react (as measured by complement fixation) with hormone preparations from other primates, and in nine such species (Fig. 3.33) the degree of these interactions was closely correlated with their phyletic relationships.

A wider survey has shown that pituitary growth hormone extracts from most vertebrates can react with monkey antiserum that is formed in response to injected rat growth hormone (Hayashida and Lagios, 1969; Hayashida, 1970, 1971, 1973). Measurements of the relative ability of such pituitary extracts to antagonize the interaction of rat growth hormone with its antisera, in radioimmunoassays, have shown that the wider the phyletic distance between the species the less effective this antagonism (or ability to react with the antisera) becomes (Fig. 3.34a). No significant interaction was seen in glandu-

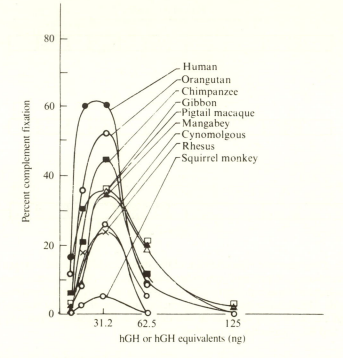

Fig. 3.33. The interactions between rabbit antiserum to human growth hormone and growth hormone from nine different species of primates. The values given are for complement-fixation curves. The equivalence points of all the curves have been aligned with the antigen concentration of human growth hormone that gave the maximum value. The degree of the immunological relationship is directly proportional to the amount of complement fixed at equivalence by that of the human and animal growth hormones. (From Tashjian, Levine, and Wilhelmi, 1965.)

lar extracts from teleost fishes. Hormones from other vertebrates, including the lungfish, have an interaction in this system that increases as the phyletic scale is ascended. It is also notable (Fig. 3.34*b, c*) that pituitary extracts of chondrichthyean, holostean, and chondrostean fishes show abilities to antagonize the reaction of rat growth hormone with its antiserum. These reactions parallel their abilities to stimulate growth in the rat tibia test.

Purified growth hormones have been prepared from members of several classes of lower vertebrates including birds, reptiles, amphibians, chondrosteans, teleosts, and elasmobranchs (Farmer, Papkoff, and Hayashida, 1974, 1976; Farmer et al., 1976, 1981; Hayashida and Lewis, 1978). Their precise amino acid sequences are unknown, but all of these hormones displayed general chemical similarities to mammalian growth hormone, suggesting that the general structure of the hormone has been conserved during evolution. Differences were observed, however, in such properties as their electrophoretic mobility, amino acid composition, immunological behavior, and biological activity, indicating that they have undergone some evolutionary changes. The teleost growth hormone from tilapia displayed only slight activity in the rat tibia growth test and although its antiserum showed some immunological cross-reactivity to that of another teleost (the perch), there was little cross-reaction with preparations from distantly related bony fish, such as the sturgeon.

Prolactin increases milk secretion in mammals, and this response can be used to measure the hormone's activity even in the low concentrations that appear in the plasma (Frantz, Kleinberg, and Noel, 1972). Pituitary gland extracts from birds, reptiles, and amphibians all promote this response but not those from fishes. Pigeons secrete a milk-like paste (pigeons' milk) from their crop sac with which they feed their young. This response is stimulated by prolactin from tetrapods *and* lungfishes but that from other fishes is ineffective. Prolactin, when injected into certain newts [*Notophthalmus (Diemictylus) viridescens*] at a particular stage in their life cycle, causes them to seek water preparatory to breeding. This is called the "eft (or newt) water-drive response" and can be initiated by prolactin from all the principal groups of vertebrates [except the cyclostomes, which seem to lack a prolactin hormone (see Bern and Nicoll, 1968)]. This response cannot be mimicked by any other pituitary hormone and has been used to demonstrate the presence of an analogous prolactin-like secretion throughout the vertebrates. Further evidence of the occurrence of this hormone in fishes has followed the discovery that certain teleost fishes, when in fresh water, usually die following removal of the pituitary gland, and this is due to excessive losses of sodium. When injected with mammalian prolactin they retain sodium and survive. Teleosts' pituitaries contain a "hormone" that also has the latter effect and which, as a reflection of its difference from mammalian prolactin, has been called "paralactin." The phyletic distribution of all these effects follows a precise pattern that is shown in Fig. 3.35.

The foregoing observations suggest two things: first, that the prolactin hormone is not identical in all vertebrates and has been subject to evolutionary change and, second, that it seems likely that it has assumed diverse biological roles.

Purified prolactins have been prepared from birds (chicken and turkey; Scanes, Bolton, and Chadwick, 1975; Burke and Papkoff, 1980) and teleost fish (tilapia, salmon, flounder, plaice, and carp; Farmer et al., 1977; Idler, Shamsuzzaman, and Burton, 1978; Ng, Idler, and Burton, 1980*a*). Turkey prolactin has a molecular weight of about 26,000, which is similar to that in mammals. It stimulates the development of the pigeon crop-sac but the dose response curves are not parallel to those found using ovine prolactin, suggesting differences in structure. The fish prolactins have usually been character-

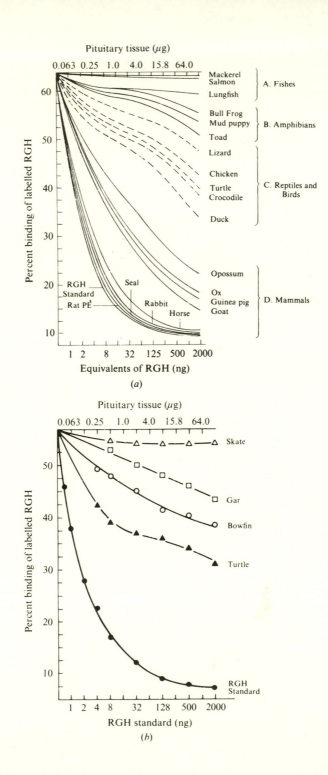

(a)

(b)

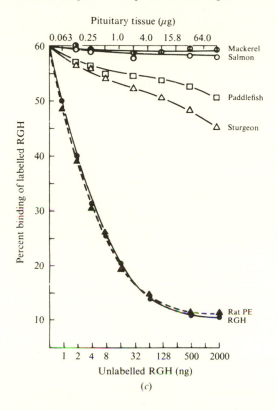

Fig. 3.34. Diagrams showing the immunochemical relationships of different vertebrate growth hormones. The curves represent the relative abilities of these preparations
to reduce the binding of ^{131}I-labelled rat growth hormone to monkey antiserum (to rat
growth hormone). Thus, rat growth hormone (RGH) nearly completely displaced the
labelled rat growth hormone, while that from the mackerel, salmon, and skate was
completely ineffective. Growth hormone from other sources fell between these ex-
tremes. (*a* and *b* from Hayashida, 1970, 1971; *c*, Hayashida and Lagios, 1969.)

ized by their abilities to stimulate sodium retention in teleosts, and they are
more effective than ovine prolactin in this respect. The prolactin from *Tilapia
mossambica* has a molecular weight of only 19,400 compared with 22,200 for
its growth hormone and about 23,000 in sheep prolactin. It contains only two
disulfide bridges, compared with three in mammalian prolactin. It was suggested that the fish prolactin may lack the disulfide bridge near the amino
terminal in the mammalian hormone and this difference could be associated
with a peptide chain which is shortened by 24 amino acids. This protracted
form of prolactin may reflect a more ancestral type that became more elongated during evolution, a situation that probably contributed to changes in its

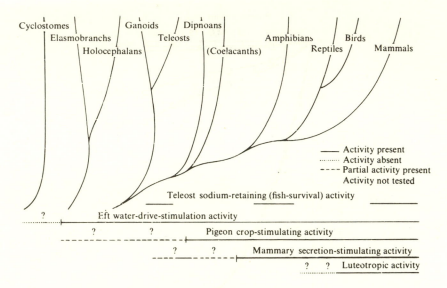

Fig. 3.35. Distribution of some of the biological activities that can be initiated by mammalian prolactin and prolactin-like hormones from other vertebrates. (From H. A. Bern, personal communication in Bentley, 1971.)

biological activity. It is noteworthy that all growth hormones also possess only two disulfide bridges, though teleost growth hormone completely lacks activity in the sodium-retention assays and the two hormones display no immunological cross-reactivity.

The precise chemical structure of prolactins from nonmammals has not yet been described. The similarities and differences in their biological effects (see Table 3.8) nevertheless confirm that, although they are basically analogous, differences exist in their structure. The "ancestral" molecule (or its receptor) may have been relatively less specific in its action than, for instance, that present in contemporary mammals. This may be reflected in the one-way specificity of mammalian prolactin, which acts in fish while fish prolactin fails to have an effect in mammals. With the origin of the tetrapods, changes occurred in the molecule that are illustrated by its ability to stimulate the pigeon crop-sac and mammary gland. This evidence is, of course, derived from extant species and, if it indeed did occur in that long-past time, the particular effects were not then of contemporary biological significance. The pigeon crop-sac and mammary gland were not to appear for many millions of years. It seems quite likely that prolactin had other roles at that time.

As described earlier, placental lactogen has both prolactin- and growth hormone-like action in mammals. When tested in nonmammals, however, this hormone behaves differently from the mammalian pituitary hormones as

Table 3.8. *Distribution of several prolactin activities in vertebrate pituitaries*

Prolactin activity	Cyclostomes	Chondrichthyes	Teleosts	Lungfish	Amphibians	Reptiles	Birds	Mammals
Osmoregulatory (in teleosts)			+/−					+
Water-drive-inducing (in efts)	−?	+	+		+	+	+	+
Growth-stimulating (?) (in tadpoles)			−		+	+	+	+
Crop-sac-stimulating (in pigeons)	−	−[a]	−[a]	+	+	+	+	+
Mammotropic (in mice)		[b]	−[b]	−	+	+	+	+
Luteotropic						−?	+?	+

[a] Partial activity has been reported from these groups; this is considered to be minimal and not fully crop-sac-stimulating in the manner seen with lungfish and amphibians.
[b] Partial activity has been reported from these groups that is distinguished from the "fully effective" response seen with tetrapods.
Source: Bern and Nicoll, 1969.

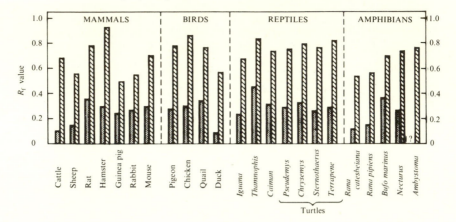

Fig. 3.36. A comparison of the electrophoretic mobilities (R_f) of prolactins and growth hormones from different species of tetrapod vertebrates. Stippled bars, growth hormone; cross-hatched bars, prolactin. (From Nicoll and Licht, 1971.)

it exhibits no eft water-drive activity, does not increase growth in tadpoles, and fails to promote sodium retention in a teleost fish, *Tilapia mossambica* (Gona and Gona, 1973; Clarke et al., 1973). Thus, although placental lactogen shares some actions with mammalian prolactin and growth hormone, these do not necessarily extend to the effects of the last two in nonmammals.

Tetrapod prolactins have been shown to exhibit differing chemical behavior. The electrophoretic mobilities (R_f), reflecting size and electrical charge, show considerable differences (Fig. 3.36) from each other and cross an almost five-fold range. No phyletic order is apparent; the ox, duck, and frog all have a similar R_f, while the rat, quail, turtle, and toad are all much higher. Interspecific differences, nevertheless, can be seen to exist. Growth hormones can also be separated in this way and, although they also exhibit differences in electrophoretic mobility, they are quite distinct from the prolactins (Fig. 3.36). The two hormones may still exhibit immunochemical similarities as seen in amphibians where the electrophoretically separated hormones both compete with rat growth hormone for binding to rat growth hormone antiserum (Hayashida, Licht, and Nicoll, 1973).

The prolactins exhibit a multitude of different biological effects. These (Bern and Nicoll, 1968; Ensor, 1978; Nicoll, 1980) have been classified into several categories including actions related to reproduction and parental care, osmoregulation, growth, metabolism, and the integument. In all, over 60 different effects have been described. It is unlikely that these all reflect physiological roles, but they do illustrate the considerable biological reactivity of the prolactin molecule. This may provide it with an adaptability, and propensity, to be utilized in a great diversity of physiological roles, only some of which have already been identified, or even applied. Prolactin would seem to be a hormone whose structure, as well as physiological role, have both evolved. Its diverse biological actions suggest that it is the most versatile vertebrate hormone.

Origin and evolution of hormones

The archetypal hormones probably evolved from molecules that possessed an innate ability to combine with proteins in the plasma membrane and cytosol. The cell components may have provided the ancestral hormone receptors. If such an interaction resulted in some advantageous change in the cell's activities then it could have been perpetuated as a result of natural selection. A number of contemporary hormones, apart from displaying a chemical kinship to each other, also exhibit similarities to other types of molecules in the body. Such a relationship has been observed when comparing the N-terminal amino acid sequences in secretin, glucagon, growth hormone, and placental lactogen with five proteolytic enzymes, chymotrypsin, bovine and human trypsin, and elastase (Adelson, 1971; Weinstein, 1972). Such observations have resulted in the suggestion that protein hormones may have evolved from enzymes. As many hormones are formed in cells with an embryonic kinship to the gastrointestinal tract such enzymes could have been enterosecretory proteins. The initial transposition from an enzyme to a hormone may have involved the duplication of a gene coding for the enzyme so that evolution could then proceed separately along two pathways, one involving the hormone. Hormones also bear structural analogies to plasma proteins, which suggests common origins. Thus, the gastrointestinal prohormones and prealbumin exhibit similarities that are frequent enough to indicate that they may have arisen from a common ancestor (Jörnvall et al., 1981). Such relationships have also been noted between the cobalamin carrier transcobalamin II and its receptor in the human placenta (Seligman and Allen, 1978), as well as vitamin B_{12}-intrinsic factor and its receptor in the pig small intestine (Kouvonen and Gräsbeck, 1979). Such observations have led to the suggestion that physiological ligands, such as hormones, and their receptors may each have evolved from a common ancestor as a result of gene duplication. One of the resulting genes may have provided an evolutionary line of hormones and the other receptors for them. Such an origin may help to account for complementary relationships of hormones and receptors.

The evolution of a hormone may be important in relation to properties other than their interactions with receptors (see, e.g., Blundell and Wood, 1975). Thus, storage and secretion of hormones by cells may depend on the abilities of the molecules to associate with each other or other substances present in endocrine glands. The attainment of adequate levels of hormones in the circulation, as well as appropriate rates for their destruction, may be related to their ability to cross permeability barriers and disperse in the body, and to interact with enzymes that may metabolize them.

Hormones that are synthesized from steroids (e.g., corticosteroids, androgens, and estrogens) or single amino acids (e.g., thyroxine and epinephrine) are the result of chemical modifications perpetrated on the parent or substrate molecule by enzymes. Their evolution will depend on changes in this metabolic machinery, such as the formation or sequestration of a particular enzyme. These types of hormones have shown considerable evolutionary conservatism in their structures, possibly reflecting the genetic complexities needed to initiate successful evolutionary change in such biosynthetic processes.

Protein and polypeptide hormones appear to have evolved largely as the result of genetically determined changes in the size and amino acid sequences of their precursors. This may involve the sequence within the active fragments that are destined to act as the hormones, as seen when comparing oxytocin and vasopressin. In addition, if other sites on the precursor are changed it may allow an enzyme to cleave novel fragments from it, and these may then become hormones.

The chemical structures of various hormones described in the earlier parts of this chapter indicate that many hormones exist as related members of single molecular "families" (e.g., oxytocin and vasopressin; TSH and the gonadotropins; growth hormone, prolactin, and placental lactogen). Thus, several structurally similar hormones may exist side by side in the same individual. This situation has probably arisen as the result of the duplication of a single gene in a distant ancestor, each new gene being retained and subsequently pursuing its own separate evolutionary pathway. Considerable homologies between strings of amino acids within individual hormones have also been observed (e.g., in growth hormone and prolactin), suggesting that duplications have occurred that have been retained within the same gene, being linked up in a tandem-like manner. Such intragenic tandem duplication may even be followed by genetic reduplication, as seen when comparing growth hormone, prolactin, and placental lactogen. These hormones all have common internal homologies in segments of their amino acids.

Several hormones consist of distinct subunits and in some of these, TSH and the gonadotropins, the synthesis of each section is controlled by a separate gene. Structural similarities between the subunits, however, suggest that each originated from the same ancestor as a result of gene duplication. The ultimate emergence of the new hormone, however, will be expected to depend on the acquisition of a process for uniting the two subunits.

Origin and evolution of hormone receptors

It is not known whether receptors originated quite separately from the hormones that combine with them (Csaba, 1980) or if they were both derived from a common ancestor, following gene duplication (Jörnvall et al., 1981). Candidates for such ancestral molecules include proteins present in the cytoplasm, such as the "determinants" that control nuclear events during embryogenesis (Gurdon, 1977) and those incorporated in the plasma membrane. The former type of molecules could have given rise to receptors for steroid and thyroid hormones and the latter for the proteins and polypeptides that initially act at the cell surface. Receptors appear to have evolved but the changes may not always be as great as observed in the hormones that bind to them.

A phyletic study of the insulin receptor (Muggeo et al., 1979) suggests that its broad general properties have been strongly conserved during evolution, probably more so than in the hormone. However, differences have been observed in its immunological behavior and affinity for insulin (Horuk et al., 1979) so that structural changes have undoubtedly occurred. These surviving changes appear to be of little physiological consequence to the animal and may involve unimportant parts of the receptor molecule. Small changes in the affinity of the insulin receptor for the hormone may be compensated for by the maintenance of higher levels of insulin or by an increase in the number of receptors. Steroid hormone receptors, on the other hand, may exhibit more profound changes in their structure. Thus, receptors for corticosteroids prepared from the tissues of teleosts have a much lower affinity for this type of hormone than such receptors from tetrapod vertebrates (Porthé-Nibelle and Lahlou, 1978). The high-affinity-type cytoplasmic receptors for corticosteroids, which have been studied in mammals and amphibians, appear to be absent in these bony fishes. The occurrence of inherited diseases that involve changes in the properties of receptors, but not the hormones, indicates that hormones and their receptors are under separate genetic control. Their mutual dependence, however, suggests that each will impose constraint on the evolution of the other, so that an evolutionary conservatism is not unexpected (see Wallis, 1975).

Changes in the physiology of receptors do not necessarily involve direct genetic changes. Hormones themselves can influence the properties of their receptors in a manner that suggests they are being adapted to their needs. Such induced changes can involve the numbers, distribution, configuration, and binding properties of the receptors. The levels of hormones during ontogeny may also influence the development of their receptors (Csaba, 1980). Thus, structural changes in receptors as a result of evolution are not the sole way in which they may adapt to the requirements of a particular hormone.

It has been observed that, whereas the hormones from "higher" vertebrates may interact with receptors and elicit a response in "lower" vertebrates, the reverse situation is rare (the "one-way" rule). There are, however, exceptions to this principle. For example, the growth hormone of the teleostean fishes does not elicit effects in mammals, but that from holostean and chondrostean fish is effective. Also, chondrichthyean thyrotropin is effective in mammals but that from mammals, or even teleosts, has no action in chondrichthyeans. If one nevertheless accepts the one-way premise as a general situation, then it can be used for conjecture about the type of changes that may have occurred in hormones and receptors during their evolution. The archetypal hormone–receptor interaction may have been of a simpler type than has occurred later in evolution, possibly reflecting a less complex physiological situation where there are fewer excitants to be distinguished. As evolution proceeded, however, the number of hormonal responses increased so that the hormone–receptor interactions, in order to be distinct, became more complex.

Examination of the structure of the hormones from species in various parts of the phyletic scale does not indicate that the structural complexity of hormones has undergone a marked progressive increase, though the information is admittedly sparse for nonmammals. If the one-way rule has any validity then it would appear that the observed differences in sensitivity to heterologous hormones by different species principally involve their receptors. The latter may, for instance, be simpler in lower vertebrates and so exhibit less discrimination toward different excitants than those receptors from more highly evolved species. Whether this situation is true or not, however, must await more extensive comparative studies on the structure of hormone receptors.

Conclusions

An examination of the chemical, biological, and immunological behavior, as well as the chemical structure, of homologous hormones from different species suggests that many of these may have been subject to an orderly evolutionary change. This possibility is also indicated by the similarities that persist between each hormone from closely, in contrast to distantly, related species and within the principal systematic groups of the vertebrates. In some instances, a genetic background for such changes has been described. It is nevertheless noteworthy that some hormones display little or no difference in their structure even when they are present in such distantly related species as a lamprey and a man. On some occasions it even appears that a completely "new" hormone that lacked a genetic homologue in its ancestors has evolved.

While the evolution of "new" hormones has potential importance in novel processes of coordination in the body, the functional significance of alterations in the structure of "old" hormones is less clear. Such changes may take place at chemical sites on the hormone molecule that apparently are not essential for its action and so have little or no effect on its functioning. However, if more important sites are involved, the hormone's activity may be altered and this could have important results in the animal. These changes may include a virtual absence of its effect, differences in the quantities of the hormone required to mediate the response, or an alteration in its relative ability to influence different processes within the same animal (specificity). As we shall see in succeeding chapters, the roles of hormones in the body have often changed completely during the course of evolution. Such a modification of a hormone function is not necessarily accompanied by an alteration in its chemical structure, but it often is; we are uncertain, however, as to how important this change may be in the transition.

4. The life history of hormones

The use of hormones for the purpose of coordination involves a complex series of physiological events. Such a life history begins with the formation of the excitant by the endocrine glands and concludes with the response of a target, or effector tissue, and the hormone's ultimate destruction or its excretion from the body. The events that determine the action of a hormone are shown in Fig. 4.1. This basic pattern persists throughout the vertebrates, though, as will be described, certain differences exist.

The formation of hormones

Although the formation of all hormones is determined at the genetic level, it can be either a relatively direct translational procedure or, alternatively, occur as a result of the prior formation of enzymes that mediate synthesis. Enzymes are also involved in the former process, however.

Translational formation of hormones

It seems likely that the sequences of amino acids in the polypeptide and protein hormones directly reflect genetic translation via messenger RNA. Even when a direct translation of genetic material into the amino acid sequence of a hormone is made, the active hormone product does not result. Many peptide hormones, including glucagon, corticotropin, MSH, the neurohypophysial peptides, and calcitonin, contain relatively few amino acids and are formed as a result of a process of disassembly from much larger protein molecules. In addition, some polypeptide hormones are not simple strings of amino acids but consist of subunits as seen in insulin, the gonadotropins, and TSH. The assemblage of these subunits into a hormone must occur subsequent to the formation of the individual parts. The formation of polypeptide hormones thus usually involves the initial formation of a parent molecule called a *pre-prohormone*. As a result of post-translational changes, involving cleavage by enzymes, the pre-prohormone is broken up to form the hormone itself. The pre-prohormone has only a transient existence (usually less than 2 minutes) and is rapidly converted to the *prohormone*. There is good evidence that most peptide hormones are formed in this way.

[144]

PROCESS   MOLECULAR MECHANISM

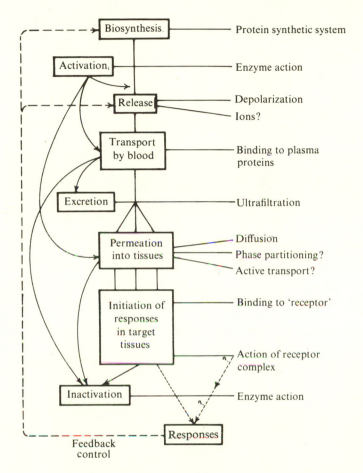

Fig. 4.1. Diagrammatic summary of the life history of a hormone commencing with its biosynthesis and concluding with the response and its inactivation. (From Rudinger, 1968. Reprinted by permission of the Royal Society.)

Such a process of hormone formation is well illustrated in the production of insulin (Steiner et al., 1974; Chan, Keim, and Steiner, 1976; Lomedico et al., 1977). The genes that are responsible for the formation of insulin have been isolated and allowed to translate their coded amino acid sequence *in vitro*. The initial product is a protein with a molecular weight of about 11,500 (compared with about 6000 for insulin). It is a linear peptide with a special 23 amino acid segment at its N-terminus (Fig. 4.2). This "signal peptide" attaches the molecule, and the ribosome on which it is being formed, to the endoplasmic

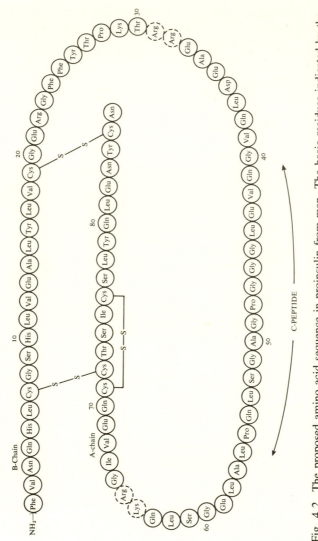

Fig. 4.2. The proposed amino acid sequence in proinsulin from man. The basic residues indicated by the broken circles have been assigned as they are known to occur in bovine and porcine proinsulin. (From Oyer et al., 1971.)

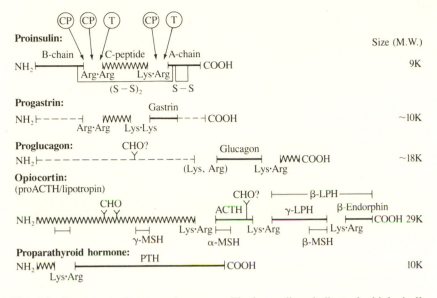

Fig. 4.3. Structures of some prohormones. The heavy lines indicate the biologically active fragments, or hormones. The dashed lines are the regions in which the amino acid sequences have not yet been determined. The parent molecules are broken down as a result of the activities of carboxypeptidase B-like (CP) and trypsin enzymes (T). These reactions occur in the regions of the arginine and lysine residues, as indicated for the proinsulin. (Based on Steiner et al., 1980.)

reticulum. The molecule is called pre-proinsulin and it only has a brief existence as it passes into the Golgi apparatus where its 86 amino acid C-terminal segment is incorporated into storage granules. This proinsulin is then broken down under the influence of a carboxypeptidase-like and tryptic enzymes that fragment the molecule in the region of certain arginine and lysine residues (Fig. 4.3). The A-chain and B-chain are released but as a result of the folding of the prohormone they remain aligned and joined to each other by two disulfide bridges. The "C"-peptide fragment is released with the hormone. Its amino acid sequence varies quite considerably in different species (Fig. 4.4). The structures of several other prohormones are shown in Fig. 4.3, and these are also converted to hormones in a similar manner to that of proinsulin.

Many polypeptide hormones are stored in granules present in the endocrine cells. These structures are bounded by membranes and are 0.1–0.4 μm in diameter. They appear to originate in the Golgi apparatus of the cell. The precursor, or prohormone, becomes associated with the granules and it seems likely that conversion to the hormone takes place in these. The granules can travel to the peripheral regions of the cell, and, in response to releasing stimuli, combine with the plasma membrane and discharge their contents into

	1	2	3	4	5	6	7	8	9	10	11	12	13	14	15	16
Human	Glu	Ala	Glu	Asp	Leu	Gln	Val	Gly	Gln	Val	Glu	Leu	Gly	Gly	Gly	Pro
Monkey	Glu	Ala	Glu	Asp	Pro	Gln	Val	Gly	Glx	Val	Glu	Leu	Gly	Gly	Gly	Pro
Pig	Glu	Ala	Glu	Asn	Pro	Gln	Ala	Gly	Ala	Val	Glu	Leu	Gly	Gly	Gly	Leu
Cattle	Glu	Val	Glu	Gly	Pro	Gln	Val	Gly	Ala	Leu	Glu	Leu	Ala	Gly	Gly	Pro

	17	18	19	20	21	22	23	24	25	26	27	28	29	30	31
Human	Gly	Ala	Gly	Ser	Leu	Gln	Pro	Leu	Ala	Leu	Glu	Gly	Ser	Leu	Gln
Monkey	Gly	Ala	Gly	Ser	Leu	Gln	Pro	Leu	Ala	Leu	Glu	Gly	Ser	Leu	Gln
Pig	Gly	—	Gly	—	Leu	Gln	Ala	Leu	Ala	Leu	Glu	Gly	Pro	Pro	Gln
Cattle	Gly	Ala	Gly	—	—	—	—	—	Gly	Leu	Glu	Gly	Pro	Pro	Gln

Fig. 4.4. A comparison of the amino acid sequences of the human, monkey, porcine, and bovine C-peptides. The solid bars indicate residues that are identical in all species. (From Steiner et al., 1972.)

the region of blood vessels. A summary of this process as it is thought to occur for insulin is shown in Fig. 4.5.

Such granules apparently furnish sites for the formation and storage of many hormones. If released into the cytoplasm of the cell the hormones may be destroyed, as has been observed for the catecholamines when they are exposed to the mitochondrial enzyme monoamine oxidase (MAO). In addition, storage granules may afford convenient vehicles in which hormones can be transported for considerable distances along nerve cells.

Some neurons form hormones by a process called *neurosecretion.* These are like ordinary nerve cells and consist of a cell body with an extended axon and they can also be depolarized and so convey electrical information. The axon, instead of terminating at another neuron or an effector tissue, like a gland or muscle, lies near a capillary into which it can discharge certain of its products (Fig. 4.6). These products may be hormones, the formation of which is initiated some distance away in the cell body. The hormones, parceled up in their granules, travel along the nerves to the peripheral sites in the axon where they can be released into the blood.

Hormones that are formed as a result of neurosecretion are those of the neurohypophysis and hypothalamus including vasopressin, oxytocin, and the various releasing hormones that control the adenohypophysis. In mammals, vasopressin and oxytocin are formed in the supraoptic and paraventricular nuclei that are situated at the base of the brain. These hormonal products pass down the axons in granules to the neural lobe. Inside the granules they are attached to protein molecules of neurophysin, the synthesis of which seems to be closely associated to that of the hormone. The neurophysin and each hormone (vasopressin and oxytocin) are synthesized from two pro-

Beta Granule Formation

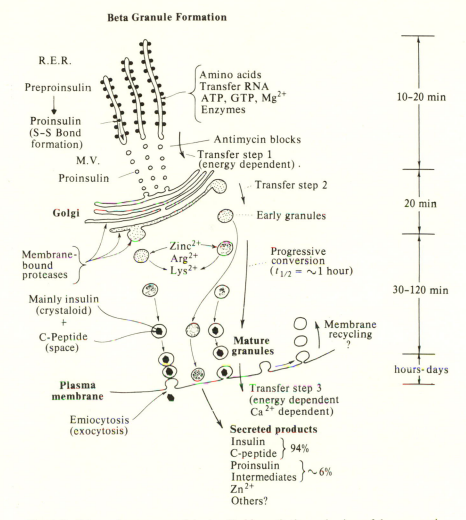

Fig. 4.5. Schematic summary of the insulin biosynthetic mechanism of the pancreatic B-cells. The time scale on the right side of the figure indicates the time required for each of the major stages in the biosynthetic process. R.E.R. = rough endoplasmic reticulum; M.V. = microvesicles. (From Steiner et al., 1974.)

hormones, each with a molecular weight of about 20,000. They have been called propressophysin and prooxyphysin (Russell, Brownstein, and Gainer, 1980). Amphibians and fishes have a single preoptic nucleus where the neurohypophysial hormones originate. The putative hormones of the urophysis in fishes (see Chapter 2) are also apparently formed by a process of neuro-secretion.

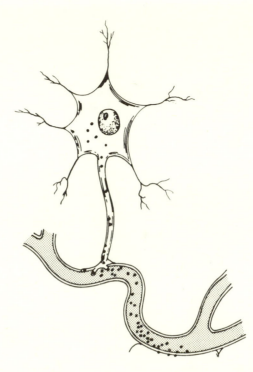

Fig. 4.6. A neurosecretory cell. The hormonal products are transported from the cell body down to the axon from which they can be released into capillaries. In contrast, ordinary nerve cells have axons that abut onto other neurons (instead of capillaries). (From R. Guillemin and R. Burgus, *The Hormones of the Hypothalamus.* Copyright © 1972 by Scientific American, Inc. All rights reserved.)

The formation of hormones by enzymically controlled synthesis

Thyroid hormones

The endocrine secretions of the thyroid, the adrenal medulla, the gonads, and the adrenal cortex are the result of biosynthetic processes controlled by enzymes. Although the enzymes themselves are the result of genetic translational processes, the hormones are synthesized in chemical reactions controlled by the enzymes. This synthesis may involve many reactions, some of which have been summarized in Fig. 3.2 (for steroid hormones) and Fig. 3.4 (for thyroid and catecholamine hormones).

Thyroid hormones contain iodine and the thyroid gland has a special ability to concentrate inorganic iodide from the blood. This ability to transport iodide actively against an electrochemical gradient is shared by some other tissues, including the intestine and salivary glands. This ability may be controlled by a single gene: In man a congenital inability to accumulate iodide in the thyroid is accompanied by a parallel deficiency at the other iodide transport sites. The accumulated iodide is oxidized to iodine, which combines with tyrosine to form the precursor of thyroxine and triiodothyronine. The latter reaction occurs with the tyrosine residue that is part of a large glycoprotein, thyroglobulin (molecular weight 660,000) which is stored extracellularly in the thyroid follicles. It is synthesized as two subunits, each with a molecular weight of about 300,000 (Van Herle, Vassart, and Dumont, 1979). Each molecule of thyroglobulin binds two of thyroxine and on the average, less than one of triiodothyronine. Thyroglobulin itself does not appear to be a particularly remarkable protein; it contains about 30 molecules of tyrosine (or about 2% by weight), and about 0.5% iodine. It nevertheless provides a site for the synthesis and storage of the thyroid hormones. The biosynthetic process for the thyroid hormones appears to be common to all vertebrates and was apparently attained early in their evolution. Nevertheless, most of our information has been derived from studies of mammals.

Thyroglobulins have been identified in thyroid tissues of species from most groups of vertebrates, even including larval cyclostomes (lampreys) where it is present in the subpharyngeal gland or endostyle (Suzuki and Kondo, 1973; Suzuki et al., 1975). These proteins exhibit many similarities with respect to their molecular size (though a few differences have been observed), as determined by centrifugation in sucrose gradients, but their amino acid constitutions may differ. In hagfish, however, a smaller protein may have assumed the role of thyroglobulin. Thyroglobulins also exhibit different immunological behavior. Antibodies to specific thyroglobulins have been prepared and these react, *in vitro*, with the homologous protein, which can be radioactively labelled. Thyroglobulins from different species may compete with this labelled protein for binding to its antibodies. The relative ability to do this suggests the degree of immunological similarity to the homologous thyroglobulin. Considerable interspecific differences have been observed in such radioimmunoassays (Torresani et al., 1973). Sheep thyroglobulin readily displaces its labelled form from anti-sheep thyroglobulin antibodies, but thyroglobulins from other mammals, such as pigs and rabbits, are much less effective. Thyroglobulin from a python and a crocodile also compete with the homologous labelled protein for such binding, but this is also much less so than that for the sheep protein. Bird thyroglobulin, from ducks, has no ability to bind with the sheep antibodies. Although it is tempting to construct phylogenetic trees with such information the paucity of species examined makes such predictions of doubtful significance. The measurements nevertheless illustrate the diversity that can occur among thyroglobulins from different species.

Catecholamines

Epinephrine (adrenaline) and norepinephrine (noradrenaline) are formed in chromaffin tissues. These hormones are present not only in the adrenal gland but also are associated with nervous tissue in other parts of the body. Norepinephrine is also formed in certain nerve endings in the sympathetic nervous tissue and the brain. The original precursor of these catecholamines is tyrosine (see Fig. 3.4), which by a series of enzymically controlled reactions is converted to 3,4-dihydroxyphenylalanine, or dopa, and thence to dopamine. These reactions occur in the cell's cytoplasm. The dopamine is accumulated by storage granules in which it is converted, under the influence of dopamine β-hydroxylase, to norepinephrine. Norepinephrine can be *N*-methylated to epinephrine under the influence of the enzyme phenylethanolamine-*N*-methyltransferase (PNMT), which, in mammals (see Chapter 3), can be induced in the presence of high concentrations of corticosteroids. There is evidence to suggest that norepinephrine and epinephrine are stored in different granules and even different cells in the adrenal medulla. This could be determined by regional differences in the access of corticosteroids to the medullary tissue, thus influencing the local levels of PNMT (Pohorecky and Wurtman, 1971).

Steroid hormones

The formation of the steroid hormones also appears to be basically the same in all vertebrates (Sandor, 1969). All of these are formed from cholesterol, which is present in high concentrations in the steroidogenic endocrine glands. This parent molecule may be formed *in situ* from acetate (see Fig. 3.2) or be accumulated from the plasma. The enzymic conversion of cholesterol to pregnenolone and progesterone is common to all of the steroidogenic endocrine glands. Sandor (1969) has suggested that the use of steroids as hormones may have been determined by a primeval mutation that invented the enzyme systems that determine this transformation of cholesterol. The conversion of progesterone to the androgen, estrogen, and adrenocorticosteroid hormones involves the successive actions of diverse enzymes that hydroxylate, oxidize, aromatize, and reduce the steroid at some of the 21 carbon positions present. The ability of a species to synthesize steroid hormones with differing structures depends on the presence, absence, or activities of the enzymes that mediate these changes. The chondrichthyean fishes that secrete 1α-hydroxycorticosterone possess an enzyme, 1α-hydroxylase (which converts corticosterone to the hormone), that has not been found in other vertebrates. The formation of aldosterone in tetrapods is determined by the presence of 18-hydroxylase (it converts corticosterone to aldosterone). Rats and mice cannot form cortisol and lack 17α-hydroxylase (that converts progesterone to 17α-hydroxyprogesterone) in their adrenal cortex. Mutations may also arise that influence the ability of a species to form certain hormones. In man, a congenital condition known as the adrenogenital syndrome is due to the complete or partial block of the 21-hydroxylating system (the conversion of progesterone to deoxycorticosterone and 17α-hydroxyprogesterone to 17α-hydroxy-11-deoxycorticosterone). Such enzymic differences determine the presence or absence of the various steroid hormones and furnish the raw materials of evolutionary change.

The steroidogenic enzymes are associated in the cytoplasm of the cell and the mitochondria. The pathways and sites of the enzymes determining the formation of adrenocorticosteroids in the frog's interrenal are shown in Fig. 4.7.

The release of hormones from the endocrine glands

Nature of the stimuli

The role of the endocrine glands in the regulation of bodily functions is dependent on the release of their secretions on appropriate occasions. Secretion is initiated upon the receipt, by the gland, of a suitable stimulus, which may increase or decrease the discharge of its hormone. The message may arrive either by way of a nerve or be carried in the blood that perfuses the tissue. The primary event that initiates this stimulus may arise either from the external environment (exteroceptive stimulus) or inside the body (interoceptive stimulus).

Exteroceptive stimuli that may affect the endocrine glands include the receipt of light, a change in temperature or of the osmotic concentration (of an aqueous environment), and the acquisition of food, water, and salts. Social situations such as the proximity of prey, a predator, a mate, or the young may evoke psychogenically mediated responses in the endocrine glands. Climatic events such as rain, temperature and even, possibly, humidity and atmospheric pressure can also influence a hormone's release. The receipt of and endocrine response to such external stimuli help the animal to maintain an equitable relationship with the events that happen around it. Exteroceptive stimuli are especially useful in providing cues that are involved in reproduction.

Interoceptive stimuli are those that result from changes in the physicochemical conditions within the body. Ultimately they may reflect the external conditions: For instance, a lack of drinking water and a hot, dehydrating environment will lead to an increase in the osmotic concentration of the body fluids. Internal stimuli include changes in the concentration of salts, such as sodium, potassium, and calcium, in the body fluids, alteration of the hydrostatic pressure of the blood vascular system, oscillations in the levels of nutrients, like glucose, amino acids, and fatty acids, as well as changes in the

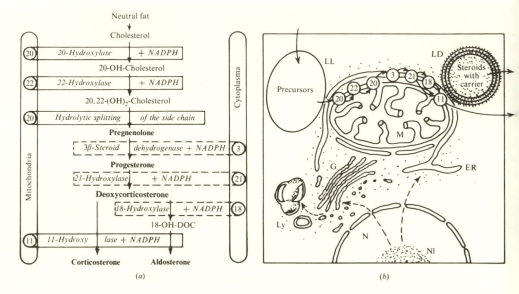

Fig. 4.7. The biosynthesis of corticosteroid hormones (*a*) in the adrenocortical tissue of a frog. This is illustrated with reference to the cell organelles (*b*). Corticosterone can also act as a precursor for aldosterone. In some species, 18-hydroxylase may be a mitochondrial rather than a microsomal enzyme. In vertebrates that form cortisol another microsomal enzyme, 17α-hydroxylase, active on progesterone and leading to 17α-OH-progesterone, deoxycortisol, and cortisol, is present. ER = endoplasmic reticulum; G = Golgi apparatus; N = nucleus; LD = electron-dense lipid droplet; LL = electron-lucid lipid droplet; Ly = lysosomes; M = mitochondrion; Nl = nucleolus. Solid arrows are pathways of steroid synthesis from precursors to steroid bound to a carrier; broken arrows indicate cellular responses activated by ACTH. (From Lofts and Bern, 1972.)

body temperature. The physiological factors influencing release of hormones are summarized in Table 4.1.

Hormones can exert tropic effects on other endocrine glands and the terminal secretions, once released, may travel back to the region where the tropic hormone originated and inhibit its further release. This last effect completes the cycle of events that closes the loop of a *negative-feedback system* that plays a vital role in regulating the endocrine system. Such a control system is well known to engineers and its action is illustrated in Fig. 4.8. The release of hormones from the median eminence, which in turn controls the formation and discharge of the tropic hormones of the adenohypophysis, is regulated in this manner. The hormones that exert the inhibitory effects in the median eminence may alter the thresholds for stimulation of the neurosecretory cells. In addition, such a negative-feedback can also act directly on the adenohypophysis as seen with the action of thyroid hormones that inhibit the release of thyro-

Table 4.1. *Principal stimuli influencing the release of hormones*

Hormone	Releasing stimuli
Aldosterone	Low plasma Na concentration, angiotensin II and III
Angiotensin	Renin
Calcitonin	Hypercalcemia
Cholecystokinin	Digestive products in the upper intestine
Cortisol and corticosterone	Corticotropin
Enterogastrone	Fats and oils in intestine
Enteroglucagon	Feeding
Epinephrine	Neural stimuli (mediated by acetylcholine)
Estrogens	FSH
FSH	External stimuli, such as light, low estrogen levels
Gastrin	Feeding (vagal reflex; local reflex from food in stomach)
GIP	Fats in intestine
Glucagon	Hypoglycemia, gastrin, CCK, high amino acids and low fatty acids in plasma, exercise
Growth hormone	Sleep, exercise, apprehension, hypoglycemia
Hypophysiotrophic hormones: CRH, TRH, etc.	Hypothalamic neuronal stimuli (dopamine and mono-amine transmitters), inhibited by negative-feedback mechanisms carried by hormones and metabolites
Insulin	Hyperglycemia and amino acids in plasma, glucagon, growth hormone, in ruminants high levels of propionic and butyric acid, vagal stimulation, CCK and secretin. Inhibition by epinephrine
LH or ICSH	External stimuli, sexual excitement (male), estrogen "surge" (female), low progesterone or testosterone levels
MSH	Light on retina, low plasma corticosteroid levels, inhibition by neural stimuli
Melatonin	Darkness (adrenergic neural stimulation)
Oxytocin	Suckling, parturition
Parathyroid hormone	Hypocalcemia
Progesterone	LH, chorionic gonadotropin, prolactin
Prolactin	Diurnal rhythm (sleep), suckling, parturition, plasma osmotic concentrations (low in fish, high in mammals?), estrogens
Renin	Low Na in plasma, hemorrhage, reduced renal blood flow, nerve stimulation (β-adrenergic), increased osmotic concentration in renal blood supply
Secretin	Acid in upper intestine
Testosterone	LH
Thyroid hormones	TSH
TSH	Low thyroxine, temperature reduction
Vasopressin, ADH	Increased osmotic concentration of plasma
$1\alpha,25\text{-}(OH)_2$-vitamin D_3	Low Ca and phosphate levels in plasma; parathyroid hormone

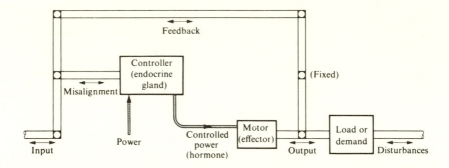

Fig. 4.8. Schematic diagram of a servosystem showing the analogies of the classic engineering control mechanism of physiological coordination by the endocrines. The double arrows indicate that the links can move to and fro. Stimuli (both input and feedback) are fed into the controller (endocrine gland in the physiological analogy) through a misalignment detector that is sensitive to changes from a certain set-point. It is conceivable that the latter may be reset in certain physiological conditions (such as hibernation). In response the controller varies its power output (or hormone secretion) to the motor or effector. The latter adjusts the physiological needs and tends to restore equilibrium (of metabolites, salts, water, other hormones, temperature, etc.), the degree of which is transmitted back to the controller via the feedback arc. (From Bentley, 1971.)

tropic hormone in this way. The interrelations of the hormones of the hypo-thalamus, adenohypophysis, and more peripheral endocrine glands are shown in Fig. 4.9.

The feedback mechanism involving the action of the peripheral endocrine secretions on the hypothalamus is called a *long-loop feedback*. There is also evidence suggesting that the adenohypophysial secretions may exert a similar action on the hypothalamus by what is termed a *short-loop feedback*. It should be noted that peripheral hormones do not necessarily initiate a negative-feedback inhibition on the hypothalamus. High estrogen levels can stimulate the release of LHRH, which initiates the events that result in ovulation. Such a *positive feedback* has also been shown in the hypothalamus of the goldfish where thyroxine stimulates the release of a thyrotropin-inhibiting hormone (Peter, 1971). The sporadic information available from experiments on nonmammals suggests that feedback control working through the hypothalamus and the adenohypophysis is widespread. More information is needed and there is some doubt as to the importance of such effects in cyclostomes (Larsen and Rosenkilde, 1971; Fernholm, 1972).

A negative-feedback inhibition of hormone secretion also results from changes in the concentrations of the products of the hormone's actions. The retention of water or sodium, as a result of the actions of antidiuretic hormone and aldosterone, respectively, reduces the further release of these hormones. Com-

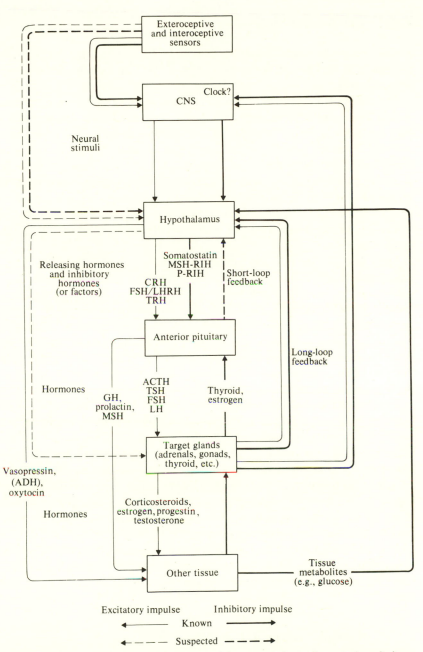

Excitatory impulse   Inhibitory impulse

←——— Known ———→

←— — — Suspected — — —→

Fig. 4.9. Factors controlling the release of hormones from the anterior pituitary (adenohypophysis). Shown is the use of negative-feedback inhibition of secreted hormones and tissue metabolites in influencing further hormonal release. As can be seen, the hypothalamus (and its associated median eminence) plays a central role in this process. Stimulatory effects are shown by thin lines and inhibitory processes by thick lines. (Based on Krieger, 1971.)

parable mechanisms exist involving glucose levels and the regulation of insulin, glucagon, and growth hormone, as well as calcium, parathyroid hormone, and calcitonin.

The endocrines, apart from influencing each other's release through tropic and feedback mechanisms, may also interact with each other and so modify their secretory activity. Epinephrine can inhibit the release of antidiuretic hormone from the neurohypophysis and that of insulin from the islets of Langerhans. This inhibition is an α-adrenergic effect that contrasts with the β-adrenergic effect that increases the release of insulin. Glucagon promotes the release of insulin and growth hormone directly; such effects do not depend on changes in blood glucose levels. Excesses of growth hormone are diabetogenic and inhibit the formation of insulin, though smaller amounts apparently stimulate its release.

Many less precise and less specific stimuli than those described in the foregoing can initiate the discharge of hormones from endocrine glands. Such stimuli may contribute to the homeostatic process, though they may also confuse it. No endocrine gland can exist and be uninfluenced by events outside what we may like to think of as its homeostatic area of influence. Nonspecific stimuli, especially if strong enough, can elicit a discharge of many hormones. This is sometimes referred to as "stress" and is particularly likely to occur in experimental situations that contribute to the confusion of the perpetrating scientists.

The tropic effects of hormones on the secretion of other endocrine glands can contribute to the processes of biological amplification in the body. An initial stimulus may only produce a change that involves a very small amount of energy while the energetic demands of the response may be relatively immense. The quantities of the excitants necessary to initiate an effect and the amount of the products formed (on a weight for weight basis) are illustrated in Fig. 4.10. We have no information as to the quantity of neural transmitters necessary to trigger a release of corticotropin-releasing hormone from the median eminence, but quantitative changes have been estimated for the subsequent physiological events. The final release of cortisol promotes the formation of 5600 μg of glycogen. The initial amount of CRH required to do this is about 0.1 μg so that the final response represents an amplification of 56,000 times.

"Pulsatile" release of hormones

Hormones are not necessarily released in a continuous stream but also in discrete bursts. This process may take place in a single large "surge" or in a pulsatile manner, one "pulse" succeeding another at regular intervals. The latter is also referred to as "episodic" release. This phenomenon did not

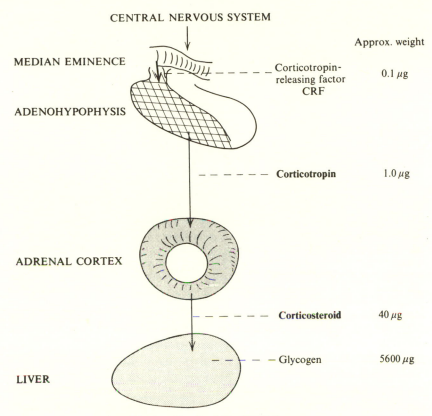

CENTRAL NERVOUS SYSTEM

Approx. weight

MEDIAN EMINENCE — — — — Corticotropin-
releasing factor
CRF — 0.1 μg

ADENOHYPOPHYSIS

— — — — — — Corticotropin — 1.0 μg

ADRENAL CORTEX

— — — — — — Corticosteroid — 40 μg

— — — — — Glycogen — 5600 μg

LIVER

Amplification 10 × 40 × 140 = 56,000 times

Fig. 4.10. An illustration of biological amplification as shown in the endocrine system. Corticotropin-releasing hormone initiates the release of corticotropin (ACTH), which leads to the deposition of glycogen in the liver.

become apparent until sensitive and convenient radioimmunoassay methods were developed for assaying hormones in small, serially collected samples of plasma. The pulses of released hormones may take place as frequently as every few minutes to as little as three or four times a day.

The rhythmical, pulsatile release of hormones can result in their attainment of stable basal, or tonic, concentrations in the plasma. When a hormone is released in such a manner it will at first be diluted, as a result of its redistribution in the body fluids, and it will then be removed from the circulation. The rate of this process is called its "clearance time." This occurs as a result of its uptake by tissues, its metabolic breakdown, and its excretion in the urine and bile (see later in this chapter). Its levels will thus decline in the intervals

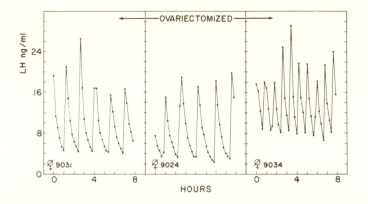

Fig. 4.11. The spontaneous pattern of release of luteinizing hormone (LH) in three ovariectomized ewes. Blood samples were collected every 12 minutes for 8 hours during the normal midanestrous period in May. It can be seen that the plasma concentrations of LH varied in a pulsatile manner from about 2 to 30 ng/ml. The time between the peak (or trough) levels reflects the times of release of the hormone from the pituitary gland and varies from about once every 50 to once every 100 minutes. The particular pattern was consistent in each sheep. The mean LH concentration in the plasma and the basal level, below which it did not drop, also varied, the highest being in the ewe that released LH most frequently. (From Karsch, 1980.)

between the pulses. The level to which they drop will depend on the factors that influence the hormone's clearance and the quantity that was originally released. If no further pulses occur, as in a surge of released hormone, the plasma concentrations will eventually be reduced to zero. A useful measure of this decline is the half-life ($t_{1/2}$) of the hormone in the circulation, which is the time it will take to reach 50% of its previous level. If a hormone is released more frequently than the time of its half-life, then it will tend to accumulate and its concentration in the plasma will rise. It will then reach a new equilibrium level after the equivalent of about four half-lives. The particular basal level attained will depend on the frequency of the pulses and the amount of hormone released on each such occasion. The basal level will appear graphically (Fig. 4.11) as an oscillating-type system with abrupt increases and declines centered around a steady median value. Such a mechanism provides a system whereby different tonic levels of hormones can be attained, and maintained, such as in daily rhythms of hormone concentrations in the plasma or in more long-term changes that accompany reproductive cycles. The factors that determine the frequency of the pulses and the amount of hormone released on each occasion are not well understood and often seek hypothetical refuge in engineering jargon, such as "pulse generator."

The testes of rams are small during the spring; these animals do not normally come into breeding condition until the autumn. When LHRH is injected

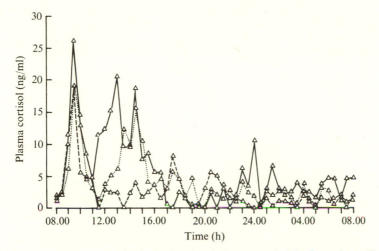

Fig. 4.12. Diurnal variation in the levels of plasma cortisol in three sheep. It can be seen that the highest concentrations were recorded during the early hours of daylight. The concentrations of cortisol in the plasma were not uniform but showed continual oscillations, suggesting that it is released in sudden "pulses". (From McNatty, Cashmore, and Young, 1972.)

into rams in the spring (Lincoln, 1979), in an episodic manner, for 60 seconds every 2 hours, the testes enlarge. These sheep then come into a reproductive condition similar to that which occurs normally in autumn.

Cyclical release of hormones

Many hormones are released periodically: at certain well-defined hours of the day (diurnal or circadian rhythms), during the reproductive cycle, or at certain seasons and times of the year (circannual rhythms). Such timing may be especially important in coordinating the events of the reproductive cycle and insuring that this occurs during the times of the year most appropriate to the survival of the young. A predictably functioning release mechanism also insures that adequate hormone levels, necessary for the animal's optimal daily activities, are available. Such release of hormones is usually controlled by centers in the brain (sometimes called "biological clocks") that are programmed by stimuli that include the length of the daily period of light, the external temperature, and changes in the seasons, as well as certain interoceptive stimuli.

The levels of corticosteroids in the blood vary in a distinct pattern during the course of the day. In mammals, this is well-known in primates including man, and dogs, rats, and mice. Release is related to the incidence of light and the corticosteroids are lowest in concentration during the night and reach a distinct peak after various periods of daylight. In man and sheep (Fig. 4.12),

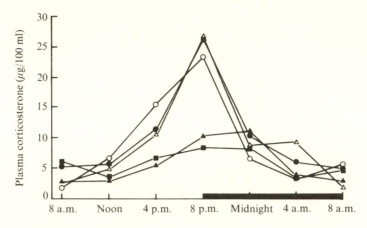

Fig. 4.13. The circadian pattern in the concentration of corticosteroids in the plasma of rats. The effects of the administration of exogenous corticosteroids to young, developing rats on the subsequent circadian periodicity in the endogenous corticosteroid levels are shown. It can be seen that the administration of dexamethasone or hydrocortisone (cortisol) on days 2 to 4 after birth suppressed the rhythmical release. However, when dexamethasone was given on days 12 to 14 after birth no effect was seen: ●, control; ○, saline, 0.1 ml, days 2–4; ■, hydrocortisone acetate, 500 μg, day 3; ▲, dexamethasone PO$_4$, 1 μg, days 2–4; △, dexamethasone PO$_4$, 1 μg, days 12–14. (From Krieger, 1972. Copyright © 1972 by the American Association for the Advancement of Science.)

this is seen in the morning hours, soon after dawn, though in laboratory rats, which are nocturnal, it is delayed until the early evening hours (Fig. 4.13). Comparable changes in the plasma corticosteroid concentrations have been observed in some teleost fishes [the channel catfish, *Ictalurus punctatus* (Boehlke et al., 1966), and the gulf killifish, *Fundulus grandis* (Srivastava and Meier, 1972)] where peak concentrations occur about 8 hours after the onset of light. In another teleost (the eel *Anguilla rostrata*), diurnal variation of plasma cortisol levels does not seem to occur (Forrest et al., 1973*a*). Prolactin and growth hormone in man are released in greatest amounts during the period of sleep (Fig. 4.14) and do not appear to be directly dependent on light.

The release of gonadotropins from the adenohypophysis may also occur at precise times of the day. In the Japanese quail (*Coturnix coturnix japonica*), the pituitary is depleted of gonadotropin over a period of 4 hours commencing about 16 hours after dawn (Follett and Sharp, 1969). This release is seen in birds kept on a long-day cycle of light, 20 hours light/4 hours of dark. When they are exposed to a short day of 6 hours light/18 hours dark such changes are not observed. This long-day photoperiodic pattern is similar to that which quail experience in nature where it is associated with the onset of breeding. Comparable precise patterns in the daily release of pituitary gonadotropins

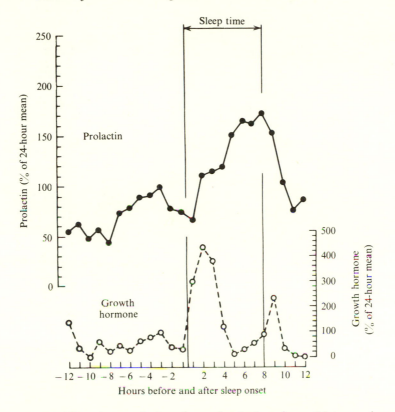

Fig. 4.14. The diurnal pattern in the release of prolactin and growth hormone in man. It can be seen that the release of both these hormones was accentuated during sleep. (From Sassin et al., 1972. Copyright © 1972 by the American Association for the Advancement of Science.)

have been observed in brook trout (*Salvelinus fontinalis*) and rainbow trout (*Salmo gairdneri*), as well as in leopard frogs (*Rana pipiens*) (O'Connor, 1972).

The pineal gland of rats and chickens (Fig. 4.15) shows an interesting cyclical pattern in activity during the day, which may be related to reproduction. The melatonin levels in the gland increase during the hours of darkness (Fig. 4.15). The enzymes *N*-acetyltransferase and hydroxyindole-*O*-methyltransferase (HIOMT) mediate the synthesis of melatonin, and the activity of the former enzyme increases at night (Binkley, 1979*a*, *b*). In rats, blinding, cutting the sympathetic nerve supply to the pineal, or placing the animals in continuous light prevents this effect. In chickens, however, extraretinal photoreceptors somewhere in the brain can mediate this rhythm.

The precise events involved in this cycle in the pineal of rats appear to be as follows (Axelrod, 1974). During the hours of darkness there is an increase in

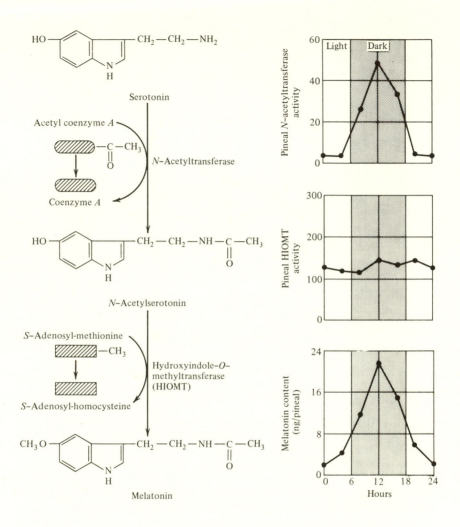

Fig. 4.15. Circadian changes in the synthesis of melatonin by the chicken pineal gland. The melatonin content of the pineal rises considerably during the hours of darkness (lower right). This change can be related to change in the biosynthetic process. The latter process is illustrated on the left side of the figure. Serotonin (5-hydroxytyptamine, 5-HT) under the influence of N-acetyltransferase gains an acetyl group. The activity of this enzyme increases during the hours of darkness (upper right). The N-acetylserotonin that is formed in this reaction is then O-methylated under the influence of HIOMT to melatonin. The latter enzyme does not appear to be rate limiting to the process, and its activity does not appear to change significantly over the period of the day (middle right). (From S. Binkley, 1979. Copyright © 1979 by Scientific American, Inc.)

the release of the neurotransmitter norepinephrine and a rise in the sensitivity of its pineal receptor. These changes result in the synthesis of the enzyme serotonin-*N*-acetyltransferase. This response is a β-adrenergic one involving cyclic AMP. The *N*-acetylserotonin that is formed from serotonin is then converted to melatonin under the influence of HIOMT. This rhythm in pineal activity is thought to be controlled by a biological clock that is situated in the hypothalamus, near the suprachiasmatic nucleus, which can be inhibited as a result of the receipt of light by the retina.

In many vertebrates, melatonin inhibits growth of the gonads and this can be correlated with the inhibitory effects of darkness on this process. It is unknown how widespread this nocturnal rhythm in the activity of the pineal is. A diurnal rhythm in pineal or plasma melatonin, however, has been observed on reptiles (Firth, Kennaway, and Rozenbilds, 1979), teleosts (Gern, Owens, and Ralph, 1978), and man (Greiner and Chan, 1978). Whether such changes are always related to the receipt of light is uncertain, however.

It has been suggested that the pineal, by responding to a biological clock, functions like the hypothalamus as a neuroendocrine-transducer that modifies the actions of other endocrine glands. It thus may provide an important pathway whereby light provides cues to the endocrine system. The pineal may be utilized to control various rhythms such as those involved in reproduction and the release of prolactin.

The activity of the endocrine glands and the release of their hormones often show profound changes that are associated with the season of the year. The diurnal patterns of release just described may be initiated at such times or be superimposed on these gross changes in activity. Such changes have most often been observed in relation to the breeding season, which, especially in animals from nonequatorial regions, usually only occurs at certain times of the year. The relationship of the size of the testes and cloacal gland of Japanese quail to the length of the day is shown in Fig. 4.16. These glands reach their maximum size in May, which corresponds to the onset of the breeding season. The morphological changes are related to the rising levels of gonadotropins and testosterone. The rates of secretion of these hormones start to increase in March, when the length of daylight is about 12 hours. Reproduction may be dictated or influenced by predictably favorable seasons or in less favored areas, like deserts, by the sudden appearance of rain. Apart from pituitary and gonadal sex hormones, cyclical changes in the activity of the thyroid gland and the adrenal cortex and medulla have been observed. The thyroid of the Japanese quail follows a pattern of activity that parallels that of the gonads (Follett and Riley, 1967). Thyroid activity (Leloup and Fontaine, 1960) is increased in fishes undergoing seasonal migrations. In the African lungfish (*Protopterus annectens*), thyroid activity is lowest during estivation at the time of seasonal drought. In toads and frogs, epinephrine attains its highest

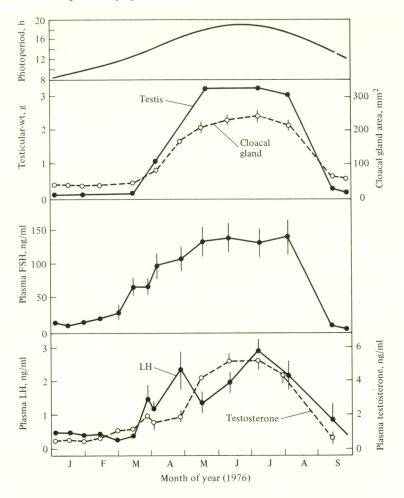

Fig. 4.16. The relationship of the growth of the testis and cloacal gland of quail to the hours of daylight (the photoperiod, upper panel) throughout the year. The lower two panels show the related changes in the plasma concentrations of luteinizing hormone (LH), follicle-stimulating hormone (FSH), and testosterone. Each point is the mean from 12 birds; the vertical lines are standard error of the mean. (From Follett and Robinson, 1980.)

concentration in the plasma during autumn and winter (Donoso and Segura, 1965; Harri, 1972). Stores of growth hormone are highest in the pituitary of the perch (*Perca fluviatilis*) in June, a few weeks prior to the summer rapid-growth period (Swift and Pickford, 1965).

The cyclical and episodic release of hormones is usually controlled through the brain. This is especially apparent when light provides the cues for these changes. External temperature may also be involved in controlling reproductive and possibly other cycles, especially in poikilotherms (see Licht, 1972). It is likely that such heat stimuli act through the central nervous system, but they could also exert more direct effects on the pituitary and gonads and even influence the rates of a hormone's metabolism. The principal photoreceptors are the eyes, but there is evidence that in some vertebrates others may exist in other parts of the brain. Temperature impinges on receptors present in the skin but may also directly influence the brain.

The diurnal rhythm of activity of glands such as the pineal and the adrenal cortex is not usually present at birth. However, in embryonic chickens the diurnal rhythm of melatonin levels in the pineal is observed before hatching, after 17–19 days of incubation (Binkley and Geller, 1975). In rats, the rhythmical release of corticosteroids from the adrenal cortex usually arises about 21 days after birth whereas in man in takes 2–8 years to appear. The adequate development of a biological clock is thought to require the presence of optimal levels of hormones. If newborn rats are injected with cortisol 2–4 days after they are born, the subsequent normal pattern of diurnal release of the corticosteroids is suppressed (see Fig. 4.13). Similarly, the normal pattern in the cyclical activity of the gonads is prevented when newborn female rats are injected with androgens. If newborn male rats are castrated soon after birth the hypothamalmus develops along the female pattern. When ovaries and a vagina are grafted into these animals they undergo cyclical changes typical of the female. Such endocrine manipulation can also contribute to behavioral abnormalities.

Systematic differences in releasing stimuli

There are many instances among the vertebrates in which the physiological roles of analogous hormones exhibit systematic differences. Such a change in the use of a hormone necessarily results in an altered responsiveness to excitatory stimuli that prompt the endocrine gland to discharge its secretion. When certain teleost fish move from fresh water to seawater they lose water and accumulate salt so that the concentration of sodium in their body fluids rises. This initiates a release of corticosteroids. Tetrapods, on the other hand, usually release such analogous hormones in response to declines in the sodium concentration of the body fluids. Prolactin and oxytocin are released during suckling in mammals. Other vertebrates lack mammary glands so that this represents a unique and phyletically novel stimulus. On the other hand, in certain teleost fish a prolactin-type hormone is released when the fish migrate from the sea into fresh water. The prolactin is then released in response to the lower calcium concentration in the fresh water. In homeotherms, thyroid hormones are discharged following a decline in the temperature of the blood flowing through certain areas in the hypothalamus but there is no indication that this happens in poikilotherms. Indeed, thyroid secretion usually increases in cold-blooded vertebrates exposed to elevated temperatures. Changes in glucose concentrations usually determine insulin release, but in ruminants (sheep and cattle) fatty acids are more important. In fishes, amino acids may be the major stimulant for secretion of insulin. Such phyletic differences in the propensity of endocrine glands to respond to certain stimuli are necessary for the evolution of a hormone's physiological role.

Conduction of stimuli to the endocrines

The conveyance of a stimulus to an endocrine gland may involve a complex series of events that take place along rather circuitous pathways. These are consistent with, and indeed are dictated by, the particular physiological requirements of the animal. The initial stimulus is usually translated into another form that may be a chemical compound or an electrical event or both. It travels to the endocrine gland, in such a modified form, by routes of varying complexity and may suffer further translation on the way. During this voyage, the stimulus may be modulated and interpolated with other information that is already available and other stimuli that also impinge on that particular communication pathway. This may take place in the brain and endocrine glands that are temporally proximal to the gland destined to receive, eventually, the final message. Such intermediary substations may involve neural areas in the brain, interconnecting endocrine glands, like the hypothalamus and the pineal, as well as the pituitary gland.

These events can be illustrated by summarily following the effects of light on reproduction. The stimulus is usually received by the eye (or in hypothalamic photoreceptors in many birds) where it is translated by the retinal receptors into electrical impulses that travel along nerves within the brain. These messages after further translation to other transmitter substances (such as acetylcholine, norepinephrine, and dopamine) eventually reach the hypothalamus and, probably also, in some species, pass through the pineal gland, which may release melatonin. The hypothalamus modulates its release of LHRH that crosses in the short portal blood vessels to the adenohypophysis. The receipt of this information will be further interpreted (in a process that probably involves the formation of cyclic AMP) in terms of an appropriate release of FSH and/or LH, which are carried in the blood to the ovaries or testes. Such an effect is illustrated by the levels of gonadotropin-releasing hormones and the gonadotropins in the hypothalamus and pituitary of the Japanese quail. There is a drop in the concentration of gonadotropin-releasing hormone that corresponds to the release of adenohypophysial gonadotropins. The gonadotroins may influence such events as ovulation and the formation and release of estrogens and progesterone as well as testosterone. These latter steroids, in turn, are carried back to the hypothalamus and pituitary gland where they provide information about the current hormone levels that will be used to modify stimuli that subsequently pass through this tissue. The process is, in detail, undoubtedly even more complex than that which has been described.

In other instances, the mechanism of the hormone's release may be simpler. The discharge of vasopressin (or antidiuretic hormone) from the neurohypophysis can occur in response to small increases in the osmotic concentration of plasma. This is thought to induce changes in osmoreceptors in the region of the supraoptic nucleus, possibly by releasing small amounts of acetylcholine, which initiates a wave of depolarization along the axons of the supraopticohypophysial tract. This results in the release of ADH from the storage granules at the terminus of the nerve. Neurohypophysial hormones may also be released in other circumstances. This release may involve non-specific stimuli, often termed "stress," that pass through higher centers in the brain to the nerve cells of the gland. Oxytocin is discharged in response to suckling in mammals; the initial receptor is in the nipple of the mammary gland from which it is transmitted along nerves to areas in the brain that initiate the release of the hormone from oxytocinergic neurons in the neurohypophysis.

Even simpler processes, not involving nerve pathways, may exist and determine the release of hormones in the body. The release of insulin can be demonstrated in isolated, perfused pieces of pancreatic tissues containing the islets of Langerhans. Elevated glucose and certain amino acid concentrations in the perfusate initiate a release of insulin. In a similar way, glucagon is discharged when the blood glucose concentration is depressed. The release of insulin from the B-cell is thought to occur as a result of the combination of glucose with a receptor (the glucoreceptor) in the cell wall (Fig. 4.17a). This event triggers a rise in the intracellular calcium concentration, possibly by increasing the permeability of the cell to this ion. The calcium initiates, and supports, the fusion of the insulin storage granules with the cell wall and the extrusion of their products into the extracellular fluid. A number of other excitants, including acetylcholine, glucagon, and some gut hormones, can also initiate a release of insulin, whereas others, such as prostaglandins and somatostatin, inhibit it. Separate receptors for these substances also appear to exist on the cell wall.

The mechanism of release of the hormones

Upon receiving a stimulus an endocrine gland may release its hormones from their storage sites. Many hormones, such as catecholamines, neurohypophysial peptides, and insulin, are spewed from their storage granules; steroid

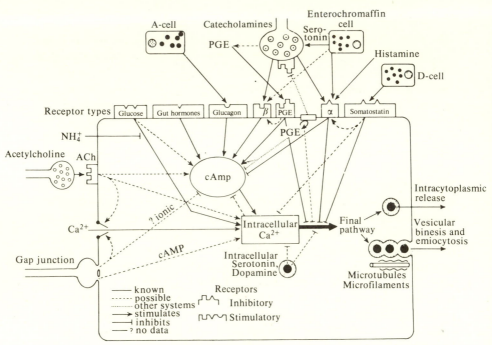

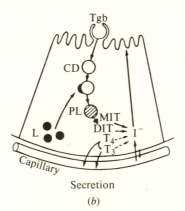

Secretion

(b)

Fig. 4.17. (*a*) Hypothetical model describing the various interacting systems that influence the release of insulin from the B-cell. (From Smith and Porte, 1976. Reproduced with permission from the *Annual Review of Pharmacology and Toxicology*. Copyright © 1976 by Annual Reviews, Inc.) (*b*) The secretion of thyroid hormones. Thyroglobulin (Tgb) is transported by endocytosis from the follicles into the thyroid cells where it appears as the colloid droplets (CD) that fuse with lysosomes (L) to form phagolysosomes (PL) in which proteolytic digestion of the thyroglobulin occurs. The iodothyronines (MIT, DIT, T_4, and T_3) are released and the active hormones pass out of the cell. Iodide (I)$^-$ is reaccumulated. (From Greer and Haibach, 1974.)

hormones are released from lipid droplets; and thyroid hormones are detached from thyroglobulins. In some instances, this is preceded by the formation of cyclic AMP as in the thyroid gland, the adrenal cortex, the ovary, and sometimes the adenohypophysis and pancreatic B-cells.

The release of hormones from intracellular storage granules has been studied in detail in the instances of the secretion of catecholamines from the adrenal medulla and vasopressin and oxytocin from the neurohypophysis (Douglas, 1972, 1974). Hormones that are stored in granules exist there as a nondiffusible complex with proteins and adenine nucleotides. Their release from the cell involves the process of emiocytosis (exocytosis or reverse pinocytosis) across the cell membranes. The nature of the events that result in such a release of hormones can be studied *in vitro* and is as follows:

i. As a result of nerve stimulation the cell membrane is depolarized and ions enter the cells. *In vitro*, this can be performed by exposing the tissue to high concentrations of potassium or by stimulating it electrically.

ii. An increase in the concentration of intracellular Ca^{2+} occurs, as a result of its uptake across the cell membrane and probably also its dissociation from binding within the cell. No release of hormone occurs *in vitro* following depolarization if the external media contain no Ca^{2+}, which thus appears to be vital for excitation–release coupling.

iii. Excitation–release coupling involves a migration of the hormone storage granules toward the cell membrane, a process that may involve the cell microtubular system.

iv. When contact between the cell membrane and the granules is made they fuse and Ca^{2+} may be important in the structural links. The entire contents of the granule, hormones, proteins, and adenine nucleotides are then extruded and pass into the capillaries. The empty granule may then be reconstituted and return to the cell cytoplasm.

Modifications of this process may occur in different endocrine glands. Insulin is released as a result of nerve stimulation, but glucose and amino acids have direct effects and these apparently also result in the admission of Ca^{2+} into the cell, or mobilization of intracellular calcium, which results in hormone release (Fig. 4.17*a*). The activation of the enzyme adenylate cyclase and the formation of cyclic AMP (as will be discussed) also appear to be involved in the release of insulin (as well as several other hormones) where it may promote an increase in intracellular calcium concentration or act more directly in the excitation–release coupling process.

Thyroid hormones are stored extracellularly, combined with thyroglobulin in the colloid of the thyroid gland follicles. Stimulation by TSH results in an activation of adenylate cyclase in the thyroid cells and the formation of cyclic AMP. In a manner that is not yet understood, this nucleotide is thought to stimulate pinocytosis during which fragments of follicular colloid are broken off and taken up by the thyroid cells; this process is thus called endocytosis. Lysosomes in the cell cytoplasm (Fig. 4.17b) that contain digestive protease enzymes combine with these ingested pieces of colloid to form phagolysosomes, which move toward the basal regions of the cell. During this migration, the colloid is digested and its components are separated so that thyroxine and triiodothyronine are freed to pass out of the cell. The iodotyrosines present are deiodinated and the iodine is either retained by the cell or, if also extruded, is subsequently reaccumulated.

The precise mechanism by which steroid hormones are released from the cell is not clear. The amounts stored in lipid droplets within the cell are relatively small compared to the large amounts of those hormones that are accumulated in granules. The synthesis and release of steroid hormones are promoted by various substances, including tropic hormones. Corticotropin and LH activate adenylate cyclase so that cyclic AMP is formed, which plays a role in the initiation of the hormone synthesis and release; these two processes have not, however, been satisfactorily separated. It is possible that an increased hormone synthesis inevitably leads to its release by a type of "overflow mechanism," but this is usually considered to be an unsatisfactory explanation. Adenylate cyclase does not mediate the release of all steroid hormones. Thus, aldosterone secretion, which occurs in response to angiotensin II and III, appears to involve calcium but not cyclic AMP.

The concentration of hormones in the blood

Hormones are normally present in the peripheral plasma at concentrations as low as 10^{-12}M and *in vitro* observations suggest that these levels may, in some instances, be similar to those necessary to stimulate the effector. However, for several reasons it is difficult to predict with certainty what the latter concentration will be. As we have seen, endocrines discharge their secretions in response to a great variety of stimuli; in some instances they may be sudden, or acute, occurrences whereas at other times they are of a more sustained, or chronic, nature. In the first case, the hormones may appear as a single surge or "spurt" of activity in the blood, such as seems to occur with the release of oxytocin in response to suckling. Thus, high concentrations of hormones may exist locally that briefly stimulate the effector and which are subsequently dissipated so that the remaining concentrations in the peripheral blood are not effective ones.

Differences in the patterns of release combined with seasonal and diurnal rhythms make it difficult to generalize as to the concentrations of hormones that are normally present in the blood. The antidiuretic hormone is normally present in the peripheral plasma of mammals at concentrations of 10^{-12} to 10^{-11} M. It is higher in small animals, such as the mouse, 2×10^{-11} M, than in large

Table 4.2. *The concentrations of vasopressin or vasotocin in the plasma of vertebrates. The animals were normally hydrated except where indicated*

	Hormone concentration (moles/liter)
Mammals	*Vasopressin*
Mouse	2×10^{-11}
Rat (i) normal hydration	1×10^{-11}
(ii) dehydrated	2.5×10^{-10}
Guinea pig	8×10^{-12}
Cat	2×10^{-12}
Rabbit	1.6×10^{-12}
Dog	2×10^{-12}
Man	4×10^{-12}
Bandicoot, wallaby, and possum (marsupials) dehydrated	1.8×10^{-10}
Amphibians	*Vasotocin*
(Anura) dehydrated	$4 \text{ to } 8 \times 10^{-10}$

Source: Robinson and MacFarlane, 1957; Heller and Štulc, 1960; Heller, 1966; Bentley, 1969.

ones such as man, 4×10^{-12} M (see Table 4.2). In the laboratory rat, ADH concentration increases about 25-fold during dehydration: from 10^{-11} to 2.5×10^{-10} M. Some dehydrated amphibians (frogs and toads) have similar circulating levels of vasotocin: about 6×10^{-10} M (Bentley, 1969). In others such as the mudpuppy and a teleost fish, the eel *Anguilla rostrata*, the hormone is not detectable (less than 10^{-11} M) under such conditions. Other polypeptide hormones like glucagon and insulin are normally present in the blood of mammals at a concentration of about 10^{-9} M.

The steroid hormones (see Idler, 1972) attain much higher concentrations in the blood than the polypeptides; thus, plasma cortisol and corticosterone levels in teleostean fish are about 10^{-7} M, though in the holostean and chondrostean fishes it is 10 times less than this. In chondrichthyeans 1α-hydroxycorticosterone usually has a concentration of 10^{-8} M in plasma. Aldosterone, compared to other corticosteroids, is present at much lower concentrations, about 10^{-10} M in mammals and 10^{-9} M in birds, though it is 10^{-8} M in amphibians. The steroid sex hormones, due to the cyclical nature of their release, are present in widely differing concentrations in the peripheral plasma: 10^{-7}–10^{-10} M.

Hormones act when present in extremely dilute solution. The levels vary somewhat with different species. The reasons for variation appear to be related to the animal's size, metabolic rate, and the hormone-binding capacity of the plasma. Receptors may also have different requirements for hormones, to which they may be bound with different strengths (affinities). It seems likely that differences in concentration may also be influenced by the body temperatures at which the animals usually function, because they affect the rate of the hormones' reactions with the effectors. Differences in the potencies of homologous hormones, as reflected in their chemical structure, may also contribute to variations in plasma levels. A mutation that initiates the formation of a less active hormone analogue could be compensated for physiologically by its release in greater quantity. An example of such an adaptation may be seen in mammals (Suiformes) that have lysine– instead of arginine–vasopressin.

Transport of hormones in the blood

Following their release from storage sites in the endocrine glands the hormones are carried, in the blood, to their various effector sites. This may be for very short distances, such as from the median eminence to the adenohypophysis, or involve much longer journeys, like from the neurohypophysis to the kidney. Some hormones have a relatively long life in the circulation and may recirculate many times before they are finally destroyed or excreted.

The blood plasma is an aqueous solution that contains high concentrations of proteins made up of several distinct components, or fractions. The hormone molecules may be dissolved in this solution or a substantial proportion can be bound to some of the proteins that are present. The binding of hormones in the blood is particularly important in the instances of the thyroid and steroid hormones.

The chemical nature of the binding of such hormones to plasma proteins is such that an equilibrium exists between the molecules that are bound and those that remain in solution. In other words, the binding is a reversible phenomenon and thus involves relatively weak chemical forces such as hydrogen and ionic bonds and Van der Waals' forces. The relative strength of these bonds, however, may vary considerably (high and low affinity) depending on the particular hormone and the nature of the binding protein. As we shall see, in some instances special proteins are present that appear to be able specifically to bind certain hormones.

There are several consequences of a hormone's binding to a protein in the plasma.

1. Hormones are usually assumed to be unable to initiate their effects when so bound. The receptor is envisaged as interacting with hormone molecules present in the aqueous phase. Their removal from solution shifts the equilibrium and may thus result in a dissociation into solution of some of the hormone that is bound to the proteins. It is implicit that the receptor has an even stronger ability than the plasma proteins to bind the hormone, and it is

even possible that more direct exchanges may occur between binding proteins and the receptors.

2. The process of the hormone's inactivation, such as can occur in the liver, and its excretion, mainly in the urine and bile, is delayed while it is in the bound form.

3. The bound hormone may constitute a circulating pool of the excitant that can extend or moderate the hormone's action. For instance, in pregnant women and guinea pigs, a plasma protein is formed that can bind testosterone and so may protect the mother from the effects of this steroid.

4. The distribution of the hormone within the body can be influenced by its binding to plasma proteins (Keller, Richardson, and Yates, 1969). A hormone–protein complex may have a special propensity to pass through the capillaries in certain vascular beds. It may thus help specifically to determine where the hormone is going.

5. Protein binding may contribute to the specificity of a hormone's action in the cell (Funder, Feldman, and Edelman, 1973). In the instance of the corticosteroids, two types of response can occur, mineralocorticoid and glucocorticoid (referring to their effects on electrolytes and intermediary metabolism), for which there are two types of receptors in the cell. Although the aldosterone receptors have a higher affinity for aldosterone than corticosterone, the normal excess of the latter hormone (in the plasma of rats) would be sufficient to negate this difference so that corticosterone would be expected, inappropriately, to occupy the aldosterone receptors. Most of the corticosterone in the plasma, unlike aldosterone, is bound to proteins in such a way that it does not obscure the effect of the aldosterone.

High proportions of hormones often exist in a bound form in the plasma. In the instance of testosterone, it is usually greater than 90% of the total present though it is less in other cases. Cortisol binding in the plasma of teleost fishes varies from 30 to 55% of the total (Idler and Truscott, 1972). Thyroxine and triiodothyronine also are substantially associated with plasma proteins; thus, in the plasma of kangaroos, more than 95% of the thyroxine and 90% of the triiodothyronine is so bound (Davis, Gregerman, and Poole, 1969).

The hormones may be bound to different protein components present in the blood plasma. In some instances, specialized proteins are present that have a high affinity (they are strongly bound) for the hormones. These include a globulin that binds cortisol (cortisol-binding globulin, or *CBG*), which is also called *transcortin*. This protein is present in most vertebrates though quantitative differences in cortisol-binding capacities of the plasma exist, indicating that interspecific variations occur (Seal and Doe, 1963). The echidna (Monotremata) lacks CBG, a deficiency that may reflect its low plasma levels of corticosteroids (Sernia, 1980). Cortisol binding is much less in the plasma of fishes than in most mammals (Idler and Truscott, 1972), and it is unlikely that the proteins are identical in the different species. Transcortin also binds corticosterone and progesterone (but not aldosterone), and in mammals its levels increase during pregnancy as a result of the action of estrogen, which promotes its formation in the liver.

Cortisol-binding globulin may exert a protective role and guard animals against excessive toxic levels of corticosteroids in plasma. The males of a small species of shrew-like marsupials, *Antechinus stuartii*, all die about 2 to 3 weeks after the start of the breeding season. These little beasts become very aggressive at this time and their plasma corticosteroid levels rise sharply (Bradley, McDonald, and Lee, 1980). This change is paralleled by a decline in the CBG concentration, a change which appears to be androgen dependent as it can be prevented by castration or increased by the injection of testosterone. The death of these marsupials is associated with gastrointestinal hemorrhage and infection, apparently reflecting the high levels of free or loosely bound corticosteroids in the plasma. The females do not experience comparable oscillations in steroids or CBG and carry on, mateless, to bear their young.

Plasma proteins that can bind sex steroids, androgens, estrogens, and progesterone with a high affinity have been identified in many vertebrates. They appear, however, to have undergone several evolutionary developments and the proteins involved are not identical. Cyclostome and elasmobranch fishes possess a plasma steroid-binding protein that binds not only androgens and estrogens but also corticosteroids, and so may serve a multiple role (Martin, 1975). Similar proteins have been identified in the plasma of teleost fish and amphibians (Ozon, 1972). A *sex hormone-binding globulin* (SHBG) is found in women (where its levels increase during pregnancy) and a protein that exhibits cross-immunoreactivity with it has also been found in other primates (Renoir, Mercier-Bodard, and Baulieu, 1980). This globulin binds both androgens and estrogens. In nonprimate eutherians an analogous protein is present, but it only binds androgens and displays no cross-immunoreactivity to the primate protein. Marsupials possess a SHBG (Sernia, Bradley, and McDonald, 1979) that also only binds androgens. The appearance of a SHBG that can bind both androgens and estrogens may thus be a specialization that, among mammals, is confined to the primates. The marsupials belonging to the order Polyprotodonta (which includes the opossums) lack (except in 1 species out of 14 examined) a SHBG (Sernia, Bradley, and McDonald, 1979) and it is also absent in rodents. The occurrence of SHBG thus appears to be quite sporadic among the mammals, possibly reflecting differences in their reproductive requirements for the sex steroids.

Many mammals possess an α-globulin that preferentially binds thyroxine, *thyroxine-binding globulin* or *TBG*, but which is absent in nonmammals (Farer et al., 1962; Tanabe, Ishii, and Tamaki, 1969). It also binds triiodothyronine but only about one-third as strongly as it does thyroxine. Although such plasma proteins strongly bind hormones, they are present in relatively small quantities so that the total amount of hormone that they can associate with is limited (low binding capacity). The plasma albumins, on the other hand, have a high binding capacity though their affinity is low. In species that lack such specialized hormone-binding proteins, or if the amount of hormone present exceeds their binding capacity, large amounts are bound to plasma albumin. In primates, and even their phyletically distant relatives, the marsupials, thyroxine is principally bound to a prealbumin (thyroxine-binding prealbumin, *TBPA*) whereas triiodothyronine combines with albumin. In some birds, reptiles, and fishes, thyroxine is bound principally to albumin and in others it is associated with both prealbumin and albumin (Tanabe, Ishii, and Tamaki, 1969).

Differences in thyroxine binding proteins exist between species. In some Australian marsupials (kangaroos), 75% of the thyroxine is bound to the prealbumin and only about 10% is associated with the postalbumin. On the other hand, in an American marsupial, the opossum, about 60% of the thyroxine is bound to the postalbumin (Davis and Jurgelski, 1973). Even among these opossums differences were observed: The postalbumin exists in two different polymorphic forms, PtA_1 and PtA_2. Six out of 177 sera examined possessed PtA_2, which has different physicochemical properties from PtA_1; for example, PtA_1 binds triiodothyronine, but PtA_2 does not. In birds plasma lipoproteins may be the principal sites for the binding of thyroid hormones (Castay, Bismuth, and Astier, 1978), but in mammals such a role for these substances is insignificant.

Peripheral activation of hormones

Some hormones are chemically altered, at sites peripheral to the endocrine gland from which they originated, in a manner that enhances their biological activity. This process is sometimes referred to as "activation" and may occur in the plasma or at tissue sites.

Angiotensin I is thus converted to its active forms, angiotensin II and III, by the action of enzymes. Not all the triiodothyronine present in the circulation is directly released from the thyroid gland. Some is formed from thyroxine in the liver and kidneys (Sterling, Brenner, and Saldanha, 1973). An integral step in the action of testosterone involves its conversion at its target site by 5α-reductase, to 5α-dihydroxytestosterone. Cortisone can also be converted peripherally to the more active steroid cortisol. Cholecalciferol (vitamin D_3) undergoes several transformations before it can exert its effects. These changes involve the formation of 25-hydroxycholecalciferol in the liver and a further hydroxylation to $1\alpha,25$-dihydroxycholecalciferol in the kidney (Lawson et al., 1971). It is also possible that some of the large protein hormones are fragmented peripherally into smaller pieces prior to their action. There is evidence that this may occur in the instance of parathyroid hormone while large amounts of proinsulin are normally present in the plasma that, possibly, may also be converted into the active hormone.

Termination of the actions of hormones

The durations of action of hormones vary: They may persist for many hours and even days, or only have a short-lived, transitory effect. This is usually in keeping with the nature of the homeostatic processes they mediate. Even a prolonged effect, however, will necessitate the renewed release of hormones because of the metabolic and excretory processes that inevitably result in their inactivation and elimination from the body.

Hormones persist in the circulation for different periods of time. The half-life of vasopressin in man is about 15 minutes, that of cortisol is about 1 hour, and that of thyroxine is nearly a week. This reflects the speed of their degradation in the body, the rate at which they may be eliminated in the urine and bile, the protection afforded them as a result of binding to proteins, and the nature of their actions at the effector site. The effects of binding are well illustrated by comparing the rates of removal (clearance) of corticosteroid hormones from the blood. In man, about 1600 liters of plasma are normally completely cleared of aldosterone each day. In the instance of cortisol, however, only about 180 liters are so purged in this time. The difference principally reflects the strong binding of cortisol to cortisol-binding globulin (transcortin) in the plasma. Considerable interspecific differences are apparent in the half-lives of hormones: arginine–vasopressin (ADH) has a half-life of about 1 minute in laboratory rats, compared to 15 to 20 minutes in man. This probably reflects the effects of size: Small animals destroy and eliminate hormones more rapidly than large ones. The precise chemical structure of the hormone may also be important; lysine–vasopressin has a half-life that is nearly twice as long in the rat as arginine–vasopressin (Ginsburg, 1968). The considerable differences that have been observed in the potency of human, porcine, and salmon calcitonin substantially reflect the differences in their degradation rates (Habener et al., 1971; DeLuise et al., 1972). In rats, porcine calcitonin is destroyed by the liver whereas that from man and the salmon is degraded by the kidney. Salmon calcitonin is much more resistant to inactivation (in rats, dogs, and man) than the mammalian hormones. Body temperature will also be expected to have an effect on the rate of inactivation of hormones so that, in cold-blooded vertebrates, the inactivation and excretion process will be modified accordingly. In the toad *Bufo marinus*, at 26°C arginine–vasotocin has a half-life of 33 minutes while in the domestic fowl (at 43°C) it is only 18 minutes (Hasan and Heller, 1968). The effects of differences in size, species, and temperature on hormone metabolism have not yet, however, been thoroughly evaluated.

The action of a hormone may be terminated in several ways.

1. In order to act it must attain a certain critical concentration in the neighborhood of its receptor site. If a hormone is released in a short burst as, for instance, usually occurs with oxytocin, the receptor will respond to a local high concentration that is sequestered like a small packet in the plasma. The response will then be terminated simply as a result of the subsequent dilution and redistribution of the hormone in the body fluids. A hormone may also be removed if it is bound or has accumulated at tissue sites. Epinephrine is readily taken up by the adrenergic nerve terminals. Less specific binding to tissues, such as that of neurohypophysial peptides to skeletal muscle, may also contribute to the removal of hormones from the circulation.

2. Small amounts of hormones may be eliminated unchanged in the urine and bile, but this usually amounts to less than 5% of the total released. The activity of hormones is generally reduced or destroyed as a result of their metabolism by enzymes in the tissues, particularly in the liver and kidneys. They are transformed in various ways and the by-products are usually then excreted in the urine and bile. The hormone's chemical structure may be altered in several ways. The catecholamines can be methylated by catechol-O-methyltransferase (COMT) or, to a lesser extent, they may be deaminated as a result of the action of monoamine oxidase (MAO). The action of the thyroid hormones is largely destroyed by the removal of iodine from the molecule by a deiodinase enzyme. Protein hormones are broken up by proteolytic enzymes. Steroid hormones (and, to some extent, thyroxine) are combined chemically with glucuronic and sulfuric acids, in a process called "conjugation" that results in increased water solubility, which enhances their chances for excretion in the urine and bile. Prior to such a conjugation considerable changes may be wrought in the chemical architecture of the steroids. Despite such chemical alterations, some hormones may still retain some of their biological activity. This is especially apparent in the gonadotropins that appear in large quantities in the urine of mammals during pregnancy. The activity of such urine in promoting ovulation and spermiation in various animals is used as a basis for pregnancy tests. Despite the retention of such biological effects, the products in the urine exhibit a different chemical behavior from that of the gonadotropins present in the pituitary gland.

Mechanisms of hormone action

Hormones exert actions on every major group of tissues in the body. Their effects are numerous and include changes in the intermediary metabolism of fats, proteins, and carbohydrates, growth and development of the tissues, changes in the permeability of membranes, and the contraction or relaxation of muscles. The precise manner in which they effect such processes is only incompletely understood.

It is generally considered that a hormone, in order to act, must eventually influence an enzyme activity. It may do this by a process of activation, promoting enzyme formation (induction) or possibly acting as, or providing, a cofactor in the chemical reaction that it promotes. As described earlier, this process is thought to be initiated as a consequence of the hormone's combination with a precise chemical moiety in the cell that is called its receptor. The exact role of this hormone–receptor unit in triggering the response is uncertain. It may provide a carrier for the hormone and so facilitate its transfer to a vital site in the cell or, conversely, the hormone may facilitate transfer of the hormone receptor to an essential site. The receptor–hormone complex itself may act as an enzyme activator (or inhibitor) or cofactor. The receptor could exist as part of an enzyme complex so that an activating effect of a hormone could be direct, but it usually seems to comprise a separate subunit.

Hormonally mediated responses to hormones may take place in successive steps (a "cascade") involving numerous enzymes. Conceivably, by acting at a single rate-limiting stage in such a process and altering the supply of an essential metabolite, a hormone could influence a complex series of metabolic events in the cell. It is also possible that a hormone may directly influence more than a single process in such a chain of events.

Hormones often initiate responses in several different tissues in the body. In such instances, the basic effects may be similar or even differ from one another. For instance, aldosterone initiates a change in membrane permeability to sodium in the kidney, colon, salivary glands, sweat glands and, in amphibians, the skin and urinary bladder. Alternatively, epinephrine mobilizes fatty acids from fat cells but glucose from muscle cells. Despite the diverse tissue sites and the nature of the processes involved, the underlying initiating mechanism is generally the same. For instance, the effects of epinephrine on fat and muscle cells are both mediated by the action of cyclic AMP (see the following section).

At the present time the mechanisms of action of hormones are thought to be affected through either of two major groups of processes in the cell (others may also exist). The nucleotide adenosine-3',5'-monophosphate, or cyclic AMP, is a vital link in many endocrine responses and its formation is promoted, or inhibited, by numerous hormones. Other hormones, by regulating genetic transcription in the cell nucleus, can control the formation of proteins and enzymes that are essential for a response. Other processes, however, may also be involved, such as calcium and prostaglandins, which may modulate these responses or, possibly, even mediate them. Not all the actions of all the

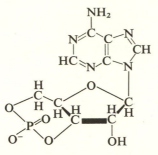

Fig. 4.18. The structural formula of adenosine-3′, 5′-monophosate (cyclic AMP).

hormones are even partially understood. It is likely that some with multiple actions may utilize several different mechanisms. Some hormones, notably insulin, do not easily fit into either of these classifications. It is possible that some hormones, including insulin, may control translational processes on the ribosome.

The role of adenosine-3′,5′-monophosphate (cyclic AMP)

Our knowledge of the part played by adenosine-3′,5′-monophosphate, or cyclic AMP, in the action of many hormones is primarily due to the work of Earl Sutherland (see Robison, Butcher, and Sutherland, 1971; Sutherland, 1972). This adenine nucleotide is chemically related to ATP, from which it is formed. Its chemical structure is shown in Fig. 4.18.

The discovery of the endocrine role of cyclic AMP was made during an investigation of the mechanism of action by which epinephrine and glucagon convert glycogen to glucose in the liver. This reaction is dependent on several enzymes but the presence of a phosphorylase is rate limiting. This enzyme exists in two forms, a relatively inactive phosphorylase *b*, which on the incorporation of phosphate is converted to the much more active form, phosphorylase *a*. This activation takes place not only in intact cells but also in broken-cell preparations upon the addition of epinephrine or glucagon to them. The phosphorylase is a soluble enzyme that can be separated in the supernatant fraction of the broken cells. This enzyme, however, cannot be activated by the hormones when alone in this solution; the presence of the particulate cell material is a necessary condition. When the latter fraction is exposed to epinephrine or glucagon a substance can subsequently be washed from the cell particles that, when added to the supernatant, activates the phosphorylase. This activating chemical was found to be cyclic AMP and it is formed from ATP as a result of the action of an enzyme, *adenylate cyclase*, which is part of the cell membrane. In the liver, this enzyme is activated by epinephrine or glucagon and in skeletal muscle preparations only by epinephrine.

Table 4.3. *Distribution and hormonal sensivity of mammalian adenylate cyclase*

Tissue	Hormone
Liver	Glucagon and epinephrine
Skeletal muscle	Epinephrine
Cardiac muscle	Catecholamines
	Glucagon
	Triiodothyronine
Kidney	Vasopressin
	Parathyroid hormone
Bone	Parathyroid hormone
	Calcitonin
Brain	Catecholamines
Adrenal	ACTH
Corpus luteum	LH and prostaglandins
Ovary	LH
Testes	LH and FSH
Thyroid	TSH
	Prostaglandins
Parotid	Catecholamines
Pineal	Catecholamines
Lung	Epinephrine
Spleen	Epinephrine
Adipose	Epinephrine
Brown adipose	Catecholamines
Platelets	Prostaglandins
Leucocytes	Catecholamines and prostaglandins
Erythrocytes	None demonstrated
Uterus	Catecholamines
Pancreas	None demonstrated
Anterior pituitary	Several
Vascular smooth muscle	None demonstrated

Source: Robison, Butcher, and Sutherland, 1971.

Adenylate cyclase has been found in many animals: mammals, birds, amphibians, and fishes, as well as many invertebrates. This enzyme is also present in bacteria, though apparently not in higher plants. Most tissues in the body contain adenylate cyclase. Apart from liver, they include muscle, kidney, heart, brain, adipose tissue, bone, and many endocrine glands (Table 4.3).

The role of adenylate cyclase and cyclic AMP in mediating the effect of hormones was first described for the actions of glucagon or epinephrine on glycolysis in the liver or skeletal muscle. A description of this process can thus be used as a prototype for its role. It should, however, be remembered that although the initiating reactions may be similar for many tissues and

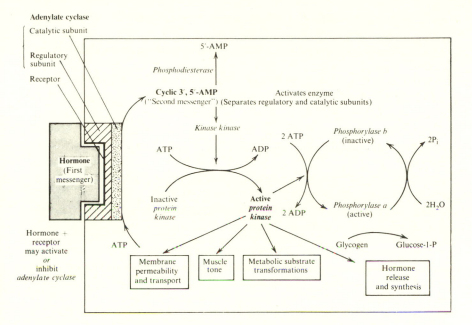

Fig. 4.19. A diagrammatic representation of the role of cyclic AMP as a "second messenger" in the mechanism of a hormone's action. A variety of protein kinases are thought to be present in different cells, which, when activated by cyclic AMP, can mediate the phosphorylation or dephosphorylation of certain proteins. This may result in diverse responses.

various hormones, the responses of the final effector (or responding system) will often differ. As summarized in Fig. 4.19, the initiating event is the interaction of the hormone and its receptor, in the cell membrane, which results in the activation (or sometimes the inhibition) of adenylate cyclase.

There is now a considerable amount of information available about the nature of the receptor–adenylate cyclase system. This complex is situated in the plasma membrane, the receptor facing outward, toward the extracellular space, the adenylate cyclase inward. These complexes have been isolated in fragments of plasma membranes and can be studied in this form. The receptors from several hormones have been separated from the enzyme and, with the aid of detergents, prepared in a soluble form. They are glycoproteins with molecular weights ranging from about 100,000 to 200,000. Such isolated receptors have been shown to be able to specifically bind hormones in a manner similar to that in the intact tissue. The peptide hormone receptors may normally form a relatively fixed unit with adenylate cyclase or they may be able to float laterally in the lipids of the plasma membrane. The latter receptors, when they interact with their hormones, can then combine with adenylate cyclase and so modulate its activity. Receptors for several different hormones can be present in a single plasma membrane, each of which, when combined with its specific hormone, can interact with a common adenylate cyclase moiety. This phenomenon is called the "mobile receptor theory" (Bennett, O'Keefe, and Cuatrecasas, 1975; De Häen, 1976). Such receptors have been separated and then grafted onto foreign cells where they can mediate a response that is foreign to the recipient, but similar to that of the donor. Thus, β-adrenergic receptors from turkey erythrocytes have been transplanted onto Friend erythroleukemia cells (Schramm et al., 1977). These "F"-cells possess adenylate cyclase but it does not normally respond to epinephrine. The transplanted β-receptors can, however, mediate its response to this type of catecholamine hormone, so that cyclic AMP is then formed.

The receptor–adenylate cyclase system contains three important sites (see Rodbell et al., 1975; Citri and Schramm, 1980):

a. A "catalytic site" (or unit) with which the substrate, ATP, can combine,

b. The "receptor site," which modulates the enzyme's activity, and

c. A "regulatory site," which combines with guanosine triphosphate (GTP), an event that is vital for the function of the enzyme. When the bound GTP is broken down the response is terminated.

All three of the units in the receptor–adenylate cyclase complex have been chemically separated from each other. The activity of the system can subsequently be restored by rejoining these pieces.

Cyclic AMP can mimic the actions of many hormones on a variety of tissues. The formation of this nucleotide, as a result of hormonal activation of adenylate cyclase, appears to mediate the actions of such hormones as (apart from epinephrine and glucagon) melanocyte-stimulating hormone, vasopressin, corticotropin, thyrotropic hormone, luteinizing hormone, and parathyroid hormone (see Table 4.4). The ultimate responses are very diverse and include glycolysis, lipolysis, changes in the permeability of membranes to water, sodium, and calcium, the dispersion of melanin in the melanophores, and the formation and release of many other hormones including thyroxine, corticosteroids, and sex steroids.

The nature of the effector response to hormones differs widely. It has been suggested, however, that in those processes that involve the action of cyclic AMP as a "second messenger," this is always the result of the phosphorylation, or dephosphorylation, of a protein and involves a *protein kinase*. Protein kinases (see Krebs, 1972) are ubiquitous cellular enzymes that can transfer phosphoryl groups onto proteins, usually at the site of a serine residue. Such kinases normally exist in an inactive form, but when they combine with substrates such as calcium and cyclic nucleotides they become active. Cyclic AMP-dependent kinases, when combined with cyclic AMP, dissociate into

Table 4.4. *Relations between the metabolic actions of some hormones and cyclic AMP*

Hormone	Actions shared by cyclic AMP
Epinephrine	Glycolysis (liver and muscle), lipolysis (fat cell)
	Heart muscle contraction
ACTH	Steroid synthesis in adrenal cortex
LH	Steroid synthesis by corpus luteum
TSH	Production and release of thyroid hormones
Growth hormone	Lipolysis (?)
Vasopressin	Increased water movement in renal tubule
MSH	Melanophore, dispersion of melanin
Glucagon	Glycolysis (liver), lipolysis (fat cell)
Parathyroid hormone	Mobilization of bone calcium

two parts, a regulatory subunit, containing the nucleotide, and an active catalytic unit. The holoenzyme is inactive:

$$R.C + cAMP \longrightarrow R.cAMP + C$$

inactive   regulatory   active catalytic
holoenzyme   subunit   subunit

Different types of proteins may be activated as a result of their phosphorylation by protein kinases. Metabolic enzymes can be activated; for instance, epinephrine can increase the conversion of glycogen to glucose and fats to glycerol and fatty acids, processes that depend, respectively, on a phosphorylase and lipase, both of which are activated by a cyclic AMP-dependent protein kinase. Changes in the configuration of proteins in the plasma membrane, and hence increased permeability to water, result from the activation by antidiuretic hormone of a protein kinase in the renal tubule.

Apart from initiating increases in levels of cyclic AMP in cells it is possible that some hormones may act by bringing about a decrease in the nucleotide's concentration. An understanding of the mechanism of action of insulin has been somewhat elusive but in some instances, especially when the basal levels of the nucleotide are elevated, it has been found to decrease the concentration of cyclic AMP (see Kono, Robinson, and Sarver, 1975). Decreases in cyclic AMP levels have also been observed in smooth muscles during contractions elicited by catecholamines. Such effects could be due to a reduction in the activity of the adenylate cyclase or an increase in the rate of its destruction.

The inactivation of cyclic $3',5'$-AMP to $5'$-AMP in cells is normally due to the action of the enzyme *phosphodiesterase*. This enzyme could provide a potential site for the action of hormones but such effects have not been described. Phosphodiesterase can be inhibited by certain drugs, notably the methylxanthines, which include caffeine.

The activation of adenylate cyclase can be accomplished by several hormones, in various tissues, and can result in diverse responses. The actions of hormones are, however, relatively specific, so how, when several utilize cyclic AMP, can a separation of their roles and effects be accomplished? In some instances more than one hormone is thought to act on a single adenylate cyclase. This seems to be so in the effects of epinephrine, glucagon, ACTH, and TSH on fat cells *in vitro*. The effects of the two latter hormones may not be physiological but reflect the relative, nonspecific reactivity of the receptors, and the adenylate cyclase, that are present. The adenylate cyclase that mediates the effects of ADH on the kidney is much more specific in its reaction with hormones. This selectivity is probably dictated by the nature of the hormone receptors. In some instances, a tissue exhibits more than one adenylate cyclase-mediated response to a single hormone. Neurohypophysial hormones increase both water and sodium transfer across amphibian skin and urinary bladder, and these effects are mediated by different receptors and effectors. In the amphibian membranes, it seems likely that two adenylate cyclases and/or two "pools" of cyclic AMP exist in the tissue, each mediating a distinct response.

Differences in the ability of various hormones to activate adenylate cyclase from different tissues or species could reflect a polymorphism in the enzyme's structure; however, it appears that differences in the associated receptor mainly determine the particular response that is manifested. Mutations that result in changes in the protein linking receptors to the catalytic component of the adenylate cyclase system have been identified, however, in the mouse and man (Farfel, Salomon, and Bourne, 1981). Cell-free (membrane fragment) preparations of adenylate cyclase exhibit different quantitative responses to hormone analogues. Adenylate cyclase preparations from rat, mouse, rabbit, and ox kidney are more readily activated by arginine–vasopressin than by lysine–vasopressin (and much less by oxytocin). On the other hand, the enzyme complex prepared from the kidneys of pigs is most responsive to lysine–vasopressin (Dousa et al., 1971). This relative sensitivity corresponds to that of the homologous hormones present in each species, whereas oxytocin does not normally act on the kidney, anyway. These differences, however, as discussed, appear to reflect the response of a receptor that need not necessarily be a permanent part of the enzyme itself. The properties of renal adenylate cyclase systems from a number of vertebrates have been compared (Table 4.5). Mammal, bird, reptile, and amphibian enzymes were all strongly activated in the presence of fluoride. The mammalian enzyme systems were also activated by neurohypophysial hormones but this response was much less prominent, or even undetectable, in the nonmammals. This lack of response

Table 4.5. *Stimulation of renal adenylate cyclase system in various vertebrate species*

Species	Neurohypophysial hormones	Parathyroid hormone	Fluoride
Rat	+ + + +	+ + + +	+ + + +
Mouse	+ + + +	+ + + +	+ + + +
Pigeon	0	+ + +	+ + + +
Alligator	0	+ +	+ + + +
Toad	±	0	+ + + +
Bullfrog	+	0	+ + + +

+ Indicates strength of response; 0, no response; ±, rudimentary response. Homologous neurohypophysial hormones were tested.
Source: Dousa et al., 1972.

could, however, merely reflect the presence of only small amounts of the appropriate enzyme in the kidneys of the nonmammals or its lack of importance in mediating the renal responses in those species. It is notable that the adenylate cyclase system obtained from amphibian urinary bladder is stimulated by neurohypophysial hormones. The renal adenylate cyclase system that is activated by parathyroid hormone (a distinct entity from that responding to the neurohypophysial hormones) was stimulated in enzyme preparations made from the kidneys of mammals, birds, and reptiles but not in those made from amphibian kidneys. This may reflect an absence of a renal effect for parathyroid hormone in the Amphibia. Except in bacteria, adenylate cyclase has not been prepared as a soluble enzyme so that the precise studies that may confirm its polymorphism are not yet feasible.

Prostaglandins are lipid molecules that exhibit many biological actions. Their distribution is ubiquitous in the tissues of the body. They are formed from some fatty acids, especially arachidonic acid. The latter is a component of phospholipids that are present in the plasma membrane, from which they are released by the action of phospholipase A_2. The prostaglandins are formed from their substrates under the influence of the prostaglandin synthetase (or cyclooxygenase) enzyme system. A large family of such substances has been identified, including prostaglandin E_1, prostaglandin E_2, and prostaglandin $F_{2\alpha}$. They have numerous effects such as the contraction or relaxation of smooth muscle, decrease in blood pressure, and changes in the activity of hormones (e.g., an inhibition of the effects of antidiuretic hormone). Prostaglandins can interact with the adenylate cyclase system to increase or decrease the formation of cyclic AMP. It has been suggested that they may normally contribute to the regulation and modulation of a variety of intracellular processes. Thus, they may interact with hormone-dependent processes, whereas hormones, in turn, can influence the levels of prostaglandins. In some instances, such as the action of prolactin on mammary glands, they may even help mediate a hormone's effect.

The role of calcium and calmodulin

Calcium plays an important role in the mechanism of action of hormones, and it has sometimes, like cyclic AMP, also been called a "second messenger" (see Rasmussen and Goodman, 1977). It may be involved in several types of reactions in cells, including changes in the activity of enzymes such as adenylate cyclase and phosphodiesterase. Calcium also acts as a coupling agent and can link the primary actions of excitants to the ultimate expression of their response. This type of role includes the process of muscular contraction and the secretion of exocrine and endocrine glands. Changes in the levels of soluble "free" intracellular calcium may occur as a result of an increase in the influx of the ion across the plasma membrane into the cell or its mobilization from intracellular stores in the sarcoplasmic reticulum, mitochondria, or plasma membrane. Such an intermediary role of calcium generally appears to result from its interaction with an intracellular binding protein (or Ca-"receptor") called *calmodulin* (Cheung, 1980). As a result of this binding the configuration of this protein changes, so that about 50% of the molecule assumes an α-helical configuration. When in this form it can activate a large number of enzymes, including phosphodiesterase, adenylate cyclase, phospholipase A_2, and Ca-activated ATPase, as well as some cell kinases.

Calmodulin is a protein, with a molecular weight of about 16,700, containing 148 amino acids. It appears to be present in all animal cells, which is consistent with its ubiquitous function, and it has been identified in species from both the animal and plant kingdoms. Its structure is remarkably conservative and displays few differences in amino acid sequences between such species as cattle and sea anemones. Antibodies to rat testis calmodulin can interact with calmodulin prepared from this coelenterate. "It is likely that calmodulin. . . will provide a link in our understanding of the interactions of calcium and cyclic nucleotides in the control of cellular metabolism" (Means and Dedman, 1980).

Hormonal effects mediated by changes in the transcription of DNA in the cell nucleus

The synthesis of specific proteins is primarily controlled by genes, which, through their coding patterns of DNA, provide a template for the formation of RNA. The latter is formed in a process that involves the action of RNA polymerase. This RNA (messenger RNA) can move to the ribosomes in the cytoplasm where it, in turn, acts as a template for the synthesis of a specific protein. This basic mechanism of genetic transcription and translation was originally proposed as the result of the work of J. Monod and F. Jacob in bacteria. Although the details differ, it basically also applies to animal cells. In 1963 Karlson proposed that some hormones may act by initiating transcription of genetic material on the chromosome in a manner analogous to that of depressors in the genome of bacteria. These observations had their origins in invertebrate endocrinology. It was observed that the giant salivary gland chromosomes of various insects often displayed a loosening, or "puffs," in the regions of the bands of DNA. The number of such puffs could be increased following the injection of the insect metamorphosis hormone ecdysone. These puffs were identified as the sites of RNA synthesis. On the basis of these observations Karlson proposed that other hormones, including vertebrate steroid and thyroid hormones, may be acting at the site of the genes on the chromatin to initiate the synthesis of the protein for which it is encoded. This was a remarkably prophetic hypothesis.

The effects of the steroid hormones, as well as the thyroid hormones and growth hormone, are most obviously manifested as increases in growth, development, and differentiation of tissues. These changes are especially clear when one observes the effects of estradiol-17β on the uterus. Such growth is associated with a marked increase in the rate of protein synthesis. More subtle changes may accompany other responses to such hormones. For instance, aldosterone, which increases the rate of sodium transfer across some membranes, may induce the formation of a permease, or some other enzyme, that increases active sodium transport. It is not always possible to identify the specific proteins formed, but progesterone is known to increase the formation of ovalbumin in the chick oviduct, whereas cortisol enhances the production of Na–K-activated ATPase in several tissues including the kidney and, in some fish, the gills and intestine.

Protein synthesis is a cytoplasmic process taking place in association with the ribosomes that follow the translation pattern provided by the messenger RNA derived from the chromosomal DNA. Although it is possible that hormones exert some direct effects on translation at the ribosome, this does not appear to occur usually. However, it has been suggested that insulin and growth hormone may have some action at this site. One of the earliest distinguishable effects of the actions of hormones that influence protein synthesis is the formation of messenger RNA by the nucleus. This process is mediated by the action of RNA polymerase II. It is now considered unlikely that this enzyme is induced or activated by hormones. Rather a perturbation of the structure of the DNA template occurs, which facilitates the action of the enzyme. The hormones could, for instance, be creating "initiation" binding sites for RNA polymerase on the chromatin. The protein synthesis by the ribosomes can be blocked by puromycin and cycloheximide, which act only

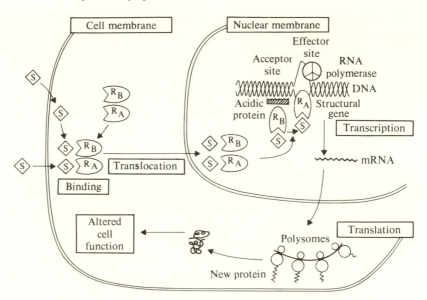

Fig. 4.20. General mechanism of action of sex steroid hormones as constructed by Chan and O'Malley (1976*a*). This scheme was based principally on information gathered from studies on the progesterone receptor in the chick oviduct. S = steroid hormone; R_A and R_B = subunits of the steroid hormone receptor. (From Chan and O'Malley, 1976. Reprinted by permission from *The New England Journal of Medicine* **294**, p. 1322.)

on the cytoplasmic translational process and do not stop the early formation of the messenger RNA. A further indication as to the nuclear site of action of hormones is their autoradiographic identification in the chromatin.

The processes by which hormones act to initiate protein synthesis, via increases in nuclear transcription of the genome, have been most completely described for the action of estradiol-17β on the uterus (see Jensen and DeSombre, 1972, 1973; O'Malley and Means, 1974) and progesterone and estrogen on the chicken's oviduct (Schwartz et al., 1976; Chan and O'Malley, 1976). These results are summarized in Fig. 4.20. Comparable mechanisms also appear to exist for progesterone, the androgens, cortisol, aldosterone, and thyroid hormone. In the instance of the latter, however, differences do exist.

The estrogenic hormone molecule, upon crossing the plasma membrane and entering the cell's cytoplasm, is bound to a protein that has been called its cytoplasmic receptor. These proteinaceous units are also called *estrophiles*. The hormone–receptor combinations can be isolated *in vitro*. They pass intact through a gel filtration column and can be separated from other cell constituents by ultracentrifugation in a sucrose gradient where they have a sedimentation coefficient of about 8 S. In the presence of salt solutions this changes to 4

S, indicating that the receptor–hormone unit can be broken into two subunits, A and B, each of which combines with a molecule of the hormone. There are about 100,000 estrophiles in each uterine cell, which is a much higher number than is present in nontarget tissues. They have a molecular weight of about 200,000. The binding for a particular hormone is relatively specific and can be prevented by substances that block their effects *in vivo*. The binding sites are 50% saturated by the hormone when its concentration is 10^{-10} M (the K_D).

At a temperature of 2° C, 75% of the estrogen is present in the 8 S complex in the cytoplasm, but if the tissue is warmed to 37° C most of it moves across into the nucleus. The estrophilic complex may be acting as a carrier for the hormone or the hormone may facilitate the transfer of one of the subunits of 8 S that could initiate the formation of messenger RNA. The estradiol-17β in the nucleus is bound in another proteinaceous form that can be separated in KCl solution and which has a sedimentation coefficient of 5 S. This complex is clearly distinguishable from the 8 S one in the cytoplasm. The 5 S nuclear complex can be freed from the chromatin by the action of DNAase, indicating that it is closely associated with DNA. The 5 S nuclear receptor has not been found in the nucleus prior to its exposure to the 8 S complex, suggesting that the latter may first be converted to 5 S, which has an *acceptor site* in the nucleus. The free hormone does not readily bind to the chromatin. The site of the combination with the chromatin is not known, but it appears to be a nonhistone acidic protein.

In the instance of progesterone, the action of this steroid on the chicken oviduct is thought primarily to involve an increase in the frequency of the initiation of transcription on the DNA template (Schwartz et al., 1976). This process may involve a perturbation of the structure of the chromatin by the A-subunit of the hormone–receptor complex. This could direct the action of the RNA polymerase. The B-subunit may act as a "binding site specifier protein" that guides the hormone–receptor complex to the appropriate site on the chromatin (O'Malley et al., 1979).

At the present time not all the actions of hormones can be accounted for by these two types of mechanism. Hormones have multiple effects and not all the effects of a single hormone can always be accounted for by a single mechanism. Thyroxine, growth hormone, and prolactin can stimulate protein synthesis in the cell and, although these actions can be prevented by inhibition of RNA polymerase, just how primary, or universal, this effect is in these hormones' actions is not clear.

The action of prolactin on the mammary gland, for instance, may involve the activation of phospholipase A_2 (Rillema, 1980). This enzyme releases arachidonic acid from lipids in the plasma membrane that is a substrate for prostaglandins. The latter are thought to initiate an increase in protein synthesis.

Triiodothyronine (Fig. 4.21) binds to chromatin and can initiate the formation of messenger RNA (see Tata, 1980). It can also, however, bind specifically to receptor sites on the inner membrane of mitochondria (Sterling, 1979). Although it is bound to a component in the cell cytosol, this complex, in contrast to steroid hormones, is not necessary for the transfer of the hormone to its receptor sites in the nucleus, or the mitochondria. An important enzyme that can be induced by thyroid hormones, and which may mediate its effects on oxygen consumption, is Na–K-activated ATPase (Edelman, 1976).

Some speculations on the evolution of the actions of hormones in cells

The two basic control mechanisms involving cyclic AMP and the regulation of genetic transcription exist in all animals as well as many microorganisms. We may thus suspect that these basic mechanisms, upon which hormones can impinge their actions, have always been present in animals, even in unicellular ones.

Cyclic AMP exists in bacteria, where it also appears to have a role in regulating cellular activities. Adenylate cyclase in bacteria is an intracellular enzyme, and it has been suggested that this represents the primitive condition. This internal site would appear to be most suitable for an enzyme that has to respond to changes in the intracellular nutrients and metabolites. In metazoan animals, adenylate cyclase appears to be confined to the cell membrane, a position that may be more apt for its interaction with metabolites and chemicals coming from other cells. These include the hormones. Such a membrane site for adenylate cyclase may thus be more opportune for intercellular cooperation and especially for interactions with molecules like the polypeptide hormones that cross cell membranes with some difficulty. It is unnecessary to postulate evolutionary changes in the structure of adenylate cyclase, only in the receptor which may be a subunit of it. The specificity of the receptor for a particular hormone has also been described. The simultaneous evolution of the complementary nature of both the hormone and its receptor is difficult to envisage. One wonders, when considering all the possible differences in structure, how they ever got together and how, at the same time, they acquired their complementary relationships (see also Chapter 3).

The transcription of genetic material plays a basic role in the life of cells. In unicellular organisms, this need only be controlled by internal accumulations of nutrients and metabolites. These may act by combining with structural units, analogous to the subunits of the 8 S cytosol receptors. With the onset of need for intercellular communication these controlling subunits may have contracted an ability to combine with materials originating outside the cell. In other words, they may have acquired, or been transformed so as to incorporate, a hormonal receptor. Steroid hormones are nonpolar materials that read-

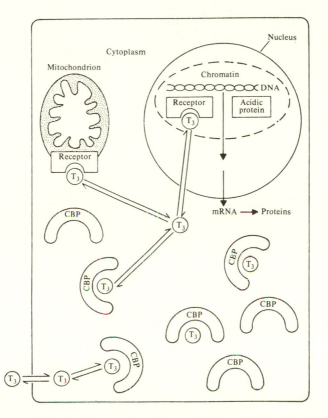

Fig. 4.21. Model showing the sequence of events that occurs when thyroid hormone enters its target cell. The unbound triiodothyronine (T_3) enters the cell by diffusion and binds to the cytosol-binding protein (CBP). The T_3–CBP complex remains in a reversible equilibrium with a very small amount of free T_3 in the cytoplasm. This unbound T_3 may interact with receptors in the nuclei or mitochondria, where it can trigger a response. (From Sterling, 1977.)

ily gain access to cells and so would appear to be well suited to such an intracellular reaction.

Conclusions

In the following chapters we will be examining the roles of hormones in coordinating different physiological processes in the body. In the present chapter we have looked at the manner by which the endocrine system itself works. Although information about nonmammals is rather sparse it appears that the underlying mechanisms of a hormone's synthesis, release, transport, mechanism of action, and its destruction are rather similar in all vertebrates.

Even when different hormones are involved, the general underlying processes involved are often similar; but major differences are often apparent between the general types of hormones, especially those made from cholesterol (steroids) and those derived from amino acids. There are a number of interspecific differences in the "life history" of particular hormones in the body and these can be related to the animal's manner of life. The natures of the stimuli that initiate a particular hormone's release are especially variable among the vertebrates and are dictated by the different physiological roles that a hormone may have assumed. In addition, quantitative differences may arise with respect to a hormone's rates of synthesis, destruction, the quantities that are stored in the gland, and the concentrations that appear in the blood. Such differences can arise at distinct stages of the life cycle of an animal, but they are also observed between various species where they can be related to such characteristics as size, rates of metabolism, and environmental factors like temperature and the availability of different nutrients, salts, and water.

5. Hormones and nutrition

Animals require a continual supply of food in order to sustain life. Such nutrients, in the first instance, are obtained from the external environment. These materials are used as an energy supply, as building blocks for growth and reproduction, and also as a source of certain essential chemicals necessary to the adequate functioning of the metabolic machinery in the body. The processes involved are thus basic to life and are regulated to a considerable extent by hormones.

Animal cells, including tissues isolated from metazoan species, can survive *in vitro*, in the absence of hormones, for extended periods of time. Except for cancer cells, their life and continual perpetuation cannot go on indefinitely. Even the more limited survival of normal tissues *in vitro* depends on an adequate supply of special nutrients that are chosen for their ability to be utilized by that particular tissue. One cannot run a gasoline engine on diesel fuel and in the same way cells can only metabolize certain forms of nutrients.

The foods that animals obtain from the environments where they live are usually chemically far more complex than can be used by their cells. The original nutrients are transformed in the body into compounds that may sometimes be immediately metabolized by the cells, or they may be converted into substances that can be stored for subsequent transformation into such compounds.

Hormones play an important role in regulating the interconversions of nutrients to metabolic substrates and their stored forms. The endocrine secretions may help to regulate the levels of nutrients by contributing to the control of their absorption from the gut, their levels in the blood, the nature and rate of their storage, their release from tissues, and their assembly into the structural elements of the body.

Animals lead diverse lives in a plethora of environmental conditions. The definitive metabolic processes are basically similar in all animals and lead to the utilization of ATP, for the supply of energy, and the building of cells. Nevertheless the physiological processes leading to these accomplishments may differ considerably. Such processes are dictated by numerous circumstances and events.

The chemical nature of the foodstuffs that animals obtain from their environments may differ greatly. In their feeding habits, animals may be carnivorous, herbivorous, or omnivorous. Even within these categories considerable differences exist in the types of food animals eat. Some animals may feed principally on invertebrates such as insects, molluscs, and worms that live in terrestrial, freshwater, or marine environments. Other animals feed on vertebrates. Plants from equally diverse situations are also used for food. The possibilities for gastronomic experiments thus appear to be endless, but only a limited number can furnish a particular species with its needs.

Animals have different patterns of feeding. Some eat almost continually, such as cattle and sheep that nibble plants hour after hour. Large predatory carnivores, like lions, snakes, and crocodiles, may only feed intermittently with days or even weeks separating their mealtimes. Circumstances, such as an unexpected drought, may inadvertently result in enforced fasting or even starvation. A dependence on body stores of nutrients for prolonged periods of time may be a fairly predictable part of an animal's life cycle, such as dictated by hibernation during winter, estivation during hot, dry summers, and migrations to more equitable regions for food and in order to breed.

The nutritive requirements of animals may differ considerably. The normal rate of metabolism of different species can differ by more than 100-fold. Warm-blooded homeothermic animals usually have a higher metabolic rate than cold-blooded poikilotherms. Even among homeotherms the basal metabolic rate differs considerably; for instance, it is about 35 times greater in the shrew than in the elephant. Factors such as size, patterns of activity, and the environmental temperatures experienced contribute to the differences in metabolic requirements of animals. Young, growing animals have special nutrient requirements, whereas breeding and care of the young alter the needs of adults.

Dominating all these differences, and dictating many of them, is the phylogeny of the species. The genetic constitution of a species determines the pattern of its nutrition and the mechanisms involved in the regulation of it. These physiological processes are presumably the result of a prolonged evolution and adaptation to environmental conditions. This is related in the diversity of the endocrine mechanisms that control the metabolism of animals.

Endocrines and digestion

Apart from catching or collecting and then eating food, the first physiological event in the nutritional process is digestion. Food is usually broken down into simpler chemical compounds prior to its absorption from the gut. This process involves the actions of acids, alkalies, and enzymes secreted by glands in the wall of the stomach and intestines as well as the exocrine, or acinar, cells of the pancreas. The orderly flow of these juices is controlled by hormones as well as by nerves (see Fig. 5.1). Indeed, the discovery of the role of *secretin* in stimulating the secretion of the exocrine pancreas into the intestine was the

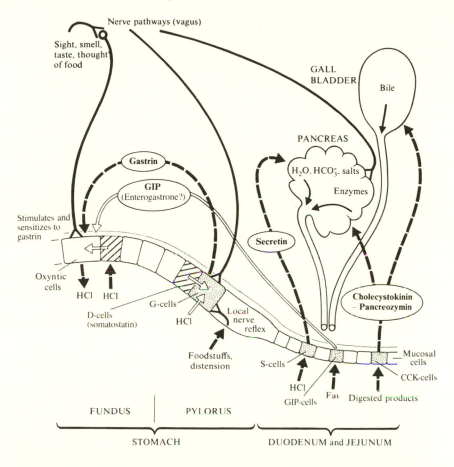

Fig. 5.1. The role of hormones in controlling gastric acid secretion, pancreatic secretion of salts and enzymes, and the contraction of the gall bladder. *Gastrin*, from the pylorus, initiates secretion of hydrochloric acid by the oxyntic cells in the fundus. The duodenal–jejunal hormones, *secretin* and *cholecystokinin–pancreozymin*, initiate the secretion of, respectively, pancreatic juice and enzymes. Gastric-inhibitory peptide (GIP), from the duodenum–jejunum, inhibits gastric acid secretion. The double open arrows indicate an inhibitory effect; the dashed ones a stimulation.

first unequivocal demonstration of the role of a hormone in the body, and it was in connection with this discovery that the term "hormone" was initially used. Bayliss and Starling performed the crucial experiment in 1902. A loop of the jejunum of an anesthetized dog was tied at both ends and was denervated. When acid (0.4% HCl) was introduced into this sac (or the duodenum), secretion of pancreatic juice was stimulated. An extract of the jejunal mucosa was made by rubbing it with sand in the presence of HCl solution.

When (after filtering) this was injected into the jugular vein of the dog, pancreatic secretion was also stimulated. As Starling remarked, "then it must be a chemical reflex." This experiment was performed at University College, London, on the afternoon of January 16, 1902, and has been summarized thus (Sir Charles Martin, see Gregory, 1962): "it was a great afternoon."

Secretin that is released as a result of the action of acid on the duodenal and jejunal mucosa stimulates the formation of a voluminous pancreatic juice, rich in bicarbonate and salt but poor in enzymes. The secretion of proteolytic, amylolytic, and lipolytic enzymes (but not water and salt) can be stimulated by the vagus nerve as a result of feeding. Another hormone, however, also assists this process. This is *cholecystokinin–pancreozymin* (it is now usually called cholecystokinin, or CCK), the role of which was not established until 40 years after that of secretin (Harper and Raper, 1943). This endocrine secretion is also formed in the upper parts of the intestine from which it is released in response to the presence of the digestive products.

Gastric secretion is also controlled by nerves and hormones. Stimulation by the vagus initiates the formation of acid by the oxyntic cells and enzymes from the chief cells. The action of the vagus on acid secretion is both direct and mediated by the release of *gastrin*, a hormone formed in the pyloric region of the stomach, which stimulates the oxyntic cells to secrete acid. The vagus also sensitizes the oxyntic cells to the action of gastrin. The complete reflex arc, which is initiated by feeding, initially involves nerve stimulation along cholinergic nerves to the oxyntic cells and the pyloric G-cells that release gastrin. This hormone then closes the reflex arc and acts on the oxyntic cells. Secretion of gastrin also results from the initiation of a local nerve reflex due to the presence of food in the stomach. Gastric acid secretion can be inhibited when fat or oils pass into the upper parts of the intestine. This stimulus apparently initiates the release of another hormone, *enterogastrone*, which inhibits secretion from the oxyntic cells. A peptide that has this effect has been isolated from extracts of the duodenum; it is called *gastric inhibitory peptide*, or *GIP* (see Chapter 3). The GIP may be identical to enterogastrone but there are other candidates, including secretin.

The presence of fats in the intestine initiates the release of bile from the gall bladder, and this also involves an endocrine reflex due to the release of cholecystokinin that contracts the gall bladder and relaxes the sphincter of Oddi (Ivy and Oldberg, 1928). This hormone has several roles in the body and, apart from its actions on the gall bladder and the endocrine and exocrine pancreas, can elicit a sensation of satiety (see later).

Secretin, gastrin, and cholecystokinin can also initiate the release of insulin or glucagon. The physiological significance of this stimulation is uncertain, but it has been suggested that such effects may contribute to the homeostasis during feeding. Such signals could mobilize insulin and glucagon in anticipation of the absorption of digested nutrients.

The presence of these humoral reflex arcs influencing digestion has been shown in several mammals, but direct evidence as to their presence in other vertebrates is lacking. From the sporadic evidence available, it seems likely that they exist. Indeed, further experiments by Bayliss and Starling in 1903 indicated that this is so with respect to the effect of secretin on the pancreas. They performed experiments on a variety of mammals, including monkey, dog, cat, rabbit, and a bird (a goose "in the process of fattening for Christmas") and confirmed the wider phyletic distribution of this humoral reflex. Secretin-like activity was also shown to be present in the duodenum of man, ox, sheep, pig, squirrel, pigeon, domestic fowl, tortoise frog, salmon, dogfish, and skate. This interesting paper by Bayliss and Starling (1903) is entitled *On the uniformity of the pancreatic mechanism in vertebrata* and must be one of the earliest contributions to comparative endocrinology.

Somatostatin is formed by the D-cells, which are present in various parts of the gastrointestinal tract including the stomach. This peptide (see Chapter 3), which is also formed in the hypothalamus and islets of Langerhans, has ubiquitous effects in the body. It can, for instance, block the release of growth hormone, insulin, and glucagon. Somatostatin can also inhibit the release of acid from the oxyntic cells and that of gastrin from the G-cells in the stomach. It appears to be released as a result of a low pH in the gastric antrum. The D-cells in the stomach lie in proximity to the G-cells and oxyntic cells so that the secreted peptide may be exerting a local (paracrine) inhibitory effect on secretion of gastrin and acid. Somatostatin has also been identified in the circulation so that it is possible that it also has a more classical type of endocrine role.

The transformation of metabolic substrates: the role of hormones

The diversity of intermediary metabolism in vertebrates

Nutrients that are utilized for the production of energy can be classified into three major groups: carbohydrates, fats, and proteins. These compounds are also incorporated into the cell structure and so are essential for growth and reproduction.

The nature of the nutrients upon which the animal's metabolism is based depends, in the first instance, on its diet. In carnivores, this consists mainly of protein and fat. The carbohydrates obtained by herbivores may consist of materials such as starches and sugars that can be broken down into simpler sugars by the digestive enzymes. The major organic constituent of most plants is cellulose, which is fermented by microorganisms present in various compartments of the gut (the cecum, the colon, or the rumen) and which produces short-chain fatty acids. These are mainly propionate, butyrate, and acetate. Such fermentation by symbiotic microorganisms is widespread in herbivorous mammals, especially in the sacculated rumen of ruminants (cattle and sheep) as well as the colon of horses and the cecum of lagomorphs, like rabbits, and the sacculated stomach of some marsupials. In nonmammals the situation is less clear but microorganisms undoubtedly aid digestion in these animals also.

Different species of animals thus show differing dietary dependencies on proteins, fats, and carbohydrates. Proteins can be broken down to their constituent amino acids, fats to fatty acids and glycerol, and carbohydrates to simple sugars, like glucose, or, with the aid of microorganisms, can give rise to fatty acids. The resulting basic subunits can be utilized directly for the production of energy or they may undergo transformations (see Fig. 5.2) into forms that can be stored and provide a readily accessible reserve. Apart from their reassembly into more complex units, interconversions of one such type of chemical compound into another may also take place in the body. Glucose can thus be readily converted in liver, adipose tissue, and the mammary glands into triglycerides (fats). Some amino acids are transformed by the process of gluconeogenesis, in the liver, to glucose whereas others are changed into fatty acids. The transformation of fatty acids to sugars is not as common, though in ruminants propionate, which is formed in large amounts, is converted to glucose. Glycerol can also be transformed into glucose.

The reserves of protein that are maintained in the body are small and in any case it is not a very suitable substrate for the storage of energy. During starvation, protein may nevertheless make an important contribution to an animal's energy requirements. Substantial amounts of glucose are stored as glycogen in the liver and muscles, but these reserves are inadequate to maintain an animal for prolonged periods of time. Triglycerides provide the most economical and convenient storage form for energy. One gram of fat furnishes 9500 calories whereas the same amount of carbohydrate and protein, respectively, supplies only 4200 and 4300 calories.

Stored fat may be dispersed widely among the tissues in the body but it usually predominates at certain sites. Adipose tissues thus exist at subcutaneous, mesenteric, perirenal, and periepididymal sites in mammals. The large fat bodies near the gonads in the abdominal cavity of frogs and toads are familiar to student dissectors. The tail of urodeles and lacertilians is also a common site for fat storage. Large quantities of fats may also be stored in the liver of poikilotherms and this is especially important in chondrichthyean fishes, though it is also seen in other vertebrates. In many fishes, fats are stored in close proximity to the muscle fibers that directly utilize their fatty acids. A characteristic type of adipose tissue called *brown fat* is present in embryonic mammals and also some adults, especially rats and hibernating species like hedgehogs, moles, bats, and squirrels. Brown fat is capable of undergoing very rapid metabolism, with considerable production of heat, particularly during arousal of hibernating animals. It has also been suggested

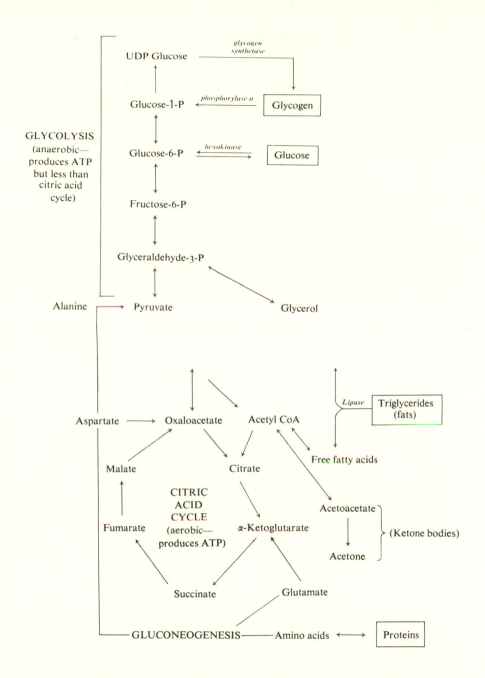

Fig. 5.2. A diagrammatic summary of the processes, enzymes, and major compounds involved in the metabolic transformations of glucose, triglycerides, and proteins. Hormones may either increase or decrease certain of these reactions, as indicated in the text and subsequent figures.

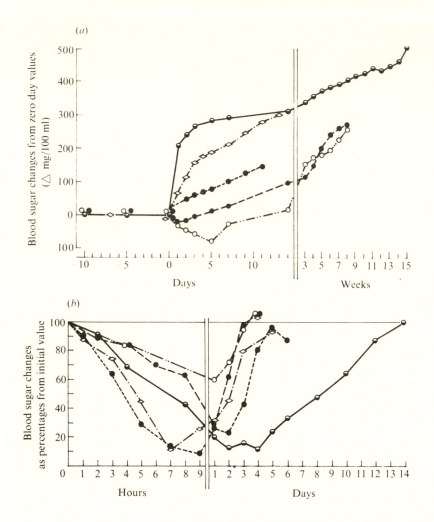

Fig. 5.3. Effects of various experimental treatments on blood glucose concentrations in amphibians and reptiles. (*a*) Changes in the blood glucose concentrations following pancreatectomy in various reptiles and amphibians. It can be seen that the elevation in the glucose occurred relatively promptly in the alligators and toads but was considerably delayed in the lizards and snakes. (*b*) Effects of injected insulin (1 unit/100 g body weight). The amphibians responded more rapidly than the reptiles, but the response was usually more prolonged in the latter. (*c*) Effects of injected glucagon (10 μg/100 g body weight). The reptiles responded more slowly than the amphibians, but the response in the former lasted for a longer time. ●—●, Alligators; ◇—◇, toads; ●———●, snakes; ○—..—○, lizards, ●---●, frogs. (From Penhos and Ramey, 1973.)

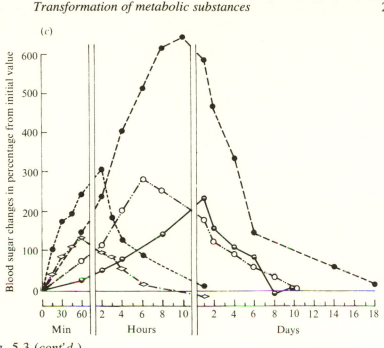

Fig. 5.3 (*cont'd.*)

that it may be concerned in the metabolic disposal of excess calorigenic nutrients, which may otherwise be converted to unnecessary stored fat.

The transformation, storage, and utilization of fats, proteins, and carbohydrates are regulated to a considerable extent by hormones. The relative differences in the availability and importance of such substrates in different species, not surprisingly, may be reflected in the animal's particular response to hormones. Quantitative, or even qualitative, differences may be observed. Some such variations in the responsiveness of different species of reptiles and amphibians to injections of insulin and glucagon, as well as pancreatectomy, are shown in Fig. 5.3. It can be seen that changes in the blood glucose levels following these treatments show considerable interspecific variability in the speed of onset and the magnitude and the duration of the responses.

Hormones that influence intermediary metabolism

When considering hormones that influence the transformation, deposition, mobilization, and utilization of fats, carbohydrates, and proteins in the body, we should be careful to distinguish between those that function physiologically and those actions that probably do not normally occur in the animals (pharmacological effects). For instance, lipolytic activity can be exhibited by at least seven pituitary hormones and also several secretions from other endocrine glands. It is possible, however, that a pharmacological action of a

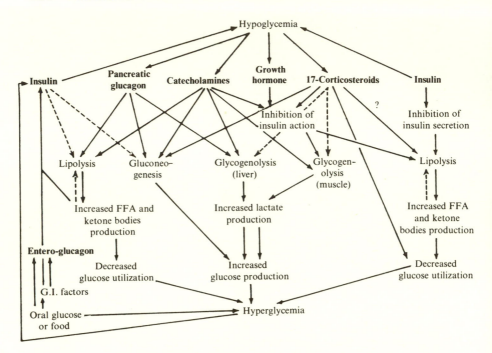

Fig. 5.4. A summary of the role of hormones in stimulating, or inhibiting, metabolic processes and transformations that control the glucose concentrations in the blood. Solid line, stimulation or increase; dashed line, inhibition or decrease. (Modified from Foà, 1972.)

hormone in one species, or tissue preparation, may reflect a physiological role in some phyletically distant species.

Insulin plays a central role in intermediary metabolism and this may be associated, especially during fasting, with the actions of the *corticosteroids*. *Glucagon* and *epinephrine* also contribute to the control system. *Growth hormone* and *thyroid hormones* modulate the processes involved in many chemical pathways. *Prolactin* (Bern and Nicoll, 1969) stimulates the formation of milk proteins in the mammary glands and in the pigeon crop-sac. Prolactin also, when injected, may have a hyperglycemic action and promote the deposition of fat in some teleost fish, amphibians, reptiles, and birds. Several hormones usually interact with each other in such metabolic processes; none can function normally in complete isolation from all the others (Fig. 5.4). As so well stated by Tepperman and Tepperman (1970), "It is virtually impossible to separate out one set of signals which control lipogenesis, another which controls gluconeogenesis and a third which controls ketogenesis. All these processes share so much of the metabolic machinery . . . ." With this reservation in mind we can consider the effects of various hormones on the processes of intermediary metabolism.

Insulin

This hormone decreases plasma glucose concentrations and, in mammals at least, also reduces free fatty acid levels. In the absence of insulin, muscle wasting occurs due to the excessive mobilization of proteins. These effects are the result of several actions, at various sites, principally the liver, muscle, and adipose tissue. *The processes mediated by insulin* are basically (see Fritz, 1972) as follows:

i. Increase in the rate of uptake of glucose across the cell membranes of skeletal and cardiac muscle, adipose tissue, and mammary gland. Organs like liver, kidney, or brain do not respond in this way to insulin.

ii. Facilitation of glycogen formation from glucose. This is partly the result of the more rapid accumulation of glucose but is also due to an increased activity of glycogen synthetase in liver, muscle, and possibly even adipose tissue.

iii. An inhibition of the mobilization of glycogen.

iv. Inhibition of gluconeogenesis from amino acids.

v. Increase in the accumulation of fatty acids across the cell membranes of adipose cells.

vi. An inhibition of lipolysis of triglycerides to form fatty acids.

vii. Increased lipogenesis, from glucose, in adipose tissue.

viii. Increased uptake of amino acids by muscle and liver cells.

ix. An inhibition, by insulin, of the mobilization of amino acids from protein [related to iv].

The effects of insulin can thus be broadly divided into its actions on the accumulation of nutrients across cell membranes and its facilitation, or inhibition, of metabolic synthesis in cells.

Variations in the responsiveness to insulin among vertebrates

The actions of insulin on intermediary metabolism have been principally studied (*in vitro* and *in vivo*) in mammals, usually in laboratory rats. The hypoglycemic action of insulin nevertheless appears to be widespread in vertebrates (see Bentley and Follett, 1965; Falkmer and Patent, 1972; DeRoos and DeRoos, 1979); it occurs among animals ranging from cyclostomes to mammals. Differences in sensitivity nevertheless are apparent, reptiles and especially birds being relatively insensitive to injected mammalian insulin. The pancreas of birds contains little insulin and this is only released "sluggishly" in response to hyperglycemia (Hazelwood, 1973) so that this hormone may not play such an important role in these as in other groups of vertebrates. Some urodeles fail to respond to injected insulin; yet others do respond (McMillian and Wilkinson, 1972). Among mammals, ruminants are less sensitive than carnivores. Such variations appear to depend on the source of the exogenous insulin, the metabolic rate and condition of the animals, and the presence of compensatory mechanisms, such as release of glucagon and epinephrine, as well as the normal diet of the animals. The hypoglycemic effect of insulin appears to be a universal one among vertebrates. Differences have been observed, however, in its ability to promote deposition of glycogen in liver or muscle or both. For instance, in lamprey liver, but not muscle, glycogen is increased by insulin injections. Muscle glycogen is increased in the skate (Chondrichthyes) and scorpion fish (Teleostei) (Leibson and Plisetskaya, 1968) where the effects on the liver are less pronounced. Insulin has unpredictable effects on plasma fatty acids; it decreases them in mammals and eels (Lewander et al., 1976), but may have no effect or even increase their concentration in birds.

The sensitivity of different species to insulin may be related to their dietary habits. It has been noted that herbivorous mammals withstand an absence of insulin far more readily than carnivorous ones (Gorbman and Bern, 1962; Fritz, 1972). Carnivores only eat periodically so that they may have a sudden large intake of nutrients that must be stored for utilization during the fast between meals. Herbivores, on the other hand, graze for long periods of the day and are continually absorbing the products, mainly fatty acids, from the large stores of digesting food in their guts. Coordination of the storage and mobilization of nutrients in the body is thus expected to be more important for periodic eaters like carnivores, than herbivores (which have aptly been called "nibblers").

Catecholamines

Epinephrine can increase the concentration of both glucose and fatty acids in the plasma. These effects are mediated in the liver and muscles, as a result of activation of a phosphorylase enzyme, and in adipose tissue by activation of a lipase. Both effects are due to the formation of cyclic AMP. The hyperglycemic effect of epinephrine is seen in species from all the main groups of vertebrates though the site of its action may differ. For instance, in lampreys, epinephrine mobilizes glycogen in liver but not muscle, while in the Chondrichthyes, glycogen from both sites is depleted, as in mammals (Bentley and Follett, 1965; DeRoos and DeRoos, 1972). The hyperlipidemic response to epinephrine has not been studied on such a broad phyletic scale, and even mammalian adipose tissue from all species studied is not uniformly responsive. Thus, mobilization of fatty acids in response to epinephrine has been shown in the rat, dog, goose, and owl but not the duck, chicken, rabbit, or pig (Prigge and Grande, 1971; Table 5.1). Epinephrine increases plasma fatty acids in the eel *Anguilla anguilla* (Larsson, 1973), but this is not seen in all fish (Minick and Chavin, 1973) and may reflect an indirect effect by inhibiting release of insulin.

Table 5.1. *Species sensitivity to hormonal stimulation of free fatty acid release from adipose tissue*

	Rat	Mouse	Rabbit	Hamster	Guinea pig	Cat	Dog	Pig	Chicken	Man
Epinephrine or norepinephrine	+ +	+ +	0 +	+ +	0 +	+	+ +	0	0	+
Corticotropin	+ +	+ +	+ +	+ +	+ +	+	0	0	+	+ +
Cortisol	+								0	+ +
Glucagon	+			+	+				+	
Thyrotropin	+ or + +		0	0	+ +	0	+	+		
MSH (α or β)	0		+ + +	0	+		+	0		
Vasopressin	0		+ +	0	+ +		0	0		

Note: + +, Strong response; +, moderate or weak response; 0, no response.
Source: Shafrir and Wertheimer, 1965.

There is some doubt, at least in mammals, as to the efficacy of the circulating concentrations of epinephrine in stimulating glycogenolysis. The concentration usually observed in the plasma appears to be too low to act physiologically on the liver. Adrenergic effects on mobilization of glucose in the liver may thus be mediated by stimulation of the sympathetic nerves. Whether this applies to all tissues, and other species, is unknown.

Glucagon

This pancreatic hormone exerts a hyperglycemic action in mammals as a result of the mobilization of liver glycogen following activation of phosphorylase. It does not have this action on skeletal muscle. Gluconeogenesis is increased. Plasma fatty acid levels are elevated by glucagon in birds and mammals but at somewhat higher concentrations than affect glucose. Lipolysis is promoted in adipose tissue from rat, rabbit, goose, duck, and fowl, but not in the dog (Prigge and Grande, 1971). Glucagon has widespread hyperglycemic effects in mammals, birds, reptiles and amphibians but is relatively ineffective on cyclostomes, chondrichthyeans, and some teleosts though it is very effective in the eel (Patent, 1970; Larsson and Lewander, 1972; Farrar and Frye, 1977). In contrast to mammals and birds, eel plasma fatty acids are unaffected by glucagon (Larsson and Lewander, 1972).

Other peptide hormones

Other hormones can also influence glucose and fatty acid metabolism. The neurohypophysial peptides (when injected), vasopressin, oxytocin, and vasotocin, exhibit hyperglycemic effects that can be demonstrated in cyclostome fishes, amphibians, birds, and mammals (see Bentley, 1966) and have also been more recently described in reptiles (LaPointe and Jacobson, 1974).

Vasotocin also increases plasma fatty acid levels in birds (John and George, 1973). The pituitary hormones MSH, ACTH, TSH, and growth hormone may all, depending on the species and the adipose tissue preparation used, facilitate mobilization of fatty acids from mammalian adipose tissue (Table 5.2; Mirsky, 1965). These actions are probably not normal physiological ones, and it is unknown whether they are widespread in nonmammals.

Adrenocorticosteroids

The steroid hormones have profound effects on intermediary metabolism (especially in fasting animals), reproductive processes, growth, and lactation. The corticosteroids increase blood glucose concentrations and promote gluconeogenesis and the deposition of glycogen in the liver. They can also promote lipolysis. In excess, corticosteroids promote muscle wasting and a negative

Table 5.2. *Hormonal effects on fat mobilization*

	Fat mobilization from adipose tissue	
Hormone	*In vitro*	*In vivo* (FFA) plasma
Glucagon	↑	↑
Epinephrine	↑	↑
Norepinephrine	↑	↑
ACTH	↑	↑
TSH	↑	
Growth hormone	↑	↑
Insulin	↓	↓
Cyclic 3′,5′,-AMP	↑	

Note: ACTH, adrenocorticotropic hormone; TSH, thyroid-stimulating hormone; ↑, increase; ↓, decrease, in rate of metabolic process indicated in column by hormone designated in row; (FFA) plasma, plasma concentration of free fatty acids.
Source: Fritz and Lee, 1972.

nitrogen balance, whereas in young animals they inhibit growth. These two effects reflect their action on protein catabolism. *The actions of corticosteroids on intermediary metabolism* can be summarized thus:

i. Increase in gluconeogenesis in the liver following a mobilization of proteins from skeletal muscle and the deamination of the amino acids that are released. This action is most important, especially during fasting.

ii. Deposition of glycogen in the liver because of an increase of the glycogen synthetase reaction.

iii. Inhibition of glycogenolysis.

iv. Reduction in peripheral oxidation and utilization of glucose.

v. Inhibition of the conversion of amino acids to proteins, and fatty acids to triglycerides.

vi. The lipolytic effects of epinephrine, glucagon, and growth hormone are enhanced by corticosteroids, which appear to have a permissive role in these responses.

Such effects appear to be widespread in the vertebrates although the information available in nonmammals is sporadic (see Chester Jones et al., 1972). Hyperglycemia in response to the injection of corticosteroids has been shown in vertebrates that range phyletically from cyclostomes to mammals. This response is associated with gluconeogenesis and elevation of tissue glycogen levels. The facilitation of gluconeogenesis is associated with increased levels of liver transaminase enzymes of teleost fishes, amphibians, birds, and mammals (though it had once been thought that this increase only occurred in the latter two groups of vertebrates) (see Janssens, 1967; Freeman and Idler,

1973). An inhibition of growth or loss in body weight has been shown in the domestic fowl and the amphibian *Xenopus laevis*, as well as two species of teleosts, *Salmo gairdneri* and *Salvelinus fontinalis* (Bellamy and Leonard, 1965; Freeman and Idler, 1973; Janssens, 1967), as well as in mammals. The actions of corticosteroids on growth and metabolism appear to be basically similar in all vertebrates.

Some differences in the general nature of responses to corticosteroids occur among noneutherian mammals. Thus, two species of kangaroos do not respond to cortisol in the usual manner (see McDonald, 1980). Plasma glucose levels are unchanged by corticosteroids in these macropodid marsupials and they also do not promote gluconeogenesis. However, this is not general to all groups of marsupials; for instance, the brush-tailed possum responds in the more conventional way. The carbohydrate metabolism of a monotreme mammal, the echidna, is also little affected by corticosteroids, but free fatty acid levels in the plasma increase markedly (see Sernia and McDonald, 1977). It seems that in these lowly mammals corticosteroids influence fat rather than carbohydrate metabolism.

Steroid sex hormones

Estrogens and androgens have widespread metabolic effects on the growth and differentiation of tissues, especially the reproductive organs. They may, however, also influence other tissues in the body, principally by promoting the formation of proteins. Androgens when administered have anabolic effects, especially on skeletal muscle (myotrophic action), promoting the formation of proteins. The magnitude of this effect depends on the species, age, and hormonal status of the animal. It is greatest in young male animals with deficient circulating androgens; it is also apparent in females but has little effect in older animals. Estrogens increase plasma lipid levels in mammals and also have an anabolic effect, but this is principally confined to the mammary glands and reproductive organs. In oviparous species, estrogens promote the formation of lipoproteins in the liver that are incorporated into the yolk of the egg. Progesterone increases the formation of avidin by the oviduct of the chicken and this is also incorporated into the egg.

Thyroid hormones, growth hormone, and prolactin

These hormones also contribute to metabolic regulation. The consumption of oxygen by homeotherms is depressed in the absence of thyroid hormones and they are also necessary for adequate growth and differentiation. The actions of several other hormones are not as pronounced in the absence of the thyroid secretion. Such reduced responses are seen for the catecholamines and corticosteroids. The thyroid gland seems to modulate the levels and activity of metabolic enzymes in cells; in the instance of epinephrine this action may involve an increase in adenylate cyclase (Krishna, Hynie, and Brodie, 1968) and β-adrenergic receptors (Williams et al., 1977). Growth hormone promotes growth and stimulates the formation of protein in cells. At least some of its actions are mediated by *somatomedins* (see Chapter 3), proteins whose formation it promotes in the liver. Growth hormone facilitates uptake of amino acids by liver and muscle cells but inhibits the action of insulin on glucose uptake and thus may have a diabetogenic effect. It has a lipolytic action on adipose tissue. Growth hormone has been shown to influence growth in mammals, birds, reptiles, amphibians, and teleost fish (see Chapter 3) so that its actions have a wide phyletic distribution. Somatomedins have been identified in many vertebrates. Prolactin influences the intermediary metabolism of the mammary glands (present in mammals) and the crop-sac of pigeons. As we shall see, prolactin may also promote fat deposition in a number of vertebrates.

Conclusions

Hormones can thus be seen to exhibit widespread actions on intermediary metabolism. In some instances, several secretions can exert similar effects though, in the animal, these may not all be physiologically equivalent. In other cases, hormones may exert opposing effects, either by acting on different processes or by a more direct inhibition. Hormones can also (see Chapter 4) directly influence one another's release and so mimic or oppose the actions of other hormones. Intermediary metabolism, although extremely complex and involving several tissues, many chemical reactions, and numerous metabolites, is a well-integrated process. This is largely the result of the actions of hormones at different types of sites (some of which are summarized in Fig. 5.5), both within the same cell and in different kinds of cells, as well as their ability to act in harmony with each other.

The underlying roles and actions of hormones in regulating intermediary metabolism appear to be basically similar in all vertebrates. The relative importance of each action of the hormone may differ, however, depending on the animal's usual life-style and the particular stage of its life cycle. Events like fasting, associated with hibernation, estivation, and migration, and reproduction (including the formation of eggs, pregnancy, lactation, and the growth, maturation, and metamorphosis of the young) are all associated with special metabolic needs and hormonally mediated effects. Some of these will be described in more detail.

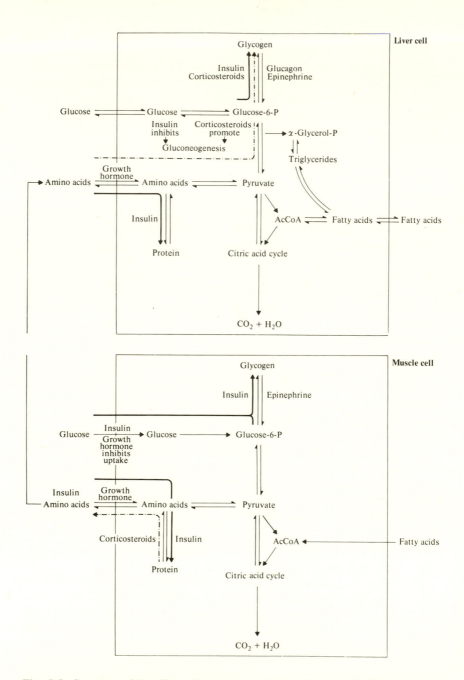

Fig. 5.5. Summary of the effects of hormones on the transfer of metabolites across the cell membranes and their metabolic transformation within liver cells, muscle cells, and fat cells. The principal metabolic transformations that are influenced by insulin are indicated by heavy solid lines and those influenced by corticosteroids, by dashed lines.

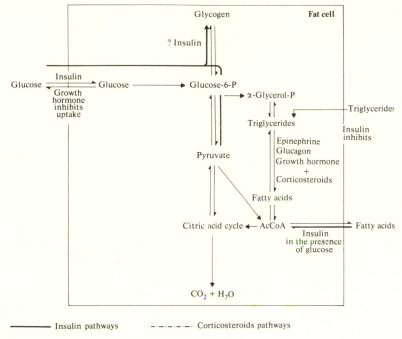

Fig 5.5 (*cont'd.*)

Hormones and calorigenesis

The production of heat in the body

The chemical reactions that are continually taking place in cells are associated with the production of heat. This can be quantitatively expressed in terms of calories; hence the term "calorigenesis." This heat represents:

i. A by-product resulting from the energy requirements of the body. This heat may result from mechanical activity of muscle contractions or from the more ubiquitous metabolic transformations of chemical substrates.

ii. In homeotherms, the production of heat energy per se may be necessary to maintain the body temperature of the animal.

All the tissues of the body contribute to the release of heat in the body and this predominates in skeletal muscle and liver. In some species, the brown fat may also be an important site of calorigenesis, especially in newborn animals and in hibernators during awakening or arousal. In the cold-adapted laboratory rat, the skeletal muscles are thought to contribute about 50% of the heat produced, the liver 25%, and brown fat only 10% (this value is greater for newborns). During arousal from hibernation, brown fat may contribute as much as 40% of the body heat of hamsters and 60% in ground squirrels (Janský, 1973).

In adult homeotherms, added heat requirements for maintaining the body temperature in cold environments are largely met by a skeletal muscle activity called shivering. The rest of the heat is produced from other metabolic activities that are continually proceeding in the body. The amount of heat produced can also be increased, in response to homeothermic need, in a manner that is independent of shivering. This is called nonshivering thermogenesis (or NST) and is greater in young animals and small species than in adult or large animals. A hamster can increase its metabolism four-fold in this manner, whereas in man the change is negligible (see Janský, 1973).

Thyroid hormones

Thyroxine and triiodothyronine have an important role in maintaining the production of heat (and the associated oxygen consumption) in mammals and birds. In the absence of the thyroid gland, the oxygen consumption of mammals declines by as much as 50% whereas an increase in the rate of thyroid secretion can stimulate metabolism by a similar amount. Thyroxine administration can nearly double the basal rate of metabolism in the golden hamster (Janský, 1973). The thyroid is thought to play a permissive, chronic role in cell metabolism rather than contributing to acute changes in metabolism. Thyroid hormones act slowly, thyroxine taking several days to exert its maximal effect. In some mammals, increased thyroid activity is associated with exposure to low temperatures (thyroidectomized rats have a limited survival in the cold) and it decreases with high temperatures. These changes are mediated by the response of the hypothalamic centers to alteration in the temperature of the blood that perfuses them (Collins and Weiner, 1968).

There are some reports that thyroid hormones increase calorigenesis in poikilotherms though it is not always possible to demonstrate this effect (indeed, a decrease has sometimes been observed). The evidence is considered to be equivocal. The injection of thyroxine into the lizard *Eumeces fasciatus* results in a 30% increase in oxygen consumption at 30°C but there is no effect at 20°C (Maher, 1965). A similar increase in oxygen consumption has been seen in the leopard frog *Rana pipiens* (McNabb, 1969) at different temperatures. Thyroid activity also varies with the body temperature of poikilotherms; it is greater at higher temperatures, which is the reverse of what happens in homeotherms. The latter's response, however, presumably reflects a homeostatic role of the thyroid in calorigenesis that would not appear to be present in cold-blooded vertebrates.

Catecholamines

Epinephrine and norepinephrine may play an important role in regulating the production of heat in the body. In homeotherms, a sudden exposure to cold conditions results in an increased production of heat. If mice are transferred from a temperature of 25°C to 0°C they normally adjust their metabolism and nearly all of them survive. When, however, they are pretreated with drugs that inhibit the β-adrenergic effects of catecholamines (for example, propranolol) they all die within 3 hours (Estler and Ammon, 1969) because of an inability to increase heat production. Normally under these conditions, this thermal response is mediated, in mice, by norepinephrine, which is released from the nerve endings of the sympathetic nervous system. Epinephrine, which is secreted by the adrenal medulla, is not usually effective (though it is so in rats), but a suggestion has been made that it also may have such an effect in some circumstances and constitutes a second line of defense.

Such an adaptational effect of catecholamines in calorigenesis can be seen in the adjustment of hypothyroid rats to a cold environment (Sellers, Flattery, and Steiner, 1974). When rats are exposed to a temperature of 4°C their urinary excretion of catecholamines increases, which probably reflects their role in increasing heat production. If such rats are thyroidectomized, they cannot then increase their production of heat, but the injection of small amounts of thyroxine allows this increase to occur so that they survive. In the hypothyroid rats, the urinary excretion of catecholamines increases even more than in the normal rats exposed to cold. It has been suggested that an increased activity of the sympathetic nervous system, including the adrenal medulla, facilitates the survival of these hypothyroid rats. It should be emphasized, however, that some thyroid hormones are essential as in the absence of the excitants (see the following section) rats do not exhibit a calorigenesis in response to catecholamines. The two types of hormones, however, appear to act in conjunction with one another in adapting the rats to cold.

Apart from laboratory rodents such as rats and mice there is little information about the role of the thyroid and catecholamines in cold adaptation of other mammals.

It is notable that epinephrine may, in addition to being calorigenic, help promote heat loss in some mammals. Sweating during exercise in monkeys is dependent on circulating epinephrine and is considerably reduced following denervation of the adrenal medulla (Robertshaw, Taylor, and Mazzia, 1973). This poses an interesting conflict of interest in the physiological role of epinephrine, its potential calorigenic effect presumably being minimal in circumstances where it also contributes to the dissipation of heat. The role of catecholamines in temperature regulation is even more diverse when one recalls that adrenergic nerves, by a peripheral vasoconstrictor action, can also promote heat conservation. This affords another interesting example of the evolution of a hormone's role.

There is little information about the effects of catecholamines on calorigenesis and oxygen consumption in nonmammals (see Harri and Hedenstam, 1972). This effect apparently cannot be demonstrated in birds (pigeon, titmouse, and gull) or in fishes. There seem to be no reports about such an action in reptiles, either. In European frogs, *Rana temporaria* injections of norepinephrine and epinephrine increase oxygen consumption by 25–35%. The response depended somewhat on the particular frogs used; it was seen in cold- and warm-adapted summer frogs but not warm-adapted winter frogs. The physiological role of catecholamines in calorigenesis of nonmammals remains in doubt.

Mechanisms of actions of hormones on calorigenesis

Calorigenesis is stimulated in tissues as a result of their mechanical activity, as in muscles, and also accompanies more general metabolic transformations. It is on the latter ubiquitous processes that the catecholamines and thyroid hormones act to stimulate thermogenesis. Their precise sites of action in the cells are uncertain. Heat may be produced as a result of an increased turnover of ATP or by an accelerated rate of mitochondrial respiration mediated by an increase in protein synthesis that may be influenced by thyroid hormones and/or permeability of cells to sodium. The formation of cyclic AMP, from ATP, as a result of the action of hormones probably results in substantial heat production (Robison, Butcher, and Sutherland, 1971). Hormones involved in this process, or released by it, increase the supply of available glucose and fatty acids that can function as substrates for the production of energy in cells. The effects of norepinephrine on oxygen consumption and the associated subcutaneous temperature of brown fat in rabbit, rat, and guinea pig are shown in Fig. 5.6. The calorigenic effects of catecholamines on brown fat can be seen dramatically *in vitro* where oxygen consumption can be increased four-fold (Table 5.3). This is the result of local mobilization and catabolism of fatty acids following the activation of lipase by cyclic AMP. It can be seen (Table 5.3) that several other hormones, glucagon, ACTH, and TSH mimic the effects of the catecholamines on brown fat, but it is unknown whether they contribute physiologically to the dramatic increases in metabolism that are observed in this tissue.

It has been suggested that a large proportion of the heat produced in the body, including that supporting homeothermy, results from the movements of sodium across the cell membrane (Edelman, 1976; Smith and Edelman, 1979). Acclimation of rats to cold has also been shown to depend on the activity of the cells' sodium "pump" mechanism (Guernsey and Stevens, 1977). Sodium

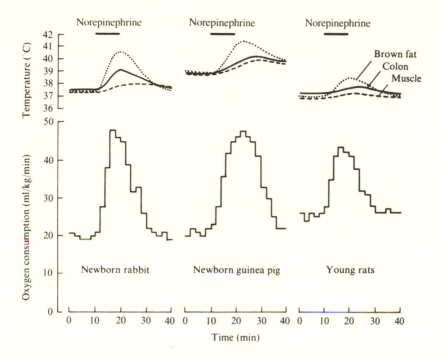

Fig. 5.6. The effects of norepinephrine (intravenous infusion) on the subcutaneous temperatures over *brown fat*, *colon*, and *muscle* and the oxygen consumption of newborn rabbits, newborn guinea pigs, and young rats. The calorigenic effects of norepinephrine are especially prominent in the brown fat while the basal rate of oxygen consumption in each species increased more than two-fold. (From Cockburn, Hull, and Walton, 1968.)

is continually entering cells but it is subsequently extruded, against an electrochemical concentration gradient, as a result of the action of the enzyme Na–K-activated ATPase (the "Na pump"), which is present in the plasma membrane. This process involves the utilization of energy and the breakdown of ATP to ADP. The latter stimulates mitochondrial respiration and ATP is reformed. The magnitude of this Na pumping activity can be influenced in two ways: (i) An increase in the rate of entry of Na across the plasma membrane into the cell will result in an increase. (ii) This process may, however, be limited by the amount of the Na–K-activated ATPase present, but the levels of this enzyme can be increased in some animals. Edelman and his collaborators have shown that thyroid hormone can increase the amount of this enzyme in several tissues, including skeletal muscle, kidney, and gut. However, such an effect was not seen in a poikilotherm, the toad *Bufo marinus* (Rossier, Rossier, and Lo, 1979). Indeed, as we have seen, poikilo-

Table 5.3. *Stimulation of oxygen consumption of slices of rat brown adipose tissue by hormones added* in vitro

Hormone	Hormone concentration (μg/ml)	Oxygen consumption (μmoles/100 mg fresh tissue per 2-hour incubation)	
		Control	Plus hormone
Norepinephrine	0.46	14.9	53.8
Epinephrine	0.50	13.9	64.7
Glucagon	1.0	13.7	35.3
ACTH	10	14.3	28.9
TSH	30	17.7	23.4

Source: Joel, 1965.

therms generally do not show substantial or predictable changes in oxygen consumption when exposed to thyroid hormones. It has thus been proposed that the evolutionary transition from poikilothermy to homeothermy may have involved the novel ability of thyroid hormones to induce the formation of Na–K-activated ATPase, thereby facilitating the ability to produce heat.

Other factors, which may also have involved the thyroid gland, are also probably involved. When the organs of two similar-sized animals, a mouse, *Mus musculus*, and a lizard, *Amphibolurus nuchalis*, were compared it was found that the homeotherm (mouse) contained a much higher proportion of mitochondria than the poikilotherm (lizard) (Else and Hulbert, 1981). The mitochondrial surface area and the cytochrome oxidase activity were much greater in the mouse. As thyroid hormones are known to be able to influence the structure and function (discussed later) of mitochondria, it was suggested that these may be another site for their action that could have contributed to the evolution of homeothermy.

The thermogenic effects of catecholamines may also be related to the activity of the Na pump (Horwitz, 1979). In this instance, however, the effect appears to be related to increases in the rate of entry of Na into the cells due to an increase in the permeability of the cell membrane to this ion. The increased amount of Na in the cells will then stimulate the activity of the Na pump and the hydrolysis of ATP.

Although the Na pump undoubtedly contributes a substantial fraction of the heat normally produced in the body, it is still contentious as to whether this process is adequate to account for enough of the calorigenesis normally associated with homeothermy.

Triiodothyronine, the form of thyroid hormone that acts in the cell, may have dual roles in stimulating calorigenesis. It can interact with sites in the nucleus of the cells and increase the formation of messenger RNA and a *de novo* synthesis of protein (Oppenheimer, 1979). This response could involve an induction of Na–K-activated ATPase. However, thyroid hormones may act at other sites in the cell (Tata, 1975; Sterling, 1979). Triiodothyronine can thus interact with receptors on the inner membrane of mitochondria and its effect is then a rapid one, resulting in an increased formation of ATP that is independent of genetic transcription (Sterling, Brenner, and Sakurada, 1980).

Responses to catecholamines, including those on calorigenesis, decline in hypothyroidism and they are enhanced in hyperthyroidism (see Lutherer, Fregly, and Anton, 1969; Fregly et al., 1979). In thyroidectomized rats catecholamines have little effect on oxygen consumption, but if thyroxine is injected their effects are increased. This effect may be due to an increase in the activity of adenylate cyclase in fat cells (Krishna, Hynie, and Brodie, 1968). Thyroid hormones have also been shown to increase the numbers of β-adrenergic receptors in cardiac muscle (Williams et al., 1977).

Corticosteroids are also necessary for optimal calorigenesis. Adrenalectomized rats (Deavers and Musacchia, 1979) and echidnas (Augee and McDonald, 1973) have considerable difficulty in thermoregulating under cold conditions. The effects of these hormones, however, appear to be relatively indirect. The peripheral blood vessels of adrenalectomized animals do not constrict as readily as those of intact animals so that the conservation of heat is impaired. As described earlier, the mobilization of fatty acids in response to catecholamines is also impaired in adrenalectomized animals.

The mechanisms involved in calorigenesis are thus complex and involve multiple effects of several hormones.

Conclusions

The calorigenic actions of hormones are most important in homeothermic vertebrates where they assist in the process of temperature regulation. The production of heat by the cells in such animals is a basic underlying characteristic, the levels of which may, within limits, be modified. Nutrients, especially proteins, obtained from the diet apparently have a direct effect on these processes, termed the "specific dynamic action of foods." The hormones act in concert with this effect. Other physiological effects mediated by nerves and hormones indirectly influence thermoregulation by facilitating or diminishing losses of heat from the body. This involves evaporation, which occurs from the respiratory tract and the skin where the control of sweat gland activity is important. The blood supply to the peripheral parts of the body, which is controlled by sympathetic nerves that release catecholamines and promote piloerection, influences heat exchanges that occur by conduction. It is unlikely that circulating epinephrine significantly alters peripheral blood flow in such circumstances. Hormones thus do not predominate but do contribute to the regulation of body temperature in mammals and birds. This effect represents an evolution of a physiological role that is not apparent in their phyletic forebears.

The storage of nutrients and their utilization during fasting

The diversity of feeding–fasting patterns in vertebrates

As described earlier, sufficient food may only be available to animals at irregular intervals of time separated by several hours, many months, or even years. In some instances, when climatic conditions are unfavorable, animals may sequester themselves in protected havens where metabolic activity is minimal. Hibernation during the winter months is well known in many small mammals (especially among the Insectivora, Rodentia, and Chiroptera) that seek refuge in burrows and allow their body temperatures to decline to levels similar to the ambient one. Bears also hibernate but maintain a much higher body temperature. Other animals become inactive during hot, dry periods of drought that result in a limited food supply and a shortage of water that may produce severe osmotic problems. In the latter instance, called "estivation," the body temperature is also similar to the ambient one, but as this is usually relatively high, metabolism would be expected to be greater than in those animals that are hibernating. Animals may survive for many months or possibly even years under these conditions. A report from Russia (Siberian Correspondent, 1973) described a live newt found in a piece of ice in Siberia that, according to carbon dating, had been entombed for nearly 100 years. African lungfish can survive for 2–3 years in a state of estivation, though more usually it is for 4–6 months between seasonal rains. Such periods of estivation are also common among amphibians that live in hot, dry deserts.

Although hibernation and estivation are associated with minimal activity and metabolic needs, other situations associated with fasting require a high expenditure of energy. Such an occasion is most dramatically seen during seasonal and breeding migrations. Birds may fly many hundreds, or even thousands, of miles from temperate regions, at the beginning of winter, to warmer tropical climes and then return again in the spring. Fishes, such as lampreys and salmon, when they become mature, migrate from the rivers, where they grew up, into the sea from which they later return in order to breed. Eels make the opposite migration from breeding grounds in the sea to rivers and then later the young return to the sea to breed. Other seagoing creatures, such as turtles and whales, also make long journeys. On many of these occasions, the animals do not feed or do so only infrequently. Reserves of nutrients are amassed in the body in preparation for the migrations during which they are expended. The endocrine glands undoubtedly have a role to play in such storage and the subsequent utilization of nutrients, but the available information is only fragmentary. Further clues can be obtained from the voluminous studies of endocrine function during normal feeding and fasting.

Endocrines and feeding

The release of hormones

Feeding, like so many other physiological processes, involves the secretion of several hormones. This begins with the release of hormones that are associated with digestion: gastrin, secretin, and cholecystokinin. These excitants, as well as enteroglucagon from the intestinal tract, promote the early release of insulin and glucagon from the islets of Langerhans. The absorbed nutrients can also influence the secretion of these hormones; glucose increases insulin release, whereas amino acids initiate the discharge of glucagon. Nerve stimulation, via the vagus, can also stimulate insulin release. Somatostatin, on the other hand, inhibits release of insulin so that the former peptide's effects may decline during feeding. However, as it has a similar action on the release of glucagon its precise role is difficult to predict, especially as all three hormones are present in the same glandular tissue. The endocrine response differs somewhat with the diet, depending on the relative amounts of protein and carbohydrate present (Table 5.4). A high carbohydrate diet is associated with high insulin and low glucagon levels whereas a predominance of protein elevates the concentrations of both of these hormones. In ruminants, the fatty acids that are absorbed from the rumen may also stimulate the release of insulin (Manns, Boda, and Willes, 1967).

Feeding behavior appears to be mainly controlled by two associated regions of the hypothalamus (see Bray, 1974). In the lateral hypothalamus there is an area, electrical stimulation of which results in increased eating; if it is destroyed feeding declines. Adjacent to this "food-seeking area" is the ventromedial nucleus and if these nerve cells are destroyed, food intake increases and body weight rises. The ventromedial nucleus thus sends appropriate signals of satiety to the food-seeking area in the lateral hypothalamus and feeding will then cease. Both of these regions of the hypothalamus respond appropriately to changes in blood glucose concentrations: High levels indicate satiety and low levels promote feeding. Hormones can have marked effects on appetite; it is decreased by glucagon and catecholamines and increased by insulin, corticosteroids, and growth hormone. Sex hormones have various effects depending on the animal's reproductive status. The sites of the hormones' actions on appetite are not clear but as insulin, glucagon, and epinephrine have pronounced effects on blood glucose levels they could be exerting indi-

Table 5.4. *Interrelationship of metabolism with the nutritional state of mammals*

	Hormonal states		Liver				Muscle			Adipose		
Nutritional states	Insulin	Glucagon	Glycolysis	Lipo-genesis	Gluconeo-genesis	Keto-genesis	Glucose uptake	Protein synthesis	Proteo-lysis	Glucose uptake	Lipo-genesis	Lipolysis
Carbohydrate-fed	+++	±	++	++	0	0	++	±	0	++	++	0
Protein-fed	++	+++	0	+	++	0	+	+	0	+	+	0
Carbohydrate- and protein-fed	++++	±	+	+++	0	0	++	++	0	+++	0	0
Fasting (low insulin)	+	++	0	0	++	++	0	0	+	0	0	++
Diabetes (absent insulin)	0	++++	0	0	++++	++++	0	0	++	0	0	++++

Notes: + to 0, either concentration of the hormone or rate of function described.
Source: Cahill, Aoki, and Marliss, 1972.

rect effects on the hypothalamic feeding centers. It is also considered likely, however, that insulin and estrogens have a direct inhibitory effect on the satiety center. Recent evidence suggests that glucoreceptors controlling feeding may also be present in the hindbrain (Ritter, Slusser, and Stone, 1981).

Signals to stop eating may originate in the gastrointestinal tract. They may result from distension and the response of endocrine cells to chemical stimuli. If a fistula is inserted into the stomach of a rat so that ingested food is lost, the animal will continue to eat (Smith, Gibbs, and Young, 1974). If food is placed directly into the duodenum of such animals, however, they then cease feeding. A number of hormones can be released as a result of chemical stimuli from this region of the gut, and it was found that injected cholecystokinin (but not secretin) also terminated feeding in these rats, as well as pigs (Anika, Houpt, and Houpt, 1981). The effects of this injected CCK can be abolished by cutting the vagi but not by lesioning the ventromedial hypothalamus (Smith et al., 1981). When antibodies to CCK are injected into the brain of sheep, feeding increases, suggesting that CCK also has an effect on eating in the central nervous system (Della-Fera et al., 1981). One of the physiological roles of CCK may thus be to signal satiety. It may act both peripherally, on the vagus, and in the brain. Cholecystokinin-like peptides have been identified in the brain so that these two effects may be distinct.

The disposal of absorbed nutrients

The nutrients that are absorbed from the digestive tract can be disposed of in the body in several ways. They may be used immediately as a source of energy. This process may be relatively direct, such as the oxidation of glucose and fatty acids. Amino acids can be transformed to glucose in the liver. The large amounts of propionate absorbed by ruminants can also be converted to glucose. Certain amino acids can also be changed into fatty acids. These products can be more readily transformed into energy. Gluconeogenesis is stimulated by glucagon and corticosteroids. This process is especially important in ruminants that must dispose of large amounts of fatty acids. As we shall see, gluconeogenesis is also fundamental for homeostasis during fasting in other species. The overall processes of oxidation of nutrients are, at least in mammals, chronically influenced by thyroid hormones but more acutely by the levels of the energy substrates themselves, as well as certain other hormones (see section on calorigenesis).

Nutrients that are not required for the immediate production of energy by the animal are stored (usually following their metabolic transformation) and can be subsequently utilized during fasting. Elevated insulin levels play a central role in this process (Fig. 5.7; see also Table 5.4). Insulin facilitates the conversion of glucose to glycogen and triglycerides, of amino acids to pro-

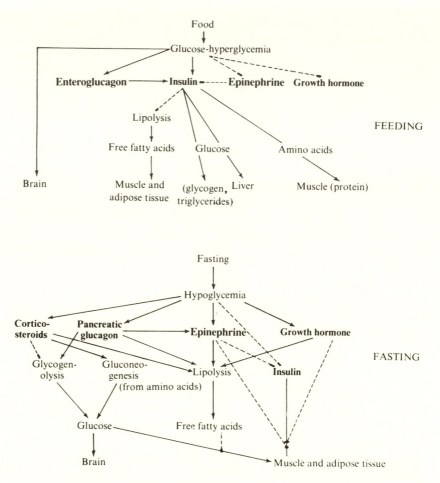

Fig. 5.7. The role of hormones in intermediary metabolism during *feeding* and *fasting*. Solid line, stimulation or increase; dashed line, inhibition or decrease. (Modified from Foà, 1972.)

tein, and of fatty acids to triglycerides. Simultaneously, insulin inhibits further gluconeogenesis and lipolysis. Other hormones may impinge their influence on these processes but usually in a negative manner. For instance, a decreased secretion of glucagon and epinephrine is favorable to lipogenesis. Growth hormone favors incorporation of amino acids into proteins. Prolactin may (see following section), on certain occasions, facilitate the deposition of fat in some species.

The principal energy store utilized by animals during periods of prolonged fasting is fat, though proteins can also undoubtedly be used. Obese animals fare better and survive longer. The immense fat reserves of migrating birds, fish, and whales are well known. The continual food collecting and feeding activities of mammals prior to winter hibernation are almost legendary. The total fat content of the body is normally controlled (by processes that are not understood) within limits that can be considered modest, both physiologically and esthetically.

Preparations for migration

Animals periodically show dramatic departures from their normal limits of fat content that may be associated with their potential requirements during a period of fasting. Humpback whales reproduce in warm equatorial waters, after a migration that takes about 5 months. They do not feed during this time but live off their fat stores, which are accumulated as a result of excessive feeding on krill in the Arctic and Antarctic oceans. The maximum accumulations of fat in several species of migratory birds may be equivalent to about 50% of the total body weight (Table 5.5). The normal fat levels in such birds are 3–10% of their body weight. This fat is contained in various parts of the body but especially in cutaneous and subcutaneous sites and can be deposited very rapidly in 1 to 2 weeks. The timing of this activity appears to be the result of changes in the length of the day and so is under photoperiodic control. Depending on the season the stimulus may be an increase (as in spring) or a decrease (as in autumn) in the hours of daylight (King and Farner, 1965). Such changes in day-length are associated, in birds, with a nocturnal restlessness and activity (*Zugunruhe*), frantic feeding, development of the gonads, and changes in the pituitary gland. It is thus reasonably suspected that the deposition of fat in these circumstances is associated with endocrine signals, but the evidence is difficult to interpret. The problem is probably largely the result of trying to define the role of single hormones when the interactions of several are involved. The effects of photoperiodic stimulation are (possibly apart from those in the pineal gland) conveyed to the endocrine system through the pituitary gland, which is suspected of playing an early role in the preparations for migration.

Prolactin, when injected, can accelerate deposition of fat in birds (white-crowned sparrows, *Zonotrichia leucophrys gambelii*) whether or not they have been subjected to photostimulation (Meier and Farner, 1964). Such injections must be given at certain times of the day and they act optimally at times related to the cyclical release of thyroxine and corticosteroids (Meier, 1970; Meier et al., 1971). Such an effect of prolactin has also been observed in teleost fishes, amphibians, and reptiles. The temporal relationship of go-nadal development and fat deposition in birds led to the early proposal that these may be related (Rowan gonadal hypothesis) and this (despite earlier

Table 5.5. *Maximum lipid deposition as indicated by fattest individual so far extracted in samples of 20 or more birds of each of seven species of families that undertake long overseas migratory flights in autumn*

Species and family	Sex	Total wet wt (g)	Total extracted lipids (g)	Fat-free wet wt (g)	Nonfat dry wt (g)	Fat index (g fat/g nonfat dry wt)
Ruby-throated hummingbird (Trochilidae)	Female	5.65	2.59	3.06	0.74	3.50
Blackpoll warbler (Parulidae)	Male	24.08	12.29	11.79	3.59	3.42
Red-eyed vireo (Vireonidae)	Male	25.37	11.02	14.35	4.36	2.53
Summer tanager (Thraupidae)	Female	37.82	21.77	16.05	7.08	3.07
Swainson thrush (Turdidae)	Male	53.11	25.94	27.17	9.02	2.88
Bobolink (Icteridae)	Male	50.26	24.67	25.59	9.04	2.73
Yellow-billed cuckoo (Cuculidae)	Female	93.50	46.54	46.96	15.48	3.01

Source: Odum, 1965.

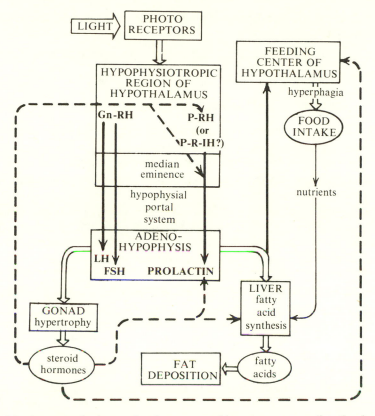

Fig. 5.8. A proposed scheme whereby nerves and hormones may mediate the changes in avian gonadal growth and deposition of fat that occur seasonally in response to photoperiodic stimulation. This scheme is based on observations in white-crowned sparrows (*Zonotrichia l. gambelii*). The dashed lines represent possible actions of the gonadal steroid hormones and the heavy solid lines neurotransmitter roles of the hormones (that of prolactin on the feeding center is uncertain). (Modified from Stetson and Erickson, 1972.)

contradictions) has recently been, at least partly, confirmed (Stetson and Erikson, 1972). When white-crowned sparrows are castrated *prior* to photostimulation the hyperphagia and fat deposition do not occur. A proposed scheme that relates some of these endocrine events is shown in Fig. 5.8. It has also been suggested that reduced levels of glucagon may promote fat deposition in migratory birds (Goodridge, 1964), an observation that is consistent with suggestions as to the causes of obesity in man.

Knowledge of the endocrine processes controlling deposition of fat is, despite a widespread applicability in animal and human nutrition, not well understood. It seems to involve several hormones but these, and the pattern of their effects, may differ depending on the species, the diet, and the physiological occasion.

Endocrines and fasting

Hormones and the utilization of stored nutrients

During fasting, the animal is dependent on its endogenous stores of nutrients for energy. These must be converted to substrates that can be metabolized, principally glucose, fatty acids, and ketone bodies, the levels of which are increased by a low insulin and a high glucagon level in the plasma (see Table 5.4).

During periods of fasting that last only a few hours, or in times of sudden acute need, such as for violent action, mobilization of glycogen stores in the liver and muscles may be sufficient. This transformation is increased by the action of glucagon (on the liver) and, on occasions, also catecholamines (see Fig. 5.7).

During longer periods of time, when no food is available, the liver and muscle stores of glycogen are usually maintained. Fatty acids are mobilized from triglycerides in adipose tissue and are used to provide energy. This may be a direct process involving β-oxidation or result in the formation of ketone bodies. These latter substances, acetoacetate, β-hydroxybutyrate, and acetone, are produced mainly in the liver, but in ruminants they are also formed by the rumen epithelium and the mammary glands. Fatty acid mobilization is favored (see Table 5.4 and Fig. 5.7) by low insulin levels, resulting from hypoglycemia as well as elevated glucagon concentrations. Catecholamines, growth hormone, and corticosteroids can also promote lipolysis. Glucose is also necessary for adequate utilization of ketone bodies in the citric acid cycle whereas certain tissues, especially brain, have a specific requirement for glucose. This substrate is obtained during starvation, as a result of gluconeogenesis, from certain amino acids as well as from propionate and glycerol. Gluconeogenesis is stimulated by glucagon and corticosteroids. As a prerequisite, amino acids must be mobilized and proteolysis is favored by low insulin levels and is promoted by corticosteroids.

Although fat is the main source of energy during prolonged fasting in vertebrates, considerable protein catabolism also occurs and indeed, in the more terminal stages of starvation, may be inevitable. This protein utilization may be more important in poikilotherms. During estivation in amphibians and lungfishes, very high concentrations of urea accumulate in the body fluids (McClanahan, 1967; Smith, 1930), indicating substantial protein catabolism (Janssens, 1964). These animals tolerate high urea concentrations (as much as 600 mM) in their body fluids, levels that would be fatal to a mammal. Aquatic species that form ammonia instead of urea would be more readily able to excrete the toxic by-products of protein catabolism during fasting. In addition, mammals that hibernate and birds that migrate are subjected to circumstances that are often associated with limited supplies of water and may (apart from the associated muscle wasting) find excessive protein catabolism an added disadvantage, because of the problem of extra nitrogen excretion.

In mammals the metabolic rate may decrease during prolonged fasting, and this adjustment is accompanied by a decline in the levels of triiodothyronine in the plasma. There is, in addition, a decrease in the number of thyroid hormone receptors in the cell nucleus (Schussler and Orlando, 1978). Such a dual effect on the activity of thyroid hormones may assure a maximal conservation of energy.

Stored nutrients may be mobilized and oxidized to produce energy at greatly contrasting speeds. A bird during a nonstop migratory flight of up to 2400 km may use up almost its entire fat reserves in 40–60 hours (Odum, 1965). Hibernating and estivating animals, on the other hand, have relatively small energy requirements so that the rate of utilization of stored nutrients is much slower than usual. No definitive information is available about the catabolic role of the endocrines in these circumstances.

Migration of fishes

High levels of corticosteroids are present in the plasma of migrating and spawning salmon, rainbow trout, eels, and sturgeon (Robertson et al., 1961; Murat, Plisetskaya, and Woo, 1981) and it seems likely that they facilitate gluconeogenesis. These steroids are also released in response to exercise in the trout, *Salmo gairdneri* (Hill and Fromm, 1968) and stress in the sockeye salmon (Fagerlund, 1967). Such stimuli could be occurring during migration. As described earlier, injected glucagon fails to increase free fatty acid levels in the blood of eels, whereas catecholamines are ineffective in the goldfish. It is nevertheless possible that they may be effective in migrating fishes. An increase in the activity of the thyroid gland occurs in migrating Atlantic salmon (*Salmo salar*) (Leloup and Fontaine, 1960) and may facilitate fasting metabolism.

Lampreys, *Lampetra fluviatilis* (on histological evidence), have an active thyroid gland at the beginning of their breeding migration from the sea into rivers (Pickering, 1972), although this activity declines as sexual maturity is approached. Sea lampreys (*Petromyzon marinus*) appear to release larger amounts of thyroxine during spawning (Hornsey, 1977). Normally, the digestive tract degenerates during the migration but this can be prevented if the lampreys are hypophysectomized, gonadectomized, or injected with estrogens (Larsen, 1969; 1980). It thus appears that gonadal hormones, either by direct action, or possibly due to a change via a negative-feedback mechanism in the pituitary or hypothalamus can influence tissue catabolism in migrating lampreys. Unlike in salmon, corticosteroids cannot be detected in the plasma of the migrating sea lamprey, *Petromyzon marinus* (Weisbart and Idler, 1970), so that their gluconeogenic role in migrating cyclostomes is doubted.

Insulin concentrations in the plasma have been measured at various times during the spawning migrations of lampreys, sturgeon, and salmon (see Murat, Plisetskaya, and Woo, 1981). High levels of this hormone are maintained during the migrations, when feeding does not occur, and it may contribute to the conservation of the stores of nutrients, especially proteins, by the fish. During spawning, plasma insulin levels decline dramatically.

Bird migration

We can only speculate about the effects of hormones in the metabolism of birds during migratory flights. Glucagon has a potent effect in increasing free fatty acid levels in the plasma of birds so that this hormone, which is present in substantial quantities in the avian islets of Langerhans, may be important on these occasions. The potent effects of injected vasotocin in elevating free fatty acid concentrations in the plasma of pigeons are also interesting. This hormone can be released in response to dehydration and stress, which may be expected to occur in migrating birds, so that (as suggested by John and George, 1973) its lipolytic action could be useful during migratory flight. No changes in plasma prolactin levels were observed during the migration of snow geese from Louisiana to Canada (Campbell, Etches, and Leatherland, 1981). It seems likely that corticosteroids may play an important role in the regulation of metabolism of migrating birds, but this possibility does not appear to have been investigated.

Hibernation in mammals

A number of animals, especially mammals, that experience long, cold winters, may undergo periods of reduced activity during which they do not feed regularly. This condition is called hibernation and it may be prolonged for 7 or 8 months. The physiological and behavioral pattern, however, varies in different species. Small animals that hibernate usually reduce their body temperatures to within a few degrees of the ambient one but they briefly "wake up" at regular intervals of about 2–10 days. Their body temperature then returns to normal. Following such arousal some species, such as chipmunks, will then eat food that they have stored during the previous summer and autumn. They then return to the state of hibernation. Other species, such as woodchucks, do not feed during these breaks in their winter "sleep." Bears hibernate in dens for periods of up to 8 months during which time they live off the large fat stores accumulated as a result of long periods of eating in the autumn. In contrast to the small mammals their body temperatures drop only slightly, about 4° or 5°C, but other bodily functions, such as basal metabolic rate, heart rate, and kidney function, decline dramatically. It would appear that hormones must play a special role in the intermediary metabolism and calorigenesis of such hibernating animals. Information is, however, some-what limited.

It is suspected that both the catecholamine hormones and the corticosteroids may contribute to the events of hibernation but this is largely hypothetical and is based on the known effects that such hormones have on intermediary metabolism and calorigenesis. In a hibernating rodent, the woodchuck *Marmota monax*, the rates of urinary excretion of catecholamines have been shown to be lowest in January, just prior to the onset of hibernation, and they may reach their highest values just before awakening in April (Wenberg and Holland, 1973). Metabolites of steroid hormones are excreted at very low levels early in hibernation but these rise steadily toward the time of arousal in spring. Indeed, it has been found that the injection of a corticosteroid hormone, cortisol, into a monotreme, the echidna *Tachyglossus aculeatus*, can delay or prevent the onset of the hibernation (referred to as "torpor") that these animals may assume when they are exposed to low environmental tempera-tures during fasting (Augee and McDonald, 1973). The rate of metabolism, blood glucose concentration, and body temperature all decline in hibernating echidnas, and it appears the actions of corticosteroids may oppose this.

Some courageous observations have been made on hibernating black bears, *Ursus americanus* (Nelson, 1980). The nitrogen metabolism of these animals is reduced to very low levels during this winter fast, the animal living princi-pally off its substantial fat reserves. During hibernation the levels of pituitary thyroid-stimulating hormone (TSH) and thyroid hormones decline, but these hormones can be stimulated by the injection of TRH. It thus appears that these animals are in a state of hypothyroidism that is due to a decrease in the activity of the hypothalamus. A decline in thyroid function has also been observed preparatory to hibernation in ground squirrels, *Spermophilus tridecemlineatus* (Hulbert and Hudson, 1976). During their long winter sojourn bears not only do not feed but they may also give birth and suckle their cubs, a rather remarkable endocrine feat.

Estivation in African lungfish

Estivation may last 2–3 years in African lungfish. During this phase of its life cycle, *Protopterus annectens* only secretes thyroid hormone at about 1/75 of the rate that it does in its normal free-swimming state (Leloup and Fontaine,

1960). The TSH content of the pituitary is similar in both conditions so that the release of TSH may be inhibited by estivation. Godet (1961) has suggested that an inhibition of the adenohypophysis precedes estivation in these lungfish. It was found that when the pituitary was removed after estivation had commenced there was no effect on the subsequent torpor. An intact pituitary gland was, however, indispensable for the fishes' survival on emergence from estivation. Many amphibians also estivate during periods of drought, but their endocrinology does not appear to have been investigated on these occasions.

Hormones and lactation

Lactation, or the secretion of nutrients by the mammary glands in order to feed the young, is an activity confined to the mammals. Somewhat analogous processes, such as the formation of pigeons' milk by the crop-sac of some birds, can occur in other vertebrates.

Milk is a nutrient solution, the composition of which differs depending on the species and the duration of the lactation. (For a complete account see Cowie, Forsyth, and Hart, 1980.) It contains fats, carbohydrates, and proteins as well as minerals and other essential dietary items. The fat content of milk varies from 0.3% in the rhinoceros to 49% in porpoises; the lactose from 0.3% in whales to 7% in man and the protein from 1.2% in man to 13% in whales. The contained energy is equivalent to 500 kcal/kg milk in the rhinoceros to 2773 kcal/kg in the reindeer (Kleiber, 1961). The nutritional requirements of the mother may thus be considerable.

The mammary glands, like adipose tissue, can make triglycerides and also lactose. The proteins present are largely synthesized by the mammary tissue though some, such as serum albumin and immunoglobulins, are transferred directly from the plasma. In order to deliver these nutrients efficiently to the young, the mammary tissue differentiates into a complex system of alveoli and ducts. These processes, morphological differentiation and the secretion of nutrients, are regulated by several hormones so that the mammary glands afford an example of the multiple actions that hormones may have on a tissue.

Figure 5.9 depicts the arrangement of the mammary gland tissues and a summary of the activities of hormones on them. The milk is secreted into the alveoli and passes down the duct prior to release from the teat or nipple. Experiments on animals with an intact pituitary, but with the ovaries removed, indicate that development of the alveoli is influenced largely by estrogens and progesterone whereas the duct system responds principally to estrogens. The neurohypophysial hormone, oxytocin, is released (see Chapter 4) as a result of a suckling stimulus on the nipple. It contracts the myoepithelial cells that surround the alveoli, and results in milk letdown (the *galactobolic*

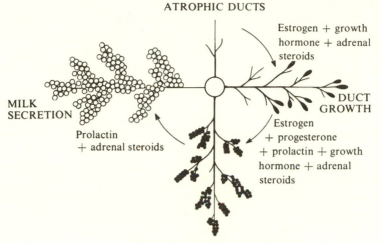

ATROPHIC DUCTS

Estrogen + growth
hormone + adrenal
steroids

MILK
SECRETION

DUCT
GROWTH

Prolactin
+ adrenal steroids

Estrogen
+ progesterone
+ prolactin + growth
hormone + adrenal
steroids

LOBULO-ALVEOLAR GROWTH

Fig. 5.9. The role of hormones in the growth of the mammary gland and the secretion of milk. In this instance the example is the *laboratory rat*, but in other species different combinations of hormones (called the "lactogenic complex") may be required (see text). (Based on Lyons, 1958, from Cowie, 1972.)

effect). Initiation of the secretion of milk (*lactogenesis*) by the alveoli is dependent on prolactin, which is released during parturition and as a result of suckling. Although these hormones are directly essential for the activities mentioned, adequate circulating levels of other hormones, including thyroid hormones, corticosteroids, and growth hormone, are also necessary. The experiments that demonstrate this are complex and involve restoration of the growth and function of the mammary glands in animals from which the pituitary and the ovaries have been removed. No single hormone appears to be effective on one function, but combinations involving ACTH or corticosteroids, prolactin, growth hormone, thyroid hormones, estrogens, and progesterone are necessary. The metabolic effects of the absence of certain of these hormones appear to inhibit or limit the actions of other hormones that act more directly.

Species differences are common. In the hypophysectomized rabbit, lactogenesis can be initiated by a single hormone, prolactin. In hypophysectomized goats, prolactin plus growth hormone plus corticosteroid plus triiodothyronine are required, and in rats and mice corticotropin plus prolactin are needed. Some strains of mice also require growth hormone. When a group of hormones is needed for normal lactation (the more usual situation), this is referred to as a *lactogenic complex*.

The maintenance of lactation is called *galactopoiesis* and the hormonal requirements of this are often difficult to distinguish from lactogenesis. Species differences again exist. Generally, the following occurs: removal of the ovaries once lactation has been established does not influence lactation; indeed, estrogens, and estrogens plus progesterone, are often used to terminate, or dry up, the secretion of milk. Removal of either the thyroid, adrenals, parathyroids, or the endocrine pancreas depresses lactation. The administration of thyroid hormones has a well-established effect in increasing milk production in dairy cows. At one time this treatment was seriously considered as a method to increase the yield of milk. The role of such hormones in maintaining lactation probably reflects their roles in maintaining the metabolic transformations of energy substrates in cells generally. It is not unexpected that such metabolic processes are necessary, directly and indirectly, for the adequate secretion of milk by the mammary tissue.

At the cellular level three important stages in the process of lactation have been distinguished (Baldwin, 1969; Turkington, 1972). These events (usually based on *in vitro* observations in rats and mice) are as follows:

i. Mammary cell proliferation. The increase in the rate of division of the mammary tissue cells has been shown *in vitro* to depend on the presence of insulin and is enhanced by estradiol-17β. This cell division is manifested as an increase in DNA synthesis. Progesterone also assists this process and may direct, or organize, it in such a manner as to allow the orderly development of the duct system.

ii. Differentiation of the mammary cells, which includes the acquisition of the enzymes necessary for the formation of the constituents of the milk. This requires (*in vitro*) the presence of insulin, cortisol, and prolactin (placental lactogen hormone is also effective). Prolactin initiates transcriptional processes and the formation of messenger RNA, which appears to mediate the formation of the enzyme lipoprotein lipase (Zinder et al., 1974). The latter regulates the uptake of triglycerides by the mammary tissue.

iii. The utilization (or expression) of the alveolar cells' capacity to produce milk protein, triglycerides, and lactose. Little is known about this process, but in order to function optimally an adequate and continuing complement of hormones is necessary. Prolactin increases the synthesis of several milk proteins by these cells. It acts on the genes to increase the transcription of three messenger RNAs coding for different casein molecules and also for α-lactalbumin (see Rosen et al., 1980; Rillema, 1980). The precise mechanism of its action is unknown, but it also appears to influence post-transcriptional events in the cell and its effects may be modified by corticosteroids, cyclic nucleotides, and prostaglandins.

The mammalian lactational process is a phylogenetically novel one that has arisen in the later stages of vertebrate evolution. Its complexities, not surprisingly, necessitate coordination by the endocrines. The hormones involved

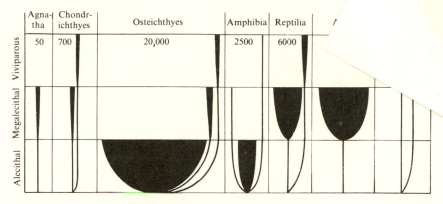

Fig. 5.10. A diagrammatic representation of the relative frequency of the occurrence in the vertebrates of megalecithal eggs (with large amounts of stored yolk) and alecithal eggs. Alecithal eggs are confined to aquatic species; the production of eggs with small amounts of yolk in terrestrial vertebrates necessitates the viviparous habit, which is also included. It should, however, be noted that megalecithal eggs and viviparity also occur among the aquatic fishes and amphibians. Numbers = no. of species. (From Browning, 1969, based on a drawing by H. A. Bern.)

probably existed prior to the mammary gland itself so that an evolution of their role in the body has occurred in this case. Certain processes occur in vertebrates that are analogous to mammalian lactation. The formation of pigeon's milk by the crop-sac of pigeons and doves, with which they feed their young, is an example. This process, which involves the proliferation of the crop-sac epithelium, can be induced by prolactin. In certain teleost fishes, mucous secretion from the skin can be stimulated by prolactin (see Bern and Nicoll, 1968). In one such fish (*Symphysodon*), the newly hatched young have been observed to feed off the surface of the parental fish, suggesting another possible "lactational" role for prolactin.

Storage of nutrients in the egg

The early development of the young is supported by nutrients stored in the egg. These materials may be sufficient for the complete embryonic development of the young (megalecithal eggs) or only support more limited differentiation (alecithal) that leads to the emergence of a free-swimming larva. An even more limited growth may precede viviparous development in the uterus. Alecithal eggs are usually confined to species that develop in a watery environment. Megalecithal eggs are common in terrestrial species as well as certain fishes including the Agnatha and Chondrichthyes (Fig. 5.10).

Estrogens, progesterone, and gonadotropins may play important roles in the growth and maturation of the egg, which are summarized in Fig. 9.21.

The "white" of the megalecithal eggs contains water, salts, and proteins, including avidin and ovalbumin, which, as we have seen, can be formed by glands in the chicken oviduct as a result of stimulation by progesterone. Progestins may also stimulate the secretion of the jelly from the oviductal glands that surrounds alecithal eggs, as in amphibians.

The yolk of eggs contains most of the nutrients required for the predicted development of the young. In the domestic fowl, about 50% of this is composed of fats and proteins. The number of eggs produced successively in one period of time (the clutch) varies considerably in different species but 20–30 are not uncommon in birds and reptiles. Under domestic conditions, a chicken may produce 200–300 eggs in a year. Even species that produce alecithal eggs must contribute considerable amounts of nutrients to the eggs, especially as such species often produce many hundreds of such eggs.

The deposition of the yolk proteins is called *vitellogenesis*. Estrogens can promote the appearance in the blood of elevated levels of lipophosphoproteins called *vitellogenins*. In the African clawed toad, *Xenopus laevis*, these molecules consist of two subunits, each with a molecular weight of about 200,000 (see Wahli et al., 1981). They bind calcium and hence their appearance in the plasma is associated with an elevation of plasma calcium concentrations. Vitellogenins are formed in the liver in a process that is normally induced in the female by estrogens. However, when these hormones are injected into males their livers also form vitellogenins. This effect of estrogens is very specific (it is not seen when other steroid hormones are injected) and is prominent in oviparous species (Urist and Scheide, 1961; Urist, 1963; Urist et al., 1972). The vitellogenic effects of estrogens are apparently absent in mammals, though the situation in the oviparous monotremes has not been investigated. The response has, however, been observed in cyclostomes, chondrichthyeans, teleosts, lungfishes, amphibians, reptiles, and birds. Its presence in lampreys and the Pacific hagfish (Pickering, 1976; Yu et al., 1981) suggests that this role of the estrogens appeared early in vertebrate evolution. (A comparable process of vitellogenesis also occurs in insects.) In *Xenopus* at least four genes appear to be involved in the synthesis of four related vitellogenins, and estrogens can apparently act at all these nuclear loci.

The vitellogenins are carried in the plasma to the ovary where they are taken up by the developing ova. Once they are so accumulated they are converted to smaller lipophosphoproteins called *phosvitin* and *lipovitellin*, which are stored to await the needs of the developing embryo. Gonadotropins, apart from stimulating the secretion of estrogens, have a separate effect on vitellogenesis as they have also been shown to increase the uptake of yolk proteins by the developing ova. This effect of these pituitary hormones has been observed in chondrichthyeans, teleosts, and amphibians (Craik, 1978; Idler and Ng, 1979; Follett and Redshaw, 1974).

Hormones and growth and development

The process of growth may place considerable nutritional demands on an animal. The development of the young is associated with rapid increases in the mass of the body and thus has special nutritional requirements. The development of the young *in utero* also contributes indirectly to the mother's nutritional requirements. In addition, growth processes also occur continually in the bodies of adult animals and involve cyclical changes in the reproductive organs as well as the replacement of cells, due to age and general wear and tear. These growth processes principally involve the formation of proteins (and thus a positive nitrogen balance) and phospholipids. The contribution of minerals, especially calcium and phosphorus, should not be forgotten.

Like lactation, successful growth requires the harmonious regulatory activities of several hormones. Those more directly involved in development and maturation of the young are growth hormone and the thyroid hormones. Androgens and estrogens also contribute to the control of the maturation of the reproductive system and, in adults, help maintain it and regulate the cyclical changes in these tissues. Androgens, as we have seen, may also exert anabolic effects and promote protein formation in muscle. A hormonal background of adequate insulin, which promotes conversion of amino acids to protein, and a not overadequate supply of corticosteroids, which facilitate proteolysis, is also important.

The process of growth consists of (*a*) an increase in the number of cells; (*b*) an enlargement in their size; (*c*) increases in their complexity and differentiation; and (*d*) replacement of cells as a result of loss due to age, wear and tear, injury, or as part of morphogenetic differentiation as in (*c*).

Growth hormone produces a striking increase in cell DNA, indicating an increase in cell number. Cell size is also increased. This effect is manifested in all tissues, including cartilage, and leads to the building of bone. Growth hormone promotes the accumulation of amino acids across the cell membrane and enhances the formation of messenger RNA, which determines the formation of specific proteins that presumably contribute to the structural elements as well as cell growth and reproduction. Growth hormone also promotes the incorporation of sulfate into cartilage. This observation led to the discovery that growth hormone was not necessarily acting directly on the tissues (see Daughaday, 1971). The incorporation of sulfate into cartilage was found to be due to a protein present in the plasma that was initially called "sulfation factor." In addition, this factor also promotes the incorporation of uridine into RNA and thymidine into DNA in the tissue. The formation of this intermediary is promoted in the liver by growth hormone and it has been renamed *somatomedin* (Daughaday et al., 1972). A family of such growth factors has been identified (see Chapter 3). It seems likely that somatomedins also medi-

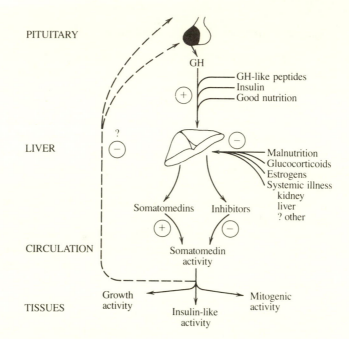

PITUITARY

GH

GH-like peptides
Insulin
Good nutrition

(+)

LIVER

?

(−)

(−)

Malnutrition
Glucocorticoids
Estrogens
Systemic illness
kidney
liver
? other

Somatomedins   Inhibitors

(+)

(−)

CIRCULATION

Somatomedin
activity

TISSUES

Growth
activity

Insulin-like
activity

Mitogenic
activity

Fig. 5.11. The role of growth hormones (GH) and the somatomedins in the process of the growth of tissues. Growth hormone stimulates the formation of somatomedins by the liver, and this process can be facilitated by insulin and good nutrition. Poor nutrition, illness, and some other hormones, such as corticosteroids and estrogens, can reduce this response or possibly result in the formation of substances that inhibit the action of somatomedins. The latter may inhibit further release of growth hormone by a negative feedback type of inhibition (Berelowitz et al., 1981, from L. S. Phillips and R. Vassilopoulou-Sellin, 1980. Reprinted by permission from *The New England Journal of Medicine* **302**, 438.)

ate other effects of growth hormone. Somatomedins persist for a longer time in the plasma than growth hormone, which thus need not be released continually in order to exert its effects. It is also interesting that somatomedins have a stronger effect in tissues from weanling rats than from fetal and old rats. The last two are known to be unresponsive *in vivo* to such effects of growth hormone (Heins, Garland, and Daughaday, 1970). The widespread actions of growth hormone in mammals, birds, reptiles, amphibians, and teleost fish have already been described (Chapter 3).

The role of growth hormone in relationship to that of other hormones under physiological conditions is summarized in Fig. 5.11. The ability of growth hormone to increase the synthesis of somatomedins in the liver is enhanced by good nutrition, insulin, and possibly the presence of related hormones such as prolactin and placental lactogen. Estrogens, excess corticosteroids, malnutrition, and some diseases reduce the formation of somatomedins. It is also suspected that factors exist in the plasma that have inhibitory effects, and a plasma protein has been identified to which somatomedins are bound. However, in the free, unbound form somatomedins increase the growth of cells and the rates of their multiplication (mitogenic effect). Deficiency of somatomedins due to malnutrition, disease, drugs, or an unfavorable genetic constitution can result in reduced growth and even dwarfism in man. African pygmies possess somatomedins but, apparently for genetic reasons, their tissues cannot respond adequately to these growth factors.

As indicated earlier (Chapter 3) prolactin and growth hormone share a number of activities, a property that may reflect their structural similarities. It is therefore not surprising to find that in some species injected prolactin may influence growth. This effect of injected mammalian prolactin has been observed in the pigeon, the lizard, *Lacerta s. sicula*, and tadpoles of the frogs *Rana temporaria* and *R. pipiens* (Bates, Miller, and Garrison, 1962; Licht and Hoyer, 1968; Frye, Brown and Snyder, 1972). Metamorphosis in the larval amphibians is prevented by injected prolactin. It is uncertain if such effects reflect another physiological role of prolactin or are merely due to similarities between its structure and activity to that of the homologous growth hormones.

Apart from their role in regulating the differentiation and maintenance of the sexual apparatus, androgens have important effects on growth. They exert anabolic effects on various tissues promoting the laying down of protein in skeletal muscle and cartilage, and they support the formation of some somatomedin-type factors. Erythropoietin is thus synthesized by the kidney and promotes the maturation of red blood cells; its formation is increased by androgens and this accounts for the higher hemoglobin levels that are present in men as compared with women. The weight of the kidneys is reduced following castration, reflecting a renotrophic action of androgens.

Apart from their calorigenic effects in homeotherms, thyroid hormones play an important role in the body, contributing to an optimal and orderly growth and development. This is most apparent in mammals that lack sufficient thyroid hormone during postnatal development. The thyroid hormones are especially necessary for adequate development of the brain, skeletal system, and reproductive organs. The disease of cretinism that results from insufficient thyroid hormones in man during infancy has been observed in other domestic animals. In newborn rats deprived of thyroid hormone, it has been shown that development of the brain is slowed but the same number of cells (DNA content) appear to develop though they remain smaller (higher concentration of DNA) than in the normal rat. The rate of multiplication of the cells is reduced in thyroid deficiency. The smaller size of the cells is reflected in neural tissue by a reduction in the development of the nerve-endings whereas the incorporation of amino acids into proteins is diminished (Balazs et al.,

1971). When tadpoles are deprived of thyroid hormone, they continue to grow but fail to metamorphose. On the other hand, if tadpoles are fed thyroid hormone they undergo an earlier metamorphosis. This latter transformation is associated with the acquisition of many adult characteristics along with the induction of certain enzymes that mediate reabsorption of the tail. The thyroid is also necessary for the maturation of the gonads in some teleost fish (Sage, 1973).

The morphological effects of thyroid hormones in mammals are also not strictly analogous to their morphogenetic effects in amphibians. Myant (1971) suggests that amphibian metamorphosis may have arisen as a special adaptation in which differentiation is suppressed during a period of rapid growth, and is then stimulated under the influence of thyroxine (see Chapter 9). The morphogenetic effect of thyroid hormones in amphibians has no true analogy in other vertebrates and appears to be a specialization. Thyroid hormones, however, appear to have a basic ability to stimulate protein synthesis in cells of homeotherms and possibly some poikilotherms like tadpoles. Such a basic effect, if exerted at strategic sites in the cell, could support multiple, different actions of the hormones on cells. The thyroid hormones thus may potentially be able to assume various functions and be especially suited to their proposed "permissive" role in cellular processes, at least in homeotherms and amphibian tadpoles.

Conclusions

The processes of nutrition, whether involving the digestion of food materials or their transformation, storage, and utilization, following absorption, or the feeding and growth of the young, are very dependent on the actions of hormones. It is especially notable that several hormones are usually involved in such processes ("multihormonal" effects) probably more often than elsewhere, and these excitants have widespread effects both with respect to the types of tissue that they act upon and the nature of the underlying biochemical events that they influence. Such ubiquitous endocrine effects are well integrated with each other by a series of feedback controls that involve the levels of metabolites and the hormones themselves. The latter indeed often directly influence each others' release. The hormonal pathways that function, or predominate, in the control of nutrition of a particular species seem to depend more on its feeding habits and the stage of its life cycle than on its phylogenetic origins. This probably reflects the common biochemical pathways underlying nutritional processes in all vertebrates. The principal species differences in responsiveness to hormones are quantitative ones, though when novel processes have evolved, such as lactation in mammals, this has been accompanied by adaptations of the roles of hormones to integrate the new systematic features.

6. Hormones and calcium metabolism

Calcium is vital for animal life. In vertebrates, this divalent ion is present at various sites, including the body fluids and structural parts of the cell (especially the mitochondria and sarcoplasmic reticulum), and in most species it is also a major component of the endoskeleton. The outer shell of the eggs of birds and many reptiles also consists principally of calcium. The physiological role of calcium appears to be the result of a rather unique set of physicochemical properties. In aqueous solution, calcium can exist in a soluble form, which is important for its mobility in the body, and yet, equally essential to its role, many of its salts, including phosphates and carbonates, have a low solubility so that a physicochemical equilibrium may exist between its solid and aqueous forms. The quantity of calcium free in solution is thus restricted in the presence of certain anions. In addition, such relatively insoluble salts in certain of their crystalline forms, principally calcium phosphate and calcium carbonate, can contribute to the mechanical support and stability of biological structures. Calcium also has a ready propensity to associate and combine with proteins. Such combinations are seen in the body fluids, where a considerable portion of the calcium is bound to serum proteins, and in cells, where it contributes to their structural stability by helping maintain essential ionic bridges at vital points in protein molecules; thus, when tissues are placed in calcium-free solutions they tend to disintegrate and the cells swell and fall apart.

Calcium plays an essential role in coordinating many events in the body that may reflect those general properties described above. Calcium stabilizes membranes and this effect can be seen in the hyperactivity of nerve fibers placed in solutions with low calcium concentrations. Such instability and the repetitive electrical depolarization of the nerve cell membrane result in tetanic contractions of the muscles they supply. Muscle contraction requires the presence of calcium; when released from the sarcoplasmic reticulum within the cell, calcium couples the initiating electrical depolarization of the cell membrane to those processes that initiate changes in the contractile proteins. In a comparable manner, calcium is necessary for the ultimate initiation of many endocrine events such as the release of hormones (see Chapter 4) and the responses of their definitive effectors.

The availability of calcium to animals varies considerably, depending on the environment and their diet. The physiological need for calcium may also

Table 6.1. *Calcium and phosphate in serum and in the environment*

Environment and representative species	Total Ca (mmole/liter)	Ionic Ca^{2+} (mmole/liter)	Total P (mmole/liter)
Environment			
Pacific ocean	10.0±0.1	10.0	0.00
Brackish water	2–5	—	—
Lake Huron	0.9±0.1	0.9±0.1	0.003
Marine invertebrate			
Nephrops	11.95	—	—
Cyclostomes			
Hagfish, *Eptatretus stoutii*	5.4±0.1	3.0±0.4	1.5±0.2
Lamprey, *Petromyzon marinus* (from fresh water)	2.60±0.1	1.74±0.2	1.3±0.1
Chimaeroid			
Ratfish, *Hydrolagus colliei*	4.8±0.3	—	2.2±0.4
Elasmobranchs			
Marine shark, *Carcharhinus leucas*	4.50±0.9	3.10±0.4	2.0±0.6
Freshwater shark, *C. leucas nicaraguensis*	3.0±0.2	1.7±0.1	1.6±1.7
Teleost			
Marine, *Paralabrax clathratus*	3.2±0.8	2.0±0.9	2.0±0.2
Freshwater, *Megalops atlanticus*	2.5±0.2	1.8±0.2	1.2±0.4
Mammal			
Man, *Homo sapiens*	2.32±0.08	1.15	—

Source: Copp, 1969.

vary a great deal. Young, growing animals, especially those that are forming large amounts of bony tissue, have much greater requirements than adults. The latter, however, periodically need more calcium for reproductive processes, as during pregnancy and lactation in mammals, and in birds and reptiles, the production of large cleidoic eggs that are covered with a shell of calcium carbonate. Even fishes and amphibians that do not have cleidoic eggs require substantial amounts of calcium for the production of their eggs. The availability of calcium in the environment varies a great deal. The concentration in seawater is even higher than that in the body fluids of vertebrates but in fresh water, little calcium is usually present (Table 6.1); thus, vertebrates living in the sea or fresh water may be expected to experience very different problems with respect to the availability of this ion. Terrestrial animals obtain most of their calcium from their diet which may include appreciable amounts obtained from certain drinking waters. In times of extra need for calcium, vertebrates, especially birds during egg laying, may consume inorganic calcium-containing minerals, such as the so-called grit fed to domestic fowl. The ultimate acquisition of calcium, whether from the external bathing solutions, drinking water, or food, takes place principally from the gut. Absorption across the intestine is regulated in relation to the animal's needs. Amphibians and fishes possess integuments that may be permeable to calcium. In the former group such exchanges appear to be quite small, except in tadpoles where a substantial uptake appears to occur across the gills (Baldwin and Bentley, 1980, 1981). The gills of fishes are also permeable to calcium and may be an important site for the accumulation of this mineral from the solutions in which they live.

The calcium that is not absorbed from the intestine, either due to a lack of physiological need for it or its presence in chemical combination that makes absorption impossible, is excreted in the feces. Additional amounts of calcium are also excreted by the kidney in the urine. Urinary losses of calcium should be viewed not only as an excretory mechanism for ridding the body of an excess of this ion but also as part of an unavoidable loss that results from the formation of urine. The calcium that is not bound to plasma proteins is filtered across the renal glomerulus, but most of this ion is subsequently reabsorbed by the renal tubules. This conservation is less effective in the presence of a high plasma calcium concentration so that extra amounts of this ion are excreted in the urine. Conversely, a hypocalcemia results in an increased renal reabsorption of calcium that is accompanied by an increased phosphate excretion. The excretion of calcium and phosphate in the urine is thus subject to physiological control.

The bony skeleton possessed by most vertebrates plays a central role in calcium metabolism. Calcium phosphate salts are an integral part of this structure and make the principal contribution to skeletal rigidity. Not all vertebrates, however, have a bony calcareous skeleton; it is absent in the cyclostome and chondrichthyean fishes that have a more elastic cartilaginous skeleton in which little calcium is deposited. In the bony vertebrates, the skeleton is the predominant quantitative site of calcium in the body though this should not be taken to reflect its relative importance, as its presence is equally vital to the soft tissues. Marine animals have a readily available and inexhaustible supply of calcium in the seawater in which they live. They thus do not need large stores of this mineral such as occurs in bones. However, species that live in fresh water and on dry land have a more limited access to calcium. The evolution of bone can thus, apart from its contribution to animal mechanics, be looked upon as an important potential contribution to the maintenance of body calcium. The calcium in the bones and that in the body fluids and soft tissues exist in equilibrium with each other. Two processes are involved in this: first, there is a relatively static *physicochemical* equilibrium that reflects the solubility (and insolubility) of the calcium salts in the body

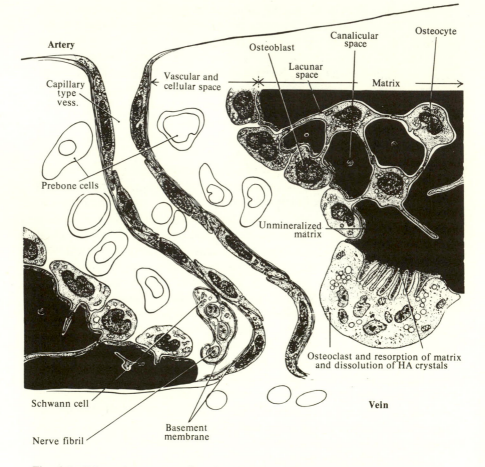

Fig. 6.1. Schematic representation of a physiological unit of bone tissue showing the osteocytes, osteoblasts, and osteoclasts. (From Doty, Robinson, and Schofield, 1976.)

and which only results in small exchanges of calcium; second, the major exchanges of calcium between the bones and body fluids are due to a *dynamic*, physiologically mediated equilibrium that results from the activities of cells in the bone, the *osteoblasts*, *osteocytes*, and *osteoclasts*.

Bone, despite its mineral-like appearance, is a living tissue (Fig. 6.1). Some fishes, however, possess acellular bone that, once formed, is much less labile than the cellular-type of bone characteristic of the tetrapods. Cellular bone has a microscopic honeycomb-like appearance due to the presence of numerous small chambers, or lacunae, each of which contains an osteocyte bone cell. Numerous fine channels (canaliculi) radiate from the lacunae to the surrounding mineralized tissue, and these provide a pathway through which the bone fluids and the dendritic-like extensions of the osteocytes maintain contact with the tissue. The osteocytes act as sentinels controlling local mineral exchanges. Bone also has an extensive network of blood vessels through which the supply of blood can be regulated according to its metabolic needs. Exchanges of bone minerals take place at any of the bone's free surfaces, the canaliculi, at the inner and outer borders of the limb bones (periosteum and endosteum), the channels through which the blood vessels pass and special tunnels, called "cutting cones," which are excavated in the tissue. Minerals are principally mobilized from the cortical regions of the shafts of the long bones, but a labile store also occurs in the medullary regions of the bones of birds where it is an especially important store during the egg-laying cycle.

Bone is formed by the osteoblast bone cells. These cells extrude collagen, which is laid down as a matrix into which calcium, phosphate, and some carbonate are subsequently deposited (called "accretion"). It is possible that the osteoblasts also play some direct role in the process of deposition but this is not essential. Minerals are deposited as two phases, initially as an amorphous form of calcium phosphate, which is subsequently changed to a structurally stronger, crystallized form resembling the mineral apatite. The osteoblast, after surrounding itself with such tissue, then is transformed into an osteocyte. The mineralized bone is not necessarily a permanent structure but can be modified and remodeled, and the calcium and phosphate may be returned to the tissue fluids. The process of "resorption" occurs more readily from the amorphous than the crystalized phase but the latter can also be broken down. Resorption of minerals from bone is associated with an increase in the activity of the osteoclasts, and the osteocytes are also involved. The process of resorption involves the secretion of enzymes by the bone cells and these facilitate the dissolution of the minerals. This process can be regulated.

Physiological control of calcium levels in vertebrates can thus be affected at four major sites: the intestine, by the control of absorption; the kidney, by the regulation of reabsorption from the glomerular filtrate; the gills, across which influx and efflux may be regulated; and the bones, which act as a storage site for calcium phosphate. The very important role of bone in calcium homeostasis cannot be overemphasized, but it should be recalled that this tissue is not present in all vertebrates. The coordination of the exchanges of calcium at these four sites is under the control of hormones. These are parathyroid hormone, from the parathyroids (which are absent in fishes), and calcitonin from the thyroid "C"-cells and the ultimobranchial bodies. To these hormones can be added another (or others) that are derived from vitamin D_3. These are $1\alpha,25$-dihydroxycholecalciferol and, possibly, $24,25$-dihydroxycholecalciferol. Prolactin from the pituitary gland may play a role in the formation of the former. It is of further interest that removal of the corpuscles of Stannius that are present in certain bony fishes (see Chapter 2) results in an elevation of plasma calcium levels, suggesting that these tissues may also secrete a hormone that regulates calcium metabolism in such fish. Estrogens aid mobilization of medullary bone in birds during egg laying.

A historical note about the discovery of parathyroid hormone and calcitonin

The discovery of the roles of the parathyroids, the mammalian thyroid C-cells, and the ultimobranchial bodies in the regulation of calcium metabolism is an interesting endocrine tale. It is not an example of a triumph of "goal-oriented" research but rather of serendipity and the persistent following up of a series of unexpected observations. Species differences in the endocrine tissues have played an important part in establishing the role of these glands.

As we have seen (Chapter 2), two types of endocrine tissue are known to be concerned with regulating calcium metabolism. These tissues are present in close morphological association with the thyroid gland. Early efforts to remove thyroid tissue surgically in man were sometimes seen to result in tetanic muscular contractions such as are associated with low plasma calcium concentrations. Closer examination revealed that in these instances the parathyroid tissues had been removed. Subsequently, other experiments in animals confirmed the importance of the parathyroids in maintaining optimal calcium levels in the blood, and extracts of these glands were shown to exhibit a hypercalcemic action. Such experiments, to demonstrate the role of the parathyroids, are relatively simple in the rat, where there are two distinct bodies of parathyroid tissue present on the surface of the thyroid gland. In another favored experimental animal, the dog, substantial amounts of parathyroid tissue (as well as C-cells) are present, embedded deeply in the thyroid gland, so that this species is not an ideal one for such experiments and has, in the past, contributed to some misinterpretations.

As related by Hirsch and Munson (1969), the rat is an ideal species, and favorite subject, for parathyroidectomy and following this operation provides an excellent preparation for the bioassay of injected parathyroid hormone. Two methods have been used to remove the rat parathyroids. A simple surgical removal of the two glands can be performed or they can be destroyed with an electrocautery. In the 1950s, the latter procedure was more popular. When, however, a comparison, partly retrospective, of blood calcium levels following each type of operation was made, the resulting hypocalcemia was observed to be much greater following electrocautery than following surgical excision. This observation probably did not initially gain the serious attention it deserved. In retrospect it has, especially when it was observed that stimulation of the thyroid gland by the electrocautery at sites removed from the parathyroid tissue also produces a drop in blood calcium concentration. The possibility was then considered that this stimulation resulted in a release of a substance from the thyroid that exhibited a hypocalcemic action. A couple of years prior to the latter observation, D. H. Copp (see Copp et al., 1962) proposed the presence of a hypocalcemic hormone in mammals that he called calcitonin, which at the time he considered to be formed by the parathyroid glands. The proposal of the presence of a new hormone was based on experimental observations involving perfusion of the thyroid–parathyroid complex in dogs. Blood containing abnormally high or low concentrations of calcium was perfused through the arteries supplying these tissues and the resulting venous outflow was then passed back into the dog, and the effects on the general, systemic, plasma calcium levels were observed. It was found that, when the perfusing blood had a low calcium concentration, the outflowing blood, when passed into the dogs' general circulation, produced a hypercalcemia such as could be accounted for by the presence of parathyroid hormone. In the opposite type of experiment, in which the thyroid–parathyroid complex was perfused with blood having a high calcium concentration, the parathyroid hormone level would be expected to decline, as indeed it does. The basic question that was asked was this: Is such a decline in parathyroid hormone sufficient to bring about a decline in the calcium concentrations from hypercalcemic to normal calcium levels? In other words, while parathyroid hormone exerts a positive effect in elevating blood calcium concentration, is the mere absence of this hormone all that is needed to adjust the calcium level in a downward direction? Copp found that this was not so. Although the venous perfusate from the thyroid–parathyroid complex that was exposed to high calcium concentrations produced a drop in calcium levels, when infused systemically, this hypocalcemia was much greater than could be produced after removing the glandular complex (Fig. 6.2). The implication was drawn that the response is a positive one involving the action of a hormone that has a hypocalcemic action; Copp called this hormone "calcitonin."

It was thought at first that calcitonin came from the parathyroids and, as already commented upon, in view of the intermixture of tissues that occurs in the thyroid region in dogs such an error was not surprising. Hirsch and Munson, on the basis of their experiments on rats, proposed the presence of a hypocalcemic hormone in the thyroid gland itself, which, in order to distinguish it from Copp's hormone, they called thyrocalcitonin. These two hormones are in fact identical and in mammals this hormone originates from the thyroid gland. By the choice of an appropriate species, this time the pig, in which the thyroid contains no parathyroid tissue, appropriate perfusion experiments of the thyroid alone, similar to those described, demonstrated the presence of calcitonin in the venous effluent blood.

The question then arose as to the site of origin of calcitonin in the thyroid

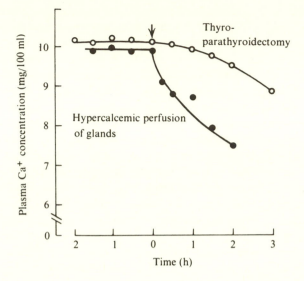

Fig. 6.2. The hypocalcemic response of dogs following perfusion of the thyroid–parathyroid gland complex with hypercalcemic solutions. In intact animals, the decline in plasma calcium concentration (at time zero) occurs promptly, but if the glandular complex is removed, the resulting hypocalcemic response is much slower. This suggests the action of a hypocalcemic hormone in the intact dogs. (From Hirsch and Munson, 1969; from data of Copp et al., 1962.)

tissue. Is it also produced by the same tissue that forms thyroxine? The answer is no. Calcitonin is formed by the parafollicular or C-cells that are present in the mammalian thyroid, and this tissue is quite distinct from that which secretes thyroxine. The presence of these secretory cells was first described by E. C. Baber in 1876, but their function was unknown until recently. Radioimmunofluorescent studies, using fluorescent antibodies to calcitonin, show quite clearly that the C-cells form this hormone.

The C-cells are present in the mammalian thyroid but not that of nonmammals. Embryologically they are derived from the ultimobranchial bodies that are present in all nonmammals except the cyclostomes. An early clue to the function of this tissue was described by Rasquin and Rosenbloom in 1954, several years before the discovery of calcitonin. It was found that Mexican cave fish *Astyanax mexicanus*, when kept for prolonged periods of time in complete darkness, suffered skeletal deformities associated with an hyperplasia of the ultimobranchial bodies. Rasquin and Rosenbloom suggested that the ultimobranchials contained a parathyroid-like hormone. Extracts of the ultimobranchial bodies obtained from chickens and subsequently from many other species, including even dogfish, showed that, when injected into rats, they exhibited a hypocalcemic effect caused by the presence of a calcitonin-like hormone that was not present in the thyroid glands of these species (Copp, Cockcroft, and Keuk, 1967*a*).

Vitamin D and 1α,25-dihydroxycholecalciferol

Cholecalciferol (see Fig. 3.1) is a vitamin (D_3) that can become a hormone (see DeLuca, 1974; Kodicek, 1974; Norman and Henry, 1974). This sterol is formed in the skin of mammals from 7-dehydrocholesterol. The reaction is a photochemical one that requires ultraviolet light, normally sunlight. The vitamin D_3 that is formed can be absorbed into the blood where it is bound to a plasma protein (Bouillon et al., 1978). It travels to the liver where it may be stored or metabolized to a more active form. Vitamin D_2, or ergocalciferol, is made from ergosterol obtained from plants. When administered to man this sterol has the same effects as vitamin D_3 and, moreover, it is usually cheaper to buy. Vitamin D_2 may even have a physiological role in plants as it can promote the growth of roots (Buchala and Schmid, 1979).

It has been recognized for more than 60 years that deficiency of vitamin D can result in a bone disease called rickets. This condition occurs in children who grow up in areas where sunlight and dietary sources of vitamin D may be deficient. The disorder results from inadequate absorption of calcium from the intestine, a process that is normally stimulated by vitamin D_3. When vitamin D is administered there is a delay of several hours before an increase in the absorption of calcium occurs from the gut (see DeLuca, 1971). There are two main reasons for this tardiness. First, the hormone must be metabolized to a more active hormonal form and, second, the action of the hormone involves genetic transcription in the epithelial cells lining the intestine.

Vitamin D_3 undergoes hydroxylation at its 25-position in the liver microsomes to produce 25-hydroxycholecalciferol (25-OHD₃). This metabolite is about twice as active as the parent compound and appears in the plasma where it is bound to a protein. It is accumulated by the kidney where it is again hydroxylated at the 1α-position to produce 1α,25-dihydroxycholecalciferol [1α,25-(OH)₂D₃, calcitriol (see Fig. 3.1)]. The reaction involves a 1α-hydroxylase that is associated with the mitochondria of the proximal segment of the nephron (Brunette et al., 1978). This enzyme has been identified in the kidneys of most groups of vertebrates ranging from teleost fishes to mammals (Henry and Norman, 1975). The conversion of 25-OHD₃ to 1α,25-(OH)₂D₃ is increased by hypocalcemia and hypophosphatemia, but it is decreased by hypercalcemia (Tanaka, Frank, and DeLuca, 1973). 1α-Hydroxylation is also promoted by parathyroid hormone and possibly also by prolactin, growth hormone, and estrogens. Some of these effects have been observed in mammals and birds, usually rats and domestic chickens. The physiological significance and the particular mechanism involved in the action of each of these hormones and minerals are, however, controversial (Fraser, 1980; Kumar, 1980). Thus, it is not clear if the effects of plasma calcium and phosphate levels are all mediated by parathyroid hormone or if the changes in the mineral composition of the plasma may exert effects of their own. Prolactin and estrogens have been shown to increase the formation of $1\alpha,25\text{-}(OH)_2D_3$ in chicks (Spanos et al., 1976*a*, *b*; Baksi and Kenny, 1977; Pike et al., 1978; Bickle et al., 1980). Such actions have not yet been unequivocally demonstrated in mammals. As these hormones are involved in reproduction it is possible that they may have a special role in assuring the additional supplies of calcium that are needed at such times. Growth hormone can increase the absorption of calcium across the intestine, especially in growing animals. In rats this effect appears to be due to an increased formation of $1\alpha,25\text{-}(OH)_2D_3$ by a facilitation of the action of the 1α-hydroxylase (Spencer and Tobiassen, 1981). Injected prolactin can increase calcium absorption across the intestine, and also mobilize calcium from bone, in vitamin D-deficient rats (Pahuta and DeLuca, 1981). Thus, this pituitary hormone may also have a direct effect on calcium absorption in the gut that is independent of the activation of vitamin D.

A number of other metabolites of vitamin D_3 have been identified, including 24,25-dihydroxycholecalciferol and $1\alpha,24,25$-trihydroxycholecalciferol (see Schnoes and DeLuca, 1980). These sterols are formed in the kidney under the influence of a 24-hydroxylase enzyme. They generally have a lower biological activity than $1\alpha,25\text{-}(OH)_2D_3$. It is considered that these and other metabolites may also have biological roles but this possibility is controversial (Fraser, 1980). The $24,25\text{-}(OH)_2D_3$ has been shown to influence bone metabolism in man and it has a longer duration of action than $1\alpha,25\text{-}(OH)_2D_3$ (Kanis et al., 1978; Ornoy et al., 1978).

A disease associated with hypercalcemia and calcification of tissues has been observed in livestock, especially cattle, which feed on certain types of plants. In Argentina, one such type of plant, *Solanum malacoxylin*, has been shown to have a vitamin D-like action (Mautalen, 1972). The active substance was shown to be $1\alpha,25(OH)_2D_3$-glycoside (Wasserman et al., 1976). This observation is one of several examples of animal hormones occurring in the plant kingdom. It has been suggested that such plants may be used to provide a commercial source of the vertebrate hormone.

Mechanisms and interactions of parathyroid hormone, calcitonin, and vitamin D_3 on calcium metabolism

Regulation of the calcium levels in the body usually depends on the interactions of three effectors: bone, intestine, and kidney. These respond to various

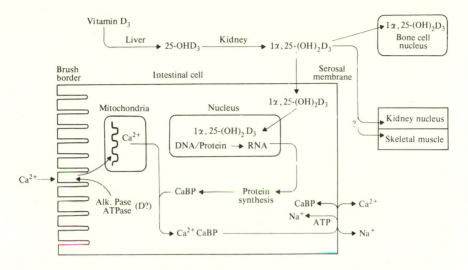

Fig. 6.3. A model illustrating the action of vitamin D_3 in promoting the absorption of calcium across the intestine. Various modifications have subsequently been proposed and these are described in the text. (From Kodicek, 1974.)

combinations of parathyroid hormone, calcitonin, and hormonal metabolites of vitamin D.

Information about the mechanism of action of calcemic hormones on transport of calcium across epithelial membranes is largely confined to the *intestine* (Fig. 6.3), though a little is also known about the kidney. In the intestine this process is mainly controlled by $1\alpha,25\text{-}(OH)_2D_3$ (see Omdahl and DeLuca, 1973; Kodicek, 1974; Wasserman and Taylor, 1976). The sterol enters the intestinal epithelial cells and combines with a specific receptor molecule in the cytoplasm, the complex then moving into the nucleus where it interacts with another type of receptor (or "acceptor") associated with the chromatin. Genetic transcription then occurs and a *de novo* synthesis of protein follows. There appear to be several such proteins formed (some are called "calcium-binding proteins," CaBP). One of these is present in the cytosol ("soluble"), another is bound to the lumenal mucosal membrane of the cells (Feher and Wasserman, 1978), and the most recently identified binds to the inner membrane of the mitochondria (Hodben, Harding, and Lawson, 1980). The first two of these proteins appear to be structurally identical. Calcium-binding proteins, which can be induced by vitamin D, have been identified in several other tissues including the kidney and bone (see later in this section) and even in nerve cells in the brain (Jande, Maler, and Lawson, 1981).

Calcium transport across the small intestine takes place in two main steps:

i. Entry into the cell, from the lumen of the gut, across the mucosal plasma membrane. The electrochemical gradient across this membrane favors the diffusion of calcium, so that the process is probably a passive one, not requiring energy. The nature of this step for calcium entry into the cell is not clear, but it may be facilitated by a calcium-binding protein in the membrane. There is also evidence for an increase in the activity of a mucosal membrane alkaline phosphatase and, possibly, a Ca-activated ATPase in response to $1\alpha,25\text{-}(OH)_2D_3$. The latter enzyme, however, does not appear to be identical to the "high-affinity" Ca-activated ATPase usually associated with active Ca transport (Ghijsen, De Jong, and Van Os, 1980). Vitamin D promotes the phosphorylation of a protein that is present in the mucosal plasma membrane of the chick intestine (Wilson and Lawson, 1981). This molecule has similar characteristics to a protein that is also induced by vitamin D. It is possible that the phosphorylation results in a change in the structural configuration of the membrane protein, which could be facilitating the entry of calcium into the cell.

ii. The normal concentration of free ionized calcium in the cytoplasm of cells is quite low, 10^{-6}–10^{-7}M. Maintenance of this level does not appear to be consistent with the passage of the large amounts of calcium that must pass through the cells of the small intestine. It appears that such calcium must be bound in some form. This sequestration could involve an uptake by the mitochondria and/or binding to the cytosol calcium-binding protein. The latter could also have a role in ferrying the calcium to and from the mitochondria. The calcium is presumably released at the opposite side of the cell in the proximity of the basal plasma membrane. It is then extruded from the cell in a process that occurs against an electrochemical gradient. This active transport appears to involve an exchange of the Ca^{2+} for two Na^+ from the extracellular fluid. The latter move down their electrochemical gradient, and the energy that is dissipated by such a movement from a high- to a low-energy state is used to energize the transport of the calcium.

Most of the calcium that appears in the glomerular filtrate in the *kidney* is reabsorbed across the walls of the nephron. This process appears to involve a Na/Ca exchange and a Ca-activated ATPase and occurs in all segments of the renal tubule (see Suki, 1979). Hormonal regulation of this reabsorption of calcium appears to be confined to the ascending loop of Henle, the distal convoluted tubule, and the cortical collecting ducts. Parathyroid hormone promotes calcium reabsorption from the latter two parts of the nephron (Agus et al., 1973; Sharegi and Stoner, 1978). This process is associated with an activation of adenylate cyclase. Another form of this enzyme that can be activated by calcitonin has been identified in the ascending loop of Henle and the distal convoluted tubule (Marx and Aurbach, 1975; Chabardès et al., 1976). Calcitonin appears to have a similar effect to that of parathyroid hormone on the kidney and increases calcium reabsorption from the glomerular filtrate; its site of action, however, appears to be the ascending loop of Henle (Quamme, 1980; Carney and Thompson, 1981). This effect could be an indirect response to lowered levels of plasma calcium. Vitamin D may also promote renal conservation of calcium but by an action on the distal renal tubule (Costanzo, Sheehe, and Weiner, 1974), an event associated with the formation of a Ca-binding protein (Christakos, Brunette, and Norman, 1981).

Phosphate excretion in the urine is also controlled by the rate of its reabsorption from the glomerular filtrate (see Dennis, Stead, and Myers, 1979). This process occurs mainly in the proximal segment of the renal tubule where it can be inhibited by parathyroid hormone. Adenylate cyclase at this site is also activated by the hormone.

The mobilization and resorption of calcium from *bone* involves changes in the activity of the osteoblast and osteoclast cells. In tissue culture the activity of the former declines while the latter increases following exposure to parathyroid hormone and vitamin D. The precise nature of the resorptive process is unknown, but it appears to involve changes in pH and the secretion of acids and enzymes that help to "solubilize" the bone (see Vaes, 1968; Raisz, 1976; Brommage and Neuman, 1979). It has been suggested that the promotion of anaerobic glycolysis and thus lactic acid formation may create an environmental pH that is favorable for resorption, but this hypothesis is contentious. In addition, hydrolase enzymes may be released from cells, especially from lysosomes, which help to break down the matrix of the bone. The synthesis and secretion of collagenase from the osteoclasts are enhanced (Sakamoto et al., 1975; Puzas and Brand, 1979). Increases in the blood supply to the bone may also contribute to the resorption of minerals.

At least three hormones appear to be directly involved, and interact, in resorption of calcium from bone.

i. Parathyroid hormone stimulates the activity of the osteoclasts and the secretion of acid and enzymes. These effects occur following the activation of adenylate cyclase in the cells and the formation of cyclic AMP. The activity of the osteoblasts declines (see Wong, Luben, and Cohn, 1977).

ii. Vitamin D_3 metabolites, especially $1\alpha,25\text{-}(OH)_2D_3$, also stimulate the activity of osteoclasts but reduce that of the osteoblasts (Wong, Luben, and Cohn, 1977). This response does not involve changes in cyclic AMP concentrations in the cells. A calcium-binding protein has been identified in bone cells as a result of the action of vitamin D_3 and it may contribute to the mobilization of the bone minerals (Christakos and Norman, 1978).

iii. Calcitonin opposes the first two effects and decreases the activities of the osteoclasts, an effect that is also mediated by cyclic AMP (see Wong, Luben, and Cohn, 1977).

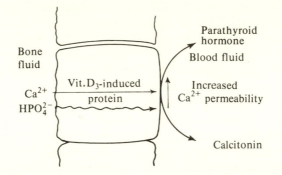

Fig. 6.4. A model for the endocrine control of calcium exchange in bone. The process of calcium resorption from bone is dependent on the presence of vitamin D_3, which apparently induces the formation of a protein that acts as a "Ca carrier" that transfers calcium from the bone fluids to the vicinity of the blood plasma. The final "jump" across the cell membrane depends on its permeability to calcium and this is increased by parathyroid hormone and reduced by calcitonin. (Modified from DeLuca, Morii, and Melancon, 1968.)

The bone cells are sometimes considered to form a barrier (see Talmage and Meyer, 1976) membrane that separates the bone fluids from the extracellular fluids. Although such a "blood–bone barrier" has theoretical attractions, the anatomical evidence for such a contiguous membrane is equivocal. Changes in the permeability of such a barrier to calcium could, however, mediate the resorption and accretion of mineral in bone. Parathyroid hormone could, for instance, increase calcium transport out of bone across such a membrane whereas calcitonin may inhibit such a process (Fig. 6.4). The $1\alpha,25\text{-}(OH)_2D_3$ may facilitate such transport, the calcium-binding protein possibly playing a role in binding intracellular calcium, as has been proposed in the transport of this mineral across the intestine. A role of hormones in promoting the secretion of bone "solubilizers" may be an integral part of such a mechanism.

Parathyroid hormone does not appear to have a direct action on the absorption of calcium from the gut. However, it is possible that calcitonin may decrease the absorption of this mineral though the evidence is equivocal (see, for instance, Swaminathan, Ker, and Care, 1974).

Cortisol and its more potent synthetic analogues, when administered to man, may cause a loss of bone mineral. The reasons for this clinical side effect of these steroids appear to be multiple. There may be a loss of the protein matrix of bone, a decreased absorption of calcium from the gut, and an increased excretion of calcium in the urine. In the gut of chicks there is a decrease in the $1\alpha,25\text{-}(OH)_2D_3$-induced calcium-binding protein (Feher and Wasserman, 1979) following the administration of cortisol.

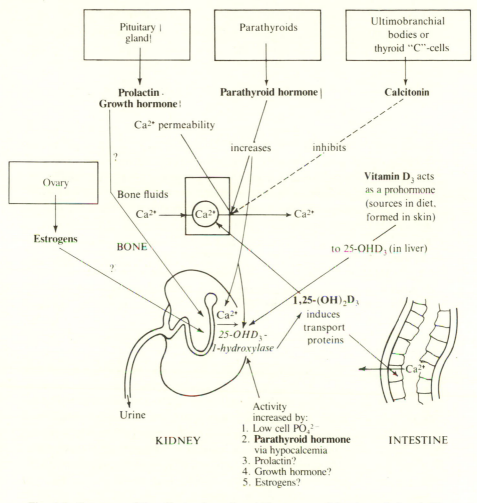

Fig. 6.5. Summary of the effects of parathyroid hormone, calcitonin, and vitamin D_3 on calcium metabolism in mammals. There are three main sites of action: the kidneys, intestine, and bone. The possibility that calcitonin acts on the kidney and changes absorption of calcium across the intestine has not been excluded.

The hormonal regulation of calcium metabolism as outlined (Fig. 6.5) is based principally on experiments using mammals and, in a few instances, birds. Calcium metabolism in nonmammalian vertebrates, however, may exhibit some interesting differences. Such variations are to be expected on quantitative grounds, as in the hen during its egg laying when relatively vast amounts of calcium are rapidly utilized. Qualitatively predictable differences arise as the parathyroids are absent in fishes whereas the cyclostomes appear to have neither these glands nor ultimobranchial bodies. The absence of a calcified bony skeleton in the chondrichthyeans and cyclostomes may also be expected to be a matter of some physiological consequence in calcium metabolism.

Phyletic differences in the role of hormones in calcium metabolism

Mammals

Mammalian calcium metabolism has been described in some detail in the preceding section along with some of the anatomical variations in the distribution of the parathyroid tissues. A variety of mammals including man, dog, rat, pig, sheep, and goat have been examined, and the regulation of calcium levels in the blood is related to the activities of the parathyroids and the thyroid C-cells. The former have a vital role to play, as mammals deprived of these tissues suffer tetanic seizures because of a hypocalcemia. The rachitic effects of vitamin D deficiency on bone are also well known. It is, however, not clear how physiologically essential the thyroid C-cells normally are in mammals. During calcium stress, when large increases in blood calcium levels occur, calcitonin undoubtedly facilitates the homeostatic adjustment of the concentration of this ion; however, the role of calcitonin in the regulation of the smaller and more usual changes in calcium concentration is uncertain. Species differences appear to exist.

This question of the normal physiological role of calcitonin in mammals has been examined in young pigs (Swaminathan, Bates, and Care, 1972). Removal of the thyroid results in a rapid rise (in 1–2 hours) of the plasma calcium concentration. This hypercalcemia is presumed to be the result of a lack of calcitonin, as it can be corrected by infusing small amounts of this hormone into the animal (Fig. 6.6). It is notable, however, that when these thyroidectomized pigs were allowed to recover, without injections of calcitonin, blood calcium levels returned to normal after 24–48 hours. This recovery is probably the result of an adjustment in the rate of secretion of parathyroid hormone. It has also been found that calcitonin aids, but is not essential, in regulating calcium metabolism in young, growing rats. In adult rats on a normal diet, the evidence to indicate that calcitonin has a physiological role is contradictory (Kumar and Sturtridge, 1973; Harper and Toverud, 1973). Such studies on rats, however, have recently been repeated using injected antibodies to calcitonin to antagonize the hormone present in the body fluids (Roos et al., 1980). These antibodies were shown to consistently increase plasma calcium concentration both before and after eating. The results are consistent with calcitonin's having a physiological role in regulating plasma calcium concentration under normal pre- and postprandial conditions.

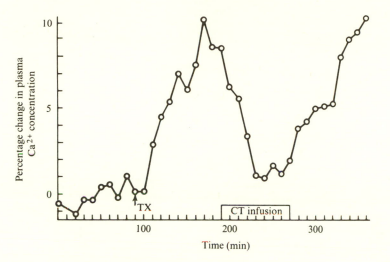

Fig. 6.6. The observed increase in plasma calcium concentration in young pigs following removal (TX) of the thyroid (and its contained calcitonin-secreting C-cells). Infusion of calcitonin (CT infusion) restored the calcium concentration to normal, but as soon as this ceased the levels climbed again. After about a day, the calcium levels returned to normal, probably as a result of other physiological adjustments, including a decline in secretion of parathyroid hormone. (From Swaminathan et al., 1972.)

Birds

Birds, especially during their egg-laying cycle, have a very high rate of calcium turnover (Copp, 1972). A domestic hen may then utilize an amount of calcium equivalent to 10% of that in its body each day. Most of this calcium is derived directly from the food. When compared on a unit body-weight basis the domestic hen during egg laying absorbs calcium across its intestine 100 times more rapidly than a man. Vitamin D, as in mammals, increases the rate of calcium absorption from the intestine of the domestic fowl (Fig. 6.7). At other times, such as during the nocturnal fast, calcium is also mobilized from the bones, principally from medullary bone rather than cortical bone. This distinction is of some endocrine importance as estrogens can influence the turnover of calcium in medullary bone whereas parathyroid hormone only acts on cortical bone. Calcium transport by the eggshell gland of quail appears to be stimulated by $1\alpha,25\text{-}(OH)_2D_3$, which is thus a novel effector organ for this hormone (Bar and Norman, 1981).

As described in Chapter 5, estrogens facilitate the formation of vitellogenins by the liver of birds. They appear in the blood, from which they are incorporated into the yolk of the developing eggs. These proteins can bind calcium and hence their presence is associated with elevation of blood calcium concentrations. This interesting hypercalcemic effect of estrogens is prominent in

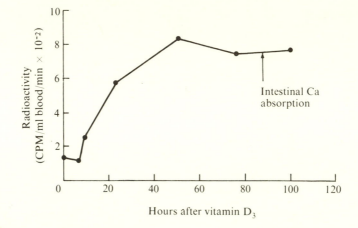

Fig. 6.7. The effect, in young chicks, of an oral dose of vitamin D₃ on calcium absorption from the intestine. The chicks were on a low-calcium diet. (Modified from Harmeyer and DeLuca, 1969.)

birds as well as other oviparous "bony" vertebrates, but is absent in mammals, which may be related to their viviparity. Such a response is especially appropriate to the needs of vertebrates that produce large megalecithal eggs that contain a lot of calcium.

Vitamin D metabolism has been extensively studied in domestic chickens. Prolactin and estrogens have hypercalcemic actions in these birds and, as described earlier in this chapter, these effects may be due to an increased formation of $1\alpha,25\text{-}(OH)_2D_3$ in the kidney. It has also been observed that ovulation in Japanese quail is associated with an increased synthesis of $1\alpha,25\text{-}(OH)_2D_3$ over the next day (Kenny, 1976). The eggs of hens raised with $1\alpha,25\text{-}(OH)_2D_3$ as their only source of vitamin D fail to hatch (Henry and Norman, 1978). If, however, these birds are also given $24,25\text{-}(OH)_2D_3$ hatchability is normal, suggesting that this metabolite of vitamin D₃ may have a special role in the nutrition of laying hens.

Both the parathyroids and the ultimobranchial bodies hypertrophy in egg-laying hens, suggesting that they are involved in regulating the calcium turnover in such birds. However, the precise contribution of each of these glands to avian calcium metabolism is still not clear.

Removal of the parathyroids results in hypocalcemia in birds and this is particularly dramatic in young, growing chicks. Injections of parathyroid hormone elevate blood calcium concentrations through an effect of cortical bone. Injected parathyroid hormone has an extremely rapid hypercalcemic action in the laying hen (Fig. 6.8); it acts six to eight times more rapidly than in the dog (Mueller et al., 1973a). This rapid initial phase of the response in

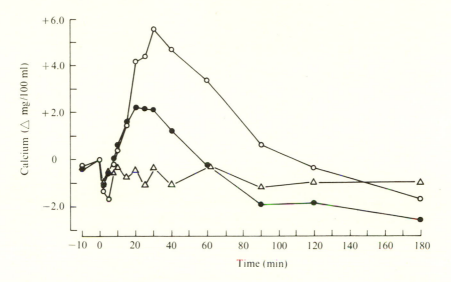

Fig. 6.8. The hypercalcemic effects of parathyroid hormone injections into female domestic fowl. The response was enhanced in those birds receiving a higher supplement of dietary calcium, suggesting that this hormone may be increasing intestinal calcium absorption or the diet decreases basal PTH levels: ○, 31 units PTH/kg, 5.00% dietary calcium; ●, 31 units PTH/kg, 2.26% dietary calcium; △, control. (From Mueller et al., 1973a.)

these birds cannot readily be related to an increased activity of the osteoclasts, suggesting that a dual mechanism of action of parathyroid hormone may exist. Parathyroid hormone preparations when injected have been shown to increase the blood supply to bone and this effect could be important for the rapid and massive mobilization of calcium that occurs during egg laying in birds. Alternatively, an inhibition of the rate of accretion of calcium into bone may be involved (Kenny and Dacke, 1974).

The parathyroid glands have also been shown to influence urinary calcium and phosphate excretion in birds. Parathyroidectomy in starlings results in a marked increase in calcium excretion, whereas phosphate excretion declines (Clark and Wideman, 1977). Injected parathyroid hormone rapidly corrects these deficiencies. It appears that in birds the kidney may play an important role in the actions of parathyroid hormone on calcium and phosphate metabolism, but clearly more species need to be studied.

Calcitonin has been identified in the blood of several species of birds including pigeon, goose, duck, domestic fowl, and Japanese quail (Kenny, 1971; Boelkins and Kenny, 1973). The plasma concentrations of calcitonin are much higher in these birds than normally observed in mammals and they can be increased by injecting calcium solutions (Fig. 6.9). In immature Jap-

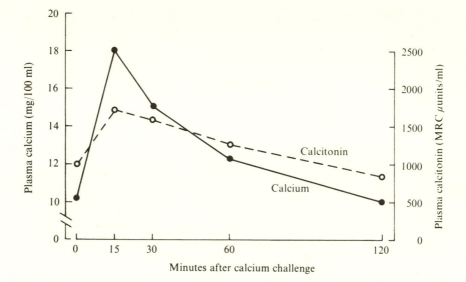

Fig. 6.9. The increase in calcitonin level in the plasma of Japanese quail given intraperitoneal injections of calcium (50 mg/kg). (Modified from Boelkins and Kenny, 1973.)

anese quail, the increase can be as great as 15-fold. It is therefore rather surprising that it has not been possible to demonstrate a physiological role of the ultimobranchial bodies in normal calcium metabolism in birds. Injections of calcitonin usually fail to elicit a hypocalcemia, but it is possible that this reflects a high basal rate of calcium turnover and a rapid compensatory release of the endogenous parathyroid hormone. It thus seems possible that calcitonin could be especially important during egg laying when it could reduce excessive oscillations in blood calcium levels and possibly even influence calcium deposition in the egg. The blood calcitonin levels, however, do not change during the egg-laying cycle of the domestic hen, nor does ultimobranchialectomy significantly influence calcium metabolism in these birds (Speers, Perey, and Brown, 1970). In addition, this operation does not significantly influence the development of the skeleton in growing chickens; thus, although some people believe that the ultimobranchial bodies may have a role in regulating calcium metabolism in birds this has not been established. Possibly, calcitonin has some other physiological role. In starlings, injected calcitonin has a diuretic effect and increases urinary sodium excretion, as is also seen in mammals (Clark and Wideman, 1980). It is considered by some that it may act as a natriuretic hormone.

Reptiles

Parathyroidectomy results in hypocalcemia and tetanic muscular contractions in several species of snakes and lizards (see Clark, 1972) and a chelonian and crocodilian (Oguro, Tomisawa, and Matuoka, 1974; Oguro and Sasayama, 1976). Such effects are, however, difficult to demonstrate in turtles though a small decline in plasma calcium concentration has been demonstrated following removal of the parathyroids of the Japanese turtle *Geoclemys reevesii* (Oguro and Tomisawa, 1972). It has also been shown, histologically, that in three species of young growing turtles parathyroid hormone increases osteocytic calcium mobilization from bones and this effect is inhibited by calcitonin (Bélanger, Dimond, and Copp, 1973). It thus seems unlikely that basic differences of a phyletic origin exist between this response in saurian and chelonian reptiles. Rather, the variation in the response may reflect the relative ease with which calcium can be normally mobilized, and it has been suggested that the bony carapace of turtles provides a calcium store that may facilitate adjustments of the calcium in the blood.

The sites of action of parathyroid hormone in reptiles are uncertain. Although bone seems to respond, the kidney may not do so. Parathyroid hormone stimulates phosphate excretion in the urine of snakes (genus *Natrix*), but neither the injection of this hormone nor parathyroidectomy could be shown to alter urinary calcium loss (Clark and Dantzler, 1972). The latter observation is in contrast to the effect of parathyroid hormone in mammals, but the experiments should be extended to other reptilian groups before any generalizations as to broad phyletic differences are made.

Despite repeated attempts (on lizards, turtles, and snakes), injections of calcitonin have not been shown to exhibit a hypocalcemic effect in reptiles. Although these observations are somewhat unexpected they conform with those in birds, which, we are told, are descended from reptiles. Reptiles, like birds, exhibit a hypercalcemic response to injected estrogens (see Clark, 1967).

Reptiles have common phyletic affinities to mammals as well as birds so that their endocrine function is of rather special interest. The reptiles share with birds the problems associated with the production of large cleidoic eggs that in many species are covered with a calcareous shell. It should thus not be surprising to observe similarities in their calcium metabolisms. Reptiles, however, are poikilotherms and this may influence the relative importance of processes that are involved in the metabolic coordination of calcium levels as the speed of the adjustments need not be as great.

Amphibians

The amphibians have a special position with regard to our understanding of vertebrate calcium metabolism, as it is in this group that the parathyroid glands first appear on the phyletic scale. On the other hand, the ultimobranchial bodies have persisted from their piscine ancestors. The information that is available is relatively sparse and is not consistent with the special interest that these animals deserve.

Parathyroidectomy reduces plasma calcium levels in anuran amphibians and this effect seems to be the result of a decreased rate of calcium resorption from bone (see Cortelyou, 1967). Injections of parathyroid hormone have a hypercalcemic action. The effects of parathyroid hormone on renal calcium excretion are not clear as parathyroid hormone, as well as parathyroidectomy, increases urinary calcium loss. It has been suggested that the response to parathyroid hormone could be complicated because of the presence of different thresholds in the sensitivity of the kidney and bone, but further observations seem to be needed to clarify this paradox. Parathyroidectomy also results in hypocalcemia in tadpoles (Sasayama and Oguro, 1975).

Parathyroidectomy has inconsistent effects in urodele amphibians (Oguro, 1973). Hypocalcemia, accompanied by tetanic convulsions, has been observed following removal of these glands in the newt, *Cyanops pyrrhogaster*, *Tylototriton andersoni*, and *Notophthalmus viridescens* (Oguro and Sasayama, 1978; Wittle and Dent, 1979). The Japanese giant salamander, *Megalobatrachus davidianus* (Oguro, 1973) is, however, unresponsive to parathyroidectomy, and this is reminiscent of early observations in three other species of urodeles in which this operation failed to induce tetany. The giant salamander appears to have parathyroid hormone present in its parathyroids, as extracts of these tissues have a hypercalcemic effect when injected into parathyroidectomized *C. pyrrhogaster*. It has thus been suggested that this salamander lacks a target-organ system for parathyroid hormone and this may also be so in some other urodeles.

Anuran amphibians (*Rana pipiens* has been principally studied) possess "lime sacs" that are novel sites for the storage of calcium in the body (see Robertson, 1969a, b). These organs are extensions of the lymph sacs, and they extend caudally along the vertebral canal and emerge between the vertebrae. They contain calcium carbonate, instead of calcium phosphate as in bone, and this exhibits a mobility that includes an added storage following the administration of calcium chloride or vitamin D. Such treatment, which produces a hypercalcemia, also results in the hypertrophy of the ultimobranchial bodies (Robertson, 1968).

Removal of the ultimobranchial bodies in *Rana pipiens* results in an initial elevation in plasma calcium concentration. After about 6 weeks, however, a hypocalcemia occurs (Robertson, 1969a, b). It is thought that these responses are the result of an excessive mobilization of calcium from the lime sacs (as well as the bones), which subsequently become depleted. Calcitonin may promote the laying down of calcium at these two sites and it may also reduce the excretion of this ion in the urine. Ultimobranchialectomy reduces the ability of tadpoles to control their body calcium levels when they are placed in solutions containing high concentrations of this mineral (Sasayama and Oguro, 1976).

Table 6.2. *Subjective analysis of amount of calcium carbonate in the paravertebral lime sacs of tadpoles* (Rana pipiens) *tabulated as estimated from X-rays*

Tadpole development stage	Tap water		High calcium	
	Normal	UBX	Normal	UBX
I–V	+	±	+	±
VI–X	+ +	±	+ +	±
XI–XVII	+ +	±	+ +	±
XVII–XXV	+ +	±	+ +	+ + +

Note: Concentration of calcium in tap water was 3 meq/liter. High-calcium water was 15 meq/liter with treatment extended for 6 weeks in all groups except in limb bud stages (I–V), which was for 2 weeks. UBX, ultimobranchialectomized; +, degree of response.
Source: Robertson, 1971.

The lime sacs are also present in tadpoles where they may contribute to the release of calcium in answer to the needs of metamorphosis. Ultimobranchialectomy in young tadpoles limits their ability to accumulate calcium in the lime sacs whether they are kept in a solution with either a high or a low calcium concentration (Table 6.2). At metamorphosis, the tadpoles in the high-calcium medium can accumulate calcium in their lime sacs despite the absence of the ultimobranchial bodies. Ultimobranchialectomized tadpoles in the low-calcium medium, however, fail to do this and as a result the bones of the adults are poorly ossified.

Tadpoles can also accumulate calcium across their integument, especially in the gills (Baldwin and Bentley, 1980). Injected calcitonin reduced such uptake of this mineral in a manner reminiscent of that observed in the gills of teleost fish (see next section). Parathyroid hormone was without effect on this process.

Although the number of species studied is limited, the Amphibia appear to utilize parathyroid hormone, calcitonin, and vitamin D_3 for the regulation of calcium metabolism. As we shall see in the next section, lungfishes, which are phyletically akin to amphibians, do not appear to respond in this manner to any of these hormones.

The fishes

A considerable diversity in the mechanisms for the regulation of calcium metabolism is not unexpected in the fishes. This may result from their great phyletic diversity, their ability to live in aqueous environments containing high (seawater) or low (most freshwater) levels of calcium, the possession of either a bony or a cartilaginous skeleton, and the presence or absence of the ultimobranchial bodies and the corpuscles of Stannius.

Paleontological evidence suggests that the ostracoderms, which were the jawless ancestors of modern fishes, had a bony dermal exoskeleton and, sometimes, also a bony endoskeleton that had the appearance of a tissue that functions as a store for calcium (see Copp, 1969; Simmons, 1971). These fishes lived in fresh water where the calcium concentration was, presumably, low so that calcium storage in the bones may have been physiologically important. When such ancestral fishes returned to the sea, where there was an unlimited quantity of calcium, they lost their bony skeleton, and this situation persists in present-day cyclostomes and chondrichthyeans. We can extend this speculation (and that is all it is!) and consider whether or not such vertebrates possessed an endocrine system for controlling calcium concentration in the body. In extant species of cyclostomes and chondrichthyeans, there is as yet no evidence for such a control mechanism. In the ancestral freshwater ostracoderms such a control could have been more important and may possibly have occurred. It is even possible that the chondrichthyean ultimobranchial bodies represent a survival from those times. The ancestral freshwater ostracoderms need not, however, have possessed such a system for controlling calcium metabolism as there is, for instance, no evidence for the endocrine control of such processes in some contemporary freshwater fishes like lampreys and the lungfishes (Urist, 1963, 1976; Urist et al., 1972).

Many cyclostome fishes live in the sea but others spend their entire life in fresh water. Lampreys migrate from the sea into fresh water where they survive for several months without feeding and produce large numbers of eggs. Cyclostomes are considered to regulate calcium levels by utilizing a so-called open system whereby calcium transfer takes place across the membranes of the body, such as in the intestine, gills, and kidney. The absence of the endocrine tissues known to influence calcium metabolism in other vertebrates, however, should not be taken to indicate a lack of such control mechanisms. Further investigation of calcium metabolism of these very interesting fish is clearly to be desired.

Although vitamin D_3 has been shown to have a role in regulating calcium metabolism in many tetrapods, there is little information to indicate that it also has such a function in fishes. Vitamin D_3 is stored in the liver of teleosts (usually more in marine than freshwater fish), but little or none of this steroid has been identified in the livers of chondrichthyeans or cyclostomes (Urist, 1963). Neither administration of vitamin D_3 to sharks and rays nor vitamin D_3 and 25-hydroxycholecalciferol to the South American lungfish had any effect on their blood calcium levels (Urist, 1962; Urist et al., 1972). There is thus, at present, little to indicate that vitamin D_3 has an endocrine role in regulating calcium levels in fishes, but we should await more extensive experiments before drawing any phyletic conclusions as to the possible evolution of its role as a hormone. The $1\alpha,25\text{-}(OH)_2D_3$ has been shown to increase plasma phosphate concentrations in eels (MacIntyre et al., 1976). It was suggested that this effect may reflect a phylogenetically early role for this hormone. Aquatic environments are usually relatively poor sources of phosphate, though seawater is rich in calcium. A hormonal control over body phosphate stores may then have been of primary importance. However, when vertebrates emerged onto dry land, calcium became in short supply and the excretion of excess phosphate was a potential problem. The phosphaturic action of parathyroid hormone may have been an important evolutionary innovation in such animals while the calcemic actions of vitamin D become more important.

The chondrichthyeans possess ultimobranchial bodies that contain a calcitonin that has a hypocalcemic action, when injected, into mammals. Injected calcitonin, however, had no effect on plasma calcium concentration in the dogfish *Squalus acanthias* nor were any other actions on renal function detectable (Hayslett et al., 1972). Possibly, calcitonin exerts other actions in chondrichthyeans that may even be unrelated to mineral metabolism. The injection of estrogens also fails to elicit a hypercalcemia in sharks (*Triakis semifasciata* and *Heterodontus francisci*) though a very small (11%) increase in calcium concentration has been observed in the dogfish *Scyliorhinus caniculus* (Urist and Scheide, 1961; Woodhead, 1969).

The bony fishes are a very diverse group and information about the regulation of calcium metabolism is almost entirely limited to teleosts. Within this group few species have been studied but considerable variability seems to exist. One of the reasons for these differences is that although all teleosteans possess a calcified endoskeleton (see Simmons, 1971), in some of these fishes, especially marine species, the bone is acellular so that the calcium deposits are relatively immobile. In other teleosts, notably those that can live in fresh water, the bone is cellular and can participate in calcium regulation. Teleosts like the eel also have large stores of calcium present in their muscles; these may be five times as great as those in tetrapods and could provide an additional site for the regulation of calcium exchange.

The injection of calcitonin into teleost fishes may or may not have a hypocalcemic action (see Chan, 1972). In the killifish *Fundulus heteroclitus*, which has acellular bone, calcitonin has no effect on plasma calcium concentration. In the eels *Anguilla anguilla* and *A. japonica*, and the freshwater catfish *Ictalurus melas*, calcitonin has been reported to decrease the blood calcium concentration but the results are equivocal (see Pang, 1973; Yamauchi et al., 1978). Calcitonin has been identified in the plasma of Japanese eels but the concentrations did not differ in fish kept in fresh water or seawater (Hirano et al., 1981). A physiological role for calcitonin in regulating plasma calcium concentrations in teleosts is usually considered to be in doubt. However, this peptide has been shown to act on bone in fishes (Lopez, Chartier-Baraduc, and Deville, 1971), and it can reduce the influx of calcium across perfused gills of salmon and eels (Milhaud et al., 1977; Milet, Peignoux-Deville, and Martelly, 1979). Receptors for calcitonin have also been identified in trout gills (Fouchereau-Peron et al., 1981).

Neither calcitonin nor parathyroid hormone has an effect on plasma concentrations of calcium in the South American lungfish *Lepidosiren paradoxa* (Pang and Sawyer, 1975). However, injected parathyroid hormone had an antidiuretic effect, whereas calcitonin elicited a diuresis. The physiological significance of these latter interesting observations is uncertain but it could reflect other possible roles for these peptides. Plasma calcitonin levels rise (Watts, Copp and Deftos, 1975) in migrating salmon during spawning, suggesting that the hormone may have a special role in reproduction.

The corpuscles of Stannius that are present in most bony fishes (see Chapter 2) also may influence calcium metabolism, and an interaction between these tissues and the ultimobranchial bodies seems to occur. M. Fontaine, in 1964, found that removal of the corpuscles of Stannius in the European eel *Anguilla anguilla* results in a marked increase (1.4-fold) in the plasma calcium concentration. This effect of Stanniectomy has been confirmed in other teleosts such as the goldfish *Carassius auratus* and the Asiatic and North American eels *A. japonica* and *A. rostrata* (see Chan, 1972). These hypercalcemic effects are accompanied by a reduced calcium excretion in the urine, and an increased osteoclastic activity that mobilizes calcium from the bone. The hypercalcemic effect of Stanniectomy can be prevented by the transplantation, or the injection of extracts, of the corpuscles of Stannius into the deficient fishes. The hormone has been called "hypocalcin," which reflects its hypocalcemic action (Pang, Pang, and Sawyer, 1974).

The activity (from histological observations) of the corpuscles of Stannius of the killifish *Fundulus heteroclitus*, appears to be greater in normal seawater than in an artificial seawater where the calcium concentration is low (see Pang, Pang, and Sawyer, 1973). Stanniectomy in these fish only results in hypercalcemia when they are bathed by solutions with a high calcium concentration; if they are in artificial calcium-poor seawater or calcium-poor fresh water, blood calcium concentrations are unaffected. Thus, the corpuscles of Stannius may play a physiological role in teleosts that live in solutions with a high calcium concentration, such as seawater, where their secretions exert a hypocalcemic action.

Although studies on the site of action of hypocalcin are incomplete, a number of observations indicate that this putative hormone may reduce the uptake of calcium across the gills. Extracts of the corpuscles of Stannius have been shown to increase the efflux and decrease the influx of calcium across the gills of eels (Milet, Peignoux-Deville, and Martelly, 1979; So and Fenwick,

1979). The gills of rainbow trout and eels contain a Ca-activated ATPase which may be able to "pump" calcium across the gills (Ma et al., 1974; Fenwick, 1976). Hypocalcin may act by inhibiting the action of this enzyme.

Both the ultimobranchial bodies and the corpuscles of Stannius appear to be able to limit changes in body calcium by a similar type of action on the gills. Such a duplication of a physiological role is rather unexpected. It should be recalled, however, that the corpuscles of Stannius are not present in all bony fishes but are confined to the Teleostei and Holostei. It has been suggested that the corpuscles of Stannius may be responsible for coarse adjustments in body calcium levels in such fishes whereas calcitonin is involved in finer regulation. (Milet, Peignoux-Deville, and Martelly, 1979).

When *Fundulus heteroclitus*, kept in artificial seawater with a low-calcium concentration, are hypophysectomized they undergo tetanic muscular contractions associated with a considerable decline in the plasma calcium levels (Pang, 1973). This response is not seen if the fish are kept in ordinary seawater with high concentrations of calcium. The effects in the low-calcium solution can be prevented by injecting the hypophysectomized fish with extracts of the pituitary or by transplanting this gland under the skin. The precise nature of the hypercalcemic "hormone(s)" that may be involved is uncertain, but it appears to be prolactin (Pang et al., 1978; Olivereau and Olivereau, 1978). This hormone is released in increased amounts in fish placed in fresh water with a low calcium concentration (Wendelaar Bonga, 1978; Wendelaar Bonga and Van der Meij, 1980). Whether or not prolactin is acting by increasing the rate of conversion of 25-OHD$_3$ to 1α,25-(OH)$_2$D$_3$, as may occur in birds, is unknown. Prolactin may have a hypercalcemic effect in teleosts that live in environments with a low-calcium concentration, like fresh water, whereas the corpuscles of Stannius may mediate the opposite, hypocalcemic, response in fish living in solutions with a high-calcium level, like seawater.

The plasma calcium levels in the female plains killifish, *Fundulus kansae*, increase three-fold in the summer compared to the winter (Fleming, Stanley, and Meier, 1964). This change in plasma calcium is not seen in the male killifish. In the female, it can, however, be imitated in winter by injecting them with estradiol, which suggests that it is a response associated with breeding. Even the male fish increase their rate of calcium uptake from the external solutions in the summer (Fleming, Brehe, and Hanson, 1973). The mechanism for the increased accumulation of calcium is not understood.

Although knowledge about the regulation of calcium metabolism in fishes is incomplete, several intrinsically interesting facts are known. We have seen that endocrine control of this process may be related to the corpuscles of Stannius, which are only present among the Osteichthyes. The secretion of the ultimobranchial bodies, calcitonin, may exert a hypocalcemic action, possibly associated with the presence of cellular bone but also involving

changes in calcium uptake across the gills. Chondrichthyeans possess calcitonin but its physiological role is unknown. Almost nothing is known about the
regulation of calcium metabolism in sharks and rays, while lampreys and hagfishes may be dependent on an "open system" that does not involve the action
of hormones. The regulation of calcium metabolism in cyclostomes and
chondrichthyeans undoubtedly will provide a very interesting area in which to
pursue this subject further. An extensive comparative account of calcium
metabolism in vertebrates has been provided by Dacke (1979).

Conclusions

Calcium is essential for the life of vertebrates, but their requirements for this
mineral, and its availability in the environment, vary considerably. This need
can be related to vertebrate phylogeny because of the systematic presence or
absence of a bony calcareous skeleton and the characteristic life of some
groups in the sea, which is rich in calcium. It is, therefore, not surprising to
observe that the role of hormones in the regulation of calcium metabolism
appears to have changed considerably during the course of evolution. In
marine fishes that lack a bony skeleton, hormonal regulation of calcium does
not seem to occur. When such an endocrine control system is present, as in
marine bony fishes, it is concerned with limiting concentration of calcium and
lowering its levels in the blood (hypocalcemic effects). The gills may be an
important effector site. This response possibly involves calcitonin, which is
present in all vertebrates except cyclostomes, and, probably in teleosts, a
secretion, hypocalcin, from the corpuscles of Stannius. Teleost fishes in fresh
water, where the calcium levels are low, may utilize a pituitary hormone,
possibly prolactin or corticotropin, to help maintain adequate concentrations
of calcium in their body fluids. In mammals and birds prolactin may also play
a role in calcium metabolism by promoting the activation of vitamin D and
possibly even by exerting direct effects on the intestine and bone. Tetrapods
have acquired a "new" hormone, parathyroid hormone, whose role is to
mediate a hypercalcemia, and bone is the major site of its action. A physiological role for calcitonin in calcium metabolism is doubtful in many species of
tetrapods. Two other hormones contribute to the regulation of calcium metabolism but apparently in distinct groups of vertebrates; vitamin D has an
hormonal function in tetrapods where it facilitates the accumulation of calcium in the body but it is uncertain whether it has such a role in fishes.
Estrogens have assumed a "special" endocrine role in many vertebrates where
they assist the deposition of calcium in the developing egg; this effect is
absent in mammals.

7. Hormones and the integument

The skin and gills of vertebrates constitute the major external interface between the animal and its environment. This integument is physiologically and anatomically a very important tissue that exhibits considerable diversity reflecting the differences that exist in the physicochemical gradients between the vertebrates and their environments. The integument may thus play a role in the animal's osmoregulation, thermoregulation, and respiration. In addition, the integument provides signs and signals that can promote social and sexual contact and can help the animal to blend in with its surroundings and so protect it from predators, or help it catch its food. Of primary importance is the skin's role as an integumental skeleton by which it contains the animal in a condition that facilitates its locomotion. The relative importance of these various roles of the integument varies in different species and the structure varies accordingly also.

In fishes and larval amphibians, the gills, which function as organs of respiration, make up a large part of the animal's external surface. Exchanges of oxygen and carbon dioxide readily occur across these highly vascularized tissues, which are also the sites of considerable movements of water and salts. Many fishes contain special cells in their gills and skin called "chloride-secreting cells" that are the site for active extrusion of salts. The endocrine control mechanisms influencing the permeability of the gills are described in Chapter 8.

The skin is the major nonbranchial interface between the animal and its environment. In its simplest form, the skin consists of two major layers of tissues: an outer epidermis, which has several strata of cells, and an inner dermis. However, such a simple arrangement does not exist in nature, as various other structures are also included in the skin that modify its properties. These structures include scales, hair, feathers, pigment cells, secretory glands, and certain sense organs. Such accessories contribute to the particular physiological properties exhibited by the integument of each species.

The skin is thus a complex tissue that has different physiological needs that depend on the species, the environment, and the stage of the animal's life cycle. The constitution of the skin is not static but undergoes continual change commensurate with the normal needs of growth and repair. In addition, rapid changes in the physiological properties of the skin also can occur, such as involve an increased blood supply and the secretion of sweat in response to the need to dissipate heat in the mammals, and an increased osmotic permeability to water, as a result of dehydration, in amphibians. Many cold-blooded vertebrates can rapidly alter the distribution of pigment in the skin so that they blend more closely with the shades and hues of their surroundings. Seasonal changes commonly occur in the integument, such as the changes in pigmentation that may be associated with breeding and alteration of the color, length, and density of fur and feathers in summer and winter. Such changes in the fur and feathers may alter the insulative properties of the integument and contribute to the animal's camouflage.

While the skin has a considerable innate ability to regulate its functions, it is also dependent on the nervous and endocrine systems with whose aid it can coordinate its activities with the rest of the body. The skin has a plentiful nerve supply that mediates its sensory functions and regulates its blood supply. The secretions of cutaneous glands are also predominantly under neural control, though circulating catecholamines exert some effects on them. Rapid changes in the distribution of pigment may be controlled by nerves, but hormones are also very important. The endocrines help in the maintenance of the nutritional and anatomical integrity of the skin as well as such processes as molting, pigmentation, and the function of certain cutaneous glands. Hormones that influence cutaneous function include several from the pituitary such as prolactin, MSH, vasotocin, ACTH, LH, and TSH, and also thyroxine, the catecholamines, corticosteroids, gonadal steroids, and melatonin. Some of the actions of these hormones are confined to relatively few species while the effects of a hormone on the skin may be quite different in one species as compared with another. The variation that is observed in the cutaneous effects of particular hormones suggests that considerable evolution has occurred in their special roles. It is also often difficult to decide whether the actions following an excess or deficiency of a hormone are the result of its direct action on a specific cutaneous effector or are due to merely a more diffuse, indirect effect such as may result from general changes in the animal's physiological and nutritional status.

Hormones and molting

The epidermis is regularly renewed as its outer layers drop off and are replaced by new cells that are formed from the underlying epithelium. This may be a more or less continuous process, as is common in mammals, or it may take place suddenly at regular intervals varying from a few days, as in many amphibians, to several months in certain lizards and snakes. The hair of mammals and feathers of birds are also subject to such periodic renewal, and this may also occur at precise times of the year such as at the onset of winter or spring or just before, or after, the breeding season (pre- and postnuptial molts). In reptiles the shedding of the epidermis is often called *sloughing*, and the shedding of the pelage in mammals, the plumage of birds, and the epidermis in amphibians is called *molting*. The molting cycle is considered to involve three processes: tissue proliferation, differentiation, and shedding, or ecdysis.

It is generally considered that the regular cyclical molting that occurs in fish, reptiles, and amphibians reflects an autonomous rhythm in the skin upon which the actions of hormones can impinge in a permissive manner (Ling, 1972). The seasonal molts that occur commonly in birds and mammals are closely allied to the external stimuli, principally the photoperiod, but they are also modified by the external temperature and the nutritional condition of the animal.

The pituitary, gonads, and the thyroid glands are the principal endocrines that influence molting in vertebrates.

Removal of the pituitary usually prevents, or considerably prolongs, the length of the reptilian and amphibian molting cycles and blocks the seasonal molts observed in many birds and mammals. This effect of hypophysectomy is the result of the absence of several hormones, a lack of TSH, with its tropic effect on the thyroid, is very important but the lack of prolactin and corticotropin may also contribute to the debility. In lacertilian reptiles, urodele amphibians, birds, and mammals the thyroid hormones accelerate the molting process; thyroidectomy has an inhibitory effect. It is interest that in ophidian reptiles (snakes) removal of the thyroid results in a decrease in the length of the sloughing cycle which is in direct contrast to what is observed in their lacertilian relatives (Chiu and Lynn, 1972; Maderson, Chiu, and Phillips, 1970). Differences in the effects of hormones on molting also occur within the Amphibia, for whereas this process is facilitated by the thyroid gland in *Ambystoma mexicanum* and *Notophthalmus viridescens* (Urodela) this is not so in *Bufo bufo* (Anura) (Jorgenson, Larsen, and Rosenkilde, 1965; Hoffman and Dent, 1977). In the latter toads, however, corticotropin and the corticosteroids (corticosterone is most active) are necessary for successful molting. To further complicate any attempts to define a phyletic uniformity, it has been found that corticotropin completely *inhibits* sloughing in the lizard *Gekko gecko* (Chiu and Phillips, 1971*a*).

Prolactin has diverse actions in vertebrates and it is especially notable that many of its effects are on the integument (see Dent, 1975) or derivatives of it, most notably the mammary glands (which are merely modified sweat glands). It has been shown that the injection of prolactin decreases the length of the sloughing cycle in the lizard *Anolis carolinensis* and *Gekko gecko* (Maderson and Licht, 1967; Chiu and Phillips, 1971*b*). Prolactin has been found to facilitate the growth and increase the appetite of lizards so that its effect could reflect their nutritional condition. In this respect, it should also be remembered that hypophysectomized lizards usually do not eat and are not in perfect health. Prolactin injections have also been shown to accelerate molting in a urodele amphibian, the red eft, *Notophthalmus viridescens* (Chadwick and Jackson, 1948), though such an action has not been demonstrated in other amphibians. This effect is related to an increase in the mitotic activity of the epidermis (Hoffman and Dent, 1977). It is interesting that, in this newt, prolactin promotes the transition (or metamorphosis) from the terrestrial form into the aquatic breeding stage when it returns to water ("water-drive effect" of prolactin), and this is associated with cutaneous changes. Whether the effect on the skin is a primary one is uncertain, as the prolactin could be stimulating some noncutaneous process concerned more generally with metabolism and metamorphic change.

The seasonal changes that occur in the pelage of mammals and the plumage of birds appear to be principally under photoperiodic control. Changes in the hours of light are transmitted via the eyes and hypothalamus to the pituitary. Such photoperiodic changes also control the gonadal cycles so that it may be difficult to separate the two events. Removal of the pituitary prevents the short-tailed weasel (*Mustela erminea*) from growing a brown spring coat (they stay white), even when they are exposed to a photoperiod that induces this growth in intact weasels (Rust and Meyer, 1969). The effects of photoperiod on the pelage and plumage appear to be mediated through the action of the gonadal steroids, corticosteroids, and thyroxine, which are in turn controlled by the hypothalamus and pituitary (Ebling and Hale, 1970). In voles (*Microtus agrestis*) an increased activity of the thyroid glands in spring may precipitate molting (Al-Khateed and Johnson, 1971; Johnson, 1977). This time of the year is also one of preparation for breeding, and therefore the gonads also hypertrophy. Both the thyroid hormones and sex hormones are thought to act together to foster the growth of the thin summer coat, the development of dense, fine hair giving way to coarse, thicker hairs. Thyroxine promotes the molt and the growth of coarse hairs, and the increased levels of sex steroids inhibit the growth of fine hairs. In the autumn the opposite occurs and falling levels of thyroxine and sex hormones promote another molt and the growth of a coat of fine hair that provides better insulation in the winter.

The precise manner in which hormones influence molting is not understood. Cyclical changes in molting in poikilotherms are usually characterized by brief periods of cellular activity and rapid cell division interspersed by periods when little activity occurs, which are referred to as the "resting phases." It is thought that thyroxine and prolactin shorten the resting phase (except in snakes!) during which time the skin's activity is reduced by lower levels of these hormones. In newts, however, prolactin has been shown to

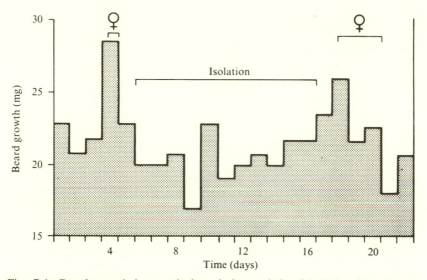

Fig. 7.1. Beard growth in man is less during periods of isolation from women. Restoration of female company results in a sudden "spurt" in the growth of the facial hair, and it has been suggested that this is the result of the release of male sex hormones, probably pituitary LH. (From Anon., 1970.)

promote active cell division in the epithelium and this is reminiscent of its action on the crop-sac epithelium in pigeons. In toads, the absence of the pituitary does not prevent the formation of new layers of epidermis (or sloughs) but prevents the shedding, or casting off, of these cells. This shedding is promoted in toads by corticotropin and corticosteroids. It is interesting that some frogs that estivate during periods of drought are protected from excessive dehydration by a cocoon composed of accumulated layers of epithelial cells (Lee and Mercer, 1967), and it seems possible that this may reflect a decline in their pituitary and adrenal function.

Although not strictly an example of molting, balding (loss of hair on the scalp) in man is known to be dependent on the activity of androgenic hormones. The sex hormones also control the development at puberty of the typical patterns in distribution of hair in men and women. A rather interesting observation (Fig. 7.1) is that the growth of the facial hair in man shows a rapid increase when female company is restored following a period of abstinence. This change probably reflects a surge in the release of LH and the secretion of testosterone due to the very thought of women. Such an effect has been observed in rats in which the presence of the female stimulates release of these hormones in the male (Graham and Desjardins, 1980). A conditioned reflex can be developed using other cues in association with the presence of the female.

Hormones and skin glands

Vertebrates possess several types of secretory glands in their skin that serve a variety of functions (Quay, 1972). These glands can be classified into two major groups: the mucous glands and the proteinaceous glands.

The proteinaceous-type glands have undergone considerable evolutionary modification and include (to name only a few) a variety of venom glands in fishes and amphibians, the uropygial (or preening) gland that is present in many birds, and the sebaceous glands, sweat glands, and mammary glands of mammals. The sweat glands play an important role in temperature regulation in mammals. The sebaceous glands are associated with hair follicles and the fatty sebum that they secrete serves to protect the hair from wetting. The sebaceous glands and the sweat glands sometimes secrete special odoriferous substances that may play an important role in territorial behavior (by defining territorial limits) and also act as sexual attractants. Such scent glands may become enlarged and congregate in distinct areas of the body. Examples of this include the "side glands" on the heads of shrews, the "anal glands" and submandibular "chin glands" in rabbits, the "ventral gland" in gerbils, and the forehead skin of roebucks.

The maturation and function of the sweat glands and sebaceous glands in mammals are influenced by hormones (Strauss and Ebling, 1970). The odoriferous scent glands are commonly observed to be larger in the male than the female whereas maturation of sebaceous and sweat glands occurs during puberty in man. It thus seems likely that they are influenced by sex hormones, and androgenic steroids are undoubtedly involved. Natural and experimental differences in the development of the submandibular chin glands in rabbits are shown in Fig. 7.2. These glands are much larger in sexually mature than immature rabbits, and they decrease in size when the testes are removed. Injections of testosterone promote the development of the chin glands while estrogens inhibit it. Progesterone also increases their weight, but this may be due to the inherent androgenic activity of this steroid hormone.

Castration results in a decrease in the development and a reduced secretion of sebum by the sebaceous glands in the skin of mammals (see Ebling, 1974). Hypophysectomy has a similar effect but the response does not merely reflect an absence of gonadotropins, which can stimulate the secretion of androgens (see Shuster and Thody, 1974). The administration of testosterone to hypophysectomized rats that have also been castrated does not completely restore the activity of the sebaceous glands. Removal of the posterior lobe of the pituitary gland has a similar effect to complete hypophysectomy. Neither oxytocin nor antidiuretic hormone stimulates the sebaceous glands in such animals, but it was found that α-MSH completely restored their activity. It was suggested that α-MSH may be acting as a sebotropic hormone. Androgens and α-MSH

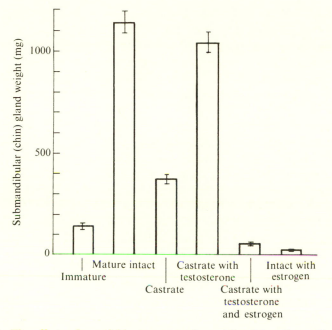

Fig. 7.2. The effects of gonadal hormones on the weight of the submandibular (chin) glands in male rabbits. It can be seen that these glands exhibit considerable increases in weight after sexual maturity, an effect that is prevented by castration. Testosterone injections overcome this effect of castration while injecting estrogens bring about an involution of the glands. (From Strauss and Ebling, 1970.)

appear to act together to enhance each other's action (synergism) on the sebaceous glands (Ebling et al., 1975; Thody et al., 1976). The androgens appear to increase their growth and size wheres α-MSH promotes the secretion of lipids.

In primates, especially man, the watery secretion of the eccrine sweat glands plays an important role in providing water that can be evaporated from the body surface and so aid in the dissipation of heat (see Chapter 8). Sweat contains dissolved salts, including sodium and potassium, the concentration of which is influenced by aldosterone; the sodium/potassium ratio declines under its influence. In addition, although the secretion of sweat is primarily under neural control in primates it is also increased, during exercise, by circulating epinephrine.

Mucous glands are present in the integument of fishes and amphibians. Their role is contentious but seems to be related to an aquatic life where it has been suggested that the mucous secretion may have a protective action on the skin as well as serving certain special functions such as the formation of a cocoon in estivating African lungfish and providing food for the young of a cichlid teleost. In the latter fish, this mucous secretion is promoted by the injection of prolactin (Egami and Ishii, 1962). Injected thyroxine increases the number of mucous cells in the skin of the guppy, *Poecilia reticulata* (Schwerdtfeger, 1979). It has also been suggested that prolactin may limit the permeability of teleost fish and urodele amphibians to water and sodium (see Chapter 8) by an action on the mucous glands, but the evidence for this is equivocal.

Knowledge about the role of hormones in regulating growth, development, and secretion of the integumental glands is incomplete. Clearly, however, some such glands do respond to hormones though the primary importance of neural secretory stimuli should not be forgotten. The functions of such glands vary considerably in different vertebrates, and it is interesting that many have attained a responsiveness to certain hormones. The nature of these effects appears to be related to the other basic functions of the hormones in the body; thus, gonadal steroids influence skin glands that are involved in sexual activities, and aldosterone acts in a manner commensurate with its role as a hormone that conserves sodium in the body.

Information about hormone effects on skin glands in nonmammals is sparse. Epinephrine stimulates chloride secretion from the skin glands of European frogs and this may reflect a direct action of such hormones in the body or a mimicking of a stimulation of the sympathetic nerves. Epinephrine, as well as vasotocin, has also been shown to stimulate the secretion of a sticky, milky-white material from the proteinaceous skin glands of the African clawed toad, *Xenopus laevis* (Ireland, 1973). The effect of the catecholamine, but not vasotocin, can be prevented by an α-adrenergic blocking drug.

The control of epidermal skin proliferation

The control of cell proliferation in the skin has elicited much interest, especially among dermatologists. Apart from the control of the normal replacement of the epithelium, the rapid but controlled processes of repair that accompany wound healing are rather remarkable. These processes, like other changes in the skin, are thought principally to involve an internal control mechanism upon which hormones may impinge their influence.

Chalones are substances that have been isolated from a number of tissues including the mammalian epidermis (see Bullough, 1971). They inhibit the mitoses of epidermal cells including melanocytes and keratinocytes. The chalones are glycoproteins that act both *in vitro* or *in vivo* and show a considerable degree of tissue specificity. It is thought that chalones form an active complex with epinephrine. As cortisol also inhibits mitosis of epidermal cells it may also contribute to this complex. Alternatively, it has been suggested (see Quevedo, 1972) that the effect of epinephrine, in inhibiting cell mitosis, may involve the formation of cyclic AMP and that chalones may be part of the cyclic AMP–adenylate cyclase system. The promotion of rapid epidermal cell growth, such as following a wound, appears to be stimulated as a result of the inactivation of the chalone, but it is unknown how this is done. It has been suggested that a local release of prostaglandins may overcome the inhibitory effects of the chalone complex.

In 1962 (see Cohen and Taylor, 1974; Cohen and Savage, 1974; Hollenberg, 1979), a substance was isolated from the submaxillary gland of adult male mice, which, when injected into newborn mice, accelerated the opening of the eyes and the eruption of the teeth. It was also shown to have powerful mitogenic activity and facilitated growth and keratinization of a number of epidermal tissue preparations, including the skin and cornea, from mammals and birds (*in vivo* and *in vitro*). This substance was thus appropriately dubbed *epidermal growth factor*, or *EGF*. It is a polypeptide containing 53 amino acids and it has a molecular weight of 6045. It has also been found in the milk of mice and the serum of pregnant women. A closely related peptide has been isolated from human urine (Cohen and Carpenter, 1975). Rather remarkably, mouse EGF has been shown to have a very similar structure to *urogastrone* (Gregory, 1975). This is a substance found in human urine that can inhibit gastric acid secretion and may promote the healing of peptic ulcers. Epidermal growth factor appears to be a member of the family of growth-promoting substances that include the somatomedins. Its actions may be more specially directed toward epithelial tissues, such as skin. Specific receptors for EGF have been identified in the plasma membrane of human epidermal cancer cells (Cohen, Carpenter, and King, 1980).

Hormones and pigmentation

The integument of most vertebrates contains pigment that makes a major contribution to what is often a very colorful appearance. Pigment may be present within the epidermis or dermis itself or color the integumental appendages, such as scales, hair, and feathers. Apart from contributing to man's aesthetic delight in contemplating nature, it seems likely that an animal's coloration may be useful to its physiology (Hadley, 1972). Appropriate pigmentations may contribute to the animal's camouflage, protect the internal organs from solar radiation, promote the absorption or reflection of heat and light, and so aid in photoreception and contribute to the synthesis of vitamin D in the skin. Integumental colors also provide signs that are important for appropriate dimorphic sexual behavior and reproduction.

Pigments of different colors are usually present in the skin in cells called chromatophores. These cells commonly contain a black or brown pigment called melanin and are called melanocytes or, if the intracellular distribution of pigment can be changed, melanophores. The yellow and red pigments (xanthines and carotenes) that also occur in the skin of vertebrates are contained in, respectively, xanthophores and erythrophores. Some chromatophores also contain pteridine platelets that reflect light, giving an iridescent appearance and are thus called iridophores. The complex and beautiful colors of many vertebrates are the result of blending the colors reflected by the various chromatophores.

Many vertebrates can alter their coloration in response to environmental and behavioral needs. Such changes may take place in a relatively slow manner, as when the total amount of pigment in the epidermis, or its appendages, changes. The result is a relatively static coloration that is attained over a period of days or weeks. This process is called *morphological color change* (which may involve melanocytes or melanophores) and is seen when we tan in the sun or when an animal changes the color of its pelage or plumage with the onset of summer or winter or in preparation for the breeding season. In addition, many cold-blooded vertebrates can rapidly change their color, a process that only takes a few minutes or at the most several hours. This relatively rapid response is called *physiological color change*. Both morphological and physiological color changes are influenced by the actions of hormones, especially the pituitary melanocyte-stimulating hormone.

The melanophores are cells with long dendritic-like extensions that radiate from a central core. In shape they resemble nerve cells from which they are derived. The melanin is contained within cellular organelles called melanosomes. Darkening and lightening of the skin, as occur in physiological color change, reflect a migration of the melanosomes in dermal melanophores so that they are widely distributed in the cell (dark color, the melanin is said to be dispersed) or they aggregate in small globs in the center of the cell (light color, the melanophore is said to have a punctate appearance) (Fig. 7.3). The grosser effects of these changes on the color of frog skin are shown in Fig. 7.4. The dispersal of the melanin in the melanophores may depend on a microtubular system in the cell, as certain drugs, for example cytochalasin B, that break such tubules also prevent dispersion of the pigment. The other chromatophores have a rather similar structure to the melanophores but contain different pigments. Iridophores that respond to MSH do so in the opposite manner to that of the melanophores; the platelets of reflecting materials aggregate so that the cell has a punctate appearance. Not all chromatophores exhibit a physiological color change response to MSH, and indeed this is not usually seen in the xanthophores and erythrophores (Bagnara, 1969; Taylor and Bagnara, 1972). In epidermal melanocytes (unlike the dermal melanophores), the pigment is relatively fixed in its position so that differences that occur usually reflect the total quantities of pigment that are present (morphological color change).

Fig. 7.3. The microscopic appearance of the dermal melanophores of the dogfish *Scyliorhinus canicula*. When the fish is maximally dark, as in 5, the melanosomes are dispersed throughout the cell, which can then be seen in outline, while when pale they are aggregated in the central region (as in 1) so that definition of the cell outline is obscured. The numbers, 1 to 5, correspond to the "melanophore index." (From Wilson and Dodd, 1973a).

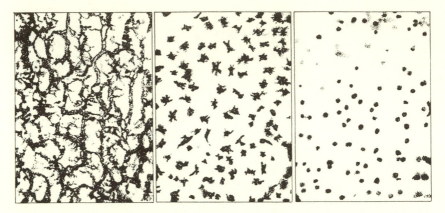

Fig. 7.4. The gross appearance of amphibian melanophores (under a low-power microscope) when the animal is (left to right): dark, intermediate, and pale in color. [From *The Pigmentary Effector System* by L. Hogben (1924). With permission of the publishers Oliver & Boyd, Edinburgh.]

The different types of chromatophores present in colorful animals, including many frogs and lizards, are arranged in layers, and changes in the distribution of pigment within these zones alter the transmission of light and the color that is perceived. The innermost layer consists of melanophores and these are overlain by the reflecting iridophores. The xanthophores form a layer closer to the surface of the skin. Changes in the density of the melanophores, which absorb light, and the iridophores, which can reflect it, thus can alter the color of a frog from a light to a dark color and influence the display of the colorful pigments in the superficially placed xanthophores. The melanophores, iridophores, and xanthophores together make up what is called a "dermal chromatophore unit."

The mechanism of release of melanocyte-stimulating hormone

As we shall see in the succeeding sections, the release of MSH plays a most important role in both physiological and morphological color change in vertebrates. A distinction between the release and the particular roles of α-MSH and β-MSH has not generally been made. The principal stimulus, especially in cold-blooded species, is the receipt of light, usually by the lateral eyes, but the pineal may also function as a photoreceptor in some species. In addition to directly influencing MSH release, in acute situations, light may also contribute to a cyclical, photoperiodic release of this hormone in some mammals that seasonally change the color of their pelage. Other stimuli that result in a release of MSH include a deficiency of adrenocorticosteroids and suckling and copulation in mammals, and increases in the osmotic concentration of the plasma in a variety of species. The latter effects may not have any physiological significance but this remains to be further explored. The intimate mechanisms that control the release of MSH from the pars intermedia are only partly understood and appear to be quite complex (see Howe, 1973; Kastin, Schally, and Kostrzewa, 1980; Hadley 1980).

When the pars intermedia is transplanted ectopically, to another part of the body away from the hypothalamus, or if its connections to this part of the brain are severed, MSH is secreted in an apparently uncontrolled manner (Iturriza, 1969; Penny, Tilders, and Thody, 1979). The regulation of hormone release thus appears to be under an inhibitory control originating in the hypothalamus.

The pars intermedia usually has a nerve supply that comes from the base of the brain that contains aminergic neurons that secrete catecholamines and neurosecretory fibers that release peptide hormones. Cholinergic nerves have also been observed. The vascular supply to the tissue shows considerable species variability, but portal vessels coming from the hypothalamus and neural lobe have been described. The controlling stimuli appear to mainly involve aminergic nerves that originate in the region of the hypothalamus. It is possible that a dual-control system is present that inhibits and promotes the release of MSH.

The neural stimuli responsible for regulating the secretion of MSH from the pars intermedia appear to be mediated by secretion of dopamine, norepinephrine, and, possibly, epinephrine and 5-hydroxytryptamine (serotonin) (see Wilson and Dodd, 1973b; Bower, Hadley, and Hruby, 1974; Olivereau, 1978; Kastin, Schally, and Kostrzewa, 1980). The inhibitory effects on release of the hormone may be mediated by dopamine and possibly an α-adrenergic effect of norepinephrine. Stimulation can also occur as a result of the action of norepinephrine or epinephrine, and this response is of the β-adrenergic type. The MSH-secreting cells from the pars intermedia of rats have been separated and studied in tissue culture (Cote et al., 1980; Munemura et al., 1980). Release of the hormone followed β-adrenergic stimuli and this effect is accompanied by an increase in the formation of cyclic AMP. Both of these effects can be antagonized by dopamine. The possible roles of the peptides MSH-R-IH and MSH-RH (see Chapter 3) are not clear; observations on their actions have not been consistent, and their physiological role is at present equivocal.

To complicate matters even further, it has been proposed (see discussion of hormones and seasonal changes in fur color) that melatonin, from the pineal gland, may mediate photoperiodic release of MSH by an action on the hypothalamus. Melatonin injections have been shown to increase release of MSH, and a physiological rise in the levels of the former, such as occurs during darkness, could result in the latter's release under normal conditions.

Studies on the mechanism of release of MSH were initially performed on amphibians but have since been extended to mammals, reptiles, and fishes. The observations have generally indicated that a similar system for the control of secretion of this hormone exists throughout the vertebrates. Thus, the evidence for an inhibitory role of dopamine and α-adrenergic stimulation appears to be consistent in species as phyletically separate as dogfish and rats. Pharmacological observations on a teleost fish, eels, and a reptile, the lizard *Anolis carolinensis*, suggest that nerves that secrete 5-hydroxytryptamine (serotonin) may stimulate the secretion of MSH (Olivereau, 1978; Levitin, 1980). Whether this mechanism is unique to these groups of vertebrates is at present unknown.

Physiological color change

Physiological color changes occur in many cold-blooded vertebrates, from the cyclostomes to reptiles. These changes in the distribution of the pigment in the skin occur in response to a variety of conditions and stimuli. Many vertebrates exhibit a diurnal rhythm in the degree of aggregation and dispersion of melanin in the melanophores. They turn pale at night and dark during the day. This change may reflect the perception of light and be mediated by receptors in the eyes and the pineal, and sometimes can result from a direct stimulation of the melanophores by light. In other instances, such as in the lizard *Anolis carolinensis*, a diurnal rhythm can even be seen when the animals are kept in complete darkness. As it is not seen in these lizards after they are hypophysectomized, it probably reflects an inherent diurnal rhythm in the activity of the pituitary gland. Superimposed on such rhythmical changes in skin color are direct, and adaptive, responses to external stimuli. These stimuli include the perception of certain light patterns due to the color and shade of the substrate on which the animal is placed (*background response*) and, to a lesser extent, the external temperature and "excitement." The latter two effects, which have been observed more commonly in lizards, may override the background response.

The first recorded observations of physiological color change are more than 2000 years old, but our understanding of the mechanism involved is quite recent. An appreciation of the role of hormones in these responses principally resulted from the pioneering studies of Hogben and Winton in the 1920s. An excellent account of the work of the Hogben school in England and that of many others, including Parker in the United States, has been given by Waring (1963). Waring joined the Hogben school in the 1930s, and his account of the processes involved in regulating color change is an ideal example of the stringent analytical approach and the application of formal logic that we should all aspire to in scientific investigations. Although physiological color change does not occur in mammals or birds the elucidation of its mechanism has contributed a great deal to our understanding of the role of hormones in physiological coordination. The following account is largely based on Waring's, but one should also consult the book by Bagnara and Hadley (1972).

Types of melanophore response

Nonvisual

a. Coordinated; this type of response may be abolished by denervation of the skin or the removal of the pituitary or the adrenals.

b. Uncoordinated, where the melanophore (or possibly a skin receptor close to it) directly responds to a stimulus. This type of response can be seen rather clearly in the horned toad, *Phrynosoma blainvilli*. If these lizards are blinded, hypophysectomized, and the pineal eye is covered, and they are then placed in a black box with no light, they become a pale color. When, however, a thin beam of light is focused on a piece of denervated skin in these lizards this darkens in comparison to the rest of the integument. A localized response to temperature can also be demonstrated in *Phrynosoma*, for when an area of the skin of a maximally dark lizard is exposed to water at 37°C it pales in that region. Similarly, maximally pale skin will darken locally at a temperature of 1°C. Chameleons also exhibit dramatic localized changes in skin color; the skin of blinded animals turns dark in light but if a certain area is shaded by an object, a lighter colored "print" or outline of this object can be seen. Such responses do not involve hormones or the ordinary nerve supply and appear to reflect a direct response of the melanophore; however, a local nerve reflex initiated from a nearby cutaneous receptor or a release of a local hormone could be involved.

The visual response

This response is the result of the reception of light by the lateral eyes or in certain species, including some larval amphibians and cyclostomes, and possibly even some lizards, the pineal. The responses may be a generalized lightening (in the dark) or darkening (in the light) of the skin or be influenced by the color of the background: the background response. When the animal is on a white substrate with overhead illumination, it may turn a pale color and if on a black background (also with overhead illumination) it may turn a dark color. These changes are called the white (or tertiary) and black (or secondary) ocular-background responses. The different effects of light in these two sets of circumstances appear to be due to the stimulation of different parts of the retina; thus, the eye of a frog in a black tank of water (Fig. 7.5) receives

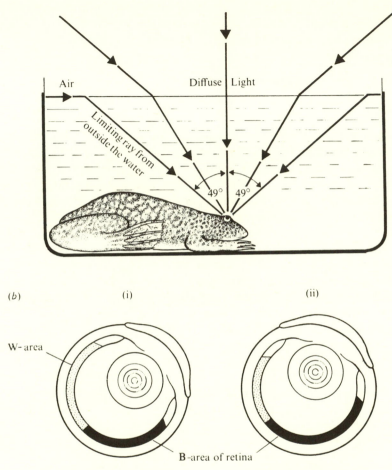

(a)

Air

Diffuse | Light

Limiting ray from outside the water

49° 49°

(b)   (i)   (ii)

W-area

B-area of retina

Axis of symmetry 35° to horizontal   Axis of symmetry 65° to horizontal

Fig. 7.5. The manner by which the reception of light initiates the dispersion of melanin in the melanophores of a frog (in this instance *Xenopus*) sitting in a tank of water with a black background and overhead illumination. As the tank has black sides all light will enter the water from above and enter the lens at an angle that is the critical angle for air and water (49°). (From Waring, 1963.) Using this data, as well as the dimensions of the eye and the refractive index of the lens (*Xenopus* has its eyes on the top of its head), Hogben provided a diagram (b, i, ii) that shows the area of the retina receiving such light rays. As these conditions result in a darkening of the skin, this has been called the B-area (for black) as opposed to the W-area (for white), which initiates skin lightening in frogs on a white background when light reaches wider areas of the retina. (From Hogben, 1942. Reproduced by permission of the Royal Society.)

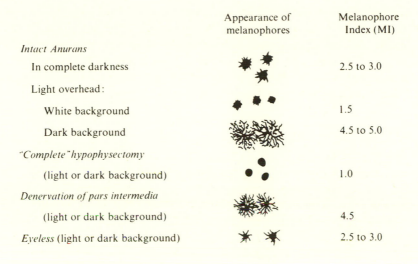

	Appearance of melanophores	Melanophore Index (MI)
Intact Anurans		
In complete darkness		2.5 to 3.0
Light overhead:		
White background		1.5
Dark background		4.5 to 5.0
"Complete"hypophysectomy		
(light or dark background)		1.0
Denervation of pars intermedia		
(light or dark background)		4.5
Eyeless (light or dark background)		2.5 to 3.0

Chart of Melanophore Index for amphibians:

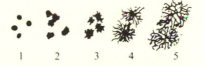

1   2   3   4   5

Fig. 7.6. The melanophore responses, mediated through the eyes, of anurans in relation to the receipt of light when on a white or a dark background. The effects of surgical changes of the pituitary on these responses have been summarized. The melanophore index (MI) in relation to the degree of dispersion, or aggregation, of melanin in the melanophores is given in the lower section. For a description and explanation of these responses the text should be consulted. (Based on Bradshaw and Waring, 1969.)

light only on the more basal parts of the retina, the "B" (for black) area, but in a white tank, where the light is reflected into the eye from all sides, the entire retina, including a "W" (for white) area, is stimulated. Such special receptor areas for light in the retina have also been found in teleost fishes and lizards.

Quantitation of the melanophore response

Early observations of vertebrate color change have been described in general subjective terms such as "pale" and "a tint rather dark than pale" that lack adequate precision for a proper scientific analysis and make comparisons of results from different laboratories almost impossible. Hogben introduced a more stringent quantitative description called the melanophore index (or MI) (Figs. 7.3 and 7.6), which has a gradation of 1, for maximally pale, with the melanin fully aggregated, to 5 when the animal is dark and the melanin is fully dispersed. In lizards, this can be translated, as in chameleons, to 1 = yellow, 3 = medium-green, and 5 = black. This simple standard of measurement allowed considerable advances to be made in the analysis of the mechanism of color change. Today, electrophotoreceptive devices are also used to quantify the melanophore responses.

Color change in amphibians

The earliest observations on the role of hormones in vertebrate color change were made on European frogs, *Rana temporaria*, and subsequently the African clawed toad, *Xenopus laevis*. With overhead illumination (see Fig. 7.6), on a white background these amphibians are pale (melanophore index about 1.5) and on a black background they are dark (MI about 4.5). When placed in complete darkness, or if they are blinded, they have an intermediate shade (MI = 2.5). In *Xenopus*, this change in melanin distribution in the melanophores is seen as a white or a black coloration but in frogs, which have overlying layers of yellow-green pigment, this appears as a pale green or yellow to a black color. When amphibians are "completely" hypophysectomized so that no pituitary tissue remains (such remnants commonly *do* remain, as in *Xenopus*), the animals become maximally pale, the MI is 1 and they cannot respond to changes in the background color. If the pars distalis is removed carefully so that the pars intermedia and pars tuberalis remain intact, the background responses are retained. This operation is relatively simple to perform in *Rana* but it is more difficult in *Xenopus* where the pars tuberalis is usually removed together with the pars distalis. This results in an inability of *Xenopus* to display a background response and it becomes permanently dark (MI = 5). The pars intermedia has a nerve supply coming down from the hypothalamus and when this is cut the anurans also have an MI of 5. Removal of the pars tuberalis appears to be associated with an interference of the hypothalamic connections to the pars intermedia and, as these are of an inhibitory nature a sustained release of its secretion, MSH, occurs.

Melanocyte-stimulating hormone (MSH) when released from the pars intermedia is carried in the blood to the melanophores where it promotes a dispersion of melanin so that the animal darkens.

The sequence of events resulting in the black background response is summarized in the following section (Fig. 7.7). Light from an overhead source, in frogs on a dark background, falls on the "B" area of the retina where it stimulates receptors that transmit messages along the optic nerve. These messages, traveling along pathways that are as yet unknown, inhibit the normal inhibitory effects of the nerves supplying the pars intermedia and this results in a release of MSH. When the frogs are on a white background, light also

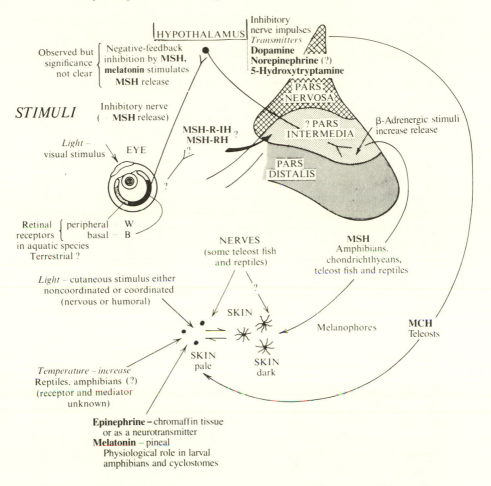

Fig. 7.7. A composite diagram summarizing the humoral and neural control of color change in vertebrates.

falls on the retina but on the receptors in the "W" area and this reduces the release of MSH. Normally, anurans kept in the dark, as well as blinded animals, have an MI of about 3, which appears to reflect a sustained, but submaximal, release of MSH. Stimulation of the "W" retinal receptors in some way inhibits this release even further, possibly by increasing the inhibitory nerve impulses to the pars intermedia.

Pallor of the skin is thus usually thought to result from a decline in the levels of MSH in the blood. In the amphibia, in contrast to some teleosts and reptiles, nerves are not involved in the aggregation of melanin in the melanophores. The possibility that a melanin-concentrating hormone, or MCH, may be present in the pituitary has been considered for over 50 years. This concept was based on the observation that an intact pars tuberalis was necessary for the occurrence of a white background in *Xenopus*. It is now, however, agreed that removal of this tissue results in the severance of the inhibitory hypothalamic connections to the pars intermedia that are necessary for maximum color change to occur (see Bradshaw and Waring, 1969). Nevertheless, sporadic observations have consistently reported that extracts of the pituitaries of teleost fish contain a substance that promotes the concentration of melanin in the melanophores, and so has a paling effect (see Baker and Ball, 1975; Rance and Baker, 1979). This biological activity has been found in the posterior lobe of the pituitary gland and it is also present in high concentrations in the hypothalamus. It does not behave chemically like an amine. This putative hormone has been called melanophore-concentrating hormone. Its levels in the hypothalamus decline in fish kept on a white background. It may be formed in neurons in the hypothalamus, the axons of which extend into the neurohypophysis, from where it is secreted. The structure of melanophore-concentrating hormone has not been chemically described, but preliminary observations indicate that it is a polypeptide containing no disulfide bonds, and it has a molecular weight between 3500 and 13,000 (Westerfield, Pang, and Burns, 1980).

The injection of epinephrine into *Rana* produces a skin lightening and an aggregation of melanin in the melanophores, which is an α-adrenergic response. On the other hand, in the skin of *Xenopus laevis* and in the spadefoot toad, *Scaphiopus couchi*, *in vitro*, catecholamines disperse melanin, which is due to an increased formation of cyclic AMP following β-adrenergic stimulation (Goldman and Hadley, 1969; Abe et al., 1969). The normal physiological importance of such actions *in vivo* is unknown.

When dried bovine pineal glands are fed to anuran tadpoles, the melanin in the melanophores on the body, but not the tail, aggregates and the animal's body pales (see Bagnara and Hadley, 1970). Under these conditions, the tadpole's internal organs can be seen clearly. This effect is due to the action of melatonin formed in the pineal gland. Normally, tadpoles, such as those of *Xenopus laevis*, pale at night and darken during the day, and this can be prevented if the pineal, but not the lateral eyes, is removed. Formation and release of melatonin occur in darkness and this appears to mediate the diurnal rhythm of color change in these tadpoles. The effect of melatonin on the melanophores is a direct one and is not mediated through the pituitary as has sometimes been suggested, as the effect of melatonin is not prevented by hypophysectomy. This physiological effect of melatonin is confined to tadpoles and does not contribute to skin lightening in adult amphibians.

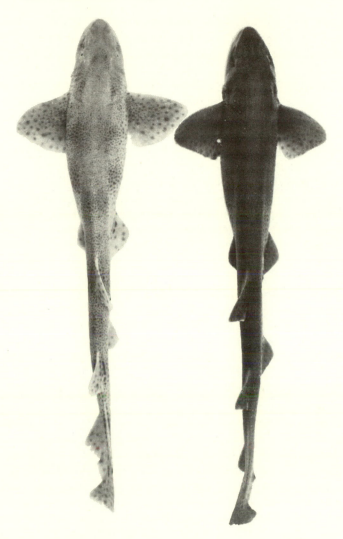

Fig. 7.8. Two dogfish, *Scyliorhinus canicula*, in their dark and pale color phases
(× 0.33). (J. F. Wilson, personal communication.)

Color change in the Chondrichthyes

Many sharks and rays exhibit dramatic changes in color depending on the
shade of the background; with overhead illumination they become dark on a
black background and pale on a white one. Two dogfish (*Scyliorhinus canicula*)
in their dark and pale phases are shown in Fig. 7.8. Waring, in 1936, found
that when he transplanted a dogfish pituitary into another dogfish that was

pale in color, it turned dark due to a dispersion of melanin in the melanophores (Waring, 1936; 1938). The release of the MSH is due to the absence of the hypothalamic neural inhibitory control mechanism present in these fish (Wilson and Dodd, 1973a). An analysis of the color change in these fish, *Squalus*, *Scyliorhinus*, and *Raja*, shows that they exhibit white- and black-background responses that are mediated humorally by MSH, just as in amphibians. Direct neural control of the melanophores does not appear to occur in the Chondrichthyes.

The background response is not seen in blinded dogfish though the fish show a slight paling in darkness that suggests the presence of a nonvisual response (Wilson and Dodd, 1973a). When kept in total darkness, the pallor exhibited by these dogfish is not seen if the pineal is removed and they become darker. The pineal may thus contribute to nonvisual color change as observed in tadpoles and cyclostomes (see later).

Color change in teleosts

Although the chondrichthyean fishes and amphibians that have been examined all have humoral control of their color change, the teleosts, which lie phyletically between these two groups, may also possess a neural coordinating mechanism. The teleosts, as has become apparent in the comparison of their other biological systems, exhibit considerable interspecific differences in the control of color change that presumably reflect the systematic diversity within this large group of fishes.

Stimulation of nerves controlling melanophores usually results in an aggregation of the melanin and a paling of the skin color in teleosts. A dispersion of melanin, in response to neural stimuli, however, may also occur in some species but the evidence for this is equivocal. I have been unable to ascertain with certainty whether the autonomic nerve fibers involved in such color changes are cholinergic or adrenergic, though the aggregating effects of injected epinephrine would tend to favor the latter; however, species differences may exist and it has been suggested that in the eel *Anguilla anguilla*, melanin dispersion is an adrenergic response while the aggregation is a cholinergic one. Alternatively, melanin aggregation may be an α-adrenergic response and dispersion a β-adrenergic one.

Responses of melanophores to neural stimuli are very rapid and may take place within several minutes. The humoral effects of MSH are, in contrast, slow and usually take 1–2 hours for their completion. This tardiness reflects the gradual buildup or removal of MSH from the circulation rather than any lethargy on the part of the melanophores themselves.

One can readily foresee the prospective biological advantages of an ability to change color rapidly as this may help protect the animal from a predator or assist it to catch food. Rapid color change may be especially important in animals that live in places where the background colors are variegated and across which the animals constantly travel in their search for food and sexual companionship. Many teleost fishes that roam about gaily colored reefs may find such rapid color change an especial advantage.

In the Teleostei, color change can be mediated by three types of mechanisms: (*a*) a humoral one, (*b*) a neural one, or (*c*) a combination of both neural and humoral processes.

Anguilla (*the eel*)

Eels exhibit black- and white-background responses but only change their color slowly, like amphibians. Hypophysectomy abolishes the full expression of these responses and as the pituitary contains a material that, when injected, disperses melanin in the melanophores, the response is considered to be predominantly a humorally mediated one.

Following hypophysectomy, the eel is not maximally pale but has a melanophore index of 1–2 and this has contributed to speculation that a melanocyte-concentrating hormone is also present in the pituitary, as, if only MSH were involved, one would expect a MI of about 1. It has been suggested that, alternatively, melanophore-dispersing nerve fibers are present.

In contrast to amphibians, hypophysectomized eels continue to exhibit a small background response; the MI is 3.5 on a black background and 1.8 on a white one. This response is abolished by the severance of cutaneous nerves that are known to innervate the melanophores; thus, although the predominant mechanism mediating color change in eels is humoral there is an underlying neural control that only becomes apparent after the pituitary is removed.

Epinephrine, when injected, readily produces an aggregation of melanin in the melanophores of eels. This hormone is produced by widely distributed chromaffin cells in teleosts, but it is unknown if it has a normal role in mediating their color changes.

Fundulus heteroclitus (*the killifish*)

Killifish exhibit the usual black- and white-background responses but these are *not* abolished following hypophysectomy. *Complete* darkening, or dispersal of melanin, does not occur following this operation, indicating that pituitary MSH may be necessary for the full expression of the black-background response.

The overriding control is nevertheless a neural one. Injections of MSH into pale fish does not disperse melanin and electrical stimulation of cutaneous nerves in dark fish evokes pallor. The injection of MSH into pale fish that have had parts of their skin denervated results in a melanin dispersion in these localized areas. Extracts of *Fundulus* pituitaries can evoke such dispersion of melanin.

In *Fundulus*, color change thus occurs in response to neural stimuli to the melanophores; there are melanin-aggregating nerve fibers, and possibly even "dispersing fibers." Underlying this mechanism, but generally overridden by it, is an ability to respond to MSH, and this hormone is necessary for a maximal darkening of the fish.

Phoxinus phoxinus *(the European minnow)*

In minnows, there is little evidence for a role of endogenous hormones in the dispersion of melanin. Black- and white-background responses are not prevented by hypophysectomy, though it has been observed that such fish cannot sustain a black coloration as readily as intact fish. Denervation of the skin abolishes the background responses and the melanin fully disperses. Nerve stimulation evokes an aggregation of melanin and aggregating nerve fibers undoubtedly exist. There is also some evidence that suggests the presence of melanin-dispersing fibers.

The injection of extracts of the pituitaries from *Phoxinus* does not disperse melanin in either the intact or denervated skin of these fish. MSH from anurans will, however, darken the denervated skin of *Phoxinus*. There is no evidence for a melanin-dispersing hormone in *Phoxinus* and indeed pituitary extracts have an opposite, aggregating effect, apparently reflecting the presence of a melanophore-concentrating hormone. As already described, melatonin has a blanching effect in a number of vertebrates due to its ability to aggregate melanin. However, the pituitary melanophore-concentrating hormone is not an amine. A role for melatonin in color change in teleosts has been suggested, but in the rainbow trout the plasma level of this hormone could not be related to such adaptation (Owens et al., 1978).

Color change, thus, in teleosts may be influenced by three types of mechanisms:

1. A predominantly humoral one that overrides a neural mechanism, as in *Anguilla*. It may involve MSH and MCH.

2. Predominantly neural coordination that overrides a humoral process but which is still important, such as in *Fundulus*.

3. A neural coordinating mechanism with no evidence for an effect of endogenous MSH, as in *Phoxinus*. The presence of an MCH has, however, not been excluded.

Color change in reptiles

The reptiles have either humoral or neural mechanisms coordinating their color changes but as yet no transitional arrangements, which involve both, have been described. Most observations, certainly the most detailed ones, have been made on lacertilians that often display very dramatic changes in color as epitomized by the chameleons.

Snakes and crocodilians also possess chromatophores, and a chelonian, *Chelodina longicolis*, has also been shown to exhibit background responses that are mediated by MSH (Woolley, 1957).

The lizard, *Anolis carolinensis*, exhibits both visual and nonvisual background responses that are abolished following hypophysectomy. The nonvisual response, which can be overridden by the visual one, may be the result of photostimulation of the pineal.

Excitement, such as results from electrical stimulation of the mouth or cloaca, results in a mottling of the skin color patterns in *A. carolinensis* because of a dispersion of melanin in some melanophores and an aggregation in others. This effect can be mimicked by the injection of epinephrine, which may normally mediate the response. The darkening appears to be due to β-adrenergic effects and the lightening to α-adrenergic ones. The melanophores are not innervated, and cutting the general nerve supply to the skin does not influence color change.

Chameleons, *Chamaeleo pumila* and *Lophosaura pumila*, exhibit rapid and dramatic changes in color that are either visual responses or are due to photoreceptors that are apparently present in the skin. The observations that have been made on these responses suggest a neural control of the melanophores; nerve stimulation results in an aggregation of melanin. It is also possible that melanin-dispersing nerve fibers are present. It thus seems that the control of color change in chameleons is a neurally coordinated process though, as no experiments seem to have been done on hypophysectomized animals, a subsidiary role of MSH cannot be completely excluded.

Color change in cyclostomes

The control of color change in the lampreys (Petromyzontoidea) and hagfishes (Myxinoidea) is intrinsically very interesting because of their lowly phyletic position on the vertebrate scale.

A background-color response, pale on a white substrate, dark on a black one, has been described in the hagfish, *Myxine glutinosa*, but the coordinating mechanism for this change is unknown.

Lampreys appear to lack a background response but exhibit a diurnal rhythm in color, dark in the day and pale at night, that in some species is mediated by the pineal gland. Young, in 1935, found that removal of the pituitary abolishes this rhythmical color change in adults and ammocoete larvae of *Lampetra planeri*, and the lamprey then becomes permanently pale in color. In the ammocoetes, pinealectomy also abolished this diurnal rhythm but, in contrast to hypophysectomy, the animals were permanently dark. These observations have recently been extended (Eddy and Strahan, 1968) to two species of Australian lampreys. These antipodean cyclostomes also exhibit a diurnal rhythm in color which stops following hypophysectomy. In larval *Geotria australis*, pinealectomy also abolishes the rhythm but in metamorphosing larval *Mordacia mordax*, the lateral eyes must be removed to see this effect.

As described earlier, the pineal is the site (especially in the dark) of formation of melatonin, which in anuran tadpoles is a very potent stimulant of melanin aggregation and so pales the skin. It has been found that the injection of melatonin into larval *Geotria* also results in skin pallor though this effect is absent in *Mordacia*. In addition, if the pineal is transplanted under the skin of *Geotria* a local paling is observed. The pineal may thus be involved in regulating the rhythmical changes in color seen in the ammocoetes of *Geotria*, both by the production and release of melatonin and by acting as a photoreceptor organ. Following pinealectomy, the ammocoetes are permanently dark, which may reflect either the lack of an inhibitory effect of melatonin on the release of MSH or, more likely, a direct antagonism to the action of MSH on the melanophores. It also seems likely, in retrospect, that the same mechanism(s) regulates color change in the ammocoetes of *Lampetra planeri*.

The involvement of melatonin in color change in some larval cyclostomes, some amphibians, and a chondrichthyean is most interesting. Melatonin does not seem to have this role in many vertebrates, but nevertheless it is present in representatives of all the vertebrate groups. The propensity of the pineal to respond to diurnal changes in light makes it a potentially valuable gland for mediating endocrine rhythms dictated by changes in the seasons. As we shall see in Chapter 9, the pineal may in this manner contribute to the control of reproductive cycles in some vertebrates.

Evolution of color change mechanisms in vertebrates

From the preceding observations on the mechanisms controlling color change, we may make some guesses about the evolution of this process in vertebrates. Information of a comparative nature about extant species from diverse phyletic groups can be interpreted in a manner that may help us to reconstruct the past. Waring has summarized the available information about vertebrate color change in Fig. 7.9, but states that he has been sternly warned "about pressing this kind of thing too far."

The original underlying mechanism coordinating color change in vertebrates would appear to have been a humoral one as this has been observed exclusively in the Chondrichthyes (elasmobranchs) and has persisted in the Anura and Chelonia, all of which can be traced back to the Triassic. Superimposed, and probably subsequent to this, has been the evolution of a neural control of the melanophores that is seen in some teleosts and reptiles. In the

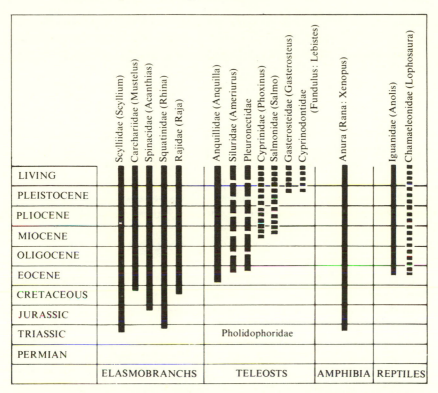

Fig. 7.9. The geological age of the different vertebrate groups in relation to the types of mechanisms (humoral, neural, or both) they utilize to coordinate their melanophore background color responses. The solid bars represent the possession of predominant humoral mechanisms; the broken bars, mixed humoral and neural mechanisms; the small squares, predominantly neural control. To this diagram could be added the Chelonia (*Chelodina oblonga*) and the Ophidia (*Crotalus*), both of which have a humoral control mechanism and are identifiable from the Triassic. It can be seen that humoral mechanisms appear to be the oldest, suggesting that neural control is a later acquisition. (From Waring, 1942.)

teleosts, there is evidence of transitional changes as some species appear to utilize both neural and humoral mechanisms. The eels (Anguillidae) are specially interesting in this respect as they are normally completely dependent on humoral control, but there is also evidence for the presence of a neural mechanism that normally does not appear to contribute to color change. We can only speculate as to whether the eel represents a stage in the evolution toward a neural control of color change or is an evolutionary regression from this development. At the bottom of the vertebrate phyletic tree are the cyclostomes that possess an MSH that helps mediate a diurnal rhythm in color change, but there is little evidence to suggest that these lowly fishes have an adaptive background color response. We can speculate even less as to whether or not the role of melatonin in controlling color changes in larval cyclostomes reflects a primeval effect of this hormone, though it might!

Morphological color change

Several hormones influence morphological changes in the color of the integument. They include MSH and sex hormones, and corticotropin and melatonin may also be involved. Such humoral effects on pigmentation are less well characterized than those of MSH on physiological color changes. Considerable interspecific differences in humoral effects on morphological pigmentation occur that make it impossible to give any generalized definition of their respective roles.

Melanin is formed from tyrosine by a complex chain of reactions that initially involves a copper-containing enzyme called tyrosinase. This enzyme may be in a soluble form in the cytoplasm of the melanocyte but when active it is attached to the melanosomes. A genetic absence of tyrosinase results in albinism though other factors, including ultraviolet light and several drugs, can also contribute quantitatively to changes in pigmentation. Increased levels of integumental melanin are associated with increased tyrosinase activity, which can be influenced by hormones.

The morphological pigmentation of the skin, as well as fur and feathers, is in the first instance the result of the formation of melanin in the epidermal melanocytes, where it is attached to the melanosomes. In the skin, each melanocyte is associated with several keratinocytes to which the melanosomes, with their attached pigment, can be transferred. This functional association is called the "epidermal–melanin unit" (Quevedo, 1972). In birds and mammals, the melanin is passed from the melanocytes associated with the feather tracts or hair follicles to the developing feathers or fur.

Dramatic changes in pigmentation associated with endocrine function have been noted in mammals that show seasonal differences in coat color, in birds that display sexual dimorphism in plumage color, and in mammals suffering from endocrine imbalances. Many monkeys display prominent changes in color of parts of their skin associated with the sexual cycle; these areas are called the "sex skin." At ovulation the buttocks enlarge and become red in color because of an engorgement of blood in the large venous sinuses. There is also an accumulation of mucopolysaccharides in the skin. This development of the sex skin is under the control of estrogens. These are only a few examples of the pigmentary changes that may occur in vertebrates and which are influenced by the action of hormones.

Hormones and seasonal changes in fur color

As described earlier, short-tailed weasels (*Mustela erminea*) change their coat color from brown to white with the onset of winter (Rust and Meyer, 1968). This change is a photoperiodic response due to changes in the length of the daylight hours and can be prevented by hypophysectomy. The latter operation results in a permanent white coat, but the growth of new brown fur can be promoted when the weasels are injected with MSH or corticotropin. In addition, if the pituitary is transplanted to the kidney, where MSH release is increased due to a lack of hypothalamic inhibition, brown fur also grows on previously white animals.

If melatonin, in a "slow-release vehicle" of beeswax, is implanted under the skin, weasels undergoing a normal spring molt from a white to brown pelage, regrow white, instead of brown, fur. Rust and Meyer (1969) suggest that this is the result of a stimulation by melatonin of the release of MSH-R-IH from the hypothalamus so that the MSH levels drop. This experimental treatment of weasels also prevents the development of the gonads and the onset of normal spring reproductive cycles, so that it is possible that an interference with the normal levels of other hormones, apart from MSH, is also contributing to the effect. Precise experiments of this kind, to determine the role of hormones in seasonal changes of coat color in mammals, have been rare but, as described in the next section, there are several other examples of pigmentary changes in coat color that can be induced by a deficiency or excess of MSH.

Hormones and morphological color change in cold-blooded vertebrates

Cold-blooded vertebrates that undergo physiological color changes have also been shown to increase the levels of melanin in their skin in response to a continual environmental "black-background" stimulation of MSH secretion. Such a change has been observed in amphibians, teleosts, and chondrichthyeans. In the goldfish, *Carassius auratus*, corticotropin increases the cutaneous levels of melanin and the activity of tyrosinase (Chavin, Kim and Tchen, 1963), an effect that cannot be mimicked by mammalian MSH. In contrast, MSH, but not corticotropin, increased cutaneous melanin synthesis in the killifish, *Fundulus heteroclitus* (Pickford and Kosto, 1957) and one must therefore be careful not to draw any general systematic conclusions about the role of such hormones in melanin synthesis in the Teleostei. As the hormone preparations used in the fishes were of mammalian origin the differences in the responses could reflect the degree of similarity of these exogenous hormones to the particular endogenous MSH present in each species of fish.

The effects of changes in the level of endogenous MSH on melanin levels in the melanophores have been observed in the dogfish *Scyliorhinus canicula* (Wilson and Dodd, 1973*b*). Removal of the neurointermediate lobe of the pituitary in this chondrichthyean resulted in an almost complete loss of melanin from the skin. In the converse experiment, when increased circulating levels of MSH were promoted by severing the inhibitory hypothalamic connections to the intermediate lobe, there was an increased concentration of melanin in the skin.

Cold-blooded vertebrates may also exhibit morphological changes in skin color and pattern that are of a rather colorful nature. Some female lizards develop colored, orange and orange-red, spots on various parts of their bodies during the period of the development of their eggs in the body; they are thus called "pregnancy spots" (Cooper and Ferguson, 1972; Ferguson and Chen, 1973; Medica, Turner, and Smith, 1973). Such changes in color have a hormonal basis that has been examined in the collared lizard, *Crotaphytus collaris* and the leopard lizard *C. wislizenii*. The injection of progesterone induces such pigmentation in ovariectomized lizards, and estrogen increases the response though it is ineffective alone. The natural levels of these hormones change during the growth of the eggs, and their circulating concentrations have been measured in such lizards and can be correlated with the development of the pregnancy spots. The injections of FSH also induces the formation of such pigmented areas in leopard lizards. It would thus appear under natural conditions that a release of gonadotropin stimulates the development of the ovarian follicle together with a release of gonadal steroids (see Chapter 9), and these directly mediate the response. The role of pregnancy spots in these lizards is uncertain, but it has been suggested that they may deter the males from inappropriate amorous advances.

Hormones and sexual dimorphism in avian plumage color

The mechanisms of the effects of hormones in mediating the seasonal sexual dimorphism in the color of plumage in weaver birds, *Steganura paradisaea*, and the nonseasonal differences in domestic brown leghorn fowl have been studied by Hall (1969).

The male weaver bird grows prominent black feathers just before (prenuptial) the breeding season. When areas of white feathers are plucked from the these birds the injection of pituitary LH results in the appearance of melanin granules in the feather tracts of these areas and a related growth of black feathers. The formation of melanin is associated with an increased activity of tyrosinase in the feather tracts. The effect of LH is direct and is not mediated through any action on the gonads as it is still seen in castrated birds.

The male house finch, *Carpodacus mexicanus*, has red or orange feathers on its crown, throat, and belly, which is in contrast to the female, in which this plumage is brown. When the colored feathers are plucked from castrated males the new, regrown feathers are of the female, brown type, which contrasts with the renewal of the colored plumes in intact birds (Tewary and Farner, 1973). It thus appears, that as in the male weaver bird, the more gaily colored plumage of the male is determined by the presence of male sex hormones. It seems, however, that these hormones are from the gonads of the finches, though an indirect action that could involve the pituitary is also possible.

The male domestic brown leghorn fowl has black feathers on its neck and breast; those in the female are pinkish brown. In this instance, the coloration of the male plumage is not hormone dependent but that of the female is due to the action of estrogens. The injection of estradiol increases melanin formation and the activity of tyrosinase. The estrogens act directly on the feather tracts as their action is a local one at the site of the injection (Greenwood and Blyth, 1935).

Mechanisms of hormone-mediated changes in integumental melanin distribution

Physiological color change is an alteration in the dispersion of melanin in the melanophores, such as that mediated by MSH and epinephrine, and this is related to the level of cellular cyclic AMP. This nucleotide is formed in the presence of MSH as a result of the activation of adenylate cyclase, which is presumably associated with the melanophores. Cyclic AMP stimulates dispersion of the melanosomes. This response requires the presence of calcium and may involve a microtubular system in the cell, but the details of this are unknown (Novales, 1972). Epinephrine, on the other hand, may inhibit adenylate cyclase and decrease the formation of cyclic AMP (α-adrenergic effect) or in some instances, as in the spadefoot toad *Scaphiopus couchi* and the lizard *Anolis carolinensis*, it also activates the enzyme (β-adrenergic effect) and so mimics the effect of MSH (Abe et al., 1969; Goldman and Hadley, 1969). The neural responses of the melanophores also appear to be mediated by changes in the levels of cyclic AMP (Novales, 1973). The aggregating nerve fibers may exert an α-adrenergic effect and dispersing fibers a β-adrenergic one. The latter effect is observed *in vitro* but its significance *in vivo* is unknown.

The increase in melanin synthesis that occurs in morphological color change appears to be due to an increase in the activity of tyrosinase. As we have seen, this enzyme is associated with the action of MSH in mammals and also LH and estrogens in birds. The precise mechanisms by which this change occurs may, however, differ for each hormone. Lee, Lee, and Lu (1972) have studied the effect of MSH on a mouse skin tumor that contains a high concentration of melanocytes (a melanoma). The increase in tyrosinase activity in response to MSH in this tumor does not appear to be the result of formation of new enzymes but rather the activation of those already present, possibly by the removal of an inhibitor. In the amphibian skin, a different mechanism may operate, and it is thought that a trypsin-like enzyme is released that activates tyrosinase, which is present in the cytoplasm. The latter enzyme is then attached to the melanosome where it initiates the conversion of tyrosine to dopa that eventually leads to the formation of melanin. In birds (Hall, 1969), the action of LH in increasing cutaneous tyrosinase activity also does not appear to involve the synthesis of a new enzyme, as its effect is not inhibited by puromycin. The effect of estradiol on pigmentation in the brown leghorn fowl, however, *is* inhibited by puromycin so that the action of this hormone may then be due to an induction of tyrosinase. The many effects of hormones on integumental pigmentation thus may be reflected in a diversity in the mechanisms by which they exert their effects. At the present time, however, it would appear that the activity of tyrosinase is central to their morphological actions.

Has there been an evolution of the role of MSH?

Although MSH has well-established effects in mediating color change in cold-blooded vertebrates, its normal role in birds and mammals is uncertain. MSH undoubtedly, in some circumstances and in certain species, stimulates the synthesis of melanin in the skin of mammals, but whether this is its normal physiological role is not clear. It is possible that MSH may exhibit such a function, especially in species like the weasel that seasonally change the color of their coat. Such an effect is, however, limited to a relatively few species and yet mammals, as well as birds, apparently possess MSH and sometimes in several molecular forms (see Chapter 3). Does it then have other physiological effects in these animals? Despite a widespread search no satisfactory answer to this has emerged.

The phyletic persistence of MSH-like activity in extracts of the pituitary gland of birds and mammals has formerly suggested that such a molecule probably endows the animal with some selective advantage and could reflect its role as a hormone. However, it is now known (see Chapter 3) that the amino acid sequence of α- and β-MSH are integral parts of the prohormone (proopiocortin) for ACTH, LPH, and the endorphins. Whether or not this parent protein is normally fragmented in such a way as to form the discrete MSH molecules will depend on the nature of the post-translational processing. In some animals, such as those with a distinct pars intermedia, this undoubtedly often occurs. However, in other instances, such as in man, the identification of MSH in extracts of the pituitary gland appears to reflect the nature of the chemical extraction process. Even plasma immunoassay procedures may cross-react (e.g., that of LPH and MSH) so that the actual identity of the peptide in the blood may be in doubt. It therefore seems that the MSH peptides may not exist as widely in nature as was formerly thought.

Despite this reservation MSH could have other roles as a hormone, apart from those involving pigmentation of the skin. The particular amino acid sequences of the MSHs appear to possess considerable innate biological activity, both when present as part of larger molecules, such as ACTH, or as individual peptides. These include an ability to activate adenylate cyclase and tyrosinase, and the mobilization of lipids. α-MSH has recently been identified in the brain of rats (Oliver and Porter, 1978). When injected into the brain of these animals it reduces the sensation of pain in a manner reminiscent of the endorphins (Walker, Akil, and Watson, 1980). MSH has also undergone tests in man for the treatment of some psychiatric conditions though with inconclusive results. It should, however, be recalled that MSH can activate tyrosinase and that this enzyme is involved in the synthesis of catecholamines, including those in the brain. Thus, one interesting hypothesis (Shuster et al., 1973) has proposed a logical metabolic pathway by which MSH may be able to influence brain function.

It is also possible that MSH may contribute to the regulation of salt metabolism in some vertebrates. There have been several accounts of the release of this peptide from pituitary glands following exposure to dehydrating conditions or salt solutions (see Howe, 1973). It has recently been shown that β-MSH and a related peptide, pro-γ-MSH, can enhance the release of corticosteroids, including aldosterone, from the mammalian adrenal cortex (Matsuoka et al., 1981; Al-Dujaili et al., 1981). (γ-MSH is a peptide that is incorporated in the proopiocortin molecule; it has little melanotropic activity and its function is unknown. It has been shown to be formed in the pituitary of dogfish [McLean and Lowry, 1981].) It has thus been of special interest to observe that the levels of plasma cortisol are four times greater in trout and eels kept on a dark, as compared to a white background (Baker and Rance, 1981). The physiological significance or possible relationship of these observations is at present not clear. MSH appears to have the potential, or prospect, to assume other physiological roles but which, if any, these may be is at present uncertain.

Conclusions

The integument is a very complex tissue that may be involved in several physiological phenomena, including osmoregulation, color change, temperature regulation, and reproduction. There are many characteristic processes involved in such mechanisms that have a definite systematic distribution (for example, sweat glands, mammary glands, and branchial chloride-secreting cells) so that when hormones are involved, as they often are, their effects follow phyletic suit. Such responses must have also arisen at distinct times during vertebrate evolution. It is interesting to observe that there is often a definite relationship between the nature of the particular hormone and the general physiological process involved; sex hormones influence sexual processes in the skin, as well as elsewhere, and adrenocorticosteroids regulate electrolyte movements in sweat glands and across the amphibian skin, as well as in the kidney (see Chapter 8). Some hormones have a special propensity to mediate processes in the integument. MSH thus influences pigmentation in nearly all groups of vertebrates commencing phyletically with the cyclostomes, and on occasions it may even promote formation of melanin in mammalian skin. Its action, however, appears to be rather "conservative" as it has no other established effect on any other types of process in the body nor for that matter in the integument either. There are, however, some theories as to other possible roles. In contrast, prolactin is "versatile" as, apart from effects at nonectodermal sites, it influences many integumental processes including molting cycles, the secretions from the mucous and mammary glands, proliferation of the crop-sac and development of the brood-patch in birds, and the control of water and salt movements across the gills of fishes. Some of these effects will be described in the next chapters.

8. Hormones and osmoregulation

About 70% of the body weight of animals is water in which are dissolved a variety of solutes, the presence of many of which is vital for life. Within the body, the solutions inside the cells differ from those that bathe the outside, and the composition of each of these solutions must be maintained so as to provide an environment with an electrolyte content and osmotic concentration suitable for life. These intra- and extracellular fluids provide the framework in which life exists.

The physicochemical properties of the body fluids in animals usually differ greatly from those of their external environment. Animals continually suffer exposure to the whims of the exoteric conditions and this will tend to change the composition of their body fluids. In addition, although the intra- and extracellular fluids have identical osmotic concentrations, there are qualitative differences in the solutes they contain, and equilibration, due to diffusion, will tend to occur. Such animals, however, maintain the gradients between their body fluids and the environment, an equilibrium that is maintained as a result of a complex pattern of physiological events. These processes involve the cells, and special tissues and organs that are concerned with osmoregulation. The integration of the functions of these homeostatic tissues relies largely on hormones. The nervous system makes little direct contribution to such regulatory processes though at the cellular level itself considerable autoregulation, independent of hormones, exists. Hormones do ultimately influence some cellular processes, of course, but they generally appear to do this in effector tissues like the kidney, gills, and gut, which are especially concerned with the overall osmoregulation of the animal. For a more complete account of the role of hormones in osmoregulation the book by Bentley (1971) could be consulted.

Animals occupy diverse osmotic environments; the major ones are the sea, fresh water like rivers and lakes, and dry land. Differences exist between the availability of water and salts within these environments and this is particularly apparent to animals that lead a terrestrial life. Water may be relatively freely available to some terrestrial species that live in areas where rainfall is high, and lakes, ponds, and rivers exist in close proximity to where they live. Other animals, however, live in dry, desert regions where water may only be available sporadically and in limited quantities. Salts are freely available to marine animals, but in fresh water the supplies are more restricted, and some terrestrial animals may occupy regions where a low salt content in the soil may be reflected in a salt-deficient diet. It is thus not surprising to find that the processes for controlling the water and solute content of the body, called osmoregulation, can differ considerably between species that habitually occupy diverse ecological situations. These differences may be manifested as a tolerance to osmotic changes but are principally seen as differences in the functions of the tissues and organs concerned in maintaining the composition of the body fluids. Not unexpectedly, the evolution of the tissues and organs concerned with osmoregulation has been accompanied by changes in the role of the endocrine glands that help integrate their functioning.

Homeostatic events that contribute to osmoregulation may involve changes in either the rates of loss of water and solutes or in the processes of their accumulation.

Osmoregulation in terrestrial environments

Terrestrial vertebrates may lose water as a result of evaporation from the skin and respiratory tract (Fig. 8.1). This loss is greatest in hot, dry air in which little water vapor is present. In homeotherms, evaporation from the respiratory tract and the skin may be increased. In the latter this may be due to the activity of the sweat glands, as a result of the need to dissipate heat. While evaporation will be the predominant route for water loss in animals living in hot, dry conditions, additional quantities pass out of the kidneys in the urine, and out of the gut in the feces. Reproductive processes, such as egg laying and lactation, will also result in increased water losses. Water is gained in the food, from which substantial quantities may be absorbed across the intestines, and as a result of drinking. In addition, amphibians, like frogs and toads, can absorb water across their permeable skin, from damp surfaces and pools of fresh water. Any excessive water that may be gained in such ways is excreted by the kidneys.

Salt losses in land-living vertebrates occur in the urine and feces and, in mammals, in the sweat gland secretions. Some birds and reptiles possess special glands in their heads called nasal salt glands, which have an ability to secrete concentrated solutions of salts. Some regulation of the losses from the sweat glands and gut occurs, and the secretion of the kidneys undergoes a rigorous process of conservation so that salts that are deficient in the body may be conserved. Additional conservation of urinary salts can occur from urine during its storage in the urinary bladder of amphibians and some reptiles. Birds and many reptiles lack a urinary bladder, but in such animals the urine passes into the cloaca and up into the large intestine where some of its contained salts and fluid can be transported back into the blood. Salts are mainly gained in the diet of terrestrial animals, and the drinking of brackish

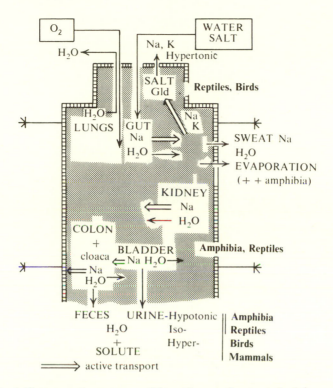

Fig. 8.1. A diagrammatic summary of the pathways of water and salt losses and gains in vertebrates living on dry land. (From Bentley, 1972.)

water may also result in salt accumulation. The latter process, however, occurs only rarely in nature. Excesses of accumulated salts can be excreted by the kidneys and in many birds and reptiles by the nasal salt glands.

Osmoregulation in fresh water

Many species of vertebrates live in fresh water, a solution that is hypoosmotic to the body fluids and which only contains small amounts of dissolved solutes. Water will thus tend to be gained by osmosis across the integument of such animals. Amphibians have a relatively permeable skin and can take up large amounts of water in this way. Fishes respire with the aid of gills, which, apart from allowing the exchange of oxygen and carbon dioxide, are also permeable to water and so are an additional route for the accumulation of water in the body (Fig. 8.2). The skin of reptiles and mammals that frequent fresh water is usually quite impermeable so that little water is accumulated

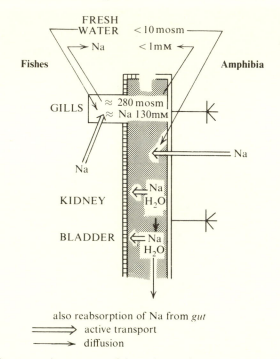

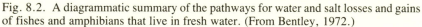

Fig. 8.2. A diagrammatic summary of the pathways for water and salt losses and gains of fishes and amphibians that live in fresh water. (From Bentley, 1972.)

through this channel and, as they breathe with the aid of lungs, they do not suffer the osmotic problems associated with the presence of gills.

Vertebrates living in fresh water may be prone to a greater salt loss than their terrestrial or marine relatives. When the integument is permeable, such as the skin in the Amphibia and the gills of fish, salt loss may be expected to occur as a result of diffusion. Such potential losses, however, are rather limited and are much smaller than may be expected on a simple physicochemical basis. In addition, salt losses may continually occur in the urine because of the necessity for the excretion of water that has accumulated by osmosis.

The solute excretion in the urine is reduced by its reabsorption from the renal tubules. In aquatic amphibians and reptiles, like turtles, and some fishes, sodium is also reabsorbed from the urine that is stored in the urinary bladder.

Salts are principally obtained from the food of aquatic vertebrates. However, additional gains of sodium chloride may be made as a result of their active transport, against electrochemical gradients, across the skin of amphibians and the gills of fish. It has also been suggested that some turtles may actively accumulate sodium across their pharyngeal and cloacal membranes during their irrigation by the external freshwater solution.

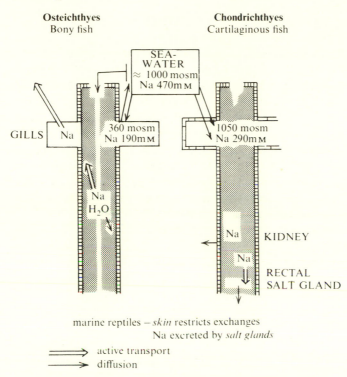

Osteichthyes
Bony fish

Chondrichthyes
Cartilaginous fish

SEA-
WATER
≈ 1000 mosm
Na 470mM

GILLS   Na   360 mosm
Na 190mM

1050 mosm
Na 290mM

Na
H₂O

Na   KIDNEY

Na

RECTAL
SALT GLAND

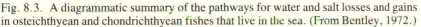

marine reptiles – *skin* restricts exchanges
Na excreted by *salt glands*

⇒ active transport
→ diffusion

Fig. 8.3. A diagrammatic summary of the pathways for water and salt losses and gains in osteichthyean and chondrichthyean fishes that live in the sea. (From Bentley, 1972.)

Osmoregulation in the sea

Most species of fishes as well as a number of reptiles and at least one frog (the crab-eating frog, *Rana cancrivora*) live in the sea. This solution is strongly hyperosmotic to the body fluids of most vertebrates. The exceptions are the hagfishes (Agnatha), the sharks and rays (Chondrichthyes), and the coelacanth, which are isoosmotic, or slightly hyperosmotic, to seawater. The crab-eating frog is also slightly hyperosmotic to the seawater in which it lives.

Most of the bony fishes (Osteichthyes) thus tend to lose water by osmosis, especially across their gill membranes. The Chondrichthyes and myxinoid Agnatha, on the other hand, may gain small amounts of fluid in this way. The sodium chloride concentration in the seawater is much higher than that in the blood of osteichthyean or even chondrichthyean fishes so that an accumulation of salt will tend to occur (Fig. 8.3).

The mechanisms utilized to maintain osmotic balance in seawater are varied (see Kirschner, 1980). Marine teleost fish drink seawater and much of the salt that is present is absorbed across the gut wall; water follows this solute by osmosis. The salt is excreted by special cells called "chloride-secreting cells," or simply "chloride cells," present in the gills and skin. Divalent ions are excreted either in the feces or urine. The gain of salts by marine chondrichthyeans is small compared to that of teleosts, and this is excreted in the urine and as a concentrated solution from a tissue unique among the vertebrates, the rectal salt gland. This salt-secreting organ is situated in the nether regions of the gut. Marine reptiles appear to have a relatively impermeable integument that restricts the gain of salt, but excesses, such as may be gained in the food, can be excreted by the cephalic salt glands that are often modified tear, or orbital, glands.

It can be seen that the osmotic problems of vertebrates differ considerably and depend on the environment where they live as well as the anatomical and physiological wherewithal that is conferred by their phylogeny. The maintenance of osmotic homeostatis is dependent on a variety of tissues and glands, some of which, like the kidney, are present in all the major phyletic groups of vertebrates, and others, like cephalic salt glands and the rectal gland, which have a more restricted distribution. The activities of many of the organs and tissues involved in osmoregulation are controlled by hormones. A summary of these is given in Table 8.1.

Osmoregulation in vertebrates is dependent on the active participation of such tissues as the kidneys, gills, skin, urinary bladder, gut, and certain salt-secreting glands. These tissues contribute to the excretion and conservation of water and salts, and their roles and physiological significance are not the same in all species.

The hormones that influence osmoregulation most directly are the neurohypophysial hormones, principally vasotocin and vasopressin (ADH), the adrenocorticosteroids, the catecholamines, and prolactin. Corticotropin and angiotensin are indirectly involved because of their roles in controlling the release of adrenocorticosteroids. Angiotensin may also have a more direct effect on some membranes while the urophysis and corpuscles of Stannius, which have putative endocrine functions, may also be involved.

Active transport and secretion of ions, especially sodium, potassium, bicarbonate, and chloride, across, or from the epithelial membranes that make up the tissues that effect osmoregulation, are basic to their physiological function. Such membranes are osmotically permeable to water, which can pass across them with an ease that may vary, depending on the membrane and the physiological conditions. The adequate functioning of these osmoregulatory tissues is ultimately dependent on their blood supply. Hormones may thus influence the activity of osmoregulatory tissues by actions at several sites:

1. Hormones may alter the processes of active sodium and chloride transport and the secretion of hydrogen ions, bicarbonate, and potassium. Cortisol and aldosterone can alter sodium and potassium movements across many

epithelial membranes. Vasotocin and ADH may also promote such transmural sodium transport. Catecholamines can increase transport and secretion of chloride ions in several tissues and also can inhibit the effects of ADH. All these hormones act directly on the cells involved.

2. Osmotic and diffusional movements of water and sodium across epithelial membranes can be changed by hormones. Vasotocin and ADH may increase the permeability to water of the renal tubule (in some species), as well as amphibian skin and urinary bladder. Prolactin can reduce the permeability of the gills of certain teleost fishes to sodium, thus limiting diffusional losses of this ion in fresh water.

3. The catecholamines and neurohypophysial peptides can alter the diameter of blood vessels and so may influence the functioning of osmoregulatory tissues by virtue of their vasoactive actions. The urine flow, especially in nonmammals, may be influenced by changes in the rate of filtration of plasma across the glomerulus and this process can be altered by these hormones. Alterations in secretion and absorption of ions across the gills of fishes may also be changed by such hormonally mediated variations in the regional blood flow.

While this chapter is principally concerned with hormones, the role of nerves in osmoregulation should also be mentioned. Neural integration is not common, though it does occur. The nasal salt-secreting glands in birds and reptiles are stimulated to secrete as a result of the stimulation of autonomic cholinergic nerves (Peaker and Linzell, 1975). Corticosteroids may contribute to the well-being of the underlying processes but do not directly stimulate secretion. The sweat glands of mammals secrete in response to the need to dissipate heat, and this usually occurs following stimulation of autonomic cholinergic or adrenergic nerves (as described in Chapter 5). However, sweat gland secretion during exercise may depend on circulating epinephrine. The vasoconstrictor tone of blood vessels is primarily dependent on the activity of adrenergic nerves that can thus, indirectly, alter the functioning of tissues. This effect is sometimes observed in the kidney of animals but is probably not a usual physiological mechanism.

The role of hormones in osmoregulation

Mammals

These vertebrates have a complement of hormones, similar to other vertebrates, that can influence osmoregulation, though the roles of such secretions may differ somewhat from those in nonmammals.

Vasopressin is unique to the mammals. It reduces urinary water losses (antidiuresis) as a result of an increased osmotic reabsorption of water from

Table 8.1. *Target organs for osmoregulatory-type responses to hormones in vertebrates*
The responses in *italics* appear to be physiological ones while others are either only pharmacological or the evidence for their normal, *in vivo* role is as yet equivocal. The responses are not necessarily present in all members of the orders of vertebrates that are indicated

Target organ	Phyletic distribution of target organ	Stimulatory hormone	Nature of response	Phyletic distribution of responsiveness
Kidney	All vertebrates	Vasotocin	*Decreased glomerular filtration rate (GFR) and increased renal tubular water reabsorption = decreased urine flow (antidiuresis)*	Amphibians, reptiles, and birds
			Decreased GFR = antidiuresis	Teleost (eels)
			Increased GFR = diuresis	Some teleosts and lungfishes
		Vasopressin	*Increased renal tubular water reabsorption = antidiuresis*	Mammals
		Prolactin	Increased urine flow (increased GFR, decreased renal tubular water reabsorption)	Some teleosts
Urinary bladder	Most vertebrates (except birds, some reptiles, and many fishes)	Vasotocin	*Increased water and Na reabsorption*	Amphibians (mostly anurans)
		Aldosterone	Increased Na reabsorption	Amphibians, reptiles, and mammals
		Prolactin	Reduced water permeability, increased Na reabsorption	Teleost (starry flounder)
		Urotensin II	Increased Na reabsorption	Teleost (goby)
Gills	Fishes and larval amphibians	Cortisol	*Increased outward Na secretion*	Marine teleosts
			Increased inward Na absorption	Freshwater teleosts
		Vasotocin	Increased inward or outward movements of Na	Freshwater (inward) or marine (outward) teleosts
		Aldosterone	Uptake across larval amphibian gills?	

Organ	Animal group	Hormone	Effect	Species
Skin	All vertebrates	Epinephrine	Decreased secretion by Cl-secreting cells	Marine teleosts
		Prolactin	*Decreased Na diffusion outward and water accumulation (inward)*	Freshwater teleosts
			Decreased Na extrusion	Marine teleosts
		Vasotocin	*Increased water and Na absorption*	Some amphibians
		Aldosterone	*Increased Na absorption*	
		Angiotensin	As above	
		Prolactin	Decreased permeability to water and Na	Urodele amphibians (?)
		Epinephrine	Increased Cl secretion	Teleosts (killifish)
Sweat glands	Mammals	Aldosterone	*Reduced Na loss in sweat*	
		Epinephrine	*Increased secretion during exercise*	
Salt glands	Some birds and reptiles (nasal and orbital glands) and chondrichthyeans (rectal glands)	Cortisol	Facilitates secretion	Birds and reptiles
		Cortisol	Reduces secretion	Chondrichthyeans
		Aldosterone	Reduces Na secretion	Lizard
		Prolactin	Increases secretion	Bird
		Vasotocin	Increases secretion	Bird
		1α-Hydroxycorticosterone	Increases Na–K ATPase	Chondrichthyeans
		Vasoactive intestinal peptide	Increases secretion	Chondrichthyeans
Salivary glands	Mammals	Aldosterone	*Decreases Na and increases K loss*	
		Epinephrine	*Dries up secretion*	
Intestine	All vertebrates	Cortisol	*Increased NaCl absorption*	Teleosts
		Aldosterone	Increased NaCl absorption in colon	Mammals, birds, amphibians
		Angiotensin	As above	Mammal (rat)
		Prolactin	Decreased salt and water absorption	Teleosts, birds
			Increased fluid absorption	Mammal (rat)

the kidney tubules. Vasopressin's phyletic forebear, vasotocin, as we have seen, has a slightly different chemical structure that confers on it a pronounced ability to contract smooth muscle such as in the oviduct and uterus and also, when injected, to promote contractions of the myoepithelial cells in the mammary glands. Injected vasotocin also has an antidiuretic action in mammals and it could thus conceivably function in such a physiological role. Vasopressin, however, lacks the prominent effects that vasotocin has on nonvascular smooth muscle contraction and so exerts a more specific action in the body. Its evolutionary perpetuation in mammals is therefore not surprising. Vasopressin does not appear to have any other physiological role, on other organs, in mammals. It can, however, exert other effects, such as increasing the blood pressure, contracting the uterus, raising blood sugar levels, and may possibly facilitate memory and learning when it is administered.

Adrenocorticosteroids play an important role in controlling sodium and potassium metabolism in mammals. The absence of the adrenal cortex in mammals quickly results in death, resulting mainly from losses of sodium and an accumulation of potassium. Aldosterone is the most effective of the adrenocortical hormones that exhibit actions on sodium and potassium metabolism in mammals, though the others, especially corticosterone, can also exert such effects. Sodium excretion from the kidney, sweat glands, and salivary glands is reduced while potassium loss is increased; there is a drop in the ratio of sodium/potassium in the secreted fluids. Aldosterone can promote sodium reabsorption from the large intestine and the mammary gland ducts (Yagil, Eltzion, and Berlyne, 1973; Dolman and Edmonds, 1975). The reabsorption of sodium from the rabbit urinary bladder is also promoted by aldosterone (Lewis and Diamond, 1976). The osmoregulatory effects of the corticosteroids in mammals are thus all directed to the same general purpose (i.e., sodium conservation and potassium excretion) and seem to involve at least five different target tissues.

Epinephrine has a less prominent role in osmoregulation. Its action in stimulating sweat gland secretion has already been mentioned. In addition, epinephrine can antagonize the release, and the effects, of ADH on the kidney. Such inhibition is an α-adrenergic action that can be demonstrated in experimental animals, though its possible physiological importance is not yet clear. It may also influence the release of renin.

Angiotensin, apart from initiating the release of aldosterone, has been shown to promote sodium reabsorption from the kidney tubule and the rat colon *in vitro* (Davies, Munday, and Parsons, 1970; Munday, Parsons, and Poat, 1971). However, the normal physiological significance of this effect is in doubt (Dolman and Edmonds, 1975). Another interesting effect of angiotensin is its ability, when injected, to promote drinking (Fitzsimons, 1972, 1979). Drinking is elicited by the sensation of thirst that arises in the brain in a number of circumstances, including a reduction in the volume of the extracellular fluids. This latter response is reduced if the kidneys are removed, suggesting that the renin–angiotensin system may be involved. Indeed, the injection of small amounts of angiotensin II into the region of the "thirst center" in the anterior diencephalon of the brain promotes drinking. These effects are mediated by the hormone's action on the subfornical organ and organum vasculosum in the brain. Receptor sites for angiotensin II have been directly observed in the latter, using a fluorescent analogue of the peptide (Landas et al., 1980). Angiotensin II also induces thirst in birds and reptiles but not amphibians (Kobayashi et al., 1979). The injection of angiotensin II into the brains of rats promotes the drinking of NaCl solutions (Bryant et al., 1980). This "Na appetite" can also be induced in sheep by the injection of ACTH, an effect that depends on the release of corticosteroids (Weisinger et al., 1980). It seems possible that these two hormones may normally interact to adjust the animals' taste for salt.

Birds

Birds possess osmoregulatory hormones that are similar to those of other nonmammalian tetrapods.

Vasotocin acts as an antidiuretic hormone comparable in its effect to vasopressin in mammals. It is released in response to dehydration (Koike et al., 1977). Vasotocin increases water reabsorption from the renal tubule of birds and in slightly higher, but still physiological, concentrations also decreases the GFR (Skadhauge, 1969). The hormone thus has two effects on the avian kidney, both of which decrease urinary water loss. Two types of nephrons have been identified in the kidney of the desert quail, *Lophortyx gambelii* (Braun and Dantzler, 1972, 1974): a mammalian-type with a loop of Henle, and a reptilian-type that lacks this tubular segment. Glomerular filtration across the reptilian-type nephron is more variable than in the mammalian-type and the former ceases functioning when excess sodium chloride is administered. The reptilian-type nephron is the site where vasotocin acts when it decreases the GFR in the desert quail.

It is interesting that vasotocin probably has another physiological role in birds, as it contracts the oviduct and so can assist oviposition. Vasotocin may thus have two physiological, but unrelated, roles and this situation may also occur in reptiles and amphibians. As described earlier (Chapter 5), vasotocin, when injected, also has a hyperglycemic and hyperlipidemic effect in birds, and it can also stimulate secretion from the nasal salt glands. The physiological spectrum of vasotocin's action could thus be even larger than just an involvement with osmoregulation, but the importance of such nonosmoregulatory effects in the body is in doubt.

The *adrenocorticosteroids*, of which birds possess aldosterone and corticosterone, reduce renal sodium loss and facilitate potassium excretion, just as in mammals. Adrenalectomy results in an excessive loss of salt in the urine of ducks (Thomas and Phillips, 1975). This effect partly reflects an insufficient reabsorption of sodium from the colon and coprodeum (Thomas and Skadhauge, 1979). Adrenalectomy in ducks also reduces the ability of the nasal salt glands to secrete hypertonic salt solutions. Corticosterone, but not aldosterone, can restore this deficiency. Corticotropin and corticosteroids also enhance salt gland secretion when they are injected into intact ducks. These effects, however, are at least partly indirect and the result of an elevated glucose concentration in the blood (see Peaker, 1971; Phillips and Ensor, 1972). Receptors for corticosterone have, however, been identified in the cytosol of cells from nasal glands of ducks, suggesting that it is the site of a specific response to this steroid (Sandor, Mehdi, and Fazekas, 1977).

The immediate stimulus for secretion of the avian salt glands is a neural one that is initiated by a hypertonicity of the plasma. Apart from corticosteroids, hormones may impinge their effects on this process. Epinephrine inhibits the salt gland's response to hypertonic saline, which probably reflects its vasoconstrictor effect. Hypophysectomy abolishes the secretory response of the salt gland and this deficiency can be partly restored by the injection of corticotropin or prolactin. Injections of prolactin into normal ducks also stimulate salt gland secretion. This action of prolactin is an interesting one for, as we shall see later, this hormone has an osmoregulatory function in some teleost fishes. It is likely, however, that its action in birds is indirect, due to its role in maintaining an optimal intake of food and water, especially following hypophysectomy (see Ensor, Simons, and Phillips, 1973).

The adrenal cortex of birds is influenced by the amount of salt in their diet and this effect can be seen in nature. It has been observed (Holmes, Butler, and Phillips, 1961) that birds living in environments near the sea or supplies of brackish water, where their salt intake may be high, have larger adrenal glands than species that habitually have fresh water to drink.

The *prolactin* stores in the pituitary are also influenced by the bird's salt intake. Herring gulls, *Larus argentatus*, given salt solutions to drink suffer a depletion in the stores of prolactin in their pituitaries, which is associated with an elevated osmotic concentration of the plasma (Fig. 8.4). Injected prolactin has been shown to decrease the absorption of sodium from the small intestine of chickens (Morley, Scanes, and Chadwick, 1981). The sodium levels in the plasma also declined, suggesting that this hormone may be enhancing the loss of sodium in the feces. It is interesting that prolactin levels in the pituitaries of rats also decline in response to dehydration (Ensor, Edmondson, and Phillips, 1972). In some teleost fishes, prolactin is released in response to hypoosmotic conditions (see later) and this hormone serves an important role in their osmoregulation.

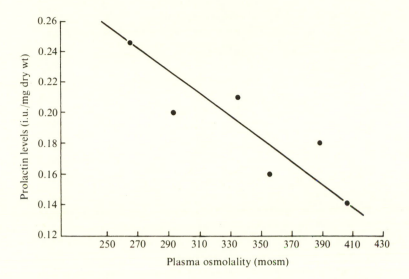

Fig. 8.4. The relationship of plasma osmolality and the storage of prolactin in the pituitary gland of juvenile herring gulls or black-backed gulls (*Larus argentatus* and *L. fuscus*). These birds were given saline solutions of different concentrations to drink. As the concentration of this drinking water was increased the plasma osmolality rose while the prolactin storage declined. (From Phillips and Ensor, 1972.)

The possible interrelations of nerves and hormones in influencing the electrolyte metabolism of birds that possess nasal salt glands are summarized in Fig. 8.5. The avian adrenal, as in mammals, can be controlled by both the adenohypophysis and a renin–angiotensin system.

Calcitonin, when injected, increased the excretion of sodium in the urine of starlings (Clark and Wideman, 1980), but the physiological significance of this effect, while intriguing, is not clear. A role for this calcemic hormone in the sodium metabolism of other vertebrates has been previously suggested, however (see Chapter 6).

Angiotensin II has been shown to promote the sensation of thirst and drinking in a number of birds including the white-crowned sparrow, Japanese quail, and several species of Australian parrots (Wada, Kobayashi, and Farner, 1975; Takei, 1977; Kobayashi, 1981). Three species of the Australian parrots belonging to the genus *Barnadius* showed differences in their sensitivity to this hormone. The twenty-eight parrot lives in quite wet areas while the Port Lincoln parrot lives in more arid zones, but it is considered to have evolved from a species from more temperate regions. The mallee ringneck parrot, on the other hand, has lived in very dry areas for a long time. The former two species were very sensitive to injected angiotensin II while the latter was quite insensitive. These observations suggested that birds originating from dry areas

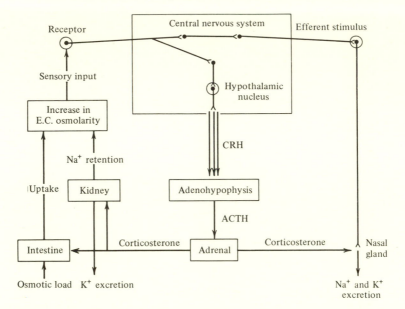

Fig. 8.5. A schematic summary of the processes thought to be involved in the control of sodium and potassium excretion in marine birds. It should be noted that not all species of birds possess a functioning nasal salt gland. (From Holmes, 1978.)

have a poor thirst and drinking response to angiotensin II as compared with those from wet regions.

Reptiles

Reptiles are poikilothermic, which profoundly influences their osmoregulation, and they represent a substantial metabolic departure from the birds and mammals. These animals live in diverse osmotic habitats including the sea, fresh water, and terrestrial situations ranging from deserts to tropical rain forests.

Reptiles possess *vasotocin* and this, when injected, can decrease urine flow by a dual action (as in birds), namely an increase of the renal tubular reabsorption of water and a decrease of the GFR (see LeBrie, 1972).

The role of the *adrenocorticosteroids* in reptilian osmoregulation is not clear. Adrenalectomy has been variously reported to cause no change in plasma electrolytes, to result in a potassium accumulation or sodium loss or, as in birds and mammals, to have both of these effects. These diverse observations may be the result of the longer periods of time that poikilotherms often take to respond to altered physiological and environmental circumstances. Injected aldosterone can increase renal tubular absorption of sodium in the water snake, *Natrix cyclopion*, but only after it has received additional sodium chloride, which presumably decreases endogenous levels of the steroid hormone (LeBrie, 1972). A similar effect has been observed in freshwater turtles (Brewer and Ensor, 1980). On the other hand, the injection of aldosterone into a saline-loaded lizard, *Dipsosaurus dorsalis*, has no effect on renal electrolyte excretion (Bradshaw, Shoemaker, and Nagy, 1972). *Dipsosaurus dorsalis*, like many birds and other reptiles, has a functional nasal salt gland, but injected corticosteroids do not initiate secretion from this gland (Shoemaker, Nagy, and Bradshaw, 1972). Indeed, the sodium secretion is abolished by aldosterone injections. It has been suggested that in the lizard salt gland, aldosterone may promote sodium reabsorption from the tubules in the glands as it does in the mammalian kidney.

When excesses of sodium chloride are administered to the lizards, *Dipsosaurus dorsalis* and *Amphibolurus ornatus*, corticosteroid levels in the plasma increase (see Bradshaw, 1972). This effect contrasts with what is seen in mammals and amphibians where a sodium deficiency is associated with increased release of corticosteroids. In these species of lizards, the increased level of corticosteroids facilitates urinary sodium losses by decreasing renal sodium reabsorption, a response that is opposite to that seen in mammals where renal sodium reabsorption is increased. The role of the adrenocorticosteroids in controlling sodium and potassium metabolism in reptiles is not completely understood. Some of the responses are reminiscent of those in other phyletic groups of vertebrates but some are quite different. Measurements of aldosterone and corticosterone concentrations in the plasma of 10 species of snakes that live either in fresh water, in the sea, or on dry land failed to show any relationship to their presumed osmotic needs (Duggan, 1981). One cannot, at present, draw any unified picture about the role of adrenocorticosteroids in the Reptilia.

Aldosterone may act at extrarenal sites in some reptiles. The urinary bladders of reptiles are the site of active sodium reabsorption from the urine that is stored there. This salt transport can be increased, *in vitro*, by aldosterone in the tortoise, *Testudo graeca*, and the freshwater turtle, *Pseudemys scripta* (Bentley, 1962; LeFevre, 1973). It is unknown whether this effect of aldosterone exists normally in these chelonians, but it could reflect an evolution of this hormone's osmotic role. As we shall see, aldosterone also stimulates sodium transport across the amphibian urinary bladder.

Angiotensin II induces drinking in reptiles (Fitzsimons and Kaufman, 1977), as is also seen in mammals and birds.

The amphibians

Osmotically, the amphibians are a very interesting group as they bridge the gap between the fishes and the amniotes. Phyletically, they represent the first terrestrial vertebrates yet they are still largely dependent on the ready availability of fresh water and have aquatic larvae. Frogs and toads (as well as newts and salamanders) do not normally drink (see Bentley and Yorio, 1979), and such behavior cannot be elicited by injecting them with angiotensin II (Kobayashi et al., 1979). The Amphibia, therefore, are a group of considerable interest both with respect to osmotic regulation and the endocrine mechanisms they utilize for this process.

Vasotocin has an antidiuretic effect in most, but not all, amphibians. Mesotocin, the other amphibian neurohypophysial hormone, when injected, can even have a diuretic effect (Pang and Sawyer, 1978; Galli-Gallardo, Pang, and Oguro, 1979). As in the birds and reptiles, vasotocin initiates both the reabsorption of water from the renal tubules and a decrease in the GFR (Sawyer, 1972*a*). The tubular response may, however, be lacking in the urodele amphibians (Brown and Brown, 1980).

The magnitude of the antidiuretic effect of the neurohypophysial hormones also varies in different species. Vasotocin, for instance, does not reduce urine flow in the African clawed toad, *Xenopus laevis*. This toad is aquatic, and consequently such a response to released vasotocin would be physiologically inappropriate and could even lead to death resulting from hyperhydration. Other aquatic amphibians, like the mudpuppy, *Necturus maculosus*, the mud eel, *Siren lacertina*, and the congo eel, *Amphiuma means*, exhibit antidiuretic effects after injections of vasotocin, but the responses are small and the amounts of hormone required are large so that the physiological significance of this effect is doubtful (Bentley, 1973). Young tadpoles do not exhibit water retention (reflecting an antidiuresis) in response to injected vasotocin, but as bullfrog (*Rana catesbeiana*) tadpoles approach metamorphosis such a response becomes increasingly apparent, though it does not reach its full expression until after metamorphosis has occurred (Alvarado and Johnson, 1966).

Vasotocin exerts several other interesting effects with respect to the osmoregulation of amphibians. The skin of amphibians is permeable to water, which moves across it by osmosis. In anurans (frogs and toads), water crosses the skin much more readily when the tissue is stimulated by vasotocin. This hormone appears to make the integument less waterproof, an action that is is also seen on the renal tubule. This direct cutaneous effect of vasotocin is equivocal in urodeles (newts and salamanders) and is not seen in some anurans, including tadpoles and *Xenopus laevis*. The ability of vasotocin to increase the skin's osmotic permeability appears to be greater in species that normally occupy dry, rather than relatively wet or damp, habitats. Different parts of the skin of each animal do not necessarily respond in a quantitatively uniform manner to vasotocin. Thus, skin from the ventral surface of several species of tree frogs (Hylidae) shows a marked increase in permeability to water, but that from the dorsal surface is unresponsive to the hormone (Yorio and Bentley, 1977).

As amphibians do not drink, this increased rate of water accumulation across the skin may aid rehydration in some species, such as those from desert areas, where water is only available sporadically.

Amphibians usually possess a large urinary bladder in which they can store water equivalent to as much as 50% of their body weight. This water can be reabsorbed in times of need and so constitutes a store that may be very useful to some species. Vasotocin, or dehydration, which releases this hormone into the blood, increases the rate of water reabsorption across the urinary bladder of many amphibians.

The crab-eating frog, *Rana cancrivora*, lives in the seawater in coastal mangrove swamps in Southeast Asia. As described earlier, this interesting amphibian maintains its blood plasma at a hypertonic concentration with respect to the external solution by retaining additional salts and urea. The skin of these amphibians is not responsive to vasotocin, but this hormone increases the permeability of the urinary bladder to urea (as well as water) and so, by permitting its reabsorption from the urine, apparently contributes to the conservation of this solute in the body (see Dicker and Elliott, 1973). A cutaneous response would be a disadvantage to such animals as it would increase the rate of water loss if they entered hyperosmotic solutions.

The tetrapod urinary bladder has no true embryological analogue in the fishes and appears to have first evolved in the Amphibia. It is interesting that some species of this group utilize it for water storage, and possibly urea conservation, with which use it has also acquired a responsiveness to vasotocin. A comparable effect on the bladder has not been demonstrated in any other vertebrate group so that this endocrine adaptation is unique.

When amphibians, in water, are injected with neurohypophysial hormones, they gain weight due to a water retention. This action is known as the "Brunn effect," or "water balance effect," and is due to a stimulation by vasotocin at the three distinct sites described above: the kidney, the skin, and the urinary bladder. A single hormone thus has multiple effects, all of which are directed to the same physiological purpose, namely the conservation of water.

As in other vertebrates, the neurohypophysial hormones can be shown, when injected *in vivo* or *in vitro*, to have other actions, including a hyperglycemic effect and an ability to contract the oviduct. However, the physiological significance of these effects is not clear.

The *catecholamines*, including epinephrine, have ubiquitous effects on tissues. One of these, which has been shown *in vitro*, is an ability to antagonize the osmotic effects of neurohypophysial hormones (Handler, Bensinger, and Orloff, 1968). Injected epinephrine has also been shown to increase cutaneous water uptake by the toad, *Bufo melanostictus* (Elliott, 1968). The action of vasotocin on the osmotic permeability of membranes is mediated by the adenylate cyclase–cyclic AMP system. Catecholamines can inhibit the formation of cyclic AMP, which is a manifestation of their α-adrenergic effects. In addition, these hormones can increase the levels of cyclic AMP, which is a β-adrenergic effect. These actions of epinephrine and norepinephrine on membrane permeability are therefore not unexpected though their physiological significance is unknown in the Amphibia. Catecholamines could, however, modulate the actions of vasotocin.

The sodium metabolism of amphibians can also be regulated by hormones that can act at several different sites in the body. Vasotocin, *in vitro*, stimulates sodium transport across the skin, from the external media to the blood, and its reabsorption from the urinary bladder. These effects, though prominent, do not seem to persist for a prolonged time and are difficult to reconcile with normal physiological regulation of sodium in the intact animal. *Aldosterone* has more persistent effects in increasing such sodium transport across the skin, the urinary bladder, and the colon in frogs and toads (Crabbé and De Weer, 1964; Cofré and Crabbé, 1965). It seems likely that aldosterone also acts on the urinary bladder of urodeles, but the evidence for this is equivocal. As this corticosteroid is released in response to sodium depletion in amphibians and is effective at low concentrations, it seems likely that it normally adjusts sodium transport at these sites. As we have seen, aldosterone stimulates sodium reabsorption from the kidney tubules in mammals, birds and possibly reptiles, but despite frequent attempts to demonstrate it, this action appears to be lacking in amphibians. This is an interesting endocrine situation, as aldosterone makes its initial phyletic appearance in the lungfishes and the amphibians, yet the important renal action of this hormone does not appear to have evolved until much later.

The *renin–angiotensin* system is present in the Amphibia and, as in other tetrapods, contributes to the release of aldosterone. The renin concentration in the kidney of salt-depleted frogs increases (Capelli, Wesson, and Aponte, 1970). When renin from frog kidney is injected back into frogs, the aldosterone (but not corticosterone) concentration in the plasma rises (Johnston et al., 1967); however, the plasma renin activity has been measured in bullfrogs and shown to *increase* following intravenous infusions of sodium chloride solutions (dehydration decreases it) which is opposite to the response observed in mammals (Sokabe et al., 1972).

The involvement of the renin–angiotensin system in the specific control of aldosterone release may be present in the Amphibia, and this effect could be one of its earliest evolutionary manifestations. There appear, however, to be differences from the physiological situation in mammals. It has even been suggested that angiotensin may constrict the efferent glomerular arterioles and so control the GFR in lower vertebrates, including amphibians.

The "water-drive" effect of injected *prolactin* in newts, *Notophthalmus viridescens* (see Chapters 3 and 7), is associated with changes in the skin, including a "thickening" and an increased secretion of mucus. It would not be surprising if this second metamorphosis, from life in a terrestrial to an aquatic environment, were associated with osmoregulatory adjustments. The injection of prolactin plus thyroxine into these newts, when they are in their terrestrial phase, has been found to result in a decrease in the the permeability of the skin to water and sodium (Brown and Brown, 1973). This change could facilitate their osmotic adaptation to an aqueous environment and, as we shall see later (see discussion of pituitary hormones, this chapter), the decrease in the permeability of the skin to sodium has some similarities to the effects of prolactin on the gills of some teleost fish. A phyletically closer analogy has been observed in another amphibian, the mudpuppy, *Necturus maculosus*, which possesses external gills. When this neotenous urodele is hypophysectomized it loses sodium at increased rates and its serum sodium concentration declines (Pang and Sawyer, 1974). The precise route for such sodium loss has not been described but, as in teleosts, it can be prevented by the injection of prolactin.

The fishes

The fishes are phyletically very diverse and contain several distinct groups that osmoregulate differently. These include the Osteichthyes (bony fishes), the Chondrichthyes (elasmobranchs, cartilaginous fishes), and the Agnatha (cyclostomes or jawless fishes, lampreys, and hagfishes). Most of the available information about the role of hormones in osmoregulation applies to a single order of the Osteichthyes, the Teleostei. The role of the gills in osmoregulation in fishes has been thoroughly summarized by Maetz (1971).

Neurohypophysial hormones

The well-known antidiuretic effect of vasotocin, which is seen in tetrapods, does not seem to occur in most fishes. Indeed, some fishes, but not all, exhibit a diametrically opposite response: The urine flow is increased. This diuresis is seen in some teleosts, like the goldfish, *Carassius auratus*, and the eel, *Anguilla anguilla*. It is also interesting that such a diuretic response is seen in all of the extant lungfishes: the African lungfish, *Protopterus aethiopicus*, and the South American lungfish, *Lepidosiren paradoxa* (Sawyer, 1972*b*), and the Australian species, *Neoceratodus forsteri* (Sawyer et al., 1976). The diuretic effect of vasotocin in fishes reflects its action in increasing the glomerular filtration rate. This effect, as suggested by Sawyer, is not very different, basically, from the mechanism with which vasotocin produces the reduction in GFR that is seen in tetrapods. In the latter, vasotocin constricts the *afferent* glomerular arteriole while, to have a diuretic action, a similar effect, but on the *efferent* arteriole, would result in the observed increase in the GFR. Thus, a vascular effector site may merely have shifted from one branch of the glomerular arteriole to the other. Not all the glomeruli may be responsive to vasotocin as there is plenty of evidence in the fishes that a change in the GFR can reflect an alteration in the number of functioning glomeruli (glomerular recruitment) rather than a change in the hemodynamics in individual nephrons.

It has also been shown that, while large doses of injected vasotocin have a diuretic action in European eels, relatively small amounts (less than 10^{-10} g/gk body weight) have an antidiuretic action as in tetrapods (Babiker and Rankin, 1973, 1978; Henderson and Wales, 1974). This response is the result of a decreased GFR and presumably reflects changes in the opposite direction from that of glomerular recruitment; tubular water reabsorption is unchanged. It thus appears that the eel afferent glomerular arteriole may also respond to vasotocin and there may be a balance between the hormonal effects at this site and the efferent arteriole, the overall response depending on the concentration of the peptide hormone. An action at the latter site may be able to override an effect on the former.

It should be emphasized that the diuretic effects of vasotocin do not occur in all fishes or even all teleosts and the antidiuretic effect has only been observed in European eels. This makes it difficult to envisage whether or not the neurohypophysial peptides really have a physiological action on the kidney or whether, on injection, the exogenous hormones are merely exerting their well-known pharmacological effects on blood vessels. Indeed, vasotocin has not even been identified in the blood of fishes. A diuretic effect of vasotocin could, however, be useful to species living in fresh water, as it may facilitate the excretion of water accumulated by osmosis, while an antidiuretic effect may be useful when the fishes are bathed in hyperosmotic solutions like seawater. In African lungfishes that undergo a period of estivation, enclosed in a cocoon in the dried mud at the bottoms of lakes and rivers where they live, the initiation of a diuresis could aid excretion of both accumulated solutes and water taken up by the fish during its "awakening" in fresh water. Indeed, African lungfishes from which the pituitary has been removed cannot survive once they are replaced in water.

Neurohypophysial peptides, when injected, have also been shown to increase the turnover of sodium chloride in some teleosts, such as eels, when they are transferred from fresh water to seawater. Such hormones may also increase active ion uptake in some fresh water teleosts. It is possible that these hormones have a direct effect on the permeability of the gill epithelium that would be analogous to some of their actions in tetrapods. Alternatively, it has been shown (Maetz and Rankin, 1969) that regional changes in the branchial blood flow occur that could mediate alterations in ion transfer; thus epinephrine, which increases the blood supply to the respiratory areas of the gills, produces a decline in salt excretion from the chloride cells. This change may be the result of a shunting of blood away from the central part of the gill filaments that contain the chloride cells. Neurohypophysial peptides (and acetylcholine) have the opposite action on the blood supply in the gills (Fig. 8.6) and so could facilitate the functioning of the chloride cells. We can thus conjecture that the vasoactive effects of the neurohypophysial peptides may be phyletically older than their direct actions on the permeability of epithelial membranes (Maetz and Rankin, 1969). Such vascular effects can be seen in *some* fishes where they mediate changes in renal and gill functions that influence osmoregulation.

Vasotocin has no effect on the urine flow in Agnathan fishes (or at least in *Lampetra fluviatilis*). The rate of sodium loss in the urine is increased, however (Bentley and Follett, 1963), while the branchial losses of sodium are unchanged. The effects of vasotocin on osmoregulation in chondrichthyean fishes do not appear to have been investigated.

Differing physiological roles of vasotocin in eliciting an antidiuresis in tetrapods and a diuresis, or antidiuresis, in some teleost fishes, would necessitate different mechanisms for regulating the release of this hormone; thus, in freshwater fish, vasotocin may possibly be released in response to a dilution, or expansion, of the body fluids, while in tetrapods (and possibly the seawater eel), it is secreted following a concentration of the extracellular fluids. The possible contrast in these two mechanisms in fish and amphibians is summarized in Fig. 8.7, though it should be noted that the release mechanism in the fish is still hypothetical.

The contrasting roles of vasotocin in altering urine flow via different mechanisms, involving the renal tubule or glomerulus, are shown in Fig. 8.8. Vasotocin can elicit a glomerular antidiuresis in nonmammalian tetrapods and a glomerular diuresis in fishes. The diuretic effect may have occurred first during evolution, reflecting an action of vasotocin on the efferent glomerular arteriole and/or the blood pressure. A change in the sensitivity may have occurred subsequently, possibly involving an increase in the number of vasotocin receptors associated with the *afferent* glomerular arteriole, resulting in a specific renovascular antidiuretic effect (see Pang et al., 1980). Vasotocin can promote water reabsorption from the renal tubules (a tubular antidiuresis) in all tetrapods (possibly with the exception of the urodeles and crocodilians), and this response may be a subsequent evolutionary adaptation.

Catecholamines

It is uncertain if the catecholamine hormones have a physiological role in the osmoregulation of fishes but epinephrine, when injected, can alter the movements of water, sodium, and chloride across the gills (see Pic, Mayer-Gostan,

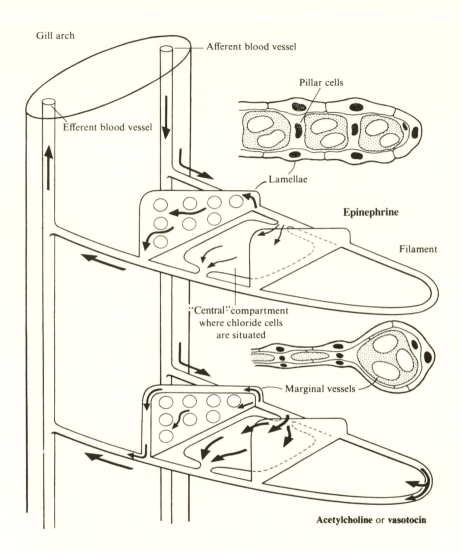

Fig. 8.6. A hypothetical model illustrating how hormones may alter the distribution of the blood flow in the gills of teleost fish and so alter their respiratory and osmoregulatory functions. Epinephrine increases the blood flow to the *lamellae* of the gills, which have a predominant respiratory function, by causing a relaxation of the *pillar cells* (PC). Acetylcholine or neurohypophysial peptides (including vasotocin and isotocin) constrict the lamellae (possibly by contracting the pillar cells) and divert blood to the central compartment. The chloride-secreting cells are situated in the interlamellar region of the central compartment so that their function is facilitated by the presence of the neurohypophysial hormones. (From Rankin and Maetz, 1971.)

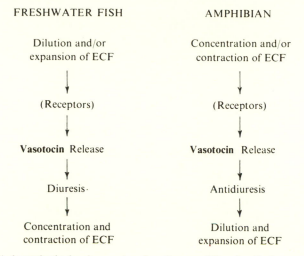

FRESHWATER FISH   AMPHIBIAN

Fig. 8.7. A hypothetical scheme showing the possible stimuli that may effect the release of vasotocin in fishes as compared to the typical tetrapod release reflex (such as exemplified in amphibians). The releasing stimuli may be diametrically opposite to each other, for while *concentration* (or contraction) of the extracellular fluid (ECF) brings about vasotocin release in amphibians, in the freshwater fishes this would need to occur in response to a *dilution* if vasotocin were to mediate a diuresis physiologically. (From Sawyer, 1972*a*.)

and Maetz, 1973). In teleosts in seawater, injected epinephrine reduces the active extrusion of sodium and chloride from the gills. It increases branchial permeability to water in either fresh water or seawater. The effects on ion movements can be prevented by α-adrenergic blocking drugs and that on water by β-adrenergic inhibitors. β-Adrenergic receptors mediate the increases of blood flow to the central lamellar regions of the gills (see the last section in this chapter) and this effect could be contributing to the osmotic change. A more direct effect on the permeability of the gill epithelium to water is, however, considered to be a more likely mechanism (Isaia, Maetz, and Haywood, 1978). The ionic responses to epinephrine also appear to be due to a direct effect on the tissue resulting from an inhibition of the activity of the chloride-secreting cells.

The skin on the head of a number of teleosts has, like the gills, been shown to contain chloride cells. These salt-secreting cells may be present on the inner surface of the operculum and the sides of the jaws. Pieces of such skin can be removed and studied *in vitro* when a transepithelial chloride transport can be observed. This active transport process can be increased or decreased by catecholamines, the particular response depending on the hormone or drug used and its concentration. Increased chloride transport follows β-adrenergic type stimulation while a decrease results from α-adrenergic effects (Zadunaisky

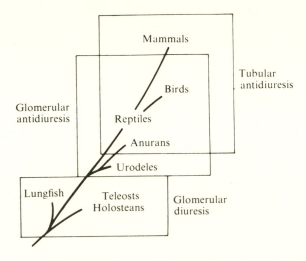

Fig. 8.8. A schematic diagram showing the phyletic distribution of the responses of the kidney to neurohypophysial hormones in vertebrates. An exception may occur in the European eel (Teleostei) where small amounts of vasotocin may also have an antidiuretic effect (see text). (From Sawyer, 1972*b*.)

and Degnan, 1980; Marshall and Bern, 1980). These effects are clearly direct ones on the tissue and do not involve changes in its blood supply.

The overall effects of injected epinephrine on fluid balance in teleosts are hypernatremia and hyperosmolarity in seawater and an accumulation of water in fresh water: Stress, such as associated with laboratory handling or forced swimming, results in elevated levels of catecholamines in the blood of teleosts and also produces disturbances in fluid balance. It thus seems likely that catecholamines may influence osmoregulation in unusual circumstances, but it is not clear whether or not they have such a role in more equitable conditions.

Adrenocorticosteroids

The corticosteroids, mainly cortisol, have a far better established effect on osmoregulation in fishes than do the neurohypophysial peptides. Again, this is predominantly seen in teleost fishes and in this group most of the experiments have been carried out on eels from Europe, *Anguilla anguilla*, North America, *Anguilla rostrata*, and Asia, *Anguilla japonica*. Apart from their ready availability and the feasibility of the performing surgical procedures on them, eels can osmoregulate in either fresh water or the sea and during their normal lives migrate between these two environments. In seawater, radioactive ion flux measurements indicate that teleosts, including eels, have a very large turnover of sodium amounting to as much as 50–60% of the total present in their body every hour, but in fresh water this is less than 1% per hour. In seawater, this ion exchange is the result of a rapid transfer of salts across the gills and the drinking of seawater, which is absorbed from the fish's gut. The excess sodium chloride is excreted across the gills by the chloride cells. The kidney plays little part in the excretion of the excess solutes in teleosts as it lacks an ability to form a hypertonic urine. The influx of sodium across the gills in seawater appears to be a passive process possibly involving exchange diffusion, the sodium that enters being exchanged for sodium leaving the body. The absorption of seawater across the gut depends largely on active transport of sodium and this is related to the presence of the enzyme Na–K-activated ATPase. This enzyme has also been localized in the chloride cells (Kamiya, 1972; Epstein, Silva, and Kormanik, 1980). The maintenance of adequate levels of Na–K ATPase in the gut, gills, and kidneys depends on the action of cortisol, which, in turn, is regulated by corticotropin from the pituitary (Epstein, Cynamon, and McKay, 1971; Pickford et al., 1970). When eels enter seawater there is an increase in the concentration of cortisol in their blood that persists for several days (Hirano, 1969; Ball et al., 1971; Forrest et al., 1973*b*). After this time, however, the steroid level declines so that it is similar to that in eels adapted to fresh water. When the euryhaline teleost *Tilapia mossambica* is transferred from fresh water to seawater a transient rise in plasma cortisol has also been observed; however, a similar change also occurs when they are moved from seawater to fresh water (Assem and Hanke, 1981). It seems that cortisol may enhance the adaptations of such fish by promoting the appropriate movements of electrolytes between the tissues and the extracellular fluids.

In teleost fishes, the corticosteroids contribute to osmoregulation, as in other vertebrates. Removal of the adrenals in freshwater eels results in a decline in plasma sodium, but in seawater there is an accumulation of this ion. This effect can be overcome by injecting cortisol (Chan et al., 1967; Butler et al., 1969; Mayer et al., 1967; Henderson and Chester Jones, 1967). It is interesting, however, that the principal steroid involved is cortisol, which, in the tetrapods, mainly influences intermediary metabolism. In teleosts, cortisol may fulfill both physiological roles.

Although the corticosteroids influence osmoregulation in the fishes, as well as the tetrapods, the mechanisms involved differ. The use of the gills is a piscine (and possibly larval amphibian) prerogative, and they are thus a site for the steroid's action that is confined to the fishes. In addition, the relative importance of the effects of corticosteroids on the gut of teleosts living in seawater may be greater than their action at this site in tetrapods or even in freshwater teleosts. The mechanisms by which the hormones facilitate sodium transport may also differ, for in the tetrapods there is little evidence to suggest that corticosteroids exert their acute effects on sodium transport by

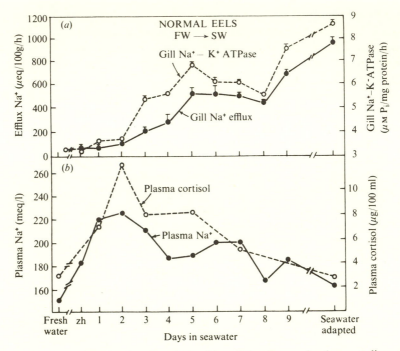

Fig. 8.9. A diagram showing the relationships of the changes in plasma sodium and cortisol concentrations (*b*), the efflux of sodium, and the branchial Na–K ATPase in North American eels (*Anguilla rostrata*) following transfer from fresh water to seawater (*a*). It can be seen that the cortisol concentrations in the plasma initially rise and this is followed by increased Na–K ATPase levels in the gills, which parallels added sodium efflux and the resulting decline in the plasma sodium concentration. (From Forrest et al., 1973*a,b*.)

increasing the levels (or inducing formation) of Na–K ATPase. Instead, it seems more likely that aldosterone increases the activity (or rate of turnover) of the sodium pump (directly or indirectly) rather than by increasing its total capacity, such as would be expected if an increase of Na–K ATPase occurred. In fishes, cortisol may still have an additional, acute type of effect like that of aldosterone in tetrapods, as there is evidence to suggest that when fish are transferred from fresh water to seawater corticosteroid-mediated changes occur independently of any alteration in the level of Na–K ATPase (see Forrest et al., 1973*a*). The temporal relationship between plasma cortisol and sodium levels, Na–K ATPase concentration in the gills, and the branchial efflux of sodium in eels during adaptation to seawater is shown in Fig. 8.9.

There is little information about the action of corticosteroids in nonteleost fish. Adrenocorticosteroids appear to have no effect on renal electrolyte losses in freshwater lampreys though they can decrease the rate of sodium loss across the gills (Bentley and Follett, 1963). Adrenalectomy results in a decrease in the salt-secreting activity of the rectal gland of the skate, *Raja ocellata* (Holt and Idler, 1975; Idler and Kane, 1976). This exocrine gland contains high concentrations of Na–K-activated ATPase, and the levels of this enzyme decline following this surgery. Corticosteroids, especially 1α-hydroxy-corticosterone, appear to help maintain the metabolic integrity of the rectal gland. Vasoactive intestinal peptide (VIP) has been shown to stimulate its secretion in the dogfish, *Squalus acanthias* (Stoff et al., 1979). This effect involves the formation of cyclic AMP and it can be inhibited by somatostatin. It is possible that VIP released into the blood may normally initiate this response though this peptide has also been identified in nerves where it may act as a type of neurotransmitter.

Pituitary hormones: corticotropin and prolactin

The pituitary has an essential role to play in influencing osmoregulation in fishes. Its effects are principally due to two of its hormones, corticotropin and prolactin. The importance of the effects of these two hormones, however, differs considerably between species of fish and depends on whether they are living in fresh water or the sea. Hypophysectomy thus may have little osmotic effect on some species though in others dramatic changes may occur, especially if they are living in fresh water.

When freshwater eels are hypophysectomized they can survive for several weeks though there is a slow depletion of their electrolytes (Fontaine, Callamand, and Olivereau, 1949; Butler, 1966). When placed in seawater, these eels cannot osmoregulate properly and they accumulate excess sodium. Hypophysectomy results in a considerable lowering of plasma cortisol levels, which can then be elevated by injecting corticotropin (Hirano, 1969); thus, a pituitary interrenal (or adrenal) control axis exists in the fishes, the interruption of which can upset osmoregulation (see Maetz, 1969). This is manifested by low levels of Na–K ATPase and reduced fluid and ion movements across the gills and gut that can be substantially corrected by injected cortisol (Epstein, Katz, and Pickford, 1967; Butler and Carmichael, 1972). This hormone, however, does not completely restore osmotic balance, suggesting that other factors may be involved.

The osmotic deficiencies resulting from surgical removal of the pituitary are also due, apart from a lack of corticotropin, to the absence of prolactin. Pickford and Phillips, in 1959, found that when killifish, *Fundulus heteroclitus*, were hypophysectomized they were able to survive in salt water but they soon died when placed in fresh water. If, however, they were injected with mammalian prolactin, their survival in fresh water was considerably prolonged. No other hormones were found to have this effect. Many species of fishes die in fresh water following hypophysectomy though others, like the goldfish, eel, and trout, can survive for considerable periods of time. The importance of the pituitary for survival in fresh water varies considerably among the teleost fishes; thus, 18 species of the order Antheriniformes were found to be unable to survive hypophysectomy if kept in fresh water though many of the order Ostariophysi survive (Schreibman and Kallman, 1969).

Death in fresh water following hypophysectomy was found in *Poecilia latipinna* to be accompanied by a considerable loss of sodium that was prevented by the injection of prolactin (Ball and Ensor, 1965, 1967). This effect has been confirmed in other species of fish and is due principally to an excessive loss of sodium across the gills (see Maetz, Mayer, and Chartier-Baraduc, 1967; Lam, 1972; Ensor and Ball, 1972). The lack of such osmotic sensitivity in some fish, like eels and goldfish, to hypophysectomy seems to be the result of the more restricted permeability of their bodies to sodium, but even in these species prolactin can be shown to decrease branchial sodium loss (efflux).

Circulating levels of prolactin are probably low in fish adapted to seawater. The importance of this is emphasized by the observation that when seawater-adapted *Tilapia mossambica* are injected with this hormone their plasma sodium concentration increases (Dharmamba et al., 1973). This treatment, if continued, would probably kill these fishes. The accumulation of sodium is the result of a reduced rate of sodium chloride secretion from the branchial chloride-secreting cells, possibly as a result of an inhibition of Na–K ATPase. The reduced activity of the chloride cells helps effect the adaptation of these fish to fresh water, but in seawater such an action would be disastrous.

Prolactin also influences the permeability of teleosts to water. The results *in vivo* have been a little contradictory as they usually indicate that the hormone increases branchial osmotic permeability but, depending on the species studied, a decrease may also be observed. In vitro observations on the gills of goldfish also suggest that prolactin reduces permeability to water (see also Ogawa, Yagasaki, and Yamazaki, 1973). In intact fish, it is often difficult to decide which effect and site of action is the primary one; thus, prolactin can increase the urine flow and this could be either a direct action on the kidney or the result of an increased branchial permeability to water. Prolactin is thought to have a direct effect on the kidney in fishes, mediated by an increased GFR or a reduced tubular reabsorption of water, or both. Apart from the kidneys, prolactin can also reduce the transfer of fluid across the fish intestine and urinary bladder (Utida et al., 1972). Prolactin may restrict the osmotic permeability of membranes at three distinct sites: the kidney tubules, the intestine, and the urinary bladder; a similar effect on the gills may also occur but this is controversial. Prolactin may have a physiological action in preventing overhydration of fish in fresh water by facilitating the excretion of water.

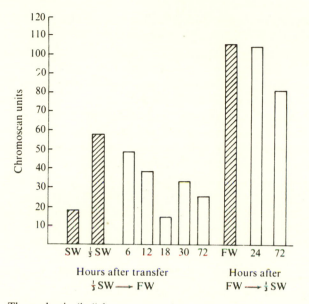

Fig. 8.10. The prolactin (in "chromoscan units") content of the pituitary glands of the teleost *Poecilia latipinna* following transfer from one-third seawater (SW) to fresh water (FW), and then to one-third seawater. After transfer from one-third seawater to fresh water the prolactin storage initially declines but subsequently increases and after 8 days reaches a level that is about six times greater than in seawater-adapted fish. Cross-hatched bars show initial control values. (From Ball and Ingleton, 1973.)

Mammalian prolactin, when injected, is effective in promoting changes in water and salt metabolism in teleost fishes, but this does not constitute proof that the endogenous hormone also acts in this way in the fishes. Teleost prolactin can elicit many of the same osmotic actions as the exogenous mammalian hormone. The amount of prolactin stored in the pituitary of the teleost *Poecilia latipinna* is six times greater in fish living in fresh water than in those in salt water, while the activity of the pituitary *eta* cells (which are thought to secrete prolactin) indicates that the hormone is being released in the freshwater fish (Ball and Ingleton, 1973). The pituitary glands of *Poecilia*, adapted to one-third seawater, contain about three times the quantity of prolactin as seawater-adapted fish (Fig. 8.10). When the one-third seawater-adapted fish are transferred to fresh water, the level of stored prolactin initially declines because of its release, but the rate of synthesis increases so that after about 8 days the amount stored in the pituitary rises to about six times the concentration seen in seawater-adapted fish. Transfer of *Poecilia* from fresh water to seawater has little immediate effect on the glandular stores, but these do show a gradual decline, presumably due to a decreased rate of synthesis. Direct measurements of prolactin concentration in the serum of tilapia have been performed by radioimmunoassay (Nicoll et al., 1981). The levels of this hormone were low in the fish kept in seawater but increased about 12 times when they were in fresh water.

Prolactin appears to have an important role in the migration of sticklebacks, *Gasterosteus aculeatus*, between the sea and rivers (see Lam, 1972). These fish normally spend the autumn and winter in the sea but in the spring migrate into rivers to breed. Sticklebacks caught in winter, in the sea, soon die if they are placed in fresh water, but if they are first injected with prolactin they survive much longer. This increased survival ability is also seen in fish in the autumn if they are kept under conditions of long day-length. It seems that, due to photoperiodic stimulation (the lengthening of the day), there is an increase in the activity of the prolactin cells, which prepares the fish for its future migration into fresh water, during which the hormone is released into the blood.

These interesting observations on the role of prolactin in teleostean osmoregulation have drawn attention to comparable effects in other vertebrates. Among fishes little is known about the role of prolactin in osmoregulation of nonteleosts. In three species of chondrichthyeans, the water uptake across the gills is decreased following hypophysectomy and is restored by injected prolactin (Payan and Maetz, 1971). This effect is similar to that observed in goldfish but is not consistent with the observations on teleostean gills *in vitro*. This effect in chondrichthyeans, as described earlier, could be the result of a primary diuretic action on the kidney rather than the gills. The possible role of prolactin in maintaining optimal hydration in birds and its effects in newts and mudpuppies have already been described and there are even some reports about its osmotic effects in mammals, including an antidiuresis. However, these results are controversial (see Baumann et al., 1977; Vorherr, Vorherr, and Solomon, 1978).

Prolactin thus has ubiquitous effects (see Ensor, 1978), including many responses that are concerned with osmoregulation. These responses appear to be mainly related to its actions on epithelial membranes, such as gills and skin. As we have seen, such effects are prominent in fishes but also occur in other vertebrates. In some instances an action can be ascribed to morphological change, such as has been observed in skin. However, more biochemically oriented effects that result in increases or decreases in permeability to water and solutes and changes in the active transport of ions also occur. The effector organs concerned include gills, skin, intestine, urinary bladder, and kidneys. The diversity of these actions is so remarkable that it is enticing to seek common mechanisms. Prolactin has been shown to influence calcium metabolism and it can exert a hypercalcemic effect in teleosts (see Chapter 6). In teleosts it is also released in response to low calcium concentrations in the external medium, such as fresh water. Calcium is known to have many effects in the body, such as limiting the permeability of membranes to ions and water, and it is involved in many metabolic processes that may influence ion transport and the release of other hormones. An effect of prolactin on calcium concentrations in body fluids could contribute to its varied actions. In the mammary glands prolactin has been shown to activate phospholipase A_2 (see Rillema, 1980), which is a calcium-activated enzyme. Such an action can have several effects. It results in a perturbation of phospholipids in the cell membrane, possibly influencing its permeability properties, and it also starts a series of events leading to the formation of prostaglandins. These fatty acids usually act as local hormones and modulate certain activities of cells, such as the adrenylate cyclase system, which is involved in the actions of several hormones, including vasotocin and epinephrine. There is also a source of experimental error. Preparations of prolactin are extracted from animal glands, usually sheep. This preparation is not a pure hormone but contains certain other constituents; for instance, it has about 1% growth hormone. More important with respect to osmoregulation, it has been shown in several laboratories to also contain vasopressin (ADH) and oxytocin (see Vorherr, Vorherr, and Solomon, 1978). These impurities could be contributing to some of the observed effects of "prolactin."

The urophysis, corpuscles of Stannius, and juxtaglomerular cells

These tissues have a glandular appearance and a putative endocrine function in some fishes. Their structure and distribution among the fishes have been described earlier (Chapter 2). There are several types of evidence suggesting that these glands may have an osmoregulatory function, including:

i. Changes in their histological appearance in different osmotic circumstances.

ii. Deficiencies in the fish's capacity to adapt to fresh water or seawater following surgical removal of the tissues.

iii. The ability of injected extracts of these glands to alter the fish's balance of water and electrolytes.

At this time no unequivocal conclusions as to the suggested osmotic importance of these tissues are available. This lack of consensus is largely the result of variations in the responses of different species of teleosts to various experimental procedures; nevertheless, the urophysis, corpuscles of Stannius, and teleostean juxtaglomerular cells contain biologically active substances and it remains possible that these materials may act as "hormones" at sites unconnected with osmoregulation.

The cichlid euryhaline fish, *Tilapia mossambica*, and the stickleback, *Gasterosteus aculeatus*, suffer an increased mortality following transfer from fresh water to saline solutions, if their urophyses are removed (Takasugi and

Bern, 1962; Ireland, 1969). This effect has not been demonstrated in all species though it is possible that the rapid regeneration of this tissue may contribute to the observed differences. The histological appearance of the urophyses of *Tilapia mossambica* suggests a depletion of the contained neurosecretory material when these fish are kept in seawater, as compared with fresh water. The urophysis of the trout, *Salvelinus fontinalis*, shows an increase in its cytological activity and it takes up radioactively labelled amino acids at an increased rate when the fish are kept in deionized water (Chevalier, 1978). The Hawaiian o'io, *Albula vulpes*, also displays such histological variations in urophysial activity (see Fridberg and Bern, 1968). The rate of electrical firing of the urophysial nerve cells has also been shown to be altered in response to osmotic changes. Both a decrease and an increase in spontaneous electrical discharge have been observed in response to hypotonicity and they depend on the particular species examined.

The biologically active materials present in the teleostean urophysis have been characterized (see Zelnik and Lederis, 1973). These substances have been separated into several components with characteristic chemical and pharmacological actions. The structure of one of these substances (urotensin II, see Fig. 3.12) has been described (Pearson et al., 1980). The urotensins are proteins and polypeptides and have been classified in the following way:

i. Urotensin I, which when injected decreases the blood pressure of rats and birds (Muramatsu and Bern, 1979).

ii. Urotensin II, which contracts various smooth muscle preparations, including the trout urinary bladder and rectum, can increase the blood pressure as well as the urine flow in eels.

iii. Urotensin III, which promotes sodium uptake across the gills of goldfish, an effect that has not been observed in other species.

iv. Urotensin IV, which increases water transfer across the toad urinary bladder (*in vitro*) and has other similarities to vasotocin with which it may be identical. Vasotocin has not been found, however, in the urophyses of all the species of teleosts that have been examined (see Gill et al., 1977; Holder et al., 1979; H. A. Bern, personal communication).

None of these substances has yet been identified in the circulation of fish, but urotensin II is discharged from the urophysis under *in vitro* conditions, which supports the possibility of its hormonal nature (Berlind, 1972*a*). The venous effluent of the urophysis passes into the renal portal blood vessels via the caudal vein, and so any urophysial secretions are in a potentially excellent situation to influence the kidney and urinary bladder.

When the euryhaline marine teleost, *Gillichthys mirabilis*, is placed in fresh water the urophysial content of urotensin IV declines but urotensin II is unchanged (Berlind, Lacanilao, and Bern, 1972). Two other proteins that are characteristically present in the urophysis of *Gillichthys* also show a considerable decline if these fish are kept in fresh water for 6 days.

The effects of crude urophysial extracts and purified preparations of urotensin II on processes associated with water and ion regulation in teleosts have been studied. Such preparations increase the blood pressure, the GFR, urine flow, and sodium excretion in North American and Japanese eels (Chan, 1975; Hirano, 1979). Urotensin II has also been shown to decrease the efflux of chloride across the head skin (*in vitro*) of a marine teleost, the goby, *Gillichthys mirabilis* (Marshall and Bern, 1979). (On the other hand, it stimulates sodium transport [*in vitro*] across the urinary bladder of this fish [Loretz and Bern, 1981]).

Somewhat sporadic evidence thus suggests the possibility that the urophysis may influence osmoregulation in fish. It may, however, have other possible endocrine functions. Urotensin II contracts the smooth muscle of the urinogenital tract, including the sperm duct in *Gillichthys mirabilis* (Berlind, 1972*b*). It has thus also been suggested that the urophysis may have a role in reproduction of fishes, such as in promoting spawning.

The probable role of the corpuscles of Stannius in calcium metabolism in teleost fishes has been described earlier (see Chapter 6). The hypercalcemia that is observed in eels following Stanniectomy is accompanied by a decline in plasma sodium concentration and a rise in potassium (M. Fontaine, 1964). It seems likely that the effects on sodium and potassium are indirect ones reflecting changes in calcium metabolism (see for instance, Pang, Pang, and Griffith, 1975). They could involve the functioning of osmoregulatory organs, such as the gills and kidneys, or endocrine glands, like the adrenocortical tissue.

A renin–angiotensin system has been identified in fishes (see Chapter 2). Renin is present not only in the kidneys but has also been identified in the corpuscles of Stannius (Chester Jones et al., 1966; Sokabe et al., 1970). Its physiological role in fishes is uncertain but by analogy to tetrapod vertebrates it could be involved in the regulation of blood pressure, kidney function (due to effects on the GFR) or, by an action on adrenocortical tissue, sodium metabolism.

When renin, prepared from the kidneys or corpuscles of Stannius of European eels, is injected into these fish it produces an increase in the blood pressure (Henderson et al., 1976). This effect presumably reflects the formation of angiotensin II. This peptide, when injected into eels and the Australian lungfish, *Neoceratodus forsteri*, also elicits a rise in blood pressure (Nishimura and Sawyer, 1976; Sawyer et al., 1976). This response is accompanied by a diuresis and natriuresis, apparently reflecting an increased GFR. (It is notable, however, that a similar vascular and renal response is also seen in the goosefish, *Lophius americanus*, which lacks glomeruli [Churchill et al., 1979].)

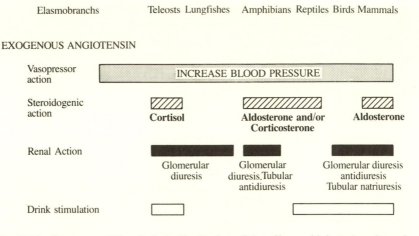

Fig. 8.11. Summary of the phyletic distribution of the effects of injected angiotensin II in vertebrates. (Adapted from Nishimura, 1978.)

The physiological significance of the vascular action of the renin–angiotensin system is not clear in fishes. The blood pressure of eels usually drops when they are transferred from fresh water to seawater (Henderson et al., 1976; Nishimura, Sawyer, and Nigrelli, 1976). Plasma renin levels are usually higher in seawater, however, so that although the release of renin may be appropriate it does not appear to have a remarkable hypertensive action.

In North American eels and toadfish (*Opsanus tau*) no relationship was observed between the concentrations of renin and cortisol in the plasma (Nishimura, Sawyer, and Nigrelli, 1976). This observation suggests that there is no physiological relationship between corticosteroid secretion and renin. However, it has been observed that injected eel renin can stimulate cortisol secretion in European eels (Henderson et al., 1976). On the other hand, angiotensin II had no effect on the concentration of aldosterone in the plasma of Australian lungfish (Blair-West et al., 1977). It is thus possible that the renin–angiotensin system may influence corticosteroidogenesis in teleosts though the evidence is equivocal. A role for this system in the control of the secretion of aldosterone, however, may not have evolved until the emergence of the tetrapod vertebrates. The effect of angiotensin on drinking in Japanese eels (Takei, Hirano, and Kobayashi, 1979) is somewhat unexpected as it is absent in amphibians. In summary, the renin–angiotensin system appears to be able to influence blood pressure throughout the vertebrates and this effect may be its primeval one. It is not altogether unexpected to find that, in view of this vascular reactivity, renin can also influence hemodynamic processes associated with glomerular filtration, and hence urine flow. The steroidogenic and dipsogenic roles, however may have been phylogenetically subsequent developments. The phyletic distribution of the effects of the renin–angiotensin system is summarized in Fig. 8.11.

Our knowledge of the nature of the processes controlling osmoregulation in fishes is incomplete, and much of the information about the role of hormones in this process is still speculative. It is thus still an exciting field of study for the comparative endocrinologist.

Conclusions

The regulation of water and salt content of vertebrates is primarily mediated by the kidney, but several other glands and tissues are also involved. Many of these "accessory" osmoregulatory organs have a distinct systematic distribution, for instance, gills in fishes, nasal salt glands in birds, and sweat glands in mammals. The osmoregulatory functions of most such organs are controlled by hormones that each tend to contribute to this process in a specific manner. Thus, the neurohypophysial hormones increase osmotic permeability whether it be in the tetrapod kidney or the amphibian skin and urinary bladder. Similarly, adrenocorticosteroids help regulate sodium metabolism by increasing transmural sodium transport in the renal tubule of mammals, the skin and urinary bladder of amphibians, the sweat, salivary, and mammary glands of mammals, and the gills of many fishes. The renin–angiotensin system has a role in controlling steroidogenesis in adrenocortical tissue and drinking in most tetrapods, but it may have other functions in fishes. Thus, the same type of hormone is often concerned (or is "utilized") with coordinating the same general physiological process, though in somewhat different ways in various groups of vertebrates. There are, however, exceptions, and systematically unique features, like the role of the gills in osmoregulation of fishes, are accompanied by what may be a unique type of action of a particular hormone, in this instance the ability of prolactin to control the permeability of the branchial (and other) epithelial membranes in teleosts. In mammals, this hormone appears to be principally concerned with the regulation of lactation though it is now suspected that it may also be capable of influencing osmoregulation in some tetrapods. The physiological significance of the effects in the latter is still in doubt and could represent a "vestigal" endocrine response.

9. Hormones and reproduction

The reproductive process is not essential for the life of the individual, though it may make it more interesting, but it is necessary for the perpetuation of the species. In many so-called lower forms of life, reproduction may be an asexual process. A notable disadvantage of this type of reproduction is a diminution in the chances of genetic variability, and the transmission of such inherited changes to other individuals, so that evolutionary adaptation is hampered. Reproduction is a complex process and this is especially true in species that occupy environments where the conditions are variable and large physicochemical changes occur. The young, developing animal is not usually as adaptable as the adult to such changes in the environment, and so must either be protected from these deviations by the parents or be produced on occasions that are most suitable to its more limited physiological capabilities. In vertebrates both conditions usually prevail; the embryo may develop to a quite advanced stage before becoming independent of the parent and it is usually produced during a season when such conditions as the temperature and food and water supply are favorable.

Reproduction in vertebrates thus involves considerable physiological coordination. The sexual process that requires the union of the sperm and ova necessitates complex physiological, social, and morphological arrangements to ensure that these gametes each ripen at a similar time, and that the two sexes then meet and effect their union. The growth and differentiation of the fertilized egg often involve complex parental care, which may occur *in utero*, within the parent itself, or in an egg that is specially produced to meet the potential needs of the embryo. Care of the young often continues for a period of time following such initial development in the egg or *in utero*. The foregoing events may not be possible, or successfully accomplished, except during certain seasons of the year when the conditions are favorable.

In vertebrates, the coordination of all the processes outlined above involves hormones and the degree of complexity of their actions directly reflects the intricacies of the reproductive processes in a particular species. The endocrine control of reproduction in man is thus more involved than in a jawless fish, like the lamprey. The basic pattern, however, is remarkably uniform and involves the endocrine secretions of the pituitary and the gonads. The influences of the hypothalamus on the pituitary gonadotropin release are vital in most groups of vertebrates, but this control seems to be lacking in the cyclostomes and chondrichthyeans. The pituitary gonadotropic hormones in vertebrates are chemically analogous but have undergone evolutionary changes in their structures. The gonadal steroids, on the other hand, are identical throughout the vertebrate series. One of the most notable endocrine differences among the vertebrates is the ability of the placenta of eutherian mammals to produce hormones that are similar to those secreted by the gonads and the pituitary. Otherwise, the evolutionary changes are largely a matter of detail. These adaptations include modifications of the gonadal ducts, such as may assist in the processes of fertilization, the production of different types of eggs, and the internal incubation of the embryo. A plethora of secondary sex characters, involving such morphological factors as size, color, and scent glands have appeared in the different vertebrate groups and these help to ensure that the sexes meet and mate in the breeding season. Also involved in the mating procedure are a multitude of different patterns of prenuptial behavior. The precise manner by which the time of breeding is controlled also varies in different vertebrates and may involve differences in the length of the gonadal cycle and adaptations to the receipt of different environmental signals, such as light, which is predominant in birds, and temperature, which seems to be more important in reptiles. Variation in the functioning of the hypothalamus, pituitary, and gonads may reflect such differences in the manner of timing of breeding.

There are several major differences in the patterns of reproduction in vertebrates that have considerable effects on the endocrine control of reproduction. These concern the manner by which fertilization is accomplished and the site where the embryo differentiates and grows.

Life in an aqueous medium, such as the sea, lakes, and rivers, provides a relatively stable physicochemical environment for ova, sperm, and the fertilized eggs. Many fishes ensure their reproduction by producing vast numbers of ova, often millions at a time that are extruded, before fertilization, into the external solution. The reproductive activities of the male are coordinated with this oviposition so that vast numbers of sperm are released among the eggs and *external fertilization* occurs. This process, which also takes place in most amphibians, has the advantage of simplicity but is only possible in aqueous situations. A general physiological corollary of external fertilization is that usually the eggs are small and contain relatively few nutrients for the support of the young. The animal thus, nutritionally, can afford to produce the vast numbers of ova necessary to assure fertilization on a scale that is adequate for the survival of a sufficient number of the young. Primarily terrestrial species, like the reptiles, birds, and mammals, must of necessity resort to *internal fertilization*. This process requires the production of fewer eggs but more intimate contact and collaboration between the sexes. Internal fertilization also occurs among fishes including some bony fishes, all of the chondrichthyeans, and some amphibians.

There are also considerable differences, again with important endocrine repercussions, between the eggs of different vertebrates and the processes that assist in their successful transposition into viable young. These differences are partly related to the nature of the fertilization process (due to the number of eggs that must be produced) and to life in a terrestrial environment. Eggs produced by most fishes and amphibians are highly prone to evaporation and so cannot readily survive on dry land. Some frogs deposit such eggs in damp burrows, but this is unusual. Birds, most reptiles, and prototherian mammals, such as the platypus, produce eggs that are covered with a protective shell that limits evaporation, and they contain large amounts of nutrients that are sufficient to sustain the young until it reaches a stage of development when it can fend for itself. Similar eggs with a horny shell and large amounts of nutritive yolk are also produced by the chondrichthyean fishes and hagfishes. The production of eggs from which the young develop in the external environment is a process called *oviparity* and the eggs are termed alecithal and megalecithal eggs, depending on the amounts of yolk nutrients that they contain. In many species, including some chondrichthyean and teleost fishes (where the ovum is often fertilized within the follicle in which it develops), amphibians, and reptiles, the eggs may be retained for prolonged periods in the oviduct during which time the young develop in a relatively protected and secluded situation. This is called *ovoviviparity*. A more intimate contact between the eggs and the wall of the oviduct or uterus may occur whereby the developing young can exchange respiratory gases and even gain fluids and nutrients. This condition is called *viviparity* and as nutrients may be gained from the parent, the eggs usually contain far less yolk. Viviparity has evolved many times in nature and is present among chondrichthyeans, teleosts (where the young are usually contained within a hollow ovary), amphibians, reptiles, and mammals. There are many endocrine variations that result from these different ways of providing for the development of the young including hormonal influences on the maturation and formation of the different types of eggs, the morphological development and physiological behavior of the oviducts, and the triggering, at the appropriate time, of the expulsion of the egg (oviposition) or young (parturition).

The eggs and young of many species receive little or no parental care once they are separated from the mother. In some species, however, they are cared for and this may be necessary for their survival. Some teleost fishes are known to deposit their eggs in specially prepared nests which they protect and over which they may circulate water. Others such as the teleosts *Tilapia mossambica* keep a brood of hatched young in the fastnesses of a large mouth that the young can emerge from or retreat to. Several species of frogs (*Pipa*

pipa, *Gastrotheca marsupiata*) and the seahorse (*Hippocampus*) keep young in a pouch, or marsupium, on their backs while other frogs incubate them in modified vocal sacs. Some snakes and most birds personally incubate their eggs, and the care with which birds feed and protect the newly hatched young is well known. Birds usually collect food, which they present to their young and often this is predigested. Pigeons and doves produce a special pasty secretion that contains a high concentration of fat and protein from their crop-sac, the so-called pigeon's milk. The formation of such a special milk secretion with which to feed the young is a characteristic systematic feature of mammals. Such processes, whether it is "tender loving parental care," "brood-iness" in birds, or lactation in mammals, are all largely controlled by hormones.

The reproductive apparatus of vertebrates

The gonads of vertebrates have a dual function, as they not only produce the germ cells but also some of the hormones that control the reproductive pro-cess; thus, the testis in addition to being the site of formation and maturation of the sperm also produces androgens, principally testosterone and andro-stenedione in the interstitial tissue (or Leydig cells) and, probably, also in the Sertoli cells. The ovary contains vast numbers of germ cells (primordial follicles) some of which, following a period of growth and maturation, ripen into ova. The follicles in which this latter process occurs are also the site of formation of estrogens. Following extrusion of the mature ovum (ovulation) from its follicle, the tissue "heals" and this may involve an invasion and proliferation of lutein cells so that a corpus luteum is formed. In many spe-cies, this structure is the site of formation of progesterone, which can also sometimes be produced by the interstitial tissue of the ovary. A more detailed description of the structure of the gonads and the hormones that they produce is given in Chapters 2 and 3.

Associated with the gonads are the duct systems through which the germ cells are delivered to the outside of the animal. Discrete gonadal ducts are absent in the cyclostomes where the eggs and sperm are shed directly into the body cavity from which they escape to the exterior through pores that are formed in the region of the cloaca.

In the teleost fishes, the ovarian ducts represent extensions of the gonadal tissue but in other vertebrates, they are modified Müllerian ducts that differen-tiate in the embryo under the influence of estrogens to form the oviducts or uterus. The oviducts and uterus are surrounded by a sheath of smooth muscle fibers that, by rhythmical contractions, can propel objects, like eggs, toward the exterior. This musculature may be relatively weak, as in the amphibians, or, as in mammals, be capable of very strong contractions and make up a major portion of the uterine wall. In mammals, this muscle layer is called the *myometrium*. The contractility of such muscle can be influenced by hormones, especially those from the neurohypophysis, which stimulate their contraction. Underlying the muscles of the gonaducts is an inner lining of cells that in mammals is referred to as the *endometrium*. This inner layer of tissue contains numerous glandular cells and may be modified in various ways so that it contributes to the well-being of the egg and, if internal fertilization occurs, the survival of the sperm and its union with the ovum. In oviparous species, segmental differences in function may occur along the length of the oviduct, such as are associated with the formation of albumin and the secretion of a hard outer shell. In amphibians, the jelly-like secretion so characteristic of clumps of frogs' spawn is secreted by glandular cells in the oviduct. In ovoviviparous and viviparous vertebrates, the lining of the oviduct is modified so as to furnish an appropriate environment for maintaining the retained egg or to allow for the implantation and development of the blastocyst and the formation of a placenta. The activity of the surrounding musculature is reduced on such occasions. The female gonaducts (oviduct or uterus and vagina) thus undergo considerable structural change during the reproductive cycle that is mediated by the action of estrogens and progesterone.

The sperm are conveyed from the testis along the vas deferens, which is also a tube surrounded by an outer layer of smooth muscle, during which time they may be mixed with secretions from certain accessory sex glands that include the prostate or prostate-like glands. The maintenance of the structure and function, and cyclical changes of the male gonaducts, their associated accessory sex glands as well as the external genitalia, such as the penis, are due mainly to the action of testosterone. In the absence of this hormone, structural and physiological degeneration of such tissue takes place.

Secondary sex characters in vertebrates

The secondary sex characters are so named because they are not primarily involved in the formation and delivery of the sperm or ova. They nevertheless may play an important part in the prenuptial and nuptial events and contribute to the behavioral and functional synchronization necessary for the fertilization of the ripened ova, the mechanical success of copulation, and the survival of the young. The secondary sex characters differ in each sex and contribute to the dimorphism of the male and female. The differences in appearance between the sexes are basically controlled genetically, and the expression of them may be influenced by the actions of sex hormones. Broad differences, such as those of size, are usually independent of the continuous action of sex hormones, while in other instances only the initial differentiation of a sexual character during early life may depend on hormones. Hormones are, however, not necessarily continuously needed to maintain such organs after their differentiation; thus, the changes in the larynx of boys at puberty, which results in a deeper voice, require the presence of testosterone though subsequent castration does not result in a return to the prepuberal soprano condition. In other instances, however, a continuous supply of hormones may be required to maintain a secondary sex character, as is seen in the instance of the penis in man and the breasts in women. Some secondary sex characters may undergo periods of development and involution that correspond to the changes in the sexual cycle and the differences in the rates of hormone production that occur during these periods. The seasonal development and subsequent shedding of the antlers of deer are well-known examples of this but there are numerous others. Dodd (1960) and Parkes and Marshall (1960) give an excellent account of these structures in cold-blooded vertebrates and birds.

In *cyclostome fish*, the endocrine control of secondary sex characters has been studied in lampreys (see Larsen, 1965, 1969, 1973). One cannot distinguish, from their external appearance, between the male and female lamprey during the early part of their autumnal breeding migration into the rivers. With the approach of spring and the onset of breeding, an anal fin appears in the female while the dorsal fins of the male heighten. Such morphological changes do not occur if the animals are hypophysectomized or gonadectomized. There is also a swelling of the cloacal region in both sexes just before breeding and the urinogenital papilla grows larger in the male. These latter characters do not, however, appear to depend on the presence of steroid sex hormones.

The *chondrichthyean fishes* also display few dimorphic secondary sex characters. In some species, the most notable difference is the presence of a pair of copulatory organs called claspers in the external cloacal region of the male. These rod-like organs develop at puberty and their size can be increased by the administration of testosterone, but this effect is not large.

The *osteichthyean fishes* show a considerable range in dimorphic sexual differences. In some species it is difficult to detect such variations, but in others there may be prominent differences in size. The male is sometimes much smaller than the female; variations in color may occur and there may be differences in the size of the fins. Such diversity may only arise, or be accentuated, during the breeding season. The dorsal fins of the bowfin, *Amia calva*, thus become a brilliant green prior to breeding while the belly turns a pale green. In the female the latter is white. The anal fins sometimes become enlarged in fishes such as *Gambusia affinis* and *Xiphophorus*, where it is called the "sword" and functions as an organ for copulatory intromission called a gonapodium. The injection of testosterone into the female fish may result in the development of secondary sex characters just like those in the male.

During the breeding season, the male South American *lungfish*, *Lepidosiren paradoxa*, develops long finger-like outgrowths from the fore- and hindlimbs.

These organs are bright red in color due to a rich blood supply, and it has been suggested that they may function as respiratory organs. The eggs of these lungfish are laid in burrows where the oxygen tension of the water is low. It has been surmised that the male, who guards this nest and fans the eggs, may secrete oxygen from these organs into the oxygen-poor water of the burrow. The growth of these so-called limb gills can be induced by the injection of testosterone (Urist, 1973), which confirms their nature as that of a secondary sex organ.

Amphibians often display prominent sexual differences during the breeding season. The bright orange coloration and the dorsal crest of the male crested newt, *Triturus cristatus*, can be induced in nonbreeding animals following the injection of testosterone. The development of the nuptial thumbpads in frogs is under hormonal control. Testosterone not only stimulates the development of these tissues but, in *Bufo fowleri*, can also promote the development of the vocal sacs in immature males and prompt them to give their characteristic mating calls or croaks.

It is usually rather difficult to distinguish between the sexes in *reptiles*. The males of many lizards, however, possess appendages about their heads and throats that have a fan-like appearance and can be erected for display. Dorsal crests are also present in some lizards. The tails of male turtles often grow longer than those of the females and aid in copulation. Some reptiles, especially snakes, possess erectile penises that differentiate from their cloacal tissues. The males in many species of snakes and lizards also possess a special secretory segment in the distal part of the renal tubule, called the "sexual segment" and the development of this is under androgenic control.

The colorful dimorphic differences in the plumage of *birds* are well known and generally (though not always) appear to be under the control of estrogens in the female; thus, the bright coloration of the male in domestic fowl, quail, and pheasants is not dependent on androgenic hormones, but the duller, more conservative plumage of the female is under estrogenic control. The injections of estrogen into male birds can stimulate the formation of female plumage. The color of the beaks of birds also is influenced by sex hormones during the breeding season. The beak of the male house sparrow is normally black in the breeding season and this coloration can be induced at other times of the year, or in castrates, by administering androgens and FSH or LH and FSH (Lofts, Murton, and Thearle, 1973). The domestic fowl possesses a red fleshy structure, called the comb, on its head. This is much more developed in roosters than in hens but it regresses following castration (caponization). The comb of the domestic fowl is extremely sensitive to the presence of androgens that induce its hypertrophy. This response in capons has been widely used to identify androgenic materials and indeed provided the first unequivocal evidence for the presence of androgenic hormones in the mammalian testis.

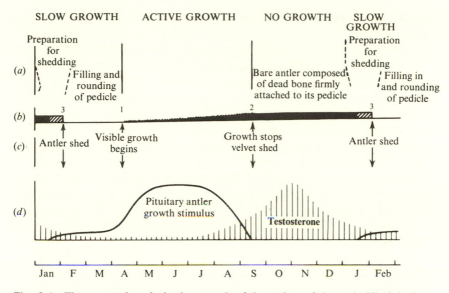

Fig. 9.1. The seasonal cycle in the growth of the antlers of the male Virginia deer (*Odocoileus v. borealis*) in relation to testosterone. This is controlled by light mediating the (hypothetical) release of pituitary hormone(s). (*a*) The three major phases; growth is slow, fast, and then ceases. (*b*) The changes in the physical size of the antlers during these periods in (*a*). (*c*) The principal events in the cycle. (*d*) The increases in the pituitary growth stimulus that rises in the spring and subsides in the autumn when it is inhibited by the rising levels of testosterone. The precise nature of the pituitary stimulus is not clear but is probably a gonadotropin(s). (Modified slightly from Amoroso and Marshall, 1960, taken from Waldo and Wislocki, 1951.)

It is unnecessary to recall the secondary sex characters in man. In other *mammals* differences in size and coloration commonly occur. The red "sex skin" of the buttocks of some female monkeys during the sex cycle has been referred to in Chapter 7. Scent glands, which are under the androgenic control, are also common in many mammals. The wild boar and male members of the cat family secrete odoriferous materials (pheromones) into the urine from special cells in the kidney and these glands are controlled by testosterone. The antlers of deer start to grow in the spring, apparently under the influence of pituitary hormones. In the autumn, when these animals breed, the antlers lose their covering of "velvet," stop growing, and come under the control of the rising testosterone levels in the blood. When the concentrations of this hormone subsequently decline the antlers are shed, usually at about the end of January (Fig. 9.1).

Finally, it should be stressed that the behavior of vertebrates during the breeding season is also a secondary sex character (a most important one) that is influenced by hormones. *Sexual behavior* (see Herbert, 1972) leads to the cohabitation of the male and female animal and the union of their genes, as provided in the sperm and ova. Subsequently, it may be utilized to foster the well-being and growth of the young. Apart from the use of attractants, such as the color of the integument, special appendages such as crests and fins, inviting smells and scents, and ethereal call sounds, animals also adopt active patterns of behavior to find and attract the opposite sex. Such solicitations become more frequent, impetuous, and insistent at the onset of the breeding seasons when ripening of the ova is imminent. Complicated courtship displays may be provided that facilitate pairing, copulation, and preparations for the future young. Hormones play an important role in such processes and influence appropriate neural events that originate in the brain. Castration and ovariectomy usually lead to a decline in such sexual behavior. Caponized roosters do not crow and strut, spayed cats do not call and will actively reject the male, castrated bulls are not aggressive and ignore the solicitations of friendly cows. This sad picture is the result of hormonal imbalance: a lack of estrogens, androgens, and progestins, and possibly a disordered secretion of pituitary gonadotropins and prolactin. The particular hormones involved vary in different species. The injections of estrogens can restore the sexual behavior of many ovarietomized members of the Carnivora but in the Rodentia progesterone is also needed. In primates androgens are needed in the female and play a major role in women. These steroids may be obtained in adequate amounts from the adrenal cortex. In the male, however, more androgens are needed and must normally be supplied by the testes. However, if sexually mature animals are castrated the masculine pattern of behavior, and even an ability to copulate, may persist for some time. Such activities can usually be fully restored by the administration of androgens. Chemical compounds that have an antiandrogenic activity, such as cyproterone, have even been utilized in attempts to reduce the sex drive in humans suffering from satyriasis.

Sex hormones can enter the brain where they interact with certain groups of neurons. They can have two types of effects;

i. In young animals they influence the development of neural circuits that mediate appropriate male and female behavior. Such effects are described elsewhere in this chapter.

ii. In sexually mature animals hormones appear to facilitate and promote the activities of neural circuits and this is manifested in what we call sexual activities.

The sites where such hormones act in the brain have been localized (see McEwen, 1981; MacLusky and Naftolin, 1981) by using radioactively labelled steroid hormones and autoradiography. Small quantities of such hormones can also be implanted in different parts of the brain and their effects on behavior noted. Thus, the sexual drive of castrated animals can often be restored by implanting crystals of hormones into the region of the hypothalamus. This part of the brain and the adjacent preoptic area and amygdala appear to be of primary importance in controlling and promoting sexually related behavior. In the female, estrogens have an important effect at such sites and in the male androgens are effective. However, it is surprising to find that estrogens are also effective in eliciting male-type behavior in the hypothalamus of the male. (The reason for this response will be described later.) Apart from mammals, such effects have also been observed in birds and reptiles (Hutchison, 1971; Crews and Morgantaler, 1979). In male canaries and frogs, injected androgens have been shown to accumulate in a region of the brain known to be associated with song or calling behavior (Kelley, 1978).

Apart from courtship and copulation, other behavioral activities that are a preliminary to breeding may be influenced by hormones. In birds this often involves the building and preparation of a nest. Nest building can be promoted in canaries that have been ovariectomized by injecting them with estrogens, though they must also be sensitized by exposure to long-day photoperiods (Hinde, Steel, and Follett, 1974). The nature of the photoperiodically induced "factor" is unknown, but the injection of LH into these birds could not mimic it. Normal nesting behavior can be induced in budgerigars if they are injected with prolactin in addition to estrogen (Hutchison, 1975). The special role that prolactin has in the incubation of eggs by birds will be described later. Ringdoves have an elaborate courtship prior to egg laying. Ovariectomy abolishes this behavior, though it can be partially restored by modest doses of estrogens. (Progesterone is also needed for nest-building behavior.) Large doses of estrogens, however, have an inhibitory effect on pre-egg laying courtship, suggesting that a blockade of the release of LHRH could be occurring, as when women take contraceptive pills. The administration of LHRH to ovariectomized ringdoves, in the presence of the high dose of estrogens, did overcome the inhibition and they then displayed normal courtship (Cheng, 1977). This effect was not due to a release of LH and could represent some other role for LHRH in behavior.

Periodicity of the breeding season; rhythms in sexual activity

Most vertebrates only breed periodically but, nevertheless, at fairly precise times of the year. In temperate zones, this more usually occurs in the spring but in some species, like deer, sheep, goats, badgers, and grey seals, it occurs in the autumn. In equatorial regions where the climate and food supply are relatively similar throughout the year, breeding may often take place at any time. Similarly, some domesticated species, like the laboratory rat, the domestic fowl, and man may breed throughout the year, a situation that appears to reflect continuously favorable circumstances.

An ability to reproduce during predictable seasons of the year clearly may be of considerable advantage as the young can then be produced at a time when such factors as the environmental temperature and the food and water supply are adequate. The chances for the survival of the young will thus be enhanced.

How is such precise timing possible? In temperate zones, the environmental conditions that prevail in a certain season are usually fairly predictable; thus, the animals can be expected to take their "cues" and make their reproductive preparations on the basis of the solar calendar. Changes in the length of the day are a direct reflection of these events so that the length of the periods of light and darkness may furnish an excellent calendar to work by. Indeed, such photoperiodic stimulation is basic for the control of the reproductive cycle in most vertebrates. The first clear indication that light influences vertebrate gonadal function was made in 1925 by a Canadian zoologist called William Rowan. He found that the gonads of the junco finch, which normally enlarge when the days grow longer in the spring, could be stimulated to grow, even in winter when the birds were subjected to artificially prolonged periods of light. Other factors, however, can also impinge on the onset of the reproductive cycle and even override it. These include temperature, the nutritional condition of the animal, and the related availability of supplies of food and water. There is also evidence for the presence of an internal, inherent rhythm in the sexual activity of some species. It is often difficult to disentangle these various factors and to decide which is predominant.

The effects of light on reproduction have been studied in many species of birds but fewer mammals and cold-blooded vertebrates. Preparations for spring breeding often commence about the end of December when the length of the daylight hours starts to increase. As shown by Rowan, these conditions can be copied in the laboratory and dramatic increases in the activity of the gonads can then be shown to occur; thus, in the Japanese quail subjected to long-day photoperiods of 20 hours light and 4 hours darkness, the weight of the testes increases from 8 mg to 3000 mg in about 3 weeks (Fig. 9.2). The subsequent substitution of a short-day photoperiod, of 6 hours light and 18 hours darkness result in a decline in the weight of the testes. The gain in testicular weight is due mainly to an increase in the length and diameter of the seminiferous tubules though increases in the activity of the Leydig and Sertoli cells also take place. Comparable increases in development also occur in the ovaries of birds.

Studies in mammals have been made on the laboratory rat, which, if kept in continuous light, suffers deficiencies in the development of its reproductive system and eventually becomes infertile. Alternate periods *per se* of light and dark also appear to contribute to gonadal stimulation in animals. The breeding cycle of the ferret has also been shown to be dependent on the length of the

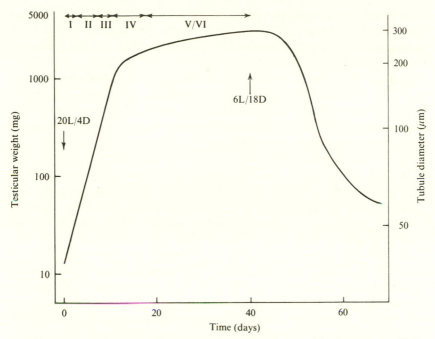

Fig. 9.2. Photoperiodically stimulated growth of the testes of the Japanese quail. For the first 40 days the birds were subjected to long, daily photoperiods of 20 h light and 4 h dark (20L/4D) and after this to short day-length of 6 h light and 18 h darkness. The diameter of the seminiferous tubules (for a given testis weight) is given on the scale on the right. At the top of the diagram (Roman numerals) the changes in the development of the sperm are given: I = spermatogonia only; II = spermatogonia dividing, a few spermatocytes; III = numerous spermatocytes; IV = spermatocytes and spermatids; and V, VI = spermatids and mature sperm. (Modified from Lofts, Follett, and Murton, 1970.)

daily period of light. Shielding the eyes from light, or cutting the optic nerves, usually abolishes the effect of such photostimulation, receptors for which appear to be present in the eye. Such ocular receptors, however, are not always vital as it has been shown, for instance, that the domestic duck still exhibits its periodic breeding behavior even after it is blinded. It has been suggested that breeding activity is due to direct photostimulation of parts of the brain, a process that may occur through their translucent skull.

The reproductive cycles of all birds or mammals do not necessarily respond to light. Such photostimulatory effects are absent in rabbits, guinea pigs, ground squirrels, and guinea fowl. These differences in response to external stimuli may reflect the effects of domestication or, possibly, in the case of guinea pigs and guinea fowl, their origin from equatorial regions where animals do not experience large changes in day-length. Tropical deer that normally breed all year round also persist in this habit after many years in Europe even though they experience cold winter conditions. Deer from equatorial regions that normally have a seasonal cycle also persist in their pattern of reproduction when moved to Europe.

Amoroso and Marshall (1960) have classified animals into those that have "a long-day" and "short-day" breeding season. Long-day animals, which breed in spring, include most birds, as well as horses, donkeys, ferrets, cats, and raccoons. Goats, deer, and sheep are short-day species that breed in the autumn when the day-length is declining.

The effects of light in stimulating development of the gonads and the timing of reproduction are not seen in the absence of the adenohypophysis or when the hypothalamic connections to the median eminence are cut. Differences in the length of the daily photoperiods of light and darkness control the release of LHRH from the median eminence, which in turn initiates the release of gonadotropins from the pituitary. The gonadotropins, FSH and LH (and sometimes also prolactin), exert their various effects on the development of the germ cells and the formation and release of the gonadal steroid hormones.

As mentioned in the foregoing, the reproductive rhythms of all animals are not responsive to light. The environmental temperature, for instance, may also play an important role. While birds and mammals often will not breed in extremely hot or cold conditions, thermal changes are usually not of great importance in determining breeding in such homeotherms. In poikilotherms, however, such effects may be more significant. Spallanzani (1784) considered that reproduction in reptiles and amphibians may be related to the environmental temperature, and this still seems to be correct though light may also contribute. Licht (1972) has carefully analyzed the role of temperature in controlling reproduction in reptiles and considers that it supplies the most important stimulus. Such stimuli could be acting at several sites.

a. A direct action on the brain could influence the release of hormones from the median eminence.

b. Temperature could be exerting a direct action on gonadotropin formation and release in the pituitary itself.

c. When the temperature is increased it has been shown that the responsiveness of the target tissues to gonadotropins increases. This is shown in Fig. 9.3 where the responses of the ovaries and oviducts of the lizard, *Xantusia vigilis*, can be seen to increase considerably at higher temperatures.

d. It is possible that changes in body temperature may indirectly alter the levels of hormones by changing the rate of their inactivation.

Temperature has also been shown to influence the reproductive cycle in fishes. In poikilotherms the effect of temperature may be of a "permissive" nature for in the presence of a low body temperature metabolism is depressed

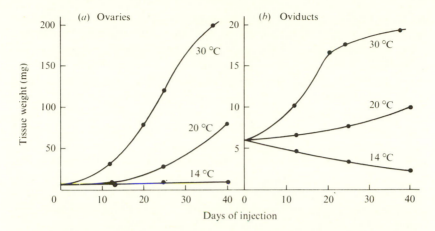

Fig. 9.3. The effects of the environmental temperature on the responses of the ovaries and oviducts of lizards (*Xantusia vigilis*) to injections (on alternate days) of ovine FSH. It can be seen that at 14° C there was little change in the weight of the tissues, there was a rather small effect at 20° C, while at 30° C the growth of the ovaries and oviduct was marked. (From Licht, 1972.)

and could be at such a low level that an action of light, or other stimulating factors, may be ineffective.

The availability of food and water can have dramatic effects on the breeding cycle. Many birds that live in the dry desert areas of Africa and Australia (so-called xerophilous species) rapidly come into breeding condition following unpredictable seasonal rains. Breeding in most amphibians, even those from temperate regions, is also finally determined by rain and the availability of water in their breeding ponds. The African toad, *Xenopus laevis*, is thought to be unresponsive to light; reproduction is determined by an optimal nutritional condition and the availability of water. Domestic animals such as sheep are often fed a special protein-rich diet to bring them into breeding condition. It should be remembered, however, that not all species breed when they are in best physiological condition as breeding may occur shortly after prolonged periods of hibernation or estivation, as seen commonly in amphibians and lungfishes, or at the end of a prolonged fast that follows a migration, such as in salmon and lampreys. At present it is not known how such a nutritional state and the availability of food and water influence breeding.

A nonestrogenic chemical compound, 6-methoxybenzoxazolinone, has been isolated from the young growing shoots of plants. When fed to montane meadow voles, *Microtus montanus*, it initiates their reproductive activity (Berger et al, 1981). This substance may provide an appropriate signal indicating that an adequate supply of herbage will most likely be available over the next few weeks. This situation is clearly favorable for successful breeding. It is possible that 6-methoxybenzoxazolinone or other chemicals like it may be utilized quite widely as "chemical triggers" for seasonal reproduction.

The breeding cycle of vertebrates is also influenced by a variety of ill-defined factors that for want of better knowledge are sometimes called psychological effects. These can be seen quite dramatically in many animals kept in captivity where they do not breed despite the fact that they otherwise appear to be in excellent physiological condition. This deficiency may be the result of the absence of certain environmental "cues" such as sufficient social contact with other members of the species and an inability to perform a ritual courtship display. Social influences can be very important (see Kelley, 1981) and it has been seen that reproduction is promoted in colonies of seabirds when the numbers grow past a critical level. The mechanism for such effects is unknown but would appear to be mediated by the central nervous system and the hypothalamus.

While the hypothalamus and median eminence usually exert a major influence in controlling reproduction, as referred to earlier, this does not appear to occur in all vertebrates. In lampreys, the pituitary is essential for reproduction but it can be transplanted to other parts of the body, away from the region of the hypothalamus and breeding can still occur (Larsen, 1973). There is similarly no evidence that the hypothalamus controls reproduction in the chondrichthyes though gonadotropins from the ventral lobe of the pars distalis are essential (Dodd, 1972*a*).

The pineal gland displays rhythmical activity that influences the synthesis of melatonin and is related to the receipt of light (see Chapter 3). Such a rhythm may be entrained to oscillations in the seasonal release of reproductive hormones. A considerable amount of research (see Reiter, 1980) has been done to explore this possibility in various vertebrates. The evidence for such a role of the pineal gland involves studying the effects of pinealectomy on normal reproductive rhythms and measurement of the effects of injected melatonin and vasotocin (another putative secretion of the pineal) on the development of the gonads and the release of pituitary hormones.

The rat pineal produces melatonin during the hours of darkness and this hormone can exert an inhibitory effect on reproduction (see Chapters 2 and 3). Pinealectomy (see Reiter and Sorrentino, 1970; Quay, 1970; Wurtman, Axelrod, and Kelly, 1968) in young rats thus hastens their sexual maturation and in adults may increase the weight of the gonads. When Syrian hamsters that are normally exposed to long-day photoperiods are placed in darkness, their testes normally decrease in weight, from 3000 mg to 500 mg and this regression can be prevented by pinealectomy (Fig. 9.4). Daily injection of melatonin can also interrupt the development of the gonads of these hamsters kept on long-day photoperiods, in a manner that mimics the effects of keeping them on short-days (Tamarkin et al., 1976).

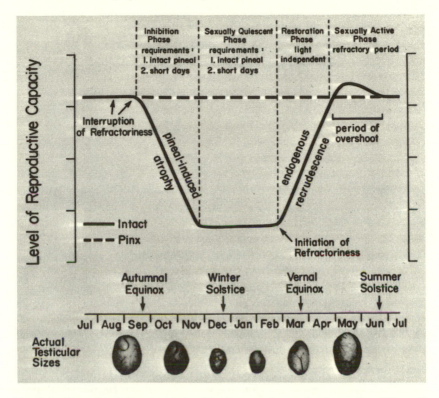

Fig. 9.4. The role of the pineal gland in the annual reproductive cycle of the male hamster. Breeding is normally initiated by the onset of days with long periods of light. Under these conditions the testes enlarge and reach their maximum size in midsummer and then, as the days get shorter, they regress and are smallest in midwinter. This regression is prevented if the hamster's pineal gland is removed. (From Reiter, 1980.)

The effects of pinealectomy on gonadal function in birds have been inconsistent, possibly reflecting surgical trauma associated with the operation (Ralph, 1970). Experiments on Japanese quail have failed to demonstrate any effect of removing the pineal on the growth of their gonads (Oishi and Lauber, 1974). Chronic administration of melatonin failed to elicit any change in the growth of the ovaries of white-throated sparrows on a long-day photoperiod (Turek and Wolfson, 1978). However, in reptiles, pinealectomy of Indian garden lizards (*Colotes versicolor*) inhibits the regression of the gonads that normally occurs when they are exposed to short-day photoperiods (Thapliyal and Haldar, 1979).

Goldfish exposed to long-day photoperiods show an increase in the weight of the gonads in the months of January to May (but not at other times), and this effect is increased more than two-fold when the fish are pinealectomized

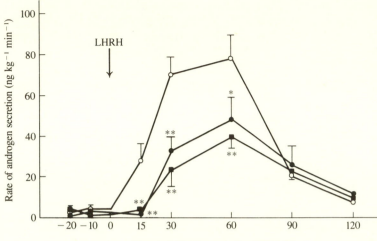

Fig. 9.5. Inhibitory effects of melatonin on the rate of androgen secretion (determined as 17-oxosteroids) by the testis of dogs following the injection of LHRH. Melatonin was injected intraarterially (● = 10 μg/kg body weight; □ = 100 μg/kg; ○ = control injection, no melatonin) 3 hours prior to the LHRH (5 μg/kg). Each point represents a mean ± S.E. *$P<0.05$; **$P<0.01$ for the statistical differences from the control injected dogs. (From Yamashita et al., 1978.)

(Fenwick, 1970). This effect has been confirmed by others. This antigonadotropic effect appears to reflect a suppression of the daily cycle of gonadotropin release (Hontela and Peter, 1980). The effects of the pineal in fish are varied, however, and progonadotropic actions have also been described. The response probably depends on the season, temperature, and the physiological condition of the fish. On the other hand, pinealectomy delays the maturation of the gonads in lampreys where pineal secretion may have a progonadotropic action! (Joss, 1973).

   Melatonin, which is secreted by the pineal, when administered to rats, mice, and weasels decreases the responsiveness of the gonads to light stimulation. This gonadal effect has also been observed in the Japanese killifish (Urasaki, 1972), the domestic fowl, and quail. The pineal may thus, through the action of secreted melatonin, exert an antigonadotropic effect. Its site of action is uncertain but is probably the brain, and this may lead to a decreased release of LHRH from the median eminence or an antagonism of its effect (Yamashita et al., 1978) (Fig. 9.5). In the female white-footed mouse, *Peromyscus leucopus*, the implantation of small quantities of melatonin into the anterior region of the hypothalamus results in a regression of the gonads (Glass and Lynch, 1981). This observation suggests that specific neural sites, where melatonin acts, are present in the preoptic and chiasmatic regions of the hypothalamus.

In sheep, prolactin is secreted in decreased amounts during the winter and this response is photoperiodic. Pinealectomy abolishes this seasonal rhythm so that prolactin levels in the plasma are high at all times of the year (Munro, McNatty, and Renshaw, 1980; Brown and Forbes, 1980). A pineal-mediated decline in prolactin secretion could be contributing to the observed regression of the gonads in animals such as Syrian hamsters (see Reiter, 1980). Vasotocin may also be secreted by the pineal, though this possibility is controversial. It does not change the release of gonadotropins from the rat pituitary *in vitro* (Demoulin et al., 1977). The injection of vasotocin into the brain of rats inhibited the preovulatory surge of LH release but not its tonic levels in the plasma (Cheesman, Osland, and Forsham, 1977). The latter observation is not consistent with a putative central role of vasotocin in inhibiting seasonal growth of the gonads.

More direct effects of melatonin on the pituitary and gonads have not been excluded. Considerable differences exist in the responses of vertebrates to pinealectomy or the administration of melatonin (the results have been called "inconsistent") so that at this time one cannot make a general statement as to the pineal's role in vertebrates. The daily rhythmical changes in the synthesis of melatonin, however, suggest that it could function as a "biological clock" mediating daily or seasonal rhythms including reproduction in some species. The pineal could thus add to or modify the role of the median eminence. It must be emphasized that at present the evidence for such an effect is equivocal.

The nature of the stimuli that control reproduction is complex and we do not yet fully understand how they exert their effects. The endocrines, in close association with the brain, principally mediate the response of the reproductive system to such stimuli. The eminent British physiologist, F. H. A. Marshall, was the first to emphasize the importance of such an interrelationship in controlling breeding. Some years ago, he summarized the situation (Marshall, 1956) as follows: "that (the) generative activity in animals occurs only as a result of definite stimuli, which are partly external and partly internal, while the precise nature of the necessary stimuli varies considerably in different kinds of animals according to the species, and still more according to the group to which the species belong."

Maturation of the gametes – the gonadal cycles

As we have seen, animals come into breeding condition at different times of the year depending on the stimuli they receive and react to, both from the external and their internal environments. If these "cues" are sufficiently appropriate and are processed correctly, then breeding will be attempted. This process involves a complex series of changes in the body that are, to a considerable extent, mediated by altering the concentrations of hormones in the blood. The sperm and the ova then mature, or ripen, in preparation for their eventual union. As these preparations are proceeding, the changing level of hormones contribute to the other physiological changes that are necessary to assure the fertilization of the ovum and, if this process is successful, the continued development of the egg and the embryo.

Such cycles in gonadal activity are relatively simple in the male when compared to those in the female. Sperm that can fertilize the ovum may be continually available for a period of many weeks, or even, as in man and feral pigeons, at all times of the year. The female, however, only produces ova available for fertilization periodically and, if not fertilized, they usually survive for less than a day. Such a periodic production of ova is an important event as it may not then occur again for many months. To mark this somewhat unique occurrence and make it clear to the male that he is at last acceptable, the female may send out various external signals and even actively seek male company. These "signs" include "calling," as in the cat, the production of a scent, as in the urine of the bitch, and the adoption of certain inviting sexual postures.

In mammals this period of sexual receptivity by the female is commonly called "heat," or by physiologists, *estrus*. The preparatory period that precedes this is proestrus, but if the animal is in a quiescent state, when no ova are being produced that are available for fertilization, it is called anestrus. The period during which the ova are being specially prepared for fertilization is called the *estrous cycle*, which varies from 4 days in the laboratory rat to 27 days in kangaroos, and (in its equivalent form, the human menstrual cycle) 28 days in women.

Many animals only experience a single estrous cycle in a year (called monoestrous) while others may have several such waves of ova production (polyestrous) spread out over several months of the breeding season or even for the entire year. Whether or not a single estrous cycle will be succeeded by another often depends on whether fertilization has occurred. If not, then there may be (though not always) another chance for successful reproduction within the overall range of the general breeding season.

Testicular cycles in vertebrates

Certain male domestic animals exhibit continual spermatogenesis and sexual readiness throughout the year, though this is not usual except in vertebrates from tropical regions and man. Seasonal breeding in a species is accompanied by a periodic maturation of the sperm (as well as the ova) along with such accessory and secondary sexual characters that facilitate its delivery on an appropriate occasion. Sperm may be available at all times during the reproductive season or mature in a single or several succeeding waves. The cystic type of spermatogenesis, where large numbers of sperm develop in unison inside envelopes that eject their contents into the seminiferous tubules, is most usual in amphibians and fishes (anamniotes) and is especially suited to those species where massive numbers of sperm are suddenly required for external fertilization. In reptiles, birds, and mammals (amniotes), sperm mature from cells in the lining of the seminiferous tubules and this may be a more or less continuous process though it may also occur in waves. This acystic spermatogenesis is thought to be more suited to internal fertilization, which may be attempted several times during a breeding season.

The maturation of the sperm may proceed in several different ways that are dictated by whether the species is a seasonal breeder and whether it is poikilothermic or homeothermic.

Postnuptial spermatogenesis is the more usual situation in seasonal poikilothermic breeders. This is illustrated by the frog *Rana esculenta*. Spermiation normally occurs in March in Northern Europe, and this is associated with a decline in testicular weight. Soon after this, however, the weight of the testis again increases and spermatogenesis proceeds throughout the summer but is halted during hibernation in winter (Fig. 9.6) though it gradually increases again in the spring. The major spermatogenetic events thus occur in the summer preceding the breeding season and following the nuptial pairing in the spring of that year. Such a pattern of testicular activity is seen in many fishes and reptiles though considerable variations can occur. In some instances, sperm may mature fully prior to the winter hibernation; in other species, spermatogenesis may be halted at some intermediate stage of development and go on later, in the spring, or again sometimes it merely slows down in winter and proceeds more slowly.

In homeotherms, *prenuptial spermatogenesis* is usual. Testicular activity following the breeding season, during the winter months, may be slight but there is a rapid increase in activity when the spring nuptials become imminent. This pattern is usual in mammals and birds that breed periodically, though some species (such as bats) may store mature sperm in the epididymis for several months, during a period of hibernation, for instance. Most birds exhibit characteristic "refractory" periods following the breeding season, when the testes fail to respond to photoperiodic stimuli or administered gonadotropins. The reptiles show a considerable diversity in testicular cycles. Chelonians usually exhibit amphibian-like postnuptial spermatogenesis, but the Lacertilia have several different testicular cycles (Fig. 9.7) and a prenuptial spermatogenesis is common.

The cyclical patterns of spermatogenesis described above are also termed *discontinuous spermatogenesis* in contrast to *continuous spermatogenesis*. The latter, apart from being present in some domestic temperate species, is common in animals that live in tropical areas where climatic conditions are

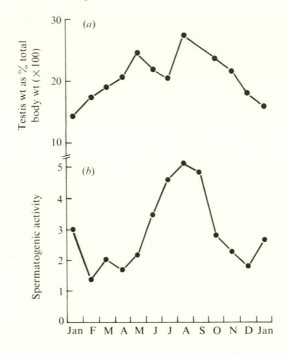

Fig. 9.6. Seasonal changes in the testicular weight (*a*) and spermatogenetic activity (*b*) in the European frog *Rana esculenta*. The decline in testicular weight that commences in May reflects spermiation during the breeding season. This sperm is that formed during the previous summer. Subsequent to this, spermatogenesis proceeds during the succeeding summer months but declines with the onset of winter hibernation. (From Lofts, 1964.)

relatively favorable at all times of the year. The frog, *Rana esculenta*, indeed has a continuous spermatogenetic cycle in warm Mediterranean regions whereas, as described previously, it has a discontinuous cycle in the more northern parts of Europe. This frog has thus been classified as a *potentially continuous breeder*, or a continuo–discontinuous type. The environmental temperature determines which pattern will persist in such species. It should be noted, however, that it is not possible to alter a discontinuous spermatogenetic cycle to a continuous one simply by raising the temperature; it will not, for instance, change in *Rana temporaria* where the tissues appear to undergo an inherent rhythm in their ability to respond to hormonal stimuli. Variations that occur in the patterns of spermatogenetic activity are illustrated in Table 9.1, which summarizes differences among amphibians from different geographic areas.

The testicular cycle (like the ovarian cycle) is controlled by the adenohypophysis. Removal of the pituitary abolishes such cyclical activity and results in a regression of the testis that involves the germ cells, in the seminiferous

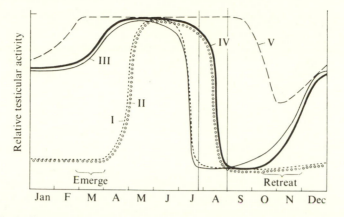

Fig. 9.7. A diagrammatic representation of the various patterns in seasonal development of the testes of lizards. I through IV represent variations in the relative testicular activity of lacertilians that live in temperate regions: V, some tropical *Anolis* lizards. "Emerge" and "retreat" indicate the times that temperate species leave and enter winter hibernation. The vertical lines indicate the period when most temperate lacertilians show no spermatogenetic activity. It can be seen that prenuptial spermatogenesis occurs in types I and II (in spring), but in types III and IV spermatogenetic activity commences in the autumn. (From Licht, 1972.)

tubules, as well as the endocrine interstitial tissue (Leydig cells). In mammals, this is thought to involve the action of FSH on the seminiferous tubules and LH on the interstitial tissue. Differences, however, occur; in reptiles an FSH-like gonadotropin appears to have a predominant ("nondiscriminating") effect on testicular function while in amphibians an LH has a comparable role (Licht, 1979). In fishes a single gonadotropin, incorporating both of these activities, may be present. While LH stimulates the production and release of testosterone, the mode of action of FSH is less certain. It is usually necessary for the full maturation of the sperm but this may not be a direct effect. Hypophysectomized mammals, birds, and fishes, in which the testes atrophy, can produce sperm following the administration of testosterone. It is surprising, however, that testosterone apparently cannot restore spermatogenetic activity in amphibians or at least in *Rana pipiens* (see Basu, 1969; Lofts, 1968).

It should be pointed out that the administration of testosterone to intact animals has often been observed to result in testicular regression and an inhibition of spermatogenesis. This paradox is apparently due to an inhibition by the androgen of the release of endogenous gonadotropins. A parallel direct effect on the seminiferous tubules has not, however, been excluded. The inhibitory effects of testosterone on spermatogenesis thus appear to reflect an overabundance of this steroid.

Table 9.1. *The types of spermatogenetic cycles exhibited by anuran amphibians that live in different geographical regions*

Species	Habitat	Remarks
Continuous type of spermatogenesis		
Bufo arenarum	S. America	
Bufo paracnemis	S. America	
Bufo granulosus d'orbignyi	S. America	
Bufo melanostictus	India, Java	Lunar periodicity from Java reported
Telmatobius schreiteri	Andes mountains (high altitude)	Low winter temperature cannot affect the cycle
Hyla raddiana andina	Andes mountains (high altitude)	
Rana erythraea	India	
Rana grahami	India	
Rana hexadactyla	India (Pondicherry)	
Leptodactylus ocellatus reticulatus	S. America	
Leptodactylus prognathus	S. America	
Leptodactylus laticeps	S. America	
Physalaemus fascomaculatus	S. America	
Pseudis paradoxa	S. America	
Pseudis mantidactyla	S. America	
Rana cancrivora	Java	
Discontinuous type of spermatogenesis		
Rana temporaria	Europe	
Leptodactylus asper	S. America	
Leptodactylus bufonis	S. America	
Phyllomedusa sauvagii	S. America	
Rana pipiens	USA	High temperature causes spermatogenetic continuity
Hyla crucifer	USA	
Pleurodema bufonina	S. America (Patagonia), Australia	
Continuo–discontinuous type of spermatogenesis		
Rana esculenta	Europe	In winter spermatogenesis goes only up to spermatid formation
Rana gracea	S. America	
Rana ocellatus typica	S. America	
Rana tigrina	India (Calcutta)	
Rana nigromaculata	Japan (Niigata), China	

Table 9.1. (*cont.*)

Species	Habitat	Remarks
Leptodactylus ocellatus typica	S. America	Two interruptions in spermatogenesis during cold winter and high summer
Variable spermatogenetic cycle as per geographic distribution		
Rana esculenta	Europe	Discontinuous cycle
ridibunda	Mediterranean region	Continuous cycle
Discoglossus pictus	Europe	Discontinuous cycle
	Mediterranean region	Continuous cycle

Source: Basu, 1969.

Testosterone is, nevertheless, usually necessary for the maturation of the sperm, but it is not yet clear whether FSH also acts by stimulating the production of an androgen. Unfortunately, the spermatogenetic effects of the administration of testosterone are often variable. It seems likely that FSH may act on the Sertoli cells to produce androgens that in turn mediate the maturation of the sperm (see Lofts, 1968).

Spermatogenesis is a prolonged and complex process that requires pre- and postnatal maturation of the gonacytes, mitotic divisions of the spermatogonia, and meiotic reduction divisions to form the spermatocytes, spermatids, and the final (spermiogenesis) differentiation of spermatozoa. Androgens, and possibly FSH, are required for certain of these steps to proceed in a normal manner, but there is considerable interspecies variation as to the stages of sperm maturation at which these hormones act. The endocrinology of gametogenesis is an important subject about which we know little. Dodd (1960, 1972*b*) has summarized what is known about this process in vertebrates. In the rat, testosterone may be necessary for early pre- and postnatal development of the gonacytes, and it also promotes the meiotic division of the spermatocytes later on. FSH is required for the maturation of the spermatids. This pattern is, however, not the same even among the mammals, and the information that is available is rather meager. In lampreys (Cyclostomata), hypophysectomy has little effect on the final stages of the maturation of the sperm. When this operation is performed in late winter or spring spermiation still occurs, but if hypophysectomy is carried out earlier, in October for instance, there is a considerable delay and spermiation may not take place at all (Larsen, 1973; 1978).

A similar slowing action on the development of the testis has been observed following hypophysectomy in the Pacific hagfish, *Eptatretus burgeri* (Patzner and Ichikawa, 1977). In chondrichthyean fishes hypophysectomy also inhibits the earlier stages (meiotic divisions of the spermatogonia) of sperm maturation. Selective removal of the ventral lobe of the pituitary of the dogfish results in a slow degeneration of the testis (Dobson and Dodd, 1977*a*, *b*). Extracts of the ventral lobe of these fish, but not mammalian gonadotropins, maintained the activity of the gonads. Seasonal changes in the development of the testes of teleost fish are also inhibited following hypophysectomy (see Peter and Crim, 1979). This effect appears to reflect the action of one or possibly two gonadotropic hormones (see Chapter 3). One such preparation ("maturational hormone") has been shown to stimulate spermatogenesis in hypophysectomized flounder (Ng and Idler, 1980; Ng, Idler, and Burton, 1980*b*).

One cannot, with the limited and varied information available, make any generalization as to which stages of spermiogenesis are hormone dependent in vertebrates. There are instances in fish, as well as reptiles, birds, and mammals, where the later stages of maturation are apparently dependent on endocrine stimulation. The situation in amphibians is especially confusing as in some instances spermatogenesis appears to be occurring at a time when the endogenous production of testosterone is low, and exogenous testosterone may exert a direct inhibitory effect on the early stages of spermatogenesis. The comparative endocrinology of gametogenesis needs, and certainly merits, further exploration.

Until recently, little direct information was available about the circulating levels of testosterone and gonadotropins in seasonally breeding animals. Changes in the concentrations of such hormones were inferred from histological examination of secondary sexual characters. The Leydig cells and their analogues, the boundary cells, show a seasonal pattern in their histological appearance (see Lofts, 1968). In the periods that precede breeding, these cells enlarge and accumulate lipids and cholesterol. These cells become depleted of these materials at the height of the breeding season, and this is thought to reflect the secretion of androgens that utilize lipids as their substrates. These histological changes are associated with breeding behavior and the development of secondary sex characters and can be imitated by the injection of gonadotropins. Such changes have been observed in fishes, amphibians, reptiles, birds, and mammals. An example of this can be seen in Fig. 9.8 where the height of the thumbpad epithelia in the frog *Rana esculenta* can be seen to decline in June when the lipid and cholesterol content of the interstitial tissue is greatest. A seasonal pattern in the ability of the testis of the cobra to convert progesterone to androgens, *in vitro*, is shown in Fig. 9.9. This androgen synthesis reaches an initial maximum during breeding in May but drops subsequently as the testes atrophy. When spermatogenesis is again initiated in the autumn the rate of progesterone to androgen conversion again increases but declines again with the onset of winter hibernation.

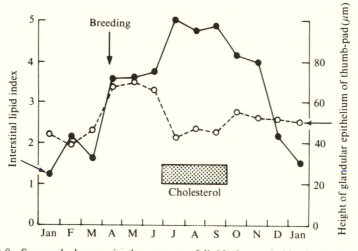

Fig. 9.8. Seasonal changes in the amount of lipid observed, histologically, in the interstitial cells of the testis of the frog *Rana esculenta*. This reaches a maximum in midsummer, after breeding has occurred, when the cellular cholesterol levels are also greatest. These changes are thought to reflect a decline in the synthesis of androgens (which utilize the lipids as substrates for their formation). This change is consistent with the decline in the development of the thumbpad, which is under androgenic control. (From Lofts, 1964.)

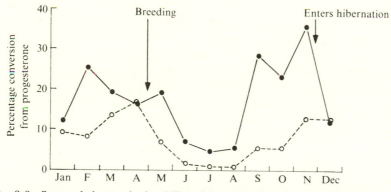

Fig. 9.9. Seasonal changes in the ability of the testis of a snake, the cobra, to convert (*in vitro*) progesterone (which acts as a substrate) to androgens. Testosterone production declines following breeding in May but rises again in late summer and autumn, during postnuptial spermatogenesis, only to decrease once more as the snakes go into winter hibernation: ●, testosterone; ○, androstenedione. (From Lofts, 1969.)

Dogfish, *Scyliorhinus canicula*, have a plasma testosterone concentration of 2 ng/ml in February and this rises to 6 ng/ml in August (Dodd, 1975). It was surprising to observe, however, that removal of the various pituitary lobes, including the ventral one, did not change the plasma level of this androgen. It is possible that in these fish the testis enjoys a relative degree of independence from the pituitary.

In the rough-skinned newt, *Taricha granulosa*, plasma androgens rise markedly in March, when breeding occurs, and then decline but rise again in August (Specker and Moore, 1980). The low levels correspond to the period of spermatogenesis and the second increase to the beginning of spermiation. The levels then decline over the winter period. The Australian lizard, *Trachydosaurus rugosa* (Bourne and Seamark, 1973) breeds in spring when the weight of its testes is about 1300 mg, compared with 180 mg in summer. During the breeding season the androgen concentration in its plasma is 33 ng/ml, but it is only 10 ng/ml at other times of the year. Similar seasonal changes in plasma androgen levels have been observed in other lizards and also in snakes and turtles (Courty and Dufaure, 1980; Bona-Gallo et al., 1980; Callard et al., 1976). A 10-fold increase in the plasma testosterone concentrations has been observed in starlings (*Sturnus vulgaris*) during the breeding season (Temple, 1974). When Japanese quail are put on a 20-hour light/4-hour dark photoperiod, as occurs at the height of summer in northern latitudes, plasma testosterone levels rise after only 4 days (Follett and Maung, 1978). A periodic decrease in the hormone-secretory interstitial, as well as the spermatogenetic, tissue occurs in all nonmammals that breed periodically. In mammals, however, there is usually permanent hormone-secretory tissue in the testis though they may also undergo seasonal changes in size.

The Sertoli tissue, which has been identified in all groups of vertebrates, has for a long time excited speculation as to its function. The histological appearance of this tissue shows changes in parallel to those of the Leydig cells (see Lofts, 1968) and spermatogenesis. An accumulation of lipids, in the Sertoli cells, follows spermiation, but these materials are depleted when spermatogenesis is occurring. In animals that normally breed continually, like laboratory rats, hypophysectomy results in an accumulation of lipids in the Sertoli cells, and this is thought to reflect a lack of their stimulation by FSH. It is now widely accepted that the Sertoli cells produce androgenic steroids that may influence spermatogenesis, and FSH probably acts to stimulate the secretion of such hormones. The cyclical changes that occur in the appearance of the Sertoli cells thus presumably reflect changes in the endogenous gonadotropin levels in the blood.

Ovarian cycles in vertebrates

The maturation of the ovum in the ovary and its extrusion (ovulation) and passage into the oviduct or uterus involve the coordinated activity of FSH, LH, sometimes prolactin, and the secretion of estrogens, progesterone and, possibly, even small amounts of androgens from the ovary. The ovarian cycle results from increases and declines in the circulating concentrations of these hormones, and this is largely the result of their interactions in stimulating or inhibiting each other's release, through a negative and positive feedback to the hypothalamus.

These hormonal rhythms have only been closely analyzed in mammals, and even these results are usually confined to more domesticated species.

Placental mammals

Three general patterns have been identified in the ovarian cycle of placental, or eutherian, mammals and these have been described (*a*) in the sheep, pigs, and cattle, (*b*) in the laboratory rat, and (*c*) in man.

Sheep

The ewe usually comes into estrus in the autumn, as a result of stimulation by shortening periods of daylight and, if pregnancy does not occur, will continually produce ova at intervals of about 16 or 17 days until the following spring. Some breeds of sheep, like the merino, may breed for longer periods of the year. The estrous cycle of the ewe lasts for about 17 days and the hormonal changes that occur are summarized in Fig. 9.10. The onset of estrus is taken as time zero in the ovarian cycle and this lasts for about 24 hours, ovulation occurring toward the end of this time. Following ovulation, the blood supply to the ruptured Graafian follicle increases and the granulosa cells luteinize to form the corpus luteum. This structure reaches a maximum size on about day 8. Luteinization of the follicle is initiated by the action of LH, and the secretion of progesterone is also stimulated by this hormone. LH is luteotropic, an effect that is seen in most mammals. In some species, including the rat, mouse, rabbit, and possibly the sheep, prolactin may also have a luteotropic effect. Sometimes the two hormones act in conjunction with each other. Progesterone secretion from the corpus luteum rises until about day 11 of the cycle and then on day 13 undergoes a precipitous decline.

Accompanying these events are the development of the Graafian follicles and the maturation of the ova. This process proceeds under the influence of FSH and the estrogens secreted by the follicular cells, which are also stimulated by LH. Thus LH appears to have a general steroidogenic effect on the ovarian tissues. During the preovulation phase of the cycle, estrogen levels are moderate but as can be seen in Fig. 9.10 they may display some periodic changes. The LH level is low but sufficient to maintain the secretion of steroid hormones. The estrogens and progesterone that are produced act on the accessory sex organs, especially the uterus and vagina, to get them into "tip-top" condition for the prospective fertilization, the implantation of the egg, and

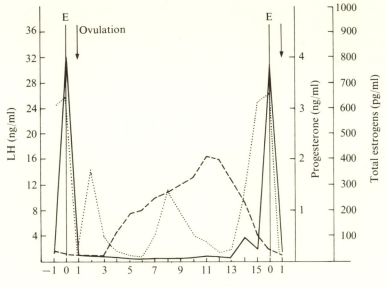

Fig. 9.10. The estrous cycle of the ewe. Estrus (E) occurs at time *zero* and is followed by ovulation. The concentrations of plasma progesterone (from the jugular), estrogens (ovarian vein), and LH (from the jugular or ovarian vein) are shown in relation to these events. It can be seen that on day 12 there is a decline in progesterone that is accompanied by climb in estrogen concentration that initiates a "surge" in release of LH resulting in ovulation. Dashed line, progesterone; dotted line, total estrogens; solid line, luteinizing hormone (jugular or ovarian vein). (From Hansel and Echternkamp, 1972.)

pregnancy. The release of LH is kept low in the preovulatory period as a result of a negative-feedback inhibition of its release that is exerted by progesterone on the hypothalamus.

Between days 13 and 16 of the cycle dramatic changes take place in the hormonal concentrations that result in ovulation. There is a rapid decline in progesterone that reflects a breakdown of the corpus luteum. LH levels are still adequate to maintain this tissue but it loses its sensitivity to the hormone. Estrogen levels then climb upward and by a positive-feedback action in the hypothalamus and pituitary bring about a massive release of LH that causes the follicle to rupture and extrude its ovum. The effect of the estrogen is reflected in a decline of stored luteinizing hormone-releasing activity in the ewe's hypothalamus (Crighton, Hartley, and Lamming, 1973).

The hormonal basis for estrous behavior has recently been questioned. It had commonly been assumed that it was solely a reflection of the actions of estrogens. The estrogen level, however, usually drops considerably just prior to estrus; the preovulatory surge is in fact rather temporary. There is now evidence to indicate that not only estrogen but also progesterone, as well as a small but significant release of an androgen, androstenedione, from the ovary, may contribute to estrous behavior.

After ovulation the LH stimulates the follicle granulosa cells to luteinize and if fertilization does not occur the cycle will then recommence. In the event of fertilization and an ensuing pregnancy, the corpus luteum, as we shall see later, will persist for a much longer time and contribute to the events of the gestational period.

The corpus luteum can thus be seen to play a commanding role in the estrous cycle and it has been called "the clock." The reason for the decline in the activity of the corpus luteum during the latter part of the estrous cycle has only recently been elucidated. It has been known for many years that, when the uterus of guinea pigs is removed, the corpus luteum persists for a much longer period of time. This effect can also be seen in the ewe, as well as the cow and sow, but not in women, the rhesus monkey, dog, badger, or marsupials (Anderson, 1973). The nonpregnant uterus, in some species, appears to produce a substance that has been called a *luteolysin* that causes the corpus luteum to atrophy. There is evidence to suggest that this may be a prostaglandin. Whatever its nature, it is interesting that this effect is not seen if the ovary is transplanted to a region that is distant from the uterus, such as the neck. The ovarian arterial blood and the uterine venous blood pass by each other in closely apposed vessels, and it seems that this special vascular arrangement functions as a unique pathway by which the luteolysin can get to the ovary before it becomes diluted, or destroyed, in the general circulation. It is important, however, to remember that this effect is not seen in all species nor has such a role for a luteolysin been accepted by all (see Nalbandov, 1973). Instead, it has been proposed that the death of the corpus luteum results from a lack of LH due to competition for this tropic hormone by newly developing ovarian follicles.

The transition of the reproductive condition of ewes from an infertile noncycling condition in summer to a series of estrous cycles, and breeding, in the autumn and winter reflects a shortening of the hours of daylight. Just how this transition may occur has been the subject of considerable research, not only on ewes (Karsch, 1980) but also on the rams (Lincoln and Short, 1980). The central endocrine change involves a rise in the plasma concentrations of LH and, as a result of this, estrogens. LH is released in a pulsatile manner (see Chapter 4), the frequency of which can vary from about twice an hour to once every 3 or 4 hours, or even less often. The levels of the LH also display oscillations around a stable baseline level, the amplitude of which partly reflects the quantity of the hormone that is released at each burst. The particular tonic concentration achieved will depend on the frequency of the

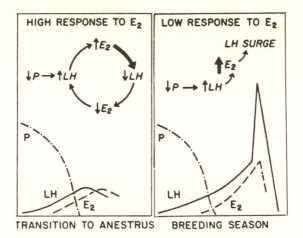

Fig. 9.11. The possible hormonal basis for the initiation of the breeding season in ewes. This event occurs in response to short day-lengths in the autumn and winter; the sheep are in anestrus during the summer. It has been proposed that the seasonal onset of the estrus cycles results from a change in the sensitivity of the brain to the inhibitory effects of estrogens on the release of luteinizing hormone (LH). In summer the sensitivity of the negative-feedback system in the hypothalamus to estrogens is high so that LH and resulting estrogen (E_2) levels are low (left panel). With the initiation of the breeding season, however (right panel), the sensitivity of the inhibitory mechanism to estrogen declines so that LH is released more readily, estrogen levels then climb and can then initiate the surge in LH release that precedes ovulation. (From Legan, Karsch, and Foster, 1977.)

pulses and their amplitude. The former appears to be more important in this instance.

During the breeding season the pituitary gland is capable of increasing the frequency of the LH release, and such changes occur during each estrous cycle. In the anestrous period, release of LH occurs infrequently and the basal level of the hormone is low. However, if such ewes are ovariectomized a rapid pulsing of LH release is quickly established and its tonic levels rise. The damping effects of the presence of the ovaries reflect the inhibitory action of estrogens on the hypothalamus, which then does not release LHRH very often. Other parts of the reproductive mechanism are basically intact during anestrus; for instance, the injection of estrogens can elicit an LH surge. The lack of cycling in the summer appears to reflect an increased sensitivity of the brain to an inhibitory action of estrogens (Fig. 9.11).

The release of LHRH from the median eminence is thought to occur as a result of periodic stimuli that originate from a "pulse generator" situated in the brain, probably in the region of the medial basal hypothalamus. Nerve impulses arising in this region may result in the release of LHRH. During the

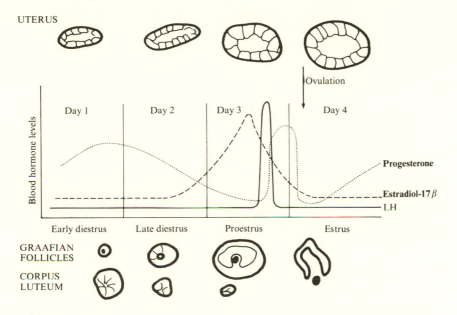

Fig. 9.12. The estrous cycle of the laboratory rat. The blood hormone levels, progesterone, estradiol-17β, and LH are shown in relation to estrus and ovulation. Also included is a representation of changes in the stages of development of the uterus, Graafian follicles, and corpus luteum. It can be seen that the corpus luteum does not persist throughout the entire estrous cycle of the rat and that the preovulatory increase in progesterone is due to the secretion of this hormone by the ovarian interstitial tissue.

estrous cycle the frequency of these pulses can be reduced by progesterone, as occurs in its luteal phase. During anestrus the frequency of the pulses is low, but this cannot be accounted for by the presence of progesterone. It appears that the activity of the pulse generator is reduced in some other way. This could be due to a direct action of the long days on its neural circuitry but, as we have seen, it also appears to involve changes in the sensitivity of the negative feedback response to estrogens. The precise nature of the actions of the long days is unknown. It is interesting, however, that cyclical changes in the development of the testes of the rams, which are also gonadotropin dependent, can be abolished by denervating the pineal gland (Lincoln and Short, 1980). The testes maintained their size and failed to regress as the days became longer. It is conceivable that the secretion(s) of this gland could be impinging in some way on the activities of the pulse generator.

The laboratory rat

The female rat has an ovarian cycle that is broadly similar to that of the ewe but there are some notable differences (see Fig. 9.12).

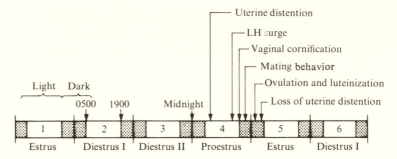

Fig. 9.13. The principal events in the rat estrous cycle in relation to the time of the day. This cycle is precisely timed on the basis of a diurnal rhythm. Ovulation can be seen to occur shortly after midnight. Other events including mating behavior, the LH "surge," and the development of the uterus, are depicted. (From Armstrong and Kennedy, 1972.)

The rat estrous cycle lasts 4 days and the events are normally regulated on the basis of a diurnal rhythm. This facilitates experiments on these animals as it is known that ovulation always occurs just after midnight and other changes can also be timed with remarkable accuracy (Fig. 9.13).

The adoption of such a rhythm is vital to the rat as the corpus luteum, which, as we have seen, acts as a "clock" in many other mammals, is small and does not persist long during its estrous cycle. Progesterone is present throughout the estrous cycle, but it does not appear that changes in its levels contribute to the timing of ovulation. Instead, there appears to be a "timed" estrogen release on the morning of proestrus sometime before 10:00 a.m. This steroid promotes LH release, as in other mammals, but in contrast to other species this is immediately followed by a marked elevation in the plasma progesterone concentration. As this precedes the formation of the corpus luteum, it presumably arises as a result of a luteotropic effect of the LH on the ovarian interstitial tissue. This progesterone is vital to the onset of estrus and subsequent changes that occur in the uterus, which becomes less distended while the endometrium becomes more glandular. Progesterone does not appear to be important for ovulation, which cannot be prevented by antiprogesterone serum.

The corpus luteum persists only if pregnancy or pseudopregnancy (as a result of copulation without fertilization) occurs; otherwise, it does not have a significant role in controlling the rat ovarian cycle.

Man

The human ovarian, or menstrual, cycle lasts for 28 days and is quite distinct from that in nonprimate mammals. A similar ovarian cycle also occurs in monkeys and apes. The corpus luteum persists for a more prolonged period of the cycle than in the rat but not for the whole of it, as in the ewe, cow, or sow. The timing of the events of the human ovarian cycle is arbitrarily taken from the initiation of menstruation. This process is due to the discharge of super-fluous structural remnants and secretions of the endometrium and contains some blood. In effect, menstruation represents the termination of the previous ovarian cycle and the life of the corpus luteum and lasts about 4 to 5 days.

The preovulatory period of the human ovarian cycle (see Henderson, 1979) thus takes place in the absence of a corpus luteum so that progesterone secretion is relatively small. 17α-OH-progesterone levels rise but this metabo-lite has little progestin activity. The follicle ripens under the influence of FSH and LH and secretes estradiol, which controls the release of this gonadotropin by a negative-feedback inhibition in the hypothalamus and pituitary. As in other mammals, a sudden surge in LH, initiated by the secreted estrogen, results in ovulation, usually on about day 16. There is also a sudden rise in FSH release at this time, which may contribute to ovulation. No precise period of estrous behavior exists in the human female. She is receptive to the male at any time of the cycle. The ruptured follicle starts to luteinize after ovulation, and progesterone secretion, due to the action of LH, rises but subsequently declines when the corpus luteum later degenerates and menstru-ation then occurs. In the event of fertilization, however, the corpus luteum persists. The reason for the premenstrual decline in the activity of the human corpus luteum is unknown and this degenerative process can proceed nor-mally following hysterectomy; apparently, it does not involve the action of a uterine luteolysin.

The role of the brain in the regulation of the primate menstrual cycle has been investigated in the rhesus monkey (Knobil, 1980). The primary initia-tor of the cycle appears to be the pulsatile release of LH, which is stimu-lated by a "pulse generator," or oscillator, that converts neural signals to endocrine ones. Lesioning experiments on the brain indicate that it is asso-ciated with the arcuate nucleus in the medial basal hypothalamus. Nerve impulses from this group of cells result in a corresponding pulsatile release of LHRH. When the arcuate nucleus is destroyed plasma gonadotropin con-centrations decline to very low levels. The infusion of LHRH increases plasma LH concentrations and when this is infused at the rate of once every hour the level returns to normal. The amounts of LH and FSH released from the gonadotrope cells in the pituitary gland in response to each pulse of LHRH are regulated by estrogens in a negative feedback manner. The ovary can thus control the length of the menstrual cycle in the presence of a constant rate of signaling by the pulse generator. However, it seems likely that the frequency of the pulses can also be regulated, probably, as in sheep, by centers higher in the brain. Neural mechanisms, as well as hormones, may be involved. Progesterone is thought to decrease the frequency of pulse generators' activity.

Ovulation

The mechanisms of initiation of ovulation may differ among the mammals though they all appear to involve a surge in the release of LH. In the examples described, ovulation takes place in response to an internal programming that controls hormone release so that ovulation is then said to be *spontaneous*. In rats and cattle, the release of LH in this type of ovulation results from a rise in plasma estrogen concentrations, which increases both the release of LHRH and the sensitivity of the pituitary gonadotrope cells to its action (Speight et al., 1980; Kesner, Convey, and Anderson, 1981). In the rhesus monkey estradiol can stimulate gonadotropin release in the absence of LHRH, provided that the pituitary gland has been suitably primed by the latter peptide hormone (Wildt et al., 1981). Estradiol may thus have a direct role in initiating the release of LH at the time of ovulation.

In other mammals, ovulation can be *induced* as a result of copulation and sexual excitement. This latter type of ovulation is known to occur in such species as the rabbit, cat, ferret, mink, and raccoon, and it is suspected that it may also sometimes occur even in women. In such species, estrogen is released from the developing ovarian follicles, which are under the influence of FSH (see Schwartz, 973). This estrogen indicates when the follicles are ripe and results in mating behavior. The latter is in contrast to spontaneous ovulators in which progesterone is also necessary. If copulation takes place, this initiates a surge of LH release, as a result of neural stimulation of the hypothalamus and pituitary, and ovulation occurs. This event is accompanied by a rise in progesterone levels and takes place several hours after coitus when the sperm are ensconced in the oviduct. In nonmammals, the situation is less clear, however; the mere presence of the male or even some substitute may be all that is necessary to initiate ovulation. Apart from gallinacious birds like the domestic fowl, as well as domestic geese and ducks, most birds do not produce eggs in the absence of the male. It has, however, been reported that some birds, such as pet parrots, will lay eggs if suitably stroked and tickled. Copulation may thus not always be necessary, and courting behavior and sexual display may be effective stimulants of ovulation.

It has been suggested that the rupture of the ovarian follicle may involve a buildup of internal hydrostatic pressure accompanied by the activation of a proteolytic enzyme that weakens its wall (see Henderson, 1979). Ovulation can be prevented by drugs that inhibit the synthesis of prostaglandins, suggesting that these fatty acid compounds may also be involved in the mechanics of ovulation. Microfilaments in the theca externa can be contracted by prostaglandins, and this effect presumably contributes to the final rupture and the release of the egg.

Following parturition, several species of eutherians, including the rabbit, ferret, mink, and raccoon, come into a postpartum heat when they copulate and this, as indicated, results in ovulation. Copulation is not always needed to precipitate ovulation in these circumstances for, as we shall see in the next section, postpartum ovulation is common in marsupials where it is a spontaneous event and occurs at a time that merely reflects an extension of the normal estrous cycle.

Delayed implantation

Pregnancy usually persists for a precise and predictable period of time. Some interesting and, at first, mystifying exceptions have, however, been encountered. Animals that conceive in the autumn and deliver their young in spring can, on some occasions, such as when the length of the daylight period is artificially increased, produce their young much earlier. There have been other instances described, especially in kangaroos, where a female has been taken into captivity and without any contact with a male has, many months later, given birth to a young one. Faced with the necessity for an explanation, some people were even forced to consider the possibility of virgin birth! The cause is, nevertheless, quite a reasonable one. In a number of mammals, especially the mustelid Carnivora (such as weasels, skunks, badgers, and sable), fur seals, roe, deer, and macropodid marsupials (kangaroos), development of the fertilized egg can sometimes cease when a blastocyst, containing about 100 cells, has been formed. This blastocyst lies dormant for a time that may extend for several months, but can be subsequently stimulated to continue development. The delay is called an *obligatory* one when it is determined by external conditions, such as light, as seen in badgers, pine marten, weasels, and the roe deer. In mustelids the activity of the corpus luteum is lost prior to implantation; increases in progesterone levels appear to trigger further development (see Wade-Smith et al., 1980). In other species, such as the mouse, rat, and macropodid marsupials it is *facultative* and controlled by more physiological events. As will be described in more detail in the later discussion on marsupials, this inhibition results from the effects of suckling and lactation. In rats and mice development appears to be initiated by estrogens but the marsupials probably utilize progesterone for this purpose (se Heap, Flint, and Gadsby, 1979).

Pregnancy

An excellent account of the role of hormones in this process is given by Heap (1972) and Heap, Perry, and Challis (1973). When the fertilized egg is retained in the oviduct or uterus and the subsequent development of the young occurs at this site, pregnancy is said to be occurring. This term is usually assumed to include the viviparous condition but may also encompass ovoviviparity. The internal incubation of the young is also called *gestation*. The condition of pregnancy appears to have reached its highest state of organization in placental mammals though little information is available about this process in nonmammals. Pregnancy is not a uniquely mammalian phenomenon as it occurs in some chondrichthyeans, teleosts, reptiles, and amphibians, though not in birds. Gestation may occur for quite long periods of time in placental mammals but this is not unique as it may extend for 2 years in some viviparous sharks and is of 1 to 4 years' duration in the ovoviviparous urodele, *Salamandra atra*.

As we have seen, the hormonal preparation of the mammalian uterus for the reception, fertilization, and implantation of the egg is initially stimulated by estrogens and progesterone, the latter usually having the subsequent dominant action, though both steroids act in simultaneous collaboration. Subsequently during pregnancy, these favorable uterine conditions need to be maintained and even modified from time to time as the fetus grows and is eventually delivered to the outside world. The necessary supplies of hormones are then not only altered qualitatively but increased quantities may also be required. These added needs have been met in various ways by the placental mammals and principally involve the function of the pituitary, the ovary, the placenta, and the uterus.

Progesterone, to use an oft-quoted phrase, is called "the hormone of pregnancy," but substantial, though usually smaller, amounts of estrogens are also used during gestation. These gonadal steroids maintain the endometrium and contribute to the considerable expansion that occurs in the myometrium during pregnancy. The hypertrophy of these muscles results from the stretching of the walls of the uterus and the induction, by estrogens, of new contractile proteins. Contractions of the uterus are not usually desirable during pregnancy, and the responsiveness of the myometrium to stimulation is reduced by progesterone. The contractile effects of oxytocin are, for instance, usually reduced by pretreatment of the uterus with progesterone while estrogens have the opposite effect and enhance the responses to this neurohypophysial hormone. Such effects have not been demonstrated in all species but are very reproducible in some, like the rabbit. A most important role of progesterone in pregnancy in placentals is the inhibition of the estrous cycle and ovulation. This effect results from a negative-feedback inhibition of the release of LH from the pituitary and may be required when the periods of gestation exceed the length of the normal estrous cycle. Corpora lutea also persist in many viviparous and ovoviviparous nonmammals and although their precise role is uncertain it is suspected that they may also have a comparable role in such animals.

The problem of how to supply the added hormonal requirements of pregnancy has been met in various ways by different species of placental mammals. Estrogens and progesterone are typically secreted by the vertebrate ovary. The corpus luteum is usually the principal ovarian source of progesterone, but in nonpregnant animals, this structure does not normally persist longer than the estrous cycle. As we shall see, this situation even occurs in pregnant Australian marsupials. The period of gestation in these animals is similar to that of their estrous cycles so that a prolongation of the life of the corpus luteum is unnecessary. In the placental mammals, which have relatively longer periods of gestation, the corpus luteum persists for a much longer time and often remains functioning throughout the entire period of pregnancy. This extended survival is the result, in some species, of an inhibition of the effects of uterine luteolysins, due to the presence of extra material in the uterus and the stimulating actions of mixtures of luteotropic hormones. These hormonal combinations may consist of FSH, LH, prolactin, and gonadotropins that may be produced by the placenta. The precise hormonal content of this so-called *luteotropic complex* differs considerably from species to species. Its function is to extend the normal lifetime of the corpus luteum and to promote the secretion of progesterone.

The production of progesterone by the ovary may be facilitated in various ways. In some species (the horse) additional corpora lutea may form, but in others (humans and cattle) only a single corpus luteum is usually present. Animals that produce several young at a time have a correspondingly greater number of corpora lutea available for the production of progesterone. During pregnancy, the secretion of progesterone by individual corpora lutea may be increased by the action of the luteotropic complex. In addition, the amount of available progesterone is the net result of its rate of production and destruction. As described in Chapter 4, proteins that bind steroid hormones are present in the plasma and when they are in such a bound condition the rate of the destruction of these hormones is reduced. During pregnancy, the formation of such steroid hormone-binding proteins in the liver may be increased, probably as a result of stimulation by estrogens. There are considerable interspecific differences in the physiological patterns that ensure adequate progesterone in pregnancy.

In some species, such as the rabbit, ovariectomy during pregnancy always results in prompt abortion. In other species, such as the sheep, guinea pig, rat, and human, this operation does not necessarily result in a loss of the fetus. The placenta in these species produces sufficient gonadal steroids to support the uterus, though the supply may be inadequate during early pregnancy. There are considerable interspecific differences in the ability of the placenta to produce hormones. The placenta of the rabbit and goat, for instance, does not produce any steroid hormones, or luteotropins, while the human placenta produces large quantities of all these materials. The pig placenta produces estrogen but not progesterone, whereas the sheep produces large amounts of progesterone.

The fetus may also contribute hormones that are involved in gestation. The fetal adrenal cortex produces large amounts of two steroids that are substrates for the progesterone and estrogens formed by the placenta. Fetal pregnenolone sulfate can be converted to progesterone while dehydroepiandrosterone sulfate is used to synthesize estrogens. These steroids then pass into either the fetal or maternal circulation where they contribute to the maintenance of pregnancy and, at the appropriate time, to its termination.

The appearance of the fetoplacental unit as a temporary endocrine organ that helps to supply the hormonal requirements of pregnancy is a fascinating physiological adaptation. Such a role has not been described in nonmammals and is controversial in marsupial mammals. It is not possible to draw any orderly phyletic line as to the distribution of this hormone-secreting tissue in placentals, and it could have evolved separately on several occasions to suit the needs of each particular species. In recent years, it has become apparent that tumorous tissues in mammals may produce a variety of hormones that normally arise from discrete endocrine glands. Perhaps there is some analogy between such tumors and the evolution of an endocrine placenta!

Parturition

The delivery of the young is a very precisely timed event that, although it undoubtedly involves hormonal changes, is not very well understood. During pregnancy, the progesterone-dominated uterus is in a relatively quiescent state and it is reasonable to suspect that removal of such a progesterone block may precipitate contractures that result in parturition. Plasma progesterone levels are quite low at birth in some animals, like sheep and ferrets, but in others, like women and guinea pigs, this is not so. Estrogen levels often rise dramatically on the approach of parturition and this may sensitize the uterus and oppose the progesterone block. In addition, under the influence of estrogens, receptors for oxytocin in the uterus increase in number just before birth (Alexandrova and Soloff, 1980). *Oxytocin* is released from the neurohypophysis and stimulates contraction of the uterus. This hormone facilitates the delivery of the young, but in some species parturition can still occur in the absence of oxytocin. Its role is therefore not always essential in mammals (see Chapter 3). *Relaxin* is released from the ovary and is also present in the placenta of rabbits and guinea pigs. Relaxin promotes the relaxation of the ligaments of the pubic symphysis to allow passage of the young through what has been described as "this triumphal arch!" This effect (on the pelvic ligaments) is detectable in most mammals but is most prominent in rabbits and guinea pigs.

Prostaglandins $F_{2\alpha}$ and E_2 have been shown to contract the uterus and soften the cervix in all mammalian species that have been examined. (These fatty acids are currently being used to promote labor and induce abortion in women.) They appear to play a determining role in normal parturition. The process of their formation prior to birth has been mainly worked out using sheep and goats, and the details of this process may differ from species to species (see Liggins et al., 1977; Liggins, 1979; Currie and Thorburn, 1977*a*). When ACTH or cortisol is injected into fetal lambs, premature parturition occurs. A rise in plasma cortisol levels has also been observed in the plasma of fetal goats prior to birth (Currie and Thorburn, 1977*b*). The corticosteroids in the fetus appear to initiate parturition by inducing the formation of a 17α-hydroxylase in the placenta. This enzyme changes the pathway of steroid synthesis so that progesterone is converted to estrogens. Prostaglandins can be formed by the placenta and the uterine endometrium, but normally this process is quite limited in pregnancy. The actions of any prostaglandins that are formed are also reduced due to the inhibitory effect of progesterone on uterine contractility. The change in placental steroid metabolism from an emphasis on the synthesis of progesterone to that of estrogens increases the contractility of the uterus and activates phospholipase A_2 in the tissues. This enzyme metabolizes membrane phospholipids and releases arachidonic acid, which is a substrate for prostaglandin synthesis.

Some species obtain a large amount of their progesterone during pregnancy from a persistent corpus luteum, rather than from the placenta. This situation occurs in goats but is more usual in small mammals. In goats it has been shown that prostaglandins secreted by the uterus pass in the uterine vein to the ovary where they have a luteolytic action. Prostaglandins can thus act at, at least, three sites during parturition: the corpus luteum and the cervix, and they also contract the uterus.

The desire to care for the young is also a hormonally mediated effect. The process of lactation from the mammary glands is normally (except it seems in man) an important part of this process in all mammals. The role of the hormones in lactation has been described in Chapter 5.

Marsupial mammals

The marsupials, which include the American opossum and large numbers of Australian mammals, such as kangaroos, have a unique form of reproduction that is accompanied by some remarkable endocrine innovations. The initiation of the study of the reproductive endocrinology of the Australian marsupials was principally due to the efforts of H. Waring on the occasion of his academic migration to Australia in 1948.

Systematically, the marsupials are distinguished from the placentals by the absence of a true placenta; only a yolk-sac placenta is present so that the physiological connection with the parent is more tenuous than in placentals. Young marsupials are born in a relatively immature state comparable to that of embryos in a quite early stage of the gestational period of placentals. The newborn young of marsupials are suckled on the teat where they undergo a considerable part of the development that would normally occur within the uterus of placentals. In many, but not all, marsupials this takes place in an external pouch or marsupium.

According to G. B. Sharman, who is well informed in these matters, marsupials and placentals probably evolved from a common oviparous ancestor. Sharman's (1970) account of marsupial reproduction should be consulted by those who wish to extend their knowledge of this process. An even more comprehensive account has been provided by Tyndale-Biscoe (1973).

Australian marsupials usually breed from midsummer to early winter: January to June in the Southern Hemisphere. This can, however, be modified so that, in periods of drought, reproduction will be inhibited while in favorable conditions it may occur at almost any time of the year. While there are many similarities between the endocrine control of reproduction in placental and marsupial mammals there are also some remarkable differences. The gonadal steroids, estrogens and progesterone, mediate the pre- and postovulatory changes in the reproductive tract during the estrous cycle of marsupials. Hypophysectomy of the tammar wallaby blocks the growth of the ovarian follicles and ovulation (see Tyndale-Biscoe and Evans, 1980). This effect appears to reflect the absence of FSH and LH, which have been identified in extracts of the animal's pituitary gland. However, in contrast to eutherians, the marsupial corpus luteum develops normally following hypophysectomy and it is not dependent on the tropic action of LH. Instead, its activity appears to be controlled by prolactin, which has an inhibitory effect. Specific receptors for prolactin have been identified in the corpus luteum of the tammar (Sernia and Tyndale-Biscoe, 1979).

In marked contrast to placentals, pregnancy in marsupials does not interfere with the concurrent estrous cycle and the maturation and ovulation of the egg. This egg is usually produced at the normal time. Certain marsupials, the macropodids or kangaroos, also display an interesting form of delayed implantation or, more correctly called in this instance, an *embryonic diapause* that differs from the process that is occasionally seen in placentals. The process of lactation in marsupials also has some rather unique features.

The estrous cycle of marsupials lasts for about 28 days and, as in placentals, consists of an initial period of follicular growth that is accompanied by the development of the uterus and vagina, at first under the influence of estrogens and then, in its secretory phase, by progesterone. Estrus follows a

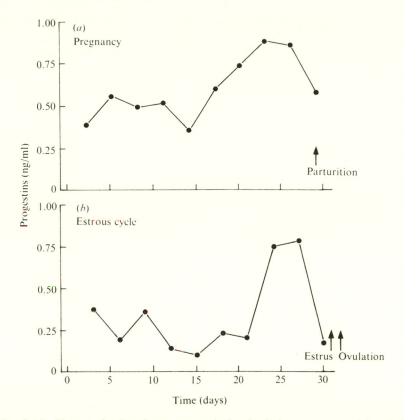

Fig. 9.14. Changes in the plasma progestin levels during pregnancy (*a*) and the estrous cycle (*b*) of a marsupial, the tammar wallaby *Macropus eugenii*. The commencement of pregnancy or the estrous cycle was initiated by removing the suckling young from the pouch, thereby initiating the development of the fertilized blastocyst (see text), or the next reproductive cycle. Progesterone levels commenced to rise on about day 15 and declined just prior to estrus and ovulation, or parturition. (From Lemon, 1972.).

sudden decline in the levels of progesterone (Fig. 9.14), reminiscent of that seen in some placentals, and lasts for several days during which ovulation occurs. The precise endocrine stimuli for ovulation have not been elucidated, but it is spontaneous and not the result of any external sexual or copulatory stimuli. Fertilization is followed by the development of the egg into a blasto-cyst and if the animal is not already lactating, pregnancy will occur. A corpus luteum is formed in the ruptured follicle and persists for the period of time that is usual in the estrous cycle; its life is *not* prolonged by the pregnancy. An extended life for the corpus luteum is not necessary in marsupials as the period of gestation is usually nearly identical to the time of the normal estrous

cycle (which continues to occur concurrently with the pregnancy!); nevertheless, the progesterone levels in pregnancy may be somewhat greater than
those in the normal estrous cycle (Fig. 9.14), which seems to reflect a hypersecretion
from the ovary. In a wallaby, the quokka, *Setonix brachyurus*, no difference
in the levels of progesterone in the nonpregnant and pregnant animals could
be detected (Cake, Owen, and Bradshaw, 1980). However, a sudden rise in the
plasma progesterone concentration was detected between the third and fourth
days of pregnancy. It was suggested that this "spike" may occur in response to
a signal from the blastocyst and serve to establish a suitable secretory condition of the endometrium. However, in the New World opossum (*Didelphys
virginiana*) no change in the blood concentrations of estradiol or progesterone
could be detected when they were pregnant (Harder and Fleming, 1981).
Thus, there may be interspecific variability in the nature of the endocrine
recognition of pregnancy among the marsupials.

Most marsupials have a quite simple placenta that is formed by a vascularization
of the chorion by blood vessels from the yolk sac (see Renfree, 1980). Such a
yolk-sac or choriovitelline placenta is quite small and, compared with that of
eutherians, provides a rather tenuous connection between the mother and
fetus. There is, however, considerable variation and the chorionic villi, which
form the attachment, can be quite well developed in some species. In bandi-
coots the allantois may also contribute to the placenta, as it does in eutherians.
In the latter the placenta plays a major role in furnishing the endocrine needs
of pregnancy but, not surprisingly, it has usually been considered unlikely
that this is so in marsupials. In several marsupials ovariectomy in mid- to late
pregnancy does not result in the death of the fetus (though parturition does not
occur) or in a regression of the uterine endometrium, which is supporting it
(Renfree, 1980). This phenomenon is also seen in eutherians where survival
of the fetus can, however, be attributed to the endocrine activities of the
placenta. Fetal membranes collected from the quokka (*Setonix brachyurus*)
and the tammar have been shown, *in vitro*, to be able to synthesize progesterone from pregnenolone (Bradshaw et al., 1975; Heap, Renfree, and Burton,
1980). The tammar placenta, however, could not form estrogens. There is
thus some circumstantial and experimental evidence that the marsupial placenta may have an endocrine role during pregnancy.

Ovulation of the egg, that ripens during pregnancy, occurs at various times
in relation to parturition. The relationships of the estrous cycle and pregnancy
are summarized in Fig. 9.15. Ovulation may occur *prior* to parturition, as in
the swamp wallaby, *Wallabia bicolor*, where the period of gestation is 35
days compared to only 32 days for the estrous cycle. In this marsupial,
preparturition ovulation is followed by copulation. If fertilization takes place,
a blastocyst develops, which, if lactation then occurs, lies dormant (see later).
In other species like *Megaleia rufa*, parturition is closely succeeded by ovula-

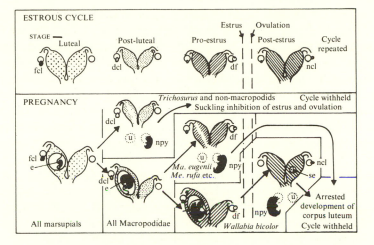

Fig. 9.15. The estrous cycle and pregnancy in marsupials. A diagrammatic summary of the size and functional relationships of the ovary and the uterus. *Estrous cycle*: fcl, functional corpus luteum; dcl, degenerating corpus luteum; df, developing Graafian follicle; ncl, new corpus luteum. *Pregnancy* (additional abbreviations): e, intrauterine embryo; npy, newborn pouch young attached to teat; se, segmenting egg. Three different patterns in the reproductive cycle of the marsupials are shown: nonmacropodids like *Trichosurus* (Australian possum) and macropodids (kangaroos), which have two types of cycle shown by *Macropus eugenii* (the tammar wallaby) and *Megaleia rufa* (the red kangaroo), and *Wallabia bicolor* (the swamp wallaby). For detailed discussion see the text. (From Sharman, 1970. Copyright © 1970, by the American Association for the Advancement of Science.)

tion, postpartum copulation, and the formation of a blastocyst. In the grey kangaroo, *Macropus giganteus*, the period of gestation is much shorter than the estrous cycle, just as seen in the Australian brush possum, *Trichosurus*, and other nonkangaroos; prescheduled future ovulation is then inhibited by the suckling stimulus provided by the young. If, however, the young is removed, ovulation follows 9 days later. In the latter part of lactation of the grey kangaroo this inhibition may decline so that ovulation and fertilization may occur, though while the young is in the pouch, the fertilized egg does not develop further than the blastocyst stage.

When the young kangaroo leaves the pouch the development of a dormant blastocyst can then proceed and pregnancy thus continues. The young kangaroos, however, remain with the mother and continue to suckle from outside the pouch; thus, the female kangaroo may have one young in the pouch, and another, much older young, "at heel." The two young then feed from different teats and the composition of the milk that each feeds on is quite different notwithstanding the fact that the endocrine secretions that are available to both glands are identical.

The delayed implantation in macropodid marsupials follows the division of the fertilized egg to a stage when 80 cells are present. This blastocyst, in contrast to the placental one, is surrounded by a shell membrane and a layer of albumin, in which state it can survive for several months. It lies in the uterus, in the branch opposite to that where the preceding pregnancy occurred. The temporary inhibition of the development of the blastocyst depends on the suckling stimulus from the young kangaroo in the pouch. Once suckling declines the blastocyst then starts to develop further. Denervation of the mammary gland has the same effect even though suckling continues (Renfree, 1979). The nature of the inhibitory stimulus is thought to result from neural stimulation of the pituitary, as a result of the suckling. Ovariectomy does not have any effect on the dormant blastocyst, but the injection of estrogen and, especially, progesterone can initiate its development. It appears that the corpus luteum of lactation, which is formed from the follicle that gave rise to the dormant blastocyst, is relatively quiescent during lactation and its subsequent development and rapid secretion initiates the succeeding pregnancy. Secreted progesterone appears to accomplish this change by initiating, and so synchronizing, both the further development of the blastocyst and the luteal phase of the uterus, the latter providing an environment that is necessary for the growth of the embryo. The nature of the inhibition of the corpus luteum is uncertain but it is interesting that the injection of oxytocin, which is normally released in response to suckling, can prevent the development of the corpus luteum and the blastocyst in kangaroos that have been deprived of their suckling young. In eutherians, prolactin is also released in response to suckling and, as described earlier, this hormone has been observed to inhibit the development of the corpus luteum in the tammar. It is considered likely that it is mainly prolactin, which is secreted in response to suckling, that is inhibiting the activity of the corpus luteum and hence the implantation of the blastocyst (Renfree, 1979).

In the tammar wallaby delayed implantation (or embryonic diapause) can also be maintained in the absence of lactation, following the weaning of a young in the spring (October in the Southern Hemisphere). The fertilized blastocyst, which usually results from mating after the birth of the just-weaned joey, then remains quiescent in the uterus until the summer solstice (December 22). At this time the dormant corpus luteum is reactivated and this results in an initiation of further development of the embryo. However, when the pineal gland is denervated the blastocyst resumes its development as soon as lactation ceases (Renfree et al., 1981). The tammar thus exhibits an embryonic diapause that can be controlled by either the stimulus of lactation or the season. The latter apparently reflects the effects of the length of day on the pineal gland.

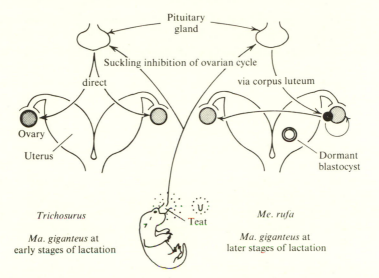

Fig. 9.16. The mechanism by which suckling may inhibit the ovarian cycle in marsupials. *Trichosurus vulpecula* and *Macropus giganteus* in early stages of lactation, return to estrus soon after the young is removed from the pouch. In these animals inhibition seems to result from a direct inhibitory effect on the ovary, mediated by the suckling stimulus. *Megaleia rufa* and *Macropus giganteus* at later stages of lactation do not return to estrus until about a month after the young is removed from the pouch. In these marsupials a functional corpus luteum is necessary for the inhibition to occur, which, as a result of suckling–pituitary stimulation, releases an "inhibitory factor." (From Sharman, 1970. Copyright © 1970 by the American Association for the Advancement of Science.)

The suppression of the normal estrous cycle that characteristically occurs in lactating marsupials also is related to the suckling stimulus (Fig. 9.16). In nonmacropodids such as *Trichosurus* (and *Macropus giganteus* in the early stages of lactation), this inhibition is due to an inhibitory action on the ovary that is probably mediated through the pituitary. In most macropodid marsupials, there is an interesting deviation from this pattern as the corpus luteum of lactation appears to be essential for the response. It is thought that, following suckling, a release of a pituitary hormone stimulates the corpus luteum, which releases a secretion that inhibits the ovary. It is uncertain what this secretion may be; if it is progesterone it presumably must be a minimal release so as to not disturb the dormant blastocyst. The nature of the pituitary stimulus is uncertain; it cannot be mimicked by the injection of eutherian LH or prolactin.

Parturition in marsupials does not appear to be such a dramatic event as it is in mammals, which bear much larger young. It nevertheless appears to be hormone dependent as it is prevented by ovariectomy. This effect may reflect the absence of estrogens, which in a sudden dramatic preparturition surge seem to contribute to the birth of the young in many placentals. The progesterone concentration in the blood also increases, but then decreases just before parturition (Lemon, 1972), and it is possible that this also has a role to play in delivery of the marsupial young. The circulating progesterone levels at parturition, however, may differ considerably, depending on the species, for, as we have seen, the length of gestation varies in relation to the estrous cycle so that the birth of the young occurs at various stages in the development of the corpus luteum. The fetus of the tammar, like that of many eutherians, has a well-developed adrenal cortex and can secrete corticosteroids (Catling and Vinson, 1976). It is thus possible that parturition in marsupials is also a prostaglandin-dependent process.

Whether or not oxytocin is also involved in parturition, as in placentals, is unknown, but it is present in the marsupial pituitary and it contracts the uterus of the wallaby, *Setonix brachyurus*, *in vitro* (Heller, 1973). This tissue is most sensitive in the late stages of pregnancy. Oxytocin also promotes milk letdown when injected into kangaroos, an effect shared with the placentals. This response may be of special importance for feeding the relatively undeveloped newly born young of marsupials. Indeed, it has been shown that the mammary gland of the agile wallaby (*Macropus agilis*) is highly sensitive to the action of oxytocin early in lactation, but this subsequently declines as the young grow larger (Lincoln and Renfree, 1981).

The reproductive pattern in marsupials shows distinct differences from that of placental mammals and appears to be well adapted to their manner of life. Contrary to some popular opinion about the "lowly" state of development of these animals, their reproduction is an extremely efficient process. The embryonic diapause of the kangaroos constitutes an excellent "insurance" for continued reproduction so that if a young is lost, or when it is weaned, another pregnancy follows with little delay.

Monotremes

These mammals are confined to Australia and New Guinea and are remarkable as they produce eggs with a keratinous shell that they care for. In the spiny anteater, *Tachyglossus*, the egg is lodged in a pouch for hatching while platypuses lay their eggs and tend them in burrows. They are monoestrous (see Griffiths, 1978). The gonads increase considerably in size prior to breeding in the late winter and spring. The platypus usually produces two eggs, and the echidna one. They are first incubated in the uterus for a period that, in the echidna, has been estimated at 9 to 27 days. Development equivalent to about 40 hours of incubation in chickens occurs at this time. Monotremes form a corpus luteum. Progesterone has been identified in the blood of the platypus and the concentration is higher in pregnant animals. Testosterone appears in high concentration in the spermatic vein of the platypus during the breeding season. In echidnas the period of incubation of the egg in the pouch is about 10 days. The young of monotremes are fed in the typical mammalian fashion, on secreted milk. In the echidna it is released in greater amounts following injection of oxytocin. Development of the mammary glands of nonbreeding or ovariectomized echidnas can be promoted by the injection of estrogens. Relatively little is known about the endocrine processes that control reproduction in these very interesting and unique animals, but there appear to be a number of similarities to those of other mammals.

Nonmammals

Precise information about ovarian cycles in nonmammals, apart from birds, is meager compared to that in mammals. Much of the available knowledge is based on morphological and histological observations on the ovaries and the accessory and secondary sexual characters, especially the oviduct. Such information is related to endocrine changes on the basis of the abilities of injected, exogenous hormones to mimic or prevent such changes. These experimental approaches, while suffering from obvious limitations have, however, demonstrated that differences indeed exist between the ovarian cycles of different nonmammalian vertebrates. With newly available radioimmunoassay procedures for measuring hormone levels in the blood, the precise role of hormones in the reproductive life of nonmammals is now being investigated more rigorously. At the present time the birds have received the most attention, but these new techniques will clearly also be extended to the reptiles, amphibians, and fishes.

There are several salient areas about which endocrine information in nonmammals promises to be especially interesting. These include:

a. The evolution and role of the corpus luteum, especially in viviparous and ovoviviparous species.

b. The mechanism by which a single gonadotropin (thought to be present in some reptiles and fishes) controls the ovarian cycle.

c. The possible physiological role of the neurohypophysial peptides in influencing the contractility of the oviducts

It is usually somewhat difficult to make a strict comparison between the ovarian cycles of mammals and nonmammals. This partly reflects a lack of information about the latter, but the timing of the events also often differs rather radically. Thus, birds usually take many weeks of preparation to come into breeding condition when ovulation becomes possible. This latter process then occurs at regular intervals of about 24 hours, which can proceed for several weeks, or in the domestic fowl, with some minor breaks, for up to 300 days of the year. This almost daily ovulation cannot be strictly compared to the entire estrous cycle of mammals but may be more synonymous with very short estrous cycles of rapidly successive periods of estrus. In other species of fish and amphibians, a single massive (sometimes referred to as "explosive") ovulation may occur, but in the meantime the egg may be held in readiness for some time; ovulation thus does not always appear to be an irrevocable event in a strictly pretimed program.

Birds

The ovarian cycle of most birds is primarily under photoperiodic control. The daily changes in light are the primary stimulus that directs the overall endocrine preparations for the breeding season (see Chapter 4). As described earlier, in quail subjected to a long-day photoperiod, a release of gonadotropin-releasing hormone, from the hypothalamus, results in a discharge of pituitary gonadotropins. Plasma LH and FSH levels have been shown to increase in Japanese quail that are photoperiodically stimulated by long day-lengths (Nicholls, Scanes, and Follett, 1973; Follett, 1976). The changes in the levels of pituitary gonadotropins indicate that a release occurs at a precise time each day, which, in the quail, is in the evening after dusk. This daily rhythmical release of gonadotropins promotes the growth of the ovary and the maturation of the follicles. Estrogens are known to stimulate the growth of the avian oviduct (progesterone may also contribute to this increase) and secondary sexual characters and are released during such preparations for reproduction.

When the bird is finally ready to breed, ovulation may begin. In domestic fowl, ducks, and geese, this is spontaneous but in other birds the presence of the male is usually necessary so that ovulation may then be said to be induced. There are several other factors that determine whether or not ovulation will occur in birds. If the newly laid eggs are continually removed from the nest some birds will continue to lay more eggs (*nondeterminate layers*). A house sparrow has thus been stimulated to produce 51 eggs in a season; ovulation apparently continued until the ovary was exhausted of suitable follicles. The domestic fowl, which can produce 300 eggs in a year, is an even more dramatic example of this phenomenon. How such birds recognize the number of eggs in the "clutch" is unknown, but it has been suggested that this may be the result of a tactile stimulus or sight. Other types of birds (*determinate layers*) produce a set number of eggs, and changing the number in the nest does not modify ovulation.

In the domestic fowl, LH, as in mammals, initiates ovulation, but the mechanism controlling the release of this hormone is not clear. The injection of progesterone promotes ovulation while estrogens delay it, effects that are in

direct contrast to those seen in mammals (Fraps, 1955). The normal rhythmical release of LH, which commences about 8 hours prior to ovulation is,
however, preceded by an increase in plasma estradiol (Fig. 9.17*b*) (see Follett
and Davies, 1979). Estrogens, however, do not directly stimulate the release
of LH in birds, though they apparently serve to sensitize the hypothalamus to
the action of progesterone. The surge of LH never *precedes* that of progesterone (Fig. 9.17*b*). Antisera to progesterone block ovulation in chickens, but
antisera to estrogens do not (Furr and Smith, 1975). The mechanism of
ovulation in birds is thus clearly different from that in mammals.

It has been shown in the domestic fowl that, as long as there is an egg in the
oviduct, further ovulation is inhibited. This effect can be mimicked by placing
an irritant, such as a piece of thread, in the oviduct. Such a condition can be
prolonged for about 3 weeks and as no regression in the ovary or oviduct
occurs secretion of FSH and estrogens is thought to be unimpaired. The
inhibition of ovulation by the presence of an egg in the oviduct is thought to
be the result of a neural stimulus that may inhibit LH release. The injection of
progesterone or LH into such birds overcomes this inhibition and promotes
ovulation.

The precise site of origin of the circulating progesterone in birds is uncertain. Birds do not possess a corpus luteum but the ovary nevertheless secretes
progesterone. This steroid may be formed by the follicles themselves, the
interstitial tissue, or the corpora atretica. The role of progesterone in the
ovarian cycle is also not clear. In mammals and other vertebrates, this steroid
inhibits the release of LH and LH-like gonadotropins by its negative-feedback
inhibition of the hypothalamus. Such an effect would be unexpected in birds
if, as suspected, progesterone stimulates the release of LH. As birds lack a
corpus luteum, LH also cannot exert its usual luteinizing and luteotropic
actions though it is possible that it may have comparable effects at other sites
in the ovary. There is need, at this stage, for a note of caution as it should be
recalled that the endocrine observations on the avian ovarian cycle are nearly
all confined to domestic species, especially *Gallus domesticus*.

The egg-laying cycle of the domestic fowl is thought to occur in the following manner (van Tienhoven and Planck, 1973). The eggs are laid in a clutch,
or "sequence," three to five in number, which are each produced at intervals
of about 26 hours. Laying can only occur during a daylight period of about 8
hours called the "open period." Ovulation usually takes place about 2 hours
after laying so that the time of the latter advances by a similar period each
day. The open period thus limits the number of eggs that can be laid before an
obligatory pause of 40 to 48 hours. Ovulation and oviposition are controlled
by an endogenous rhythm lasting for 26 to 28 hours, the commencement of
which is normally timed according to the photoperiod. Depending on such
stimulation the length of the cycle can be retarded or advanced by about 2

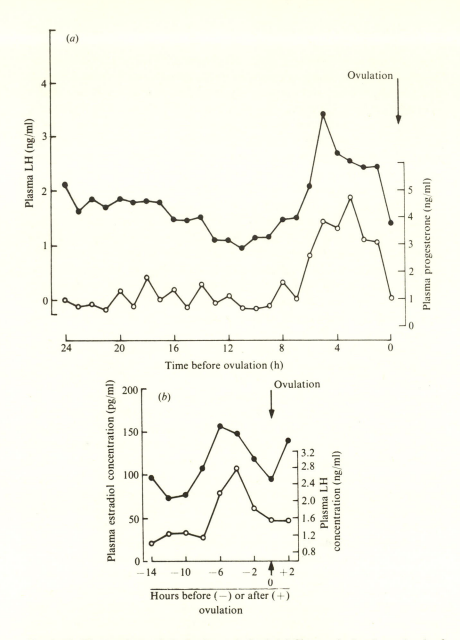

Fig. 9.17. The ovarian cycle in the domestic fowl. (*a*) Changes in the plasma levels of LH (●) and progesterone (○) during the ovulatory cycle. The relationship between the rises in the levels of progesterone and LH is not clear; the release of LH apparently does not initiate a release of progesterone as it never precedes it. (From Furr et al., 1973.) (*b*) Changes in the levels of LH (○) and estradiol (●). The rise in the level of estradiol occurs about 2 hours before that of LH. (From Senior and Cunningham, 1974.)

hours and so may vary from 24 to about 30 hours. In continual light, other periodic events, such as the times of feeding and fluctuations in temperature, can be used to initiate the egg-laying cycle. Such external stimuli appear to sensitize regions of the hypothalamus, which, in response to stimulation by circulating progesterone, and possibly estrogens, secretes LH-releasing hormone. LH is then released from the pituitary, which results in ovulation 4–8 hours later. Subsequent formation of the egg takes place in the oviduct, and the timing of the events there seems to depend on photoperiodic stimulation working in conjunction with the ruptured follicle. Removal of the latter tissue from the ovary results in a retention of the egg in the oviduct. Normally, oviposition occurs 13 to 14 hours after ovulation. It is possible that contractions of the oviduct that occur during oviposition are assisted by vasotocin, which is released from the neurohypophysis.

The production of eggs for human consumption is an important agricultural and economic industry that involves maintaining many millions of domestic hens. The costs of feeding them and providing light and heat are so large on a national basis that even quite small decreases in such costs may be of considerable economic importance. Thus, any advance in the time of onset of the age of first laying or increases in the clutch size, by decreasing the interval between oviposition and the next ovulation, may potentially provide more and cheaper eggs. Artificial changes in the periods of light to which the birds are exposed each day have effected some important advances (see Morris, 1979; Wilson and Cunningham, 1980). Sexual maturity in hens can be promoted by increasing the length of the daylight hours; for instance, pullets raised on a 14 hours light and 10 hours dark (14L:10D) schedule mature and start laying 7–10 days earlier than those kept on a 6L:18D photoperiod. If pullets initially raised on a short-day protocol are subsequently changed to a long-day one, sexual maturation can be advanced by as much as 7 weeks. The particular decrement of time to maturity depends on when the change is made. Changing the lighting period of laying hens from 6L:18D to 14L:10D can enhance their egg laying compared with birds kept constantly on the latter schedule. This change is associated with an increase in the plasma LH concentrations. Light is power, and money, so that any reduction in its use may be an economic advantage. It has been found that periods of exposure to light need not be continuous and that intermittent exposures may be as effective as continual light. Thus, when hens on short days (8L:16D) are provided with a 1- to 10-minute pulse of light each hour for the first 8 hours of darkness, advances in sexual maturity and onset of egg laying can occur. Such changes have been shown to increase plasma LH concentrations (Wilson and Cunningham, 1980).

Most birds incubate their eggs. In many instances both the male and female share periods on the nest and tend the young. The desire to incubate the eggs

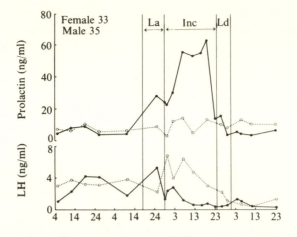

Fig. 9.18. Prolactin and luteinizing hormone (LH) concentrations in the plasma of mallards (*Anas platyrhynchos*) at different times of breeding. Values for a duck (●) and a drake (○) are shown. The vertical lines indicate the period of egg laying (La), incubation (Inc), and the leading of the young (Ld). Prolactin levels are greatly increased, whereas LH levels decline in the ducks when they are incubating the eggs. The drakes do not share this chore and do not exhibit these responses. (From Goldsmith and Williams, 1980.)

is called "broodiness" and is influenced by hormones. Preparations for incubating the eggs include building a nest and the development of a "brood patch" on the abdomen. In canaries and budgerigars this morphological change includes a loss of feathers, which appears to involve the actions of estrogens and prolactin. Subsequently, there is an increase in the vascularity, which is estrogen dependent, and an increase in tactile sensitivity involving estrogen and, probably, progesterone (Hutchison, Hinde, and Steele, 1967; Hutchison, 1975). In ringdoves progesterone may play a more important role than other hormones in initiating broodiness (Stern and Lehrman, 1969).

Prolactin levels have been measured in the plasma of a wild population of mallards over the breeding season (Goldsmith and Williams, 1980) (Fig. 9.18). In the ducks, the prolactin concentrations increase about three-fold during incubation of the eggs and decline following the hatching of the ducklings. The drakes do not incubate the eggs, and relatively low levels of prolactin persist in the plasma over the breeding season. Plasma LH declines markedly in the ducks at the end of egg laying, but this change is not seen in the drakes until much later in the season. This decline in plasma LH has been observed in several species during incubation and in ringdoves, both sexes of which share incubation, it occurs simultaneously in both the male and female (Silver, Goldsmith, and Follett, 1980). A similar reciprocal rise in plasma prolactin and decline in LH have been observed in broody domestic white rock hens (Bedrak, Harvey, and Chadwick, 1981). It is thus possible that prolactin may be interfering with the release of the gonadotropin. Ringdoves feed their young like pigeons, with a secretion formed by the crop-sac epithelium. This process is stimulated by prolactin (see Chapter 5), and a release of this hormone has been observed following the exposure of either sex to the squabs (Buntin, 1979).

Birds thus show some interesting deviations and novelties in the use of hormones for integrating their reproductive processes. An evolution of the role of certain hormones has clearly occurred in this interesting offshoot from a reptilian stock.

Reptiles

The Reptilia contain oviparous, ovoviviparous, and viviparous species. Unlike in birds, a distinct corpus luteum is formed following ovulation and although the available evidence indicates that it secretes progesterone (Highfill and Meade, 1975a, b; Colombo and Yaron, 1976), it seems unlikely that this hormone contributes to the maintenance of pregnancy. A pituitary LH and FSH have been identified in some reptiles, but in the snakes and lizards only a single gonadotropin appears to be present (Licht, 1979). The development of the ovary, its secretion of steroid hormones, and ovulation appear to be controlled by an FSH or a nonspecific gonadotropin. This is in contrast to mammals and birds, where there are two functioning gonadotropins, and amphibians, where an LH-like hormone appears to be predominant.

The changes that occur during the ovarian cycle of the ovoviviparous lizard, *Sceloporus cyanogenys*, are summarized in Fig. 9.19. The ovary starts to grow in October or November and this is accompanied by the development of the oviduct. These changes can be prevented by hypophysectomy. The gonadotropin stimulates gonadal growth and the secretion of estrogen. Implantation of small pellets of estrogens into the region of the median eminence reduces the growth of the oviduct. This effect is probably the result of a lower rate of ovarian estrogen secretion, due to the inhibition of gonadotropin release by a negative-feedback inhibition. Ovulation is also prevented by such an estrogen implant. Mammalian FSH has been shown to promote ovulation in several species of lizards (Licht, 1970), and the endogenous gonadotropin no doubt also had this effect.

Following ovulation in *Sceloporus cyanogenys* the corpus luteum develops and this is correlated with a three-fold increase in the circulating progesterone concentration. This elevated hormone level persists in the plasma until parturition, when it declines. A similar pattern in circulating progesterone levels has also been observed in the viviparous snake, *Natrix sipedon* (Chan, Ziegel, and Callard, 1973). Direct evidence that reptilian corpora lutea can produce

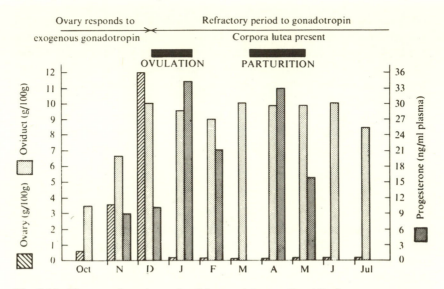

Fig. 9.19. The annual ovarian cycle of the ovoviviparous lizard *Sceloporus cyanogenys*. The ovaries and oviducts start to develop (under the influence of gonadotropin) in October and ovulation may occur in December to January. Gestation lasts for about 12 weeks and the young are delivered in late March to mid-May. Corpora lutea persist during pregnancy and the plasma progesterone levels rise, but then decline in late summer following parturition. (From Callard et al., 1972.)

progesterone has been obtained following *in vitro* incubation of such tissue obtained from the snapping turtle, *Chelydra serpentina* (Klicka and Mahmoud, 1972). In *Sceloporus cyanogenys* the circulating progesterone levels are reduced following hypophysectomy, suggesting that there is some pituitary control over this hormone, but in pregnant lizards relatively high concentrations persist so that if a luteotropic effect is present it is apparently not vital (Callard et al., 1972). In addition, the implantation of pellets of progesterone into the region of the median eminence results in a depression of the circulating progesterone concentration (Callard and Doolittle, 1973). This suggests the presence of a negative-feedback inhibition of the release of a tropic hormone and is accompanied by a decrease in the growth of the ovary and oviduct. The injection of progesterone has also been found to inhibit ovulation in the turtle, *Chrysemys picta* (Klicka and Mahmoud, 1977). The hypothalamic control of gonadotropin release can thus be influenced by estrogens and progesterone in a manner suggesting that a mechanism is present that is similar to that in mammals.

Corpora atretica, formed by the dissolution of unovulated follicles, may also contribute to the control of reptilian ovarian cycles. At the end of the summer breeding season the lizard *Anolis carolinensis* becomes refractory to the effects of photoperiodic stimulation, and this can be related to the presence of corpora atretica in the ovary. In addition, these lizards, at this time, respond poorly to injected gonadotropins, but the response to these hormones could be increased five-fold if the corpora atretica were removed. The nature of the latter's inhibitory effect on the female reproductive system is unknown (Crews and Licht, 1974).

Progesterone, as we have seen, plays an important role in maintaining pregnancy in mammals, but there is no conclusive evidence to indicate that this occurs in reptiles. Ovariectomy or hypophysectomy does not affect the course of pregnancy in a variety of viviparous and ovoviviparous species of reptiles (see Yaron, 1972). Progesterone, nevertheless, attains high concentrations during a reptilian pregnancy; so what is its function? Callard and Doolittle (1973) have suggested that its action in reptiles may be to inhibit gonadal growth during gestation, and this may represent a "more primitive" role than the regulation of the uterine environment that is seen in mammals. Corpora lutea are also formed following ovulation in oviparous reptiles, and it has been found that when this tissue is removed in gravid *Scelopordus undulatus* (an oviparous lizard) an earlier oviposition occurs, suggesting that progesterone may control the period of egg retention in such reptiles (Roth, Jones, and Gerrard, 1973). A viviparous snake, *Thamnophis elegans*, was also found to deliver its young somewhat earlier than expected following lutectomy (Highfill and Mead, 1975b).

In reptiles, as also in birds, estrogen, possibly in conjunction with growth hormone, stimulates the production of lipophosphoproteins by the liver and these are incorporated into the egg. Prolactin, when injected, has been shown to exert an antigonadal effect, but the significance of this inhibition is unknown *in vivo*. It is nevertheless interesting that these actions of estrogens and prolactin are shared with the birds.

Despite the valuable experiments that have already been performed, much additional and more precise information is needed before an adequate account of the role of hormones in the reptilian ovarian cycle can be given.

Amphibians

Most amphibians are oviparous though there are a few species that have ovoviviparous and even viviparous habits. An excellent account of the ovarian cycle in amphibians is given by Redshaw (1972) and Jorgensen, Hede, and Larsen (1978). Complete maturation of the oocytes usually takes several years while formation of the yolk, in species that live in temperate zones, commences in the summer preceding spawning (Fig. 9.20). The ovarian cycle is only partly influenced by photoperiod; other factors such as environmental temperature, the animal's nutritional condition, as well as endogenous rhythms,

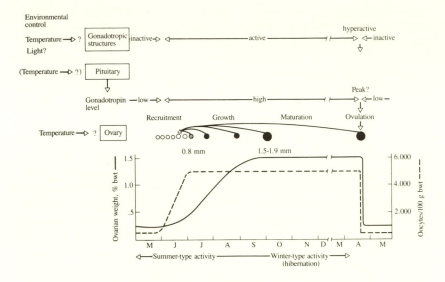

Fig. 9.20. The ovarian cycle of toads, *Bufo bufo bufo*, in Denmark. The oocytes are recruited and the ova develop during the preceding summer and remain quiescent during hibernation in the winter. Ovulation occurs the following spring, in April. The ovarian cycle appears to be under the control of environmental factors such as temperature and light and requires adequate nutrition in the summer period. Arrow shown pointing backward from the larger to the smaller oocytes indicates an inhibiting effect of the former on further recruitment of oocytes for the final maturational process. (From Jorgenson, Hede, and Larsen, 1978.)

may all impinge on this process. The development of the ova is controlled by the adenohypophysis and the hypothalamus. The LHRH content of the hypothalamus is low in the nonbreeding season of the female African toad *Xenopus laevis*, but it rises in the spring when the ovaries enlarge (King and Millar, 1979*b*).

Hypophysectomy or transplantation of the pituitary, so that it is no longer in contact with the hypothalamus, interrupts oogenesis. Two distinct gonadotropins, an FSH and an LH, have been identified in amphibians, but their respective roles in ovarian development do not appear to parallel those in mammals (Licht, 1979). The LH appears to have a predominant effect on the development of the oocytes, steriodogenesis, and ovulation. This situation is in contrast to that in reptiles where, as just described, an FSH-like hormone appears to be most active on all of these processes.

Estrogens are produced by the ovarian follicles and these contribute to the development of other sexual characters, including the oviduct, as well as vitellogenesis.

Progesterone is apparently also produced by the ovary, though the precise site of its formation is uncertain as most amphibians (except for ovoviviparous and viviparous species) do not form a distinct corpus luteum after they ovulate. Whether or not either or both gonadal steroid hormones exert a negative-feedback inhibition on the release of gonadotropin is unknown.

Ovulation can be readily promoted in amphibians by the injection of gonadotropin. This hormone may be obtained from amphibian pituitaries, but exogenous hormones from other species are also effective. The latter hormonal effects are particularly well known as they are the basis for a convenient pregnancy test for women. Human chorionic gonadotropin (hCG) that is secreted in the urine during pregnancy induces ovulation in frogs and toads. Both mammalian LH and amphibian LH are more effective than FSH in promoting the growth of the ovarian follicles and ovulation (Licht, 1979). Ovulation is often, though not always, induced at the time of sexual pairing. *Xenopus* seems to be on the verge of ovulation for prolonged periods of time while in *Rana temporaria* the eggs are stored in the oviduct, from which they are expelled when mating occurs.

The precise hormonal events that initiate ovulation and oviposition are not known. The injection of hCG into the toad *Bufo bufo* results in ovulation in about 24 hours and this is preceded by a release of a progesterone-like material into the blood (Thornton, 1972). It has also been show, *in vitro*, that progesterone can induce ovulation. In addition, the injection of progesterone hastens ovulation while estrogen retards it, a situation reminiscent of that which occurs in birds. It is to be hoped that the use of radioimmunoassay procedures will allow the measurement of amphibian gonadotropin, progesterone, and estrogens so that the natural events that occur during the ovarian cycle in amphibians can be more directly observed.

The oviduct undergoes a distinct annual cycle in *Bufo bufo* and attains its greatest size in the autumn (Jorgenson and Vijayakumar, 1970). A decline in the weight of the oviduct takes place during spawning in April. This decline is due to a loss of secretory contents that coat the eggs with a "jelly." The secretion of this jelly, like that of avidin in the fowl oviduct, is controlled by progesterone.

A diagrammatic summary of the role of hormones in controlling different ovarian functions in amphibians is given in Fig. 9.21.

Some very interesting observations have been made on the effects of progesterone on gestation in a viviparous frog *Nectophrynoides occidentalis* (Xavier and Ozon, 1971; Zuber-Vogeli and Xavier, 1973; Xavier, 1974). This frog lives in West Africa where it is subjected to periods of seasonal drought during which it estivates in burrows. Following ovulation in October the fertilized eggs are retained in the oviduct where development proceeds until parturition the following June. In November, these pregnant frogs estivate and do not emerge until April (Fig. 9.22). Corpora lutea are formed following ovulation that apparently secrete progesterone. When the ovaries from preg-

PITUITARY GONADOTROPIN

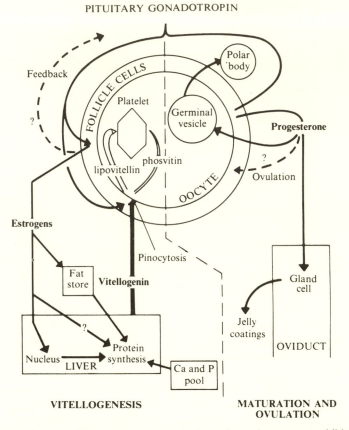

VITELLOGENESIS   **MATURATION AND
                                        OVULATION**

Fig. 9.21. The role of hormones in the production of eggs in anuran amphibians. The pituitary gonadotropin controls the release of estrogens and progesterone by the ovary. *Left*. The control of the process of *vitellogenesis*. Estrogen, formed by the follicular cells, induces the formation of vitellogenin in the liver, which is incorporated into the yolk of the developing oocytes. The uptake process is stimulated by the gonadotropin. *Right*. The *maturation of the oocyte* and ovulation. These processes are controlled by the pituitary gonadotropin, which mediates the formation of a "maturation agent" that seems to be progesterone. The progesterone also stimulates the oviductal glands to secrete the "jelly" with which the eggs are coated. This summary is principally based on experiments on *Xenopus laevis* and *Bufo bufo*. (From Follett and Redshaw, 1974.)

nant frogs are incubated *in vitro* with the progesterone substrate, pregnenolone, they convert this steroid to progesterone. This ability to form progesterone declines as gestation progresses (Fig. 9.23). Following parturition, pregnenolone is converted to other steroids by the ovarian tissue and this process increases until ovulation again takes place. The preovulatory period is also the time when production of estrogens is thought to increase. If ovariectomy is per-

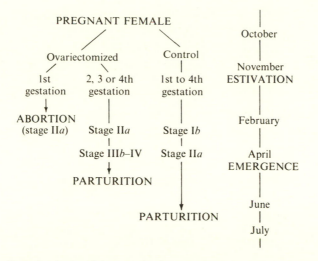

Fig. 9.22. Gestation in the viviparous frog, *Nectophrynoides occidentalis*, in relation to seasons. Normally these frogs ovulate and become pregnant in October. The dry season commences in November when they estivate. They emerge again, with the onset of rain, in April. The young are born in June. If these frogs are ovariectomized early in pregnancy, they may either abort, if the animals are young (and this is their first pregnancy) or, if they are large and it is the second, third, or fourth time of gestation, the development of the young is accelerated and they are born much earlier than usual. Ovarian progesterone is thought to delay the development of the young during the period of estivation. (From Zuber-Vogeli and Xavier, 1973.)

formed early in gestation (see Fig. 9.22) of young frogs, during their first pregnancy, abortion occurs. In more mature frogs, however, development of the young is accelerated following ovariectomy and parturition takes place about 3 months earlier than usual. The implantation of progesterone into these frogs toward the end of gestation reduces the rate of growth of the embryos. Progesterone thus appears to slow the rate of development of the embryos during the prolonged period of estivation so that the young are delivered at a more appropriate and favorable time of the year. This is indeed a novel role for progesterone.

Fishes

Sporadically distributed information is available about the control of ovarian function in fishes. This is not unexpected considering the enormous numbers of piscine species and the wide phyletic gaps that separate them. In addition, fish are usually somewhat diffident scientific collaborators. Fish may or may not exhibit seasonal changes in breeding behavior (see Peter and Crim, 1979). When the latter does occur it may, depending on the species, be promoted by

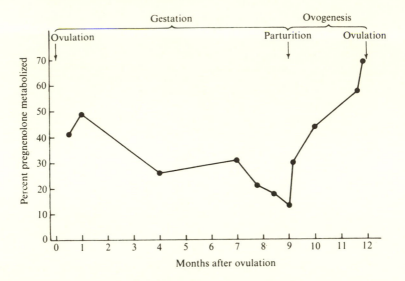

Fig. 9.23. The ability of the ovarian tissue in the viviparous frog *Nectophrynoides occidentalis* to metabolize pregnenolone (*in vitro*) at different stages of its ovarian cycle. During gestation the principal steroid produced from the pregnenolone is progesterone but after parturition 17-hydroxyprogesterone, androstenedione, and testosterone are also formed. (From Xavier and Ozon, 1971.)

either long-day or short-day photoperiods, and there is an important interaction of this with the environmental temperature. In some instances, warmth increases growth of the gonads but in others cold is more beneficial to ovarian growth. Some fish, such as lampreys and salmon, migrate up rivers from the sea in the autumn and breed the following spring and summer. Preparations for these events occur in the interim period.

Most fishes are oviparous but some species have evolved ovoviviparous and viviparous methods of reproduction. In teleosts the last two processes are usually rather different from the *in utero* development common to many other vertebrates. The young teleosts thus may develop *in situ* in the follicle or be incubated in the hollow central cavity of the ovary. In some teleosts successive broods (superfetation) may develop in such follicles that are then delivered in waves. It would not be unexpected if such dramatic differences in the procedures for reproduction resulted in endocrine adaptations.

In teleosts the development of the oocytes and vitellogenesis are dependent on the pituitary. It is controversial as to whether one or two gonadotropins are present (see Chapter 3). The injection of homologous gonadotropins can promote ovulation in teleosts. The roles of estrogens and progesterone in this process are, however, unknown. The endogenous release of gonadotropin in teleosts is under the influence of the hypothalamus; there may be a stimulating

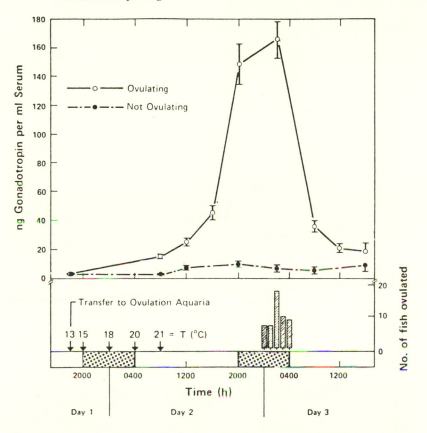

Fig. 9.24. Changes in the serum gonadotropin concentrations at the time of ovulation in goldfish. The latter event was promoted by exposing the fish to a long day (16L:8D) and moving them to a warmer aquarium, at 20° C. It can be seen that ovulation occurred in the dark early morning hours of the third day and in these fish (O - O) it was preceded by a large rise in serum gonadotropin concentration. No change in the level of this hormone was observed in the fish that did not ovulate (● -·- ●). Each point represents a mean ± S.E. (From Stacey, Cook, and Peter, 1979.)

hormone, possibly an LHRH, and also an inhibiting hormone (Peter and Crim, 1979). Gonadotropin levels in the serum of the goldfish have been shown to rise at the time of ovulation (Fig. 9.24). This process can be promoted by transferring the fish on a photoperiod of 16 hours light and 8 hours dark from an aquarium at 13°C to one at 20°C. The gonadotropin level started to rise in the daylight period and reached a peak at the beginning of the dark period, during which ovulation occurred. Plasma cortisol concentrations also increased, succeeding the rise in gonadotropin (Cook, Stacey, and Peter, 1980). The role of different steroids in the development of the ovarian follicles and ovulation is not well defined in teleosts, but cortisol may be involved.

The early development of the oocytes appears to be independent of hormones. Later development, including the uptake of the yolk proteins, can be stimulated by gonadotropin(s). The gonadotropic maturational hormone may do this by stimulating the synthesis of estrogens, progestins, and androgens while the vitellogenic hormone probably directly increases the uptake of the yolk proteins (Ng, Idler, and Burton, 1980*b*; Ng and Idler, 1980).

In cyclostome and chondrichthyean fishes, the pituitary, via the action of a gonadotropin, also controls the development of the ova. However, this process does not appear to be under hypothalamic control, though it is possible that it may be influenced by secretions that arrive via the systemic circulation. (The ventral lobe of the chondrichthyean pituitary gland has no direct vascular connection to the hypothalamus.) In lampreys, ovulation has been found to occur as long as 2 weeks after hypophysectomy (Larsen, 1973; 1978). Transplantation of the pituitary to other sites in the body does not adversely influence ovarian development. It is rather a mystery how cyclical control of reproduction is regulated in such vertebrates, which appear to be the only ones that lack a direct hypothalamic mechanism for controlling the function of the pituitary gland.

The role of corpora lutea in fish is contentious. Many teleosts possess so-called preovulatory corpora lutea, which are formed as a result of atresia of unovulated follicles and are more aptly called corpora atretica. Postovulatory corpora lutea, as well as corpora atretica, are present in many chondrichthyeans and there is considerable speculation as to whether they contribute to successful gestation in ovoviviparous and viviparous species (see Chieffi, 1967; Dodd, 1972*a*). Hypophysectomy does not interrupt pregnancy, at least for the first 3 months, in the viviparous shark, *Mustelus canis*, suggesting that an adenohypophysial control of progesterone secretion is not vital. It is, however, unknown whether progesterone contributes to gestation in ovoviviparous and viviparous fishes. Hisaw, in 1959, stated that "the elimination of yolk during follicular atresia and material from ruptured follicles at ovulation is a primitive function of corpora lutea and endocrine functions such as luteinization of the granulosa by pituitary luteinizing hormone and secretion of progesterone in response to pituitary luteotropic hormone as seen in mammals, are more recent adaptations." This interesting idea has, however, not been unquestionably accepted.

In many teleost and chondrichthyean fishes, the thyroid gland displays an increased activity during the breeding season. The latter is associated with many physiological and environmental changes so that it is difficult to be certain whether the endocrine events are primarily related to reproduction. Sage (1973) considers that it is likely that the thyroid is involved in the reproductive process in fishes as it is necessary for gonadal maturation in some species. Such a role for thyroxine could reflect a primeval endocrine use of this hormone.

Oviposition and parturition in nonmammals – a role for the neurohypophysial hormones?

The passage of eggs or young along the female gonaducts and their exit into the outside world may be assisted by rhythmical contractions of the muscles that surround these ducts. Such muscles are usually nonstriated (smooth) muscles (though striated muscle may also be present in some fishes), which have an inherent ability to contract even in the absence of nerves or hormones. The rate and pattern of contractility of smooth muscles can nevertheless be modified by such stimuli. As we have seen in mammals, oxytocin can promote contractions and aid in the process of parturition. It should be remembered, however, that smooth muscle readily reacts to local stimuli, and this may include the presence of an egg or fetus, so that oviposition or parturition can occur even in the absence of neural or hormonal stimuli.

The neurohypophysial hormones have a special ability to contract mammalian uterine smooth muscle, and this response has also been shown to occur (mainly *in vitro*) in many other species of birds, reptiles, amphibians, and fishes (Heller, 1972). Such effects can usually be elicited by low concentrations of such peptide hormones that represent only a small fraction of those stored in the neurohypophysis.

It is interesting that in poecilid teleosts that are ovoviviparous, the ovary contains smooth muscle and this is even more sensitive to the effects of vasotocin than the oviduct. The young in such fishes develop within the follicles so that their expulsion may conceivably be promoted by the contraction of these ovarian muscles.

Such contractile responses of the oviduct and teleost ovary to vasotocin can be seen to change at different stages of the sexual cycle and, as in mammals, the sensitivity is greatest at the time of parturition or oviposition. This can be seen in the ovary of *Poecilia* (Fig. 9.25), an effect that can be mimicked by the implantation of gonadal steroids into these fish. The oviduct, *in vitro*, of the mudpuppy, *Necturus maculosus*, also shows seasonal changes in sensitivity to vasotocin (Fig. 9.25*b*); this tissue may be as much as 100-fold more sensitive to this neurohypophysial hormone in May compared with its sensitivity in December and January.

Observations *in vitro*, however, must be interpreted with considerable caution as the responses may not take place *in vivo* and even if they do occur following injections of hormones they may not reflect a normal physiological response. In other words, we must be careful to distinguish the pharmacologi-

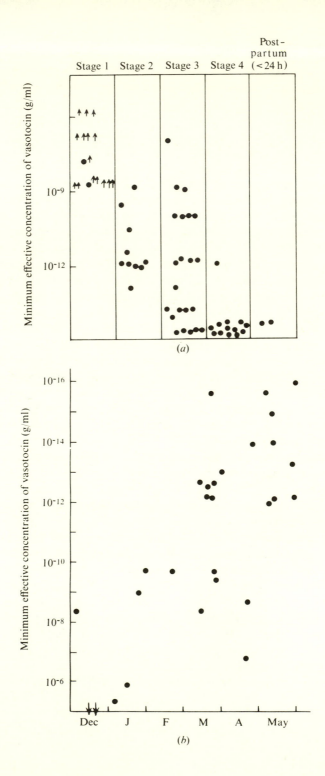

(a)

(b)

cal from the physiological effects of these hormones on the oviduct and uterus. The problem of the evolution of the roles of the neurohypophysial hormones in vertebrates has intrigued endocrinologists for many years. As we have seen (Chapter 3), such hormones are present in all vertebrates. While the neurohypophysial hormones have a fairly well-characterized role in osmoregulation of tetrapods, their role in fishes is not at all clear. The neurohypophysial hormones can contract smooth muscle of the oviduct and uterus in representatives of all the main groups of vertebrates, and it is tempting to speculate that these effects may be of physiological significance and even reflect a more ancient role for such hormones.

The evidence for this interesting endocrine hypothesis has been summarized by H. Heller (1972), who over a long scientific lifetime made many fascinating contributions to the comparative endocrinology of the neurohypophysis of which the present interest is a more recent outcome. There have been sporadic reports, spread over many years, about the ability of injected neurohypophysial hormones to promote oviposition and parturition in different vertebrates. This effect has been observed in the domestic fowl, several species of lizards, and the ovoviviparous fire salamander. The injection of neurohypophysial hormones also stimulates spawning-like behavior in the killifish, *Fundulus heteroclitus*. Vasotocin has been found to be 10 times as effective as isotocin (Pickford and Strecker, 1977). It is interesting that stored isotocin disappears from the pituitaries of the female, but not the male, killifish during the breeding season in June (Sawyer and Pickford, 1963). This change probably reflects a release of this neurohypophysial hormone that could be involved in the reproductive process. The precise physiological significance of such observations is uncertain.

Observations, *in vitro*, indicate that vasotocin is the most active of the neurohypophysial peptides in stimulating the contractility of the oviduct in nonmammals. An interesting exception is the oviduct of the dogfish, *Scyliorhinus canicula*, in which another neurohypophysial peptide present in these fish is the most active (valitocin or aspargtocin?). It would thus seem that an evolution of the receptor sensitivity in the gonaducts can occur that is related to the nature of the homologous hormones that are present. In mammals, oxytocin

Fig. 9.25. Periodic changes in the responsiveness of the ability of the ovoviviparous ovary of a teleost fish *Poecilia*, and the oviduct of a urodele amphibian *Necturus maculosus*, to contract (*in vitro*) to vasotocin. (*a*) *Poecilia* (the guppy); minimum effective concentration of vasotocin required to produce a contraction of the ovary during successive stages of gestation. The arrows indicate a lack of response; it can be seen that the preparations were most sensitive just prior to parturition (stage 4). (From Heller, 1972.) (*b*) *Necturus* (the mudpuppy); minimum effective concentration of vasotocin required to contract the oviduct at different seasons of the year. It can be seen that this was least in winter but progressively increased and was greatest in April and May. (From Heller, Ferreri, and Leathers, 1970.)

has a greater uterotonic effect than vasotocin, which is not an homologous hormone in this group of vertebrates. Such an evolution in the sensitivity of the gonaducts to the homologous neurohypophysial peptides also leads us to suspect that they may have a physiological role. We must not, however, carry this sort of thing too far!

If the neurohypophysial hormones do indeed contribute to the physiology of oviposition and parturition in nonmammals (and even ovulation in teleosts) it is probably just that – a contribution. The domestic hen can still lay an egg following neurohypophysectomy and ovoviviparous poecilid fish still produce young following hypophysectomy. As in mammals, such hormones could, however, facilitate parturition as well as oviposition.

In teleost fishes, it is possible that another neurosecretory hormone may be involved in eliciting contractions of the gonaducts. As described in Chapter 8, urotensin II, from the urophysis, exhibits an ability to contract smooth muscles from the urinary bladder and gonaducts of teleosts. This substance may have a hormonal role in the reproduction of these fishes, such as in assisting the process of spawning. Thus, a decline in the stored urotensin II in the urophysis of the white sucker, *Catostomus commersoni*, has been observed following spawning in the male, but not the female, fish (Lederis, 1973).

Hormones and the evolution of viviparity – a summary

Viviparity, or giving birth to live young, has undoubtedly evolved separately in many different vertebrates (see Sharman, 1976; Amoroso, Heap, and Renfree, 1979). Most people consider that this process has reached its highest level of physiological organization in eutherian mammals but, apart from marsupials, viviparity is also found among the reptiles, amphibians, and fishes. In the latter it is especially common in the Chondrichthyes, and it displays some interesting diversity in teleosts. The young in this group may, for instance, be retained not only in the oviducts but also in cavities in the ovaries and even in the ovarian follicles. In amphibians they may be retained in enlarged vocal sacs of male frogs and special pouches in the skin. It is questionable, however, whether such refuges constitute true viviparity, though they do appear to conform to the definition in the *Oxford English Dictionary*. Suitable contact between the maternal and embryonic tissues can be made at a number of sites in the body; abnormal pregnancies in women have even occurred that involve the development of the fetus in the peritoneal cavity. The separate evolution of viviparity on so many occasions suggests that it can be advantageous to the survival of a species. A reduction in the number of eggs produced and the quantity of yolk proteins clearly may confer some saving in energy. The chances of survival of a particular young would appear to be enhanced though this is largely a "numbers game": Lots of eggs also favor the ultimate survival of more individuals. Speed of development may be increased due to an assured and optimal supply of nutrients, for a longer period of time, and the maintenance of a suitable temperature. In homeotherms the role of the latter may be obligatory for survival, but even poikilotherms can, to some extent, regulate their body temperature, usually as a result of their behavior. Birds resort to incubating their eggs; however, this method of maintaining temperature usually restricts their movements.

Viviparity usually involves a relatively long period of commitment on the part of the female parent, and also sometimes the male, to succor the young. The period of pregnancy varies in different species; usually when it is prolonged the young are born in a better developed and more independent condition. This effect is especially apparent when one compares the newborn of marsupials with those of eutherians. Postnatal care of the young is not unique to viviparous species but it is a common accompaniment of this condition. It may involve the evolution of special arrangments for postnatal feeding, such as lactation by the mammary glands in mammals.

The morphological nature and degree of closeness of association of the embryo and parent vary in different species. In some, the egg may have adequate stores of yolk for its development and it appears to be merely incubated inside the parent. Such examples of ovoviviparity occur widely in nonmammals and in the monotremes. The sustenance of the young *in utero*, however, usually involves supplying nutrients and providing a "sink" for the excretion of some products of metabolism. Thus, blood systems show an association that allows molecular exchanges to occur between mother and embryo.

Prolonged periods of gestation and care of the young require a number of physiological adjustments by the parent and even the offspring. Many of these require the intervention of hormones. (It has been said that the embryo develops in "a sea of hormones.") They include:

i. The preparation and maintenance of the epithelium lining the oviduct and uterus for the reception and sustenance of the embryo, events that initially require estrogens, and then, mainly, progesterone.

ii. Control of the motility of the muscle that envelops the gonaducts; there is a relative immobility during gestation, but an increase in contractility at the time of birth. Progesterone reduces such activity while estrogens enhance it. At birth the latter steroids may enhance the formation of prostaglandins, which contract the uterus. Neurohypophysial hormones may also contribute to the initiation of contractions. A relaxation of, or softening of, the connective tissue of the cervix occurs under the influence of relaxin and prostaglandins.

iii. Care of the young following their birth involves hormonal regulation of parental behavior. The development of the mammary glands and the initiation and maintenance of lactation take place under the influence of estrogens, progesterone, placental lactogen, prolactin, and oxytocin.

iv. Concurrent reproductive activity is limited during gestation; this involves the hormonal suppression of further reproductive cycles and ovulation or, in some species, a delay in the implantation of the blastocyst. These effects involve actions of progesterone and prolactin.

Viviparity has been accompanied by the evolution of several important endocrine mechanisms:

i. *Internal fertilization* is essential for viviparity to occur and involves the evolution of suitable morphological structures to aid the intromission of the sperm directly into the female genital tract. Such penile structures are usually secondary sex characters, which come under the control of androgens.

ii. Changes in the *morphology of the female reproductive tract* may be necessary. For instance, the "plugging" role of the cervix is important in maintaining pregnancy and facilitating internal fertilization. Its activities are controlled by estrogens, progesterone, and relaxin.

iii. The life of the *corpus luteum*, which is an important source of progesterone, can be prolonged during pregnancy in eutherians. In marsupials it stays in a quiescent condition during lactation. It is suspected that it may also have a role to play in pregnancy in nonmammals, especially reptiles.

iv. The *placenta* may evolve endocrine mechanisms for the production of gonadotropins, which aid in the prolongation of the life of the corpus luteum, and progesterone, which supplements or even supplants its role. Estrogens are also produced that contribute to the process of birth. The endocrine association of the placenta and fetus (the fetoplacental unit) supplements its activities and facilitates the timing of parturition.

v. A prolonged contact with the maternal circulation may, potentially, be expected to influence the processes of *embryonic development*. The maternal plasma contains high concentrations of hormones, including sex steroids, which may affect the sexual differentiation of the reproductive apparatus and behavior. The mammalian fetus can, however, protect itself by forming a protein (α-fetoprotein) that binds, and so blocks, the actions of estrogens.

The role of hormones in maturation and development of the young

Reproduction can only be considered complete when the young have attained independence from the parents and are themselves able to propagate the species. The growth, differentiation, and maturation of the young, either in the egg or *in utero*, is a most remarkable process the intricacies of which were, it seems, at least until recently, more appreciated by biologists 50 years ago than in the succeeding period of time. The role of chemical substances in coordinating and directing such embryological processes is reminiscent of the action of hormones though such inductor substances are not usually classified as such and will not be dealt with here. There has, however, been speculation as to whether inductors that influence sexual differentiation are indeed sometimes identical to the gonadal steroid hormones.

It was appreciated more than 100 years ago that endocrine secretions could affect development (see Jost, 1971). The earliest observations were on the effects of congenital thyroid deficiency, which is called cretinism in man. Such a thyroid deficiency results in inadequate development of the nervous system, skeleton, and the reproductive organs. It seems that these hormones are, however, more important in early postnatal life than fetal development. Other hormones may be necessary earlier in embryonic development. In encephalectomized fetal rats and anencephalic human fetuses, the prenatal development of the adrenal cortex is retarded, apparently reflecting an absence of pituitary stimulation due to the lack of hypothalamic control. The parathyroids are also functional in the fetal rat, and removal of these glands results in hypocalcemia. The stage of embryonic development when such hormones may become important varies a great deal. It should also be recalled that hormones such as thyroxine and the steroid hormones may cross the placenta in viviparous species so that the maternal endocrines may contribute to the development of the fetus. In extreme situations, such transfer of hormones across the placenta can result in fetal abnormalities such as masculinization of female human babies born to mothers to whom large doses of progestins have been administered during pregnancy.

The effects of exogenous progestins on sexual differentiation were to a considerable extent a predictable phenomenon. The observations of Lillie in 1917 on the freemartin effect in cattle led to considerable speculation as to the role of fetal sex hormones in sexual differentiation. Lillie observed that in some cows bearing twin fetuses of opposite sexes the sexual differentiation of the genetic female was changed so that it was born with testes and a male gonaduct system. The external genitalia remained female in character and the animal was infertile. This effect has been attributed to the passage across the interdigitating fetal membranes of materials, possibly androgens, that effect the change. This interpretation as to the role of male sex hormones has not gone unchallenged and it has more recently been suggested that other factors may be involved.

These interesting observations, nevertheless, resulted in experimental testing of the ability of sex hormones to modify sexual differentiation in various other developing vertebrates. These experiments involve various techniques such as gonadectomy, parabiotically joining, thus crossing, the circulation of embryos of the opposite sex, cross-grafting of fetal gonads, and the administration of gonadal steroid hormones. Such manipulations often resulted in

considerable changes, even to the extent of reversing the recipient's predetermined genetic sex.

In mammals, gonadectomy of the genetic male, either *in utero* or, in some
species like rats, soon after they are born, results in subsequent sexual development as the female phenotype. Removal of the gonads of the young female
rat, however, makes no difference to the expressed phenotype, which remains
feminine. Thus, in mammals the development of the normal female is said to
be "neutral" as it apparently does not require further physiological intervention by the gonads. The male type, however, is said to be "inducible" as the
development of this phenotype involves the testis and the actions of its hormones. It has been said that the male is merely an induced female. This
pattern does not apply to all vertebrates. Thus, when embryonic female ducks
are gonadectomized they subsequently develop as phenotypic males. In this
species it is the female that is inducible, and this is controlled by the ovaries.
This situation applies to birds, reptiles, and many amphibians and reflects the
chromosomal constitution of the gametes. In mammals the males are heterozygous (XY) while in many nonmammals the females are heterozygous. It is
the Y-chromosome that apparently exerts a positive effect and determines the
differentiation of the embryonic gonad and subsequent sexual differentiation
of the other organs.

Before describing some of these effects further, we should recall the patterns of sexual differentiation in embryonic life (see Jost, 1979; Short, 1979).
The primordial gonad has the potential to develop into either a testis or an
ovary and is divided into two distinct regions: a cortex, which is the definitive
ovary, and a medulla, which may become the testis. This segmentation of the
primordial gonad applies to all vertebrates except cyclostomes and teleosts
where a single structure is present.

Sexual differentiation and development occur in two main phases:

i. The development of the primordial gonad into a testis or ovary. This is
predetermined by the genetic constitution of the animal. In man, this process
is completed at about 8 weeks after conception.

ii. Subsequent development of the genetic phenotype then occurs in which
hormones produced by the differentiated gonads then play an important primary role in the heterogametic sex (males in mammals and females in many
nonmammals). This process occurs both before birth, when the differentation
of the reproductive system occurs, and also after birth, when this process is
continued. These effects include both morphological and behavioral development.

The nature of the active role of the Y-chromosome in effecting the appropriate differentiation of the embryonic gonad has been the subject of some
fascinating research. It has been postulated that it involves the formation of a
substance (or substances) called an "organizer(s)." This presumed gene prod-

uct has been somewhat elusive, but one interesting candidate has been identified (see Wachtel et al., 1975; Haseltine and Ohno, 1981). There is a strain of
mice in which the females reject skin grafts from the male, but not from other
female mice. This response has been shown to be associated with the presence
of an antibody, resulting from the action of an antigen (histocompatibility-H)
from the male tissue. This antigen's formation is related to the presence of the
Y-chromosome; it is thus called the H–Y antigen. This gene product is a cell
surface antigen that has been identified in many other species but usually only
in the heterogametic sex (see Wachtel, Koo, and Boyse, 1975). It does not
appear to be species specific. It has been proposed that the H–Y antigen is a
primary determinant of the differentiation of the gonads of mammals into a
testis or, in many nonmammals, into an ovary. However, it is widely considered that other genes on the X- and Y-chromosomes, as well as the auto-
somes, are all necessary for the complete differentiation of the gonads.

The differentiation of the gonad into a testis or ovary ushers in the next
phase of sexual differentiation when sexual dimorphism becomes apparent. In
mammals this process is controlled by hormones that the testis can now
produce (see Wilson, George, and Griffin, 1981). The primordial genital tract
consists of the Wolffian and Müllerian ducts and the urogenital sinus and
tubercle (Fig. 9.26). The Sertoli cells of the newly formed testis secrete the
anti-Müllerian hormone that initiates the regression of the Müllerian ducts.
This hormone has not yet been completely characterized but it appears to be a
glycoprotein with a molecular weight of about 70,000 (see Josso, Picard, and
Tran, 1977). It is only present during fetal life and for a brief period after
birth. Monoclonal antibodies raised to fetal testicular bovine anti-Müllerian
hormone antagonize the activity of the bovine hormone but not the anti-
Müllerian hormone from rat testes (Vigier, Picard, and Josso, 1982). Subsequently, the Leydig cells of the testis produce testosterone and this steroid
promotes the development of the Wolffian duct into the vas deferens, the
epididymis, and seminal vesicles. The differentiation of the external genitalia
depends on the later acquisition of the ability of the urogenital sinus and
tubercle to convert testosterone to 5α-dihydrotestostrone. This event results in
the formation of the urethra and the penis. The early development of the
female reproductive apparatus appears to be independent of ovarian hormones;
the Müllerian duct goes on to develop into the oviducts and uterus while the
Wolffian duct degenerates. Subsequent growth and development of the secondary female sex characters, however, do involve estrogens. The administration of steroids with androgenic activity, such as progestins, to the developing
female can result in the acquisition of some male characters, for example, the
enlargement of the clitoris in the newborn to penile proportions.

The hormonally controlled pattern of sexual differentiation is not the same
in all vertebrates; in species where the female is heterozygous (XY) the

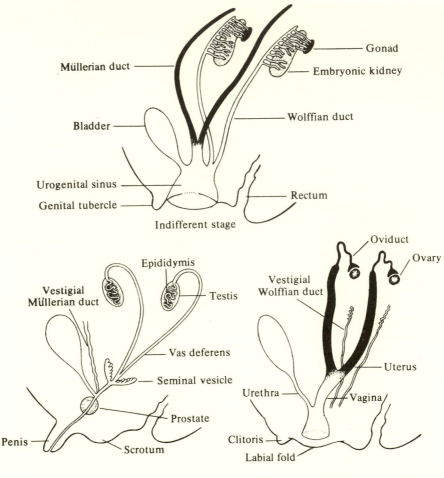

Fig. 9.26. Differentiation of the sexual organs in the embryonic mammal. The appropriate development of the genetic male is dependent on the presence of androgens. This pattern can be changed by removal or antagonism of the natural hormones or in the female by injecting hormones more typical of the opposite sex. (From Frye, 1967, by permission of Macmillan Publishing Co., Inc., New York. Copyright © 1967.)

development of the gonaducts may be different. In birds and reptiles the Wolffian duct persists, as an excretory duct, in both sexes, and this process is independent of testicular or ovarian hormones (see Price et al., 1975). The regression of the Müllerian ducts appears to be promoted by testicular or even ovarian hormones (in birds the right duct regresses in the female), but this process has not been completely characterized.

It is possible to distinguish such effects of hormones on sexual differentiation pharmacologically by administering a specific androgen antagonist, cyproterone. When this steroid is injected into male fetal rats (Elgar, Neumann, and Berswordt-Wallrabe, 1971), differentiation of the Wolffian ducts and the vesicular and prostate glands is inhibited, confirming their normal dependence on androgens. The differentiation of the gonads and the regression of the embryonic Müllerian ducts are, however, not affected.

The appropriate patterns of sexual behavior are determined early in life (during the "critical period") either before or shortly after birth. In mammals the female pattern is predetermined but the male one, for its expression, must be induced under the direction of secreted androgens and estrogens. In birds, reptiles, and some amphibians the opposite occurs: The female pattern must be subsequently developed. If male rats are castrated within the first few days of birth, the pattern of their subsequent sexual development then proceeds in a female manner (see Dorner, 1979; Brawer and Naftolin, 1979). The release of gonadotropins subsequently shows the female type cycling activity, and a positive-feedback response to injected estrogens can even be demonstrated. These genetically male rats also assume female mating postures. If such rats are given androgens in adult life they display sexual interest, but this more often involves members of the same sex. On the other hand, if young female rats are exposed to androgens during the critical period they will develop many aspects of male behavior. When young female marmoset monkeys are injected with testosterone they then display, before puberty, typically male "rough and tumble" behavior when they play (Abbott and Hearn, 1979). After puberty their behavior is somewhat unpredictable and displays some aspects that are both male and female. There has been considerable speculation regarding the effects of hormonal imbalance during the critical period of sexual development in man (see Dorner, 1979; Ehrhardt and Meyer-Bahlburg, 1981; Rubin, Reinisch, and Haskett, 1981). Thus, it has been suggested that an insufficiency of androgens during this period in genetic males may contribute to subsequent homosexual behavior. In birds, a male pattern of behavior is predetermined (the sex chromosomes in the male are homozygous, XX, instead of heterozygous, XY, as in mammals). One can alter this type of behavior to the female one by injecting sex steroids into the eggs (Adkins, 1978).

The cellular mechanisms of such effects on sexual differentiation of behavior involve the interaction of the sex steroids with particular sites in the brain (see Lieberburg et al., 1978; Maclusky and Naftolin, 1981; McEwen, 1981). By the use of autoradiography it is possible to localize the sites of accumulation of administered androgens and estrogens. There are several such sites in the brain but an especially important one in newborn rats is associated with the hypothalamus. Differentiation of the male pattern of behavior can be induced by androgens and also, surprisingly, by estrogens. Not all androgens are effective; 5α-dihydrotestosterone (DHT), which is normally the active form of testosterone, is ineffective in inducing the development of the male pattern of behavior. It is also interesting that the actions of both estrogens and androgens on this process can be blocked by antiestrogen drugs. The steroid that is active in the hypothalamus appears to be estradiol; androgens are only effective when they can be converted (aromatized) to estrogens. (Dihydrotestosterone cannot be aromatized to estrogens.) An aromatase enzyme system has been identified in the hypothalamus of members of all the major classes of vertebrates except cyclostomes (Callard, Petro, and Ryan, 1978). In mammals the special actions of estrogens in inducing male-type sexual behavior pose potential problems to the female fetuses. They may be exposed to maternal estrogens, which would tend to masculinize them. It appears, however, that they are protected from their action by a plasma protein (α-fetoprotein), only present in early life, which can then bind estrogens and so prevent their action in the brain.

The actions of sex steroid hormones in determining the ultimate pattern of sexual behavior apparently involve the development of a pattern of neural circuitry, which then persists throughout reproductive life. Such effects may be of a morphological nature and involve the development of dendritic connections between neurons. Organ cultures of embryonic, undifferentiated, hypothalamic tissue (Toran-Allerand, 1978) show morphological changes when exposed to estrogens and androgens, which are consistent with this possibility.

Sex reversal (or sex inversion) is a particularly interesting phenomenon that has never been observed in mammals, but in some teleost fishes it occurs normally during the life cycle of many species (see Reinbloth, 1970, 1972). This sex change may be from a male into a female (*protandry*) or more commonly a female into a male (*protogyny*). The stimuli that result in such sex inversions are not well understood, but the removal of the male can initiate a protogynic sex change. The female fish then may become males that produce normal sperm. The injection of testosterone can mimic this transformation in wrasses (*Thalassoma bifasciatum*), but Reinbloth does not consider this to be proof that this is the normal mechanism mediating such a sex transformation. It is, however, an attractive hypothesis for the endocrinologist. Such a possibility is supported by the observation that in the protogynous symbranchid teleost *Monopterus albus*, the rice field eel, the gonads switch their steroid syntheses from a predominance of estrogens, in their female phase, to androgens in their male phase (Chan and Phillips, 1969).

True hermaphrodites, which can simultaneously produce both sperm and eggs and even be self-fertilizing, have been described most commonly among the teleost fishes. Hermaphrodism is the normal situation in a fascinating oviparous teleost, *Rivulus marmoratus*, which lives in tidal–pluvial zones on the coasts of Florida and the West Indies (see Harrington, 1968). The differentiation of males can be promoted when embryonic development proceeds at temperatures below 20°C, a situation that does not normally occur in this fish's native habitat. It is not known whether such sexual differentiation is mediated by endocrine changes under these conditions. The normal balance of sex hormones in hermaphroditic vertebrates constitutes rather an endocrine puzzle as it is difficult to conceive how different sex hormones can exist, be controlled, and act simultaneously within the same animal. This is an intriguing problem for the comparative endocrinologist.

The control of metamorphosis

Many fishes and amphibians exist in two or more distinct morphological and physiological forms during their life cycles. The transformation from one type to another is called metamorphosis. "True" metamorphosis is considered to occur in preparation for life in a different habitat, such as fresh water (as opposed to the sea), or an aquatic compared to a terrestrial existence. These changes may involve a migration that is sometimes associated with sexual maturation and reproduction.

An interesting account of metamorphosis in fishes is given by Barrington (1968). Lampreys (Cyclostomata) undergo a prolonged period of larval development in fresh water that lasts for about 5½ years. The metamorphosis of the ammocoete to the adult, which migrates to the sea, lasts for several weeks. Subsequently, lampreys again return to the rivers to breed and then undergo physiological changes, such as reflected by their inability to osmoregulate in seawater. The environmental and physiological events that determine these metamorphic changes are unknown. Experimental manipulation of the lampreys' thyroid physiology, by placing them in solutions containing either thyroid hormones or antithyroid substances, has yielded inconclusive results.

Many teleost fish also undergo a metamorphosis. The best-known examples are seen in eels and salmon. The leptocephalus larval eel is transformed into the elver, but little is known about the mechanism of this change that takes place in the sea and precedes migration into rivers. Salmon spawn in fresh water and the young parr, upon reaching a certain size, are transformed into smolt which migrate to the sea. Considerable endocrine changes occur at this time and these can be seen histologically in the pituitary and thyroid glands (Fontaine, 1954). The uptake of radioiodine by the thyroid increases early in metamorphosis and the levels of 17-hydroxycorticosteroids in the plasma of the parr are five times greater than in the smolt. It is not possible, however, to decide whether such endocrine changes initiate metamorphosis or merely occur as a part of the general maturation process.

The possibility that changes in the endocrine glands, especially the thyroid, may initiate metamorphosis in fishes arose from the observation that thyroid hormones can initiate metamorphosis in most amphibians. The profound and dramatic morphological changes that accompany the metamorphosis of anuran tadpoles into adult frogs and toads have been a source of wonder to biologists for a long time. The transformation from a purely aquatic animal, with no limbs or lungs, into a terrestrial beast that breathes and hops about on four legs is also accompanied by many physiological and biochemical changes. The larval life and metamorphosis of amphibians may be relatively short, from several weeks in desert-dwelling species where water is available for only a short time to as long as 3 years in bullfrogs. The factors that determine the time of metamorphosis are not clear; they are partly genetic but they can also be modified by the environment. Bullfrog tadpoles from the tropical southern parts of the United States may metamorphose before the beginning of the first winter after hatching, while those in northern areas may endure three winters before this change occurs. Possible environmental factors that influence the time of metamorphosis include nutrition, temperature, the salinity and acidity of the water where they live, the relative proximity of other tadpoles ("crowding"), and some experiments even suggest that light may stimulate this process (see Dent, 1968). One can foresee that such factors could exert their effects through the activation of endocrine glands, in this instance the thyroid through its hypothalamic and pituitary control mechanisms.

The feeding of thyroid gland extracts can produce metamorphosis in tadpoles far earlier than it would normally occur. Conversely, the administration of antithyroid drugs prolongs, or even prevents, metamorphosis. Natural metamorphosis in tadpoles is accompanied by a sudden increase in thyroid gland activity, as indicated by histological changes and an increase in the rate of uptake of radioactive iodine. There can be no doubt that the activity of the thyroid gland determines metamorphosis in tadpoles but this is only a part of the endocrine story. The thyroid in tadpoles, like that in other vertebrates, is under the control of TSH from the adenohypophysis. The injection of TSH into tadpoles thus also results in a premature metamorphosis. Hypophysectomized tadpoles do not metamorphose but grow larger and larger and attain "giant" proportions. The hypothalamus and median eminence, which are usually the next sites in the chain of thyroid control, are also involved in metamorphosis for when the tadpole pituitary is transplanted to the tail, metamorphosis is prevented. Such tadpoles are dark in color due to an uncontrolled release of MSH, and they grow more rapidly than usual.

The latter effect on growth is an important clue that probably, at least partly, reflects a lack of the hypothalamic inhibition of the release of prolactin. Injections of prolactin into bullfrog tadpoles have been shown to antagonize thyroxine-induced tail resorption and delay metamorphosis (Nicoll et al.,

1965; Etkin and Gona, 1967). The same effect has been achieved by grafts or extracts (Kikuyama, Yamamoto, and Mayumi, 1980) of tadpole pituitaries so that this effect probably also exists physiologically. Antisera prepared to amphibian prolactin can accelerate metamorphosis (Clemens and Nicoll, 1977). Prolactin is thought to antagonize the peripheral actions of thyroxine and also exerts an inhibitory (goitrogenic) effect on the thyroid gland. Such effects are apparently confined to these amphibians and do not occur, for instance, in mammals.

Etkin (1970), after a careful assessment, has provided a description of how metamorphosis is normally regulated in tadpoles. This synthesis is summarized in Fig. 9.27 and is based on the observations already described that have been correlated with histological changes that occur in the hypothalamus, pituitary, and thyroid gland. Metamorphosis is divided into three stages: (*a*) a period of rapid growth called premetamorphosis; (*b*) a time of reduced growth but increased differentiation called prometamorphosis; and (*c*) metamorphic climax when there are "explosive changes"; the tail is resorbed and the frog emerges and takes up a terrestrial existence. Premetamorphosis is characterized endocrinologically as a time when thyroid hormone secretion is low, which reflects the anatomical immaturity of the hypothalamus–pituitary axis and a low rate of TSH secretion. This condition is stabilized further by the presence of large amounts of prolactin, possibly reflecting an immaturity of the inhibitory influence of the hypothalamus (P-R-IH), which further inhibits any progress toward metamorphic change. With the progressive maturation of the hypothalamus–pituitary axis toward the beginning of prometamorphosis, TRH gradually increases and releases TSH. Antisera to TSH can thus slow metamorphosis (Eddy and Lipner, 1976). The thyroxine concentration thus rises progressively during prometamorphosis. The hypothalamus continues to mature under the influence of thyroxine (called a "positive feedback"), and when this is complete and the portal blood supply to the adenohypophysis is finally established, there is a massive stimulation of the thyroid, via TRH and TSH, and metamorphic climax ensues. During this last period the levels of thyroxine are thought to be declining (due to the operation of the adult negative-feedback inhibition of TRH release).

The unequivocal proof of this theory of tadpole metamorphosis will depend on direct measurements of the hormone levels in tadpoles, a difficult, but with modern techniques, not an impossible task. The predicted increases in the levels of thyroxine and triiodothyronine in the plasma of bullfrog tadpoles have been demonstrated (Regard, Taurog, and Nakashima, 1978). The concentration of TRH in the brain of *Xenopus* tadpoles has also been shown to rise prior to metamorphosis (King and Millar, 1981).

Newts and salamanders also undergo a metamorphic transformation but this is not as dramatic as in frogs and toads. Larval urodeles possess limbs, and the

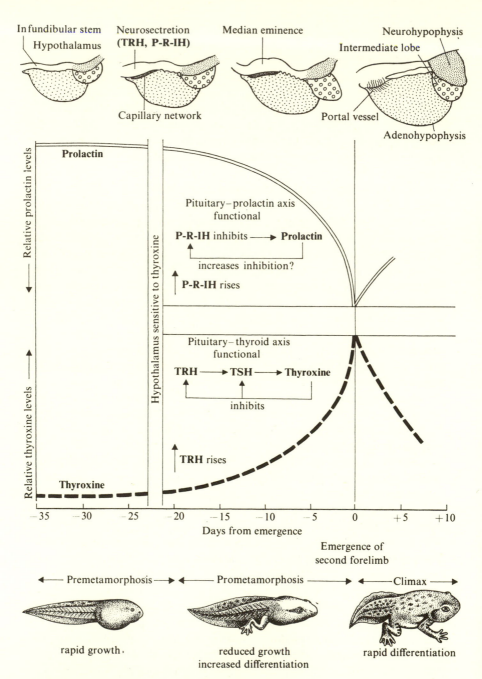

Fig. 9.27. Role of hormones in growth and development of the tadpole. *Top:* Development of the pituitary, especially in relation to establishing its connections to the hypothalamus. *Bottom:* Changes in activity and hormone concentrations of the pituitary–prolactin and pituitary–thyroid axes. Major events are acquisition by the hypothalamus of thyroxine sensitivity, hastening its development, and onset of definitive roles of TRH and P-R-IH, so that TSH and thyroxine secretion increases while prolactin decreases. (Based on Etkin, 1970.)

most prominent morphological change is often the loss of the external gills. Other, less obvious changes nevertheless also occur. This metamorphosis is under pituitary–thyroid control and can be inhibited by prolactin (Gona and Etkin, 1970). A number of species of urodeles are neotenic; they retain their larval characters and never attain the normal adult morphology but can breed while in the larval form. Well-known examples of neoteny in urodeles are seen in axolotls (*Ambystoma mexicanum*) and mudpuppies (*Necturus maculosus*).

It caused some comment in the biological world when it was found that feeding thyroid glands to axolotls caused them to metamorphose into adult salamanders. The same effect can be elicited by TSH, indicating that the natural disability of these urodeles is due to a pituitary or hypothalamic rather than a thyroid deficiency. These interesting experiments (see Dent, 1968) involved grafting of pituitaries from metamorphosing tiger salamanders (*Ambystoma tigrinum*) into axolotls, which then metamorphosed themselves. The converse experiment, transplanting axolotl pituitaries into hypophysectomized tiger salamanders, never resulted in metamorphosis of the recipients, which thereafter remained as larvae. The injection of thyroxine into the hypothalamus of neotenic tiger salamanders results in their metamorphosis (Norris and Gern, 1976). It was suggested that this hormone removes a hypothalamic block of the endocrine system, which is present in the neotenic animal. This response does not, however, appear to involve changes in TRH levels (Sawin et al., 1978). It is possible that a decline in plasma prolactin concentration is involved in these salamanders. Other neotenous urodeles, like the mudpuppy, do not metamorphose when given thyroid hormone; their tissues are not responsive to this hormone.

An interesting departure from the usual amphibian pattern is shown by the eastern spotted newt, *Notophthalmus (Diemictylus) viridescens*. This animal undergoes two metamorphoses during its life cycle. The first is from the aquatic larva to the adult, or red eft, that lives a terrestrial existence for 1–3 years, but undergoes a second metamorphosis when it returns to water to breed. This transformation is accompanied by morphological changes in the skin, and the newt also loses its tongue. The second metamorphosis is thought to be due to a rise in prolactin levels that is called the "eft water-drive effect" of this hormone (this has been referred to earlier). This change in the prolactin activity is accompanied by a decline in thyroid hormone (Gona, Pearlman, and Etkin, 1970).

The dramatic effects of thyroxine on the development of amphibians have contributed considerably to our ideas about the role of the thyroid in mammalian development. It is now generally considered, however, that the two processes are not analogous and that the "morphogenetic" effect of thyroid hormone in amphibians represents a special application of these hormones to their physiology. The aquatic and terrestrial phases in an amphibian's life cycle can be viewed as separate processes each of which has undergone evolutionary modification consistent with life in each environment. The transformation of one form to another has no parallel in other tetrapods. The use of the thyroid hormone, and probably also prolactin, to control such metamorphosis is an outstanding example of how adaptable the endocrine system is to the needs of evolutionary change.

Conclusions

Apart from nutrition, reproduction probably involves the most complex processes of humoral coordination that exist in vertebrates. It can also be considered as being under "multihormonal control." Reproduction, and the preparations for this event, involve a variety of tissues, organs, and several distinct, accurately timed physiological events (for instance, impregnation, ovulation, and parturition) and usually take a prolonged period of time to reach fruition. Hormones are especially suited to such needs for coordination. The nature of the reproductive process displays considerable morphological and physiological variation and certain types of mechanisms, such as viviparity in mammals and oviparity in birds, predominate in certain systematic groups, in which they are a feature, and involve special endocrine mechanisms. There are, however, many examples of what may be parallel evolution, such as viviparity in fishes and reptiles in which the special roles of hormones may have evolved independently but have a similar end result.

The diversity in reproductive processes does not appear to have involved many changes in the structure of the hormones themselves. The gonadal steroid hormones have the same structure in all vertebrates, though those from the pituitary show distinct differences that are probably of more consequence as regards limiting their action to a certain species than reflecting any functional predilection to coordinate novel processes. It is, however, of special note that pituitary gonadotropin appears to exist as a single molecule in some vertebrates, especially the fishes, but in most tetrapods it now seems likely that two distinct hormones, one with LH and the other FSH activity, have emerged. This dichotomy presumably allows for the operation of a more specific and precise control mechanism, but the physiological differences that, it would appear, must result are unknown. Another notable endocrine novelty has emerged in eutherian mammals and is the ability of the placenta to act as an additional site for steroid hormone synthesis and to form two pituitary-like hormones, chorionic gonadotropin and placental lactogen. There is, however, considerable interspecific variability in the endocrine function of the eutherian placenta, and it is suspected that this may have arisen on several separate evolutionary occasions and been perpetuated according to the requirements of the particular species involved.

Hormones contribute directly to the control of differentiation and growth of the embryo and behavioral and physiological processes concerned with the care of the young. The control of metamorphosis in amphibians is a unique and dramatic example of how endocrines can influence development, but they also contribute more ubiquitously, especially to sexual differentiation, in all groups of vertebrates. The phylogenetically novel process of lactation in mammals is a clear example of the evolution of endocrine function, though hormones may also contribute to the process of parental care of the young in a variety of other vertebrates.

References

Abbott, D. H. and Hearn, J. P. (1979). The effects of neonatal exposure to testosterone on the development of behaviour in female marmoset monkeys. In *Sex, Hormones, and Behaviour*, Ciba Foundation Symposium 62, pp. 299–316, Amsterdam: Excerpta Medica.

Abe., K., Robison, G. A., Liddle, G. W., Butcher, R. W., Nicholson, W. E., and Baird, C. E. (1969). Role of cyclic AMP in mediating the effects of MSH norepinephrine and melatonin on frog skin color. *Endocrinology* **85**, 674–682.

Acher, R. (1978). Molecular evolution of neurohypophyseal hormones and neurophysins. In *Neurosecretion and neuroendocrine activity. Evolution, structure and function* (edited by W. Bargmann, A. Oksche, A. Polenov, and B. Scharrer), pp. 31–43, New York: Springer-Verlag.

—— (1980). Molecular evolution of biologically active polypeptides. *Proc. Roy. Soc.* **B210**, 21–43.

Acher, R., Chauvet, J., and Chauvet, M. T. (1972). Phylogeny of the neurohypophysial hormones. Two new active peptides isolated from a cartilaginous fish, *Squalus acanthias. Eur. J. Biochem.* **29**, 12–19.

Adelson, J. W. (1971). Enterosecretory proteins. *Nature, Lond.* **229**, 321–325.

Adkins, E. K. (1978). Sex steroids and the differentiation of avian reproductive behavior. *Amer. Zool.* **18**, 501–509.

Agus, Z. S., Gardner, L. B., Beck, L. H., and Goldberg, M. (1973). Effects of parathyroid hormone on renal tubular reabsorption of calcium, sodium and phosphate. *Amer. J. Physiol.* **224**, 1143–1148.

Al-Dujaili, E.A.S., Hope, J., Estivariz, F., Lowry, P. J., and Edwards, C.R.W. (1981). Circulating human pituitary pro-γ-melanotropin enhances the adrenal response to ACTH. *Nature, Lond.* **291**, 156–159.

Alexandrova, M. and Soloff, M. S. (1980). Oxytocin receptors and parturition. 1. Control of oxytocin receptor concentration in the rat myometrium at term. *Endocrinology* **106**, 730–735.

Al-Khateeb, A. and Johnson, E. (1971). Seasonal changes of pelage in the vole (*Microtus agrestis*). 1. Correlation with changes in the endocrine glands. *Gen. Comp. Endocr.* **16**, 217–228.

Alvarado, R. H. and Johnson, S. R. (1966). The effects of neurohypophysial hormones on water and sodium balance in larval and adult bullfrogs (*Rana catesbeiana*). *Comp. Biochem. Physiol.* **18**, 549–561.

Amoroso, E. C., Heap, A. B., and Renfree, M. B. (1979). Hormones and the evolution of viviparity. In *Hormones and Evolution* (edited by E. J. W. Barrington), pp. 925–989. New York: Academic Press.

Amoroso, E. C. and Marshall, F.H.A. (1960). External factors in sexual periodicity. In *Marshall's Physiology of Reproduction* (edited by A. S. Parkes), vol. I (Pt. 2), pp. 707–831. London: Longmans.

Anderson, L. L. (1973). Effects of hysterectomy and other factors on luteal function. In *Handbook of Physiology*, Sect. 7 *Endocrinology*, vol. II, *Female reproductive system* (Pt. 2), pp. 69–86. Washington, D.C.: American Physiological Society.

Anika, S. M., Houpt, T. R., and Houpt, K. A. (1981). Cholecystokinin and satiety in pigs. *Amer. J. Physiol.* **240**, R310–R318.

Anon. (1970). Effects of sexual activity on beard growth in man. *Nature, Lond.* **226**, 869–870.

Ariëns Kappers, J. (1965). Survey of the innervation of the epiphysis cerebri and the accessory pineal organs of vertebrates. *Prog. Brain Res.* **10**, 87–151.

(1970). The pineal organ: An introduction. In *The Pineal Gland* (edited by G.E.W. Wolstenholme and J. Knight), pp. 3–25. London: Churchill Livingstone.

Armstrong, D. T. and Kennedy, T. G. (1972). Role of luteinizing hormones in regulation of the rat estrous cycle. *Amer. Zool.* **12**, 245–255.

Assem, H. and Hanke, W. (1981). Cortisol and osmotic adjustment of the euryhaline teleost, *Sarotherodin mossambicus*. *Gen. Comp. Endocr.* **43**, 370–380.

Augee, M. L. and McDonald, I. R. (1973). Role of the adrenal cortex in the adaptation of the monotreme *Tachyglossus aculeatus* to low environmental temperature. *J. Endocr.* **58**, 513–523.

Aurbach, G. D., Keitmann, H. T., Niall, H. D., Tregear, G. W., O'Riordan, J.L.H., Marcus, R., Marx, S. J., and Potts, J. T. (1972). Structure, synthesis, and mechanism of action of parathyroid hormone. *Rec. Prog. Hormone Res.* **28**, 353–392.

Axelrod, J. (1974). The pineal gland: a neurochemical transducer. *Science* **184**, 1341–1348.

Axelrod, J., Wurtman, R. J., and Snyder, S. H. (1965). Control of hydroxyindole *O*-methyltransferase activity in the rat pineal gland by environmental lighting. *J. Biol. Chem.* **240**, 949–954.

Azukizawa, M., Kurtzman, G., Pekary, A. E., and Hershman, J. M. (1977). Comparison of the binding characteristics of bovine thyrotropin and human chorionic gonadotropin to thyroid plasma membranes. *Endocrinology* **11**, 1880–1889.

Baber, E. C. (1876). Contributions to the minute anatomy of the thyroid gland of the dog. *Proc. Roy. Soc.* **24**, 240–241.

Babiker, M. M. and Rankin, J. C. (1973). Effects of neurohypophysial hormones on renal function in the freshwater- and sea-water-adapted eel (*Anguilla anguilla* L.) *J. Endocr.* **57**, xi–xii.

(1978). Neurohypophysial hormonal control of kidney function in the European eel (*Anguilla anguilla* L.) adapted to sea-water or fresh-water. *J. Endocr.* 76, 347–358.

Bach, J-F. (1977). Thymic hormones: biochemistry, and biological and clinical activities. *Ann. Rev. Pharmacol. Toxicol.* **17**, 281–291.

Bach, J-F., Dardenne, M., Papiernik, M., Barois, A., Levasseur, P., and Le Brigand, H. (1972) Evidence for a serum-factor secreted by the human thymus. *Lancet* **2**, 1056–1058.

Bagnara, J. T. (1969). Responses of pigment cells of amphibians to intermedin. *Colloq. Int. C.N.R.S.* **177**, 153–158.

Bagnara, J. T. and Hadley, M. E. (1970). Endocrinology of the amphibian pineal. *Amer. Zool.* **10**, 201–216.

(1972). *Chromatophores and Color Change*. Englewood Cliffs, N.J.: Prentice-Hall.

Baker, B. I. and Ball, J. N. (1975). Evidence for a dual pituitary control of teleost melanophores. *Gen. Comp. Endocr.* **25**, 147–152.

Baker, B. I. and Rance, T. A. (1981). Differences in concentrations of plasma cortisol in the trout and the eel following adaptation to black or white backgrounds. *J. Endocr.* **89**, 135–140.

Baksi, S. N. and Kenny, A. D. (1977). Vitamin D₃ metabolism in immature Japanese quail: effects of ovarian hormones. *Endocrinology* **101**, 1216–1220.

Balazs, R., Cocks, W. A., Eayrs, J. T., and Kovacs, S. (1971). Biochemical effects of thyroid hormones on the developing brain. In *Hormones in Development* (edited by M. Hamburgh and E.J.W. Barrington), pp. 357–359. New York: Appleton-Century-Crofts.

Baldwin, G. F. and Bentley, P. J. (1980). Calcium metabolism in bullfrog tadpoles (*Rana catesbeiana*). *J. Exp. Biol.* **88**, 357–365.

   (1981). A role for skin in Ca metabolism of frogs? *Comp. Biochem. Physiol.* **68A**, 181–185.

Baldwin, R. L. (1969). Development of milk synthesis. *J. Dairy Sci.* **52**, 729–736.

Ball, J. N. and Baker, B. I. (1969). The pituitary gland: anatomy and histophysiology. In *Fish Physiology* (edited by W. S. Hoar and D. J. Randall), vol. II, *The Endocrine System*, pp. 1–110. New York: Academic Press.

Ball, J. N., Chester Jones, I., Forster, M. E., Hargreaves, G., Hawkins, E. F., and Milne, K. P. (1971). Measurement of plasma cortisol levels in the eel *Anguilla anguilla* in relation to osmotic adjustments. *J. Endocr.* **50**, 75–96.

Ball, J. N. and Ensor, D. M. (1965). Effect of prolactin on plasma sodium in the teleost, *Poecilia latipinna*. *J. Endocr.* **32**, 269–270.

   (1969). Specific action of prolactin on plasma sodium levels in hypophysectomized *Poecilia latipinna* (Teleostei). *Gen. Comp. Endocr.* **8**, 432–440.

Ball, J. N. and Ingleton, P. M. (1973). Adaptive variations in prolactin secretion in relation to external salinity in the teleost *Poecilia latipinna*. *Gen. Comp. Endocr.* **20**, 312–325.

Bar, A. and Norman, A. W. (1981). Studies on the mode of action of calciferol. XXXIV. Relationship of the distribution of 25-hydroxyvitamin D₃ metabolites to gonadal activity and egg shell formation in the quail. *Endocrinology* **109**, 950–955.

Barajas, L. (1979). Anatomy of the juxtaglomerular apparatus. *Amer. J. Physiol.* **237**, F333–F343.

Bargmann, W. (1943). Die Epiphysis cerebri. *Handbuch der mikroskopischen Anatomie des Menschen*. Vol. VI (4). Berlin: Springer-Verlag.

Barrington, E.J.W. (1942). Blood sugar and the follicles of Langerhans in the ammocoete larva. *J. Exp. Biol.* **19**, 45–55.

   (1962). Hormones and vertebrate evolution. *Experientia* **18**, 201–210.

   (1968). Metamorphosis in lower chordates. In *Metamorphosis, A Problem in Developmental Biology* (edited by W. Etkin and L. I. Gilbert), pp. 223–270. New York: Appleton-Century-Crofts.

Barrington, E.J.W. and Dockray, G. J. (1970). The effect of intestinal extracts of lampreys (*Lampetra fluviatilis* and *Petromyzon marinus*) on pancreatic secretion in the rat. *Gen. Comp. Endocr.* **14**, 170–177.

   (1972). Cholecystokinin-pancreozymin-like activity in the eel *Anguilla anguilla*. *Gen. Comp. Endocr.* **19**, 80–87.

Basu, S. L. (1969). Effects of hormones on the salientian spermatogenesis *in vivo* and *in vitro*. *Gen. Comp. Endocr. Suppl.* **2**, 203–213.

Bates, R. W., Miller, R. A., and Garrison, M. M. (1962). Evidence in the hypophysectomized pigeon of a synergism among prolactin, growth hormone, thyroxine, and prednisone upon weight of the body, digestive tract, kidney and fat stores. *Endocrinology* **71**, 345–360.

Baumann, G., Marynick, S. P., Winters, S. J., and Loriaux, D. L. (1977). The effect of osmotic stimuli on prolactin secretion and renal water excretion in normal man and in chronic hyperprolactinemia. *J. Clin. Endocr. Metab.* **44**, 199–202.

Bayliss, W. M. and Starling, E. H. (1902). The mechanism of pancreatic secretion. *J. Physiol., Lond.* **28**, 325–353.

(1903). On the uniformity of the pancreatic mechanism in vertebrata. *J. Physiol., Lond.* **29**, 174–180.

Bedarkar, S., Turnell, W. G., Blundell, T. L., and Schwabe, C. (1977). Relaxin has conformational homology with insulin. *Science* **270**, 449–451.

Bedrak, E., Harvey, S., and Chadwick, A. (1981). Concentrations of pituitary, gonadal and adrenal hormones in serum of laying and broody white rock hens (*Gallus domesticus*). *J. Endocr.* **89**, 197–204.

Bélanger, L. F., Dimond, M. T., and Copp, D. H. (1973). Histological observations on bone and cartilage of growing turtles treated with calcitonin. *Gen. Comp. Endocr.* **20**, 297–304.

Bellamy, D. and Leonard, R. A. (1965). Effect of cortisol on the growth of chicks. *Gen. Comp. Endocr.* **5**, 402–410.

Bennett, V., O'Keefe, E., and Cuatrecasas, P. (1975). Mechanism of action of cholera toxin and the mobile receptor theory of hormone receptor-adenylate cyclase interactions. *Proc. Natl. Acad. Sci., USA* **72**, 33–37.

Bentley, P. J. (1962). Studies on the permeability of the large intestine and urinary bladder of the tortoise (*Testudo graeca*) with special reference to the effects of neurohypophysial and adrenocortical hormones. *Gen. Comp. Endocr.* **2**, 323–328.

(1966). Hyperglycaemic effect of neurohypophysial hormones in the chicken, *Gallus domesticus*. *J. Endocr.* **34**, 527–528.

(1969). Neurohypophysial function in Amphibia: hormone activity in the plasma. *J. Endocr.* **43**, 359–369.

(1971). *Endocrines and Osmoregulation. A Comparative Account of the Regulation of Water and Salt in Vertebrates.* New York: Springer-Verlag.

(1972). Introductory remarks. Symposium on endocrinology and osmoregulation. *Fedn. Proc.* **31**, 1583–1586.

(1973). Osmoregulation in the aquatic urodeles *Amphiuma means* (the congo eel) and *Siren lacertina* (the mud eel). Effects of vasotocin. *Gen. Comp. Endocr.* **20**, 386-392.

Bentley, P. J. and Follett, B. K. (1963). Kidney function in a primitive vertebrate, the cyclostome *Lampetra fluviatilis*. *J. Physiol., Lond.* **169**, 902-918.

(1965). The effects of hormones on the carbohydrate metabolism of the lamprey *Lampetra fluviatilis*. *J. Endocr.* **31**, 127–137.

Bentley, P. J. and Yorio, T. (1979). Do frogs drink? *J. Exp. Biol.* **79**, 41–46.

Berde, B. and Boissonnas, R. A. (1968). Basic pharmacological properties of synthetic analogues and homologues of the neurohypophysial hormones. In *Neurohypophysial Hormones and Similar Polypeptides* (edited by B. Berde), pp. 802–870. Berlin: Springer-Verlag.

Berelowitz, M., Szabo, M., Frohman, L. A., Chu, L., and Hintz, R. L. (1981). Somatomedin-C mediates growth hormone negative feedback by effects on both hypothalamus and pituitary. *Science* **312**, 1279–1281.

Bergland, R., Blume, H., Hamilton, A., Monica, P., and Paterson, R. (1980). Adrenocorticotropic hormone may be transported directly from the pituitary to the brain. *Science* **210**, 541–543.

Bergland, R. M. and Page, R. B. (1979). Pituitary-brain vascular relations: a new paradigm. *Science* **204**, 18–24.

Berlind, A. (1972*a*). Teleost caudal neurosecretory system: release of urotensin II from isolated urophyses. *Gen. Comp. Endocr.* **18**, 557–571.

(1972*b*). Teleost caudal neurosecretory system: sperm duct contraction induced by urophysial material. *J. Endocr.* **52**, 567–574.

Berlind, A., Lacanilao, F., and Bern, H. A. (1972). Teleost caudal neurosecretory system: effects of osmotic stress on urophysial proteins and active factors. *Comp. Biochem. Physiol.* **42A**, 345–352.

Bern, H. A. (1972). Comparative endocrinology – The state of the field and art. *Gen. Comp. Endocr. Suppl.* **3**, 751–761.

Bern, H. A. and Nicoll, C. S. (1968). The comparative endocrinology of prolactin. *Rec. Prog. Hormone Res.* **24**, 681–713.

(1969). The zoological specificity of prolactin. *Colloq. Int. C.N.R.S.* **177**, 193–202.

Bewley, T. A. and Li, C. H. (1970). Primary structures of human pituitary growth hormone and sheep pituitary lactogenic hormone compared. *Science* **168**, 1361–1362.

Bikle, D. D., Spencer, E. M., Burke, W. H., and Rost, C. R. (1980). Prolactin but not growth hormone stimulates 1,25-dihydroxyvitamin D_3 production by chick renal preparations *in vitro. Endocrinology* **107**, 81–84.

Binkley, S. (1979*a*). Pineal rhythms *in vivo* and *in vitro. Comp. Biochem. Physiol.* **64A**, 201–206.

(1979*b*). A time keeping enzyme in the pineal gland. *Scientific Amer.* **240**, 66–71.

Binkley, S. and Geller, E. B. (1975). Pineal enzymes in chickens: development of daily rhythmicity. *Gen. Comp. Endocr.* **27**, 424–429.

Blair-West, J. R., Coghlan, J. P., Denton, D. A., Gibson, A. P., Oddie, C. J., Sawyer, W. H., and Scoggins, B. A. (1977). Plasma renin activity and blood corticosteroids in the Australian lungfish *Neoceratodus forsteri. J. Endocr.* **74**, 137–142.

Blair-West, J. R., Coghlan, J. P., Denton, D. A., Nelson, J. F., Orchard, E., Scoggins, B. A., Wright, R. D., Myers, K., and Junquera, C. L. (1968). Physiological, morphological and behavioural adaptation to a sodium deficient environment by native Australian and introduced species of animals. *Nature, Lond.* **217**, 922–928.

Blair-West, J. R., Coghlan, J. P., Denton, D. A., Funder, J. W., Scoggins, B. A., and Wright, R. D., (1971). The effect of the heptapeptide (2–8) and the hexapeptide (3–8) fragments of angiotensin II on aldosterone secretion. *J. Clin. Endocr. Metab.* **32**, 575–578.

Blair-West, J. R. and Gibson, A. (1980). The renin-angiotensin system in marsupials. In *Comparative physiology: primitive mammals* (edited by K. Schmidt-Nielsen, L. Bolis, and C. R. Taylor), pp. 297–306. Cambridge University Press.

Bloom, S. R. and Polak, J. M. (1978). Gut hormone overview. In *Gut hormones* (edited by S. R. Bloom), pp. 3–18. London: Churchill Livingstone.

Blum, J. J. (1967). An adrenergic control system in *Tetrahymena. Proc. Natl. Acad. Sci., USA* **58**, 81–88.

Blundell, T. L. and Wood, S. P. (1975). Is the evolution of insulin Darwinian or due to selectively neutral mutation? *Nature, Lond.* **257**, 197–203.

Boehlke, K. W., Church, R. L., Tiemeier, O. W., and Eleftheriou, B. E. (1966). Diurnal rhythm in plasma glucocorticoid levels in the channel catfish (*Ictalurus punctatus*). *Gen. Comp. Endocr.* **7**, 18–21.

Boelkins, J. N. and Kenny, A. D. (1973). Plasma calcitonin levels in Japanese quail. *Endocrinology* **92**, 1754–1760.

Bona-Gallo, A., Licht, P., MacKenzie, D. S., and Lofts, B. (1980). Annual cycles in levels of pituitary and plasma gonadotropin, gonadal steroids, and thyroid activity in the chinese cobra (*Naja naja*). *Gen. Comp. Endocr.* **42**, 477–493.

Bouillon, R., Van Baelen, H., Rombauts, W., and de Moor, P. (1978). The isolation and characterization of the vitamin D-binding protein from rat serum. *J. Biol. Chem.* **253**, 4426–4431.

Bourne, A. R. and Seamark, R. F. (1973). Seasonal changes in testicular function in the lizard *Tiliqua rugosa*. *J. Endocr.* **57**, x.

Bower, A., Hadley, M. E., and Hruby, V. J. (1974). Biogenic amines and control of melanophore stimulating hormone release. *Science* **184**, 70–72.

Bradley, A. J., McDonald, I. R., and Lee, A. K. (1980). Stress and mortality in a small marsupial (*Antechinus stuartii*, Macleay). *Gen. Comp. Endocr.* **40**, 188–200.

Bradshaw, S. D. (1972). The endocrine control of water and electrolyte metabolism in desert reptiles. *Gen. Comp. Endocr. Suppl.* **3**, 360–373.

Bradshaw, S. D., McDonald, I. R., Hahnel, R., and Heller, H. (1975). Synthesis of progesterone by the placenta of a marsupial. *J. Endocr.* **65**, 451–452.

Bradshaw, S. D., Shoemaker, V. H., and Nagy, K. A. (1972). The role of adrenal corticosteroids in the regulation of kidney function in the desert lizard *Dipsosaurus dorsalis*. *Comp. Biochem. Physiol.* **43A**, 621–635.

Bradshaw, S. D. and Waring, H. (1969). Comparative studies on the biological activity of melanin-dispersing hormone (MDH). *Colloq. Int. C.N.R.S.* **177**, 135–151.

Bray, G. A. (1974). Endocrine factors in the control of food intake. *Fedn. Proc.* **33**, 1140–1145.

Brawer, J. R. and Naftolin, F. (1979). The effects of oestrogen on hypothalmic tissue. In *Sex, Hormones and Behaviour*, Ciba Foundation Symposium 62, pp. 19–33. Amsterdam: Excerpta Medica.

Brewer, K. F. and Ensor, D. M. (1980). Hormonal control of osmoregulation in the Chelonia. 1. The effects of prolactin and interrenal steroids in freshwater chelonians. *Gen. Comp. Endocr.* **42**, 304–309.

Braun, E. J. and Dantzler, W. H. (1972). Functions of mammalian-type and reptilian-type nephrons in kidney of desert quail. *Amer. J. Physiol.* **222**, 617–629.

(1974). Effects of ADH on single-nephron glomerular filtration rates in the avian kidney. *Amer. J. Physiol.* **226**, 1–28.

Bromer, W. W. (1972). Chemistry of glucagon and gastrin. In *Handbook of Physiology*, Sect. 7, *Endocrinology*, vol. I, *Endocrine pancreas*, pp. 133–138. Washington, D.C.: American Physiological Society.

Brommage, R. and Neuman, W. F. (1979). Mechanism of mobilization of bone mineral by 1,25-dihydroxyvitamin D_3. *Amer. J. Physiol.* **237**, E113-E133.

Brooks, C.J.W., Brooks, R. V., Fotherby, L., Grant, J. K., Klopper, A., and Klyne, W. (1970). The identification of steroids. *J. Endocr.* **47**, 265–272.

Brown, P. S. and Brown, S. C. (1973). Prolactin and thyroid hormone interactions in salt and water balance in the newt *Notophthalamus viridescens*. *Gen. Comp. Endocr.* **20**, 456–466.

Brown, S. C. and Brown, P. S. (1980). Water balance in the California newt, *Taricha torosa*. *Amer. J. Physiol.* **238**, R113–R118.

Brown, W. B. and Forbes, J. M. (1980). Diurnal variations of plasma prolactin in growing sheep under two lighting regimes and the effect of pinealectomy. *J. Endocr.* **84**, 91–99.

Browning, H. C. (1969). Role of prolactin in regulation of reproductive cycles. *Gen. Comp. Endocr. Suppl.* **2**, 42–54.

Brownstein, M. (1977). Neurotransmitters and hypothalamic hormones in the central nervous system. *Fedn. Proc.* **36**, 1960–1963.

Brunette, M. G., Chan, M., Ferriere, C., and Roberts, K. D. (1978). Site of 1,25(OH)$_2$vitamin D$_3$ synthesis in the kidney. *Nature, Lond.* **276**, 287–289.

Bryant, R. W., Epstein, A. N., Fitzsimons, J. T., and Fluharty, S. J. (1980). Arousal of a specific and persistent sodium appetite in the rat with continuous intercerebroventricular infusion of angiotensin II. *J. Physiol., Lond.* **301**, 365–382.

Buchala, A. J. and Schmid, A. (1979). Vitamin D and its analogues as new class of plant growth substances effecting rhizogenesis. *Nature, Lond.* **280**, 230–231.

Bullough, W. S. (1971). The actions of chalones. *Ag. Actions* **2**, 1–7.

Buntin, J. D. (1979). Prolactin release in parent ring doves after brief exposure to their young. *J. Endocr.* **82**, 127–130.

Burgers, A.C.J. (1963). Melanophore-stimulating hormones in vertebrates. *Ann. N.Y. Acad. Sci.* **100**, 669–677.

Burke, W. H. and Papkoff, H. (1980). Purification of turkey prolactin and the development of a homologous radioimmunoassay for its measurement. *Gen. Comp. Endocr.* **40**, 297–307.

Burzawa–Gerard, E. (1974). Separation et reassociation des sous-unités de l'hormone gonadotrope d'un poisson Téléostéen, la carpe (*Cyprinus carpio L.*). *C. R. Acad. Sci., Paris* **279**, 1681–1684.

Burzawa-Gerard, E., Dufour, S., and Fontaine, Y-A. (1980). Relations immunologiques entre les hormones glycoprotéiques hypophysaires de poissons et de Mammifères ainsi qu'entre leurs sous-unités α et β. *Gen. Comp. Endocr.* **41**, 199–211.

Burzawa-Gerard, E. and Fontaine, Y. A. (1972). The gonadotropins of lower vertebrates. *Gen. Comp. Endocr. Suppl.* **3**, 715–728.

Burzawa–Gerard, E., Goncharov, B. F., and Fontaine, Y-A. (1975). L'hormone gonadotrope hypophysaire d'un poisson Chondrostéen, L'Esturgeon (*Acipenser stellatus Pall.*). *Gen. Comp. Endocr.* **27**, 289–295.

Butler, D. G. (1966). Effect of hypophysectomy on osmoregulation in the European eel (*Anguilla anguilla L.*). *Comp. Biochem. Physiol.* **18**, 773–781.

(1969). Corpuscles of Stannius and renal physiology in the eel (*Anguilla rostrata*). *J. Fish Res. Bd. Can.* **26**, 639–654.

Butler, D. G. and Carmichael, F. J. (1972). (Na$^+$.K$^+$)-ATPase activity in eel (*Anguilla rostrata*) gills in relation to changes in environmental salinity: role of adrenocortical steroids. *Gen. Comp. Endocr.* **19**, 421–427.

Butler, D. G., Clarke, W. C., Donaldson, E. M., and Langford, R. W. (1969). Surgical adrenalectomy of a teleost fish (*Anguilla rostrata* LESUEUR): effect on plasma cortisol and tissue electrolyte and carbohydrate concentrations. *Gen. Comp. Endocr.* **12**, 503–514.

Cahill, G. F., Aoki, T. T., and Marliss, E. B., (1972). Insulin and muscle protein. In *Handbook of Physiology*, Sect. 7 *Endocrinology*, vol. I, *Endocrine pancreas*, pp. 563–577. Washington, D.C.: American Physiological Society.

Cake, M. H., Owen, F. J., and Bradshaw, S. D. (1980). Difference in concentration of progesterone in plasma between pregnant and non-pregnant quokkas (*Setonix brachyurus*). *J. Endocr.* **84**, 153–158.

Callard, G. V., Petro, Z., and Ryan, K. J. (1977). Identification of aromatase in the reptilian brain. *Endocrinology* **100**, 1214–1218.

Callard, G. V., Petro, Z., and Ryan, K. J. (1978). Conversion of androgen to estrogen and other steroids in the vertebrate brain. *Amer. Zool.* **18**, 511–523.

Callard, I. P., Callard, G. V., Lance, V., and Eccles, S. (1976). Seasonal changes in testicular structure and function and the effects of gonadotropins in the fresh-water turtle, *Chrysemis picta*. *Gen. Comp. Endocr.* **30**, 347–356.

Callard, I. P. and Doolittle, J. P. (1973). The influence of intrahypothalamic injections of progesterone on ovarian growth and function in the ovoviviparous iguanid lizard *Sceloporus cyanogenys*. *Comp. Biochem. Physiol.* **44A**, 625–629.

Callard, I. P., Doolittle, J., Banks, W. L., and Chan, S.W.C. (1972). Recent studies on the control of the reptilian ovarian cycle. *Gen. Comp. Endocr. Suppl.* **3**, 65–75.

Campbell, R. R., Etches, R. J., and Leatherland, J. F. (1981). Seasonal changes in plasma prolactin concentration and carcass lipid levels in the lesser snow goose (*Anser caerulescens caerulescens*). *Comp. Biochem. Physiol.* **68A**, 653–657.

Capelli, J. P., Wesson, L. G., and Aponte, G. E. (1970). A phylogenetic study of the renin-angiotensin system. *Amer. J. Physiol.* **218**, 1171–1178.

Carney, S. and Thompson, L. (1981). Acute effect of calcitonin on rat renal electrolyte transport. *Amer. J. Physiol.* **240**, F12–F16.

Carter, D. A. and Baker, B. I. (1980). The relationship between opiate concentration and cellular activity in the pars distalis of neurointermediate lobe of the eel pituitary. *Gen. Comp. Endocr.* **41**, 225–232.

Castay, M., Bismuth, J., and Astier, H. (1978). Thyroxine-binding proteins of duck and chicken sera. *Gen. Comp. Endocr.* **35**, 491–495.

Catling, P. C. and Vinson, G. P. (1976). Adrenocortical hormones in the neonate and pouch young of the tammar wallaby, *Macropus eugenii*. *J. Endocr.* **69**, 447–448.

Chabardès, D., Imbert-Teboul, M., Montégut, M., Clique, A., and Morel, F. (1976). Distribution of calcitonin-sensitive adenylate cyclase activity along the kidney tubule. *Proc. Natl. Acad. Sci. USA* **73**, 3608–3612.

Chadwick, C. S. and Jackson, H. R. (1948). Acceleration of skin growth and molting in the red eft of *Triturus viridescens* by means of prolactin injections. *Anat. Rec.* **101**, 718.

Chan, D.K.O. (1972). Hormonal regulation of calcium balance in teleost fish. *Gen. Comp. Endocr. Suppl.* **3**, 411–420.

(1975). Cardiovascular and renal effects of urotensins I and II in the eel, *Anguilla rostrata*. *Gen. Comp. Endocr.* **27**, 52–61.

Chan, D.K.O., Chester Jones, I., Henderson, I. W., and Rankin, J. C. (1967). Studies on the experimental alteration of water and electrolyte composition of the eel (*Anguilla anguilla* L.). *J. Endocr.* **37**, 297–317.

Chan, L. and O'Malley, B. W. (1976). Mechanism of action of the sex steroid hormones. *New England J. Med.* **294**, 1322–1328.

Chan, S. J., Keim, P., and Steiner, D. F. (1976). Cell-free synthesis of rat preproinsulins; characterization and partial amino acid sequence determination. *Proc. Natl. Acad. Sci., USA* **73**, 1964–1968.

Chan, S.T.H. and Phillips, J. G. (1969). The biosynthesis of steroids by the gonads of the ricefield eel, *Monopterus albus* at various stages during natural sex-reversal. *Gen. Comp. Endocr.* **12**, 619–636.

Chan, S.W.C. and Phillips, J. G. (1971). Seasonal variations in production *in vitro* of corticosteroids by the frog (*Rana rugulosa*) adrenal. *J. Endocr.* **50**, 1–17.

Chan, S.W.C., Ziegel, S., and Callard, I. P. (1973). Plasma progesterone in snakes. *Comp. Biochem. Physiol.* **44A**, 631–637.

Channing, C. P., Licht, P., Papkoff, H., and Donaldson, E. M. (1974). Comparative activities of mammalian, reptilian and piscine gonadotropins in monkey granulosa cell cultures. *Gen. Comp. Endocr.* **22**, 137–145.

Chauvet, M. T., Codogno, P., Chauvet, J., and Acher, R. (1979). Comparison between MSEL- and VLDV-neurophysins. Complete amino acid sequences of porcine and bovine VLDV-neurophysins. *FEBS Letters* **98**, 37–40.

Chauvet, M. T., Hurpet, D., Chauvet, J., and Acher, R. (1980). Phenypressin (Phe^2-Arg^8-vasopressin), a new neurohypophysial peptide found in marsupials. *Nature, Lond.* **287**, 640–641.

Chavin, W. (1976). The thyroid of the Sarcopterygian fishes (Dipnoi and Crossoptergii) and the origin of the tetrapod thyroid. *Gen. Comp. Endocr.* **30**, 142–155.

Chavin, W., Kim, K., and Tchen T. T. (1963). Endocrine control of pigmentation. *Ann. N.Y. Acad. Sci.* **100**, 678–685.

Cheesman, D. W., Osland, R. B., and Forsham, P. H. (1977). Suppression of the preovulatory surge of luteinizing hormone and subsequent ovulation in the rat by arginine vasotocin. *Endocrinology* **101**, 1194–1202.

Cheng, M-F. (1977). Role of gonadotrophin releasing hormones in the reproductive behaviour of female ring doves (*Streptopelia risoria*). *J. Endocr.* **74**, 37–45.

Chester Jones, I., Bellamy, D., Chan, D.K.O., Follett, B. K., Henderson, I. W., Phillips, J. G., and Snart, R. S. (1972). Biological actions of steroid hormones on nonmammalian vertebrates. In *Steroids in Nonmammalian Vertebrates* (edited by D. R. Idler), pp. 414–480. New York: Academic Press.

Chester Jones, I., Henderson, I. W., Chan, D.K.O., Rankin, J. C., Mosley, W., Brown, J. J., Lever, A. F., Robertson, J.I.S., and Tree, M. (1966). Pressor activity in extracts of the corpuscles of Stannius from the European eel (*Anguilla anguilla* L.). *J. Endocr.* **34**, 393–408.

Cheung, W. Y. (1980). Calmodulin plays a pivotal role in cellular regulation. *Science* **207**, 19–27.

Chevalier, G. (1978). *In vivo* incorporation of ^{3}H leucine and ^{3}H tyrosine by caudal neurosecretory cells of the trout *Salvelinus fontinalis* in relation to osmotic manipulations. A radioautographic study. *Gen. Comp. Endocr.* **36**, 223–228.

Chieffi, G. (1967). The reproductive system of elasmobranchs: developmental and endocrinological aspects. In *Sharks, Skates and Rays* (edited by P. W. Gilbert, R. F. Mathewson, and D. P. Rall), pp. 553–580. Baltimore: Johns Hopkins University Press.

Chiu, K. W. and Lynn, W. G. (1972). Observations on thyroidal control of sloughing in the garter snake, *Thamnophis sirtalis*. *Copeia* 1972 (no. 1), 158–163.

Chiu, K. W. and Phillips, J. G. (1971*a*). The effect of hypophysectomy and of injections of thyrotrophin into hypophysectomized animals on the sloughing cycle of the lizard *Gekko gecko* L. *J. Endocr.* **49**, 611–618.

(1971*b*). The role of prolactin in the sloughing cycle in the lizard *Gekko gecko* L. *J. Endocr.* **49**, 625–634.

Chiu, K. W., Wong, C. C., Lei, F. H., and Tam, V. (1975). The nature of thyroidal secretions in reptiles. *Gen. Comp. Endocr.* **25**, 74–82.

Christakos, S., Brunette, M. G., and Norman, A. W. (1981). Localization of immunoreactive vitamin D-dependent calcium binding protein in chick nephron. *Endocrinology* **109**, 322–324.

Christakos, S. and Norman, A. W. (1978). Vitamin D$_3$ induced calcium binding protein in bone tissue. *Science* **202**, 70–71.

Churchill, P. C., Malvin, R. L., Churchill, M. C., and McDonald, F. D. (1979). Renal function in *Lophius americanus*: effects of angiotensin II. *Amer. J. Physiol.* **236**, R297–R301.

Citri, Y. and Schramm, M. (1980). Resolution, reconstitution and kinetics of the primary action of a hormone receptor. *Nature, Lond.* **287**, 297–300.

Clark, N. B. (1967). Influence of estrogens upon serum calcium, phosphate and protein concentrations of fresh-water turtles. *Comp. Biochem. Physiol.* **20**, 823–834.

(1972). Calcium regulation in reptiles. *Gen. Comp. Endocr. Supply.* **3**, 430–440.

Clark, N. B. and Dantzler, W. H. (1972). Renal tubular transport of calcium and phosphate in snakes: role of parathyroid hormone. *Amer. J. Physiol.* **223**, 1455–1464.

Clark, N. B. and Wideman, R. F. (1977). Renal excretion of phosphate and calcium in parathyroidectomized starlings. *Amer. J. Physiol.* **223**, F138–F144.

(1980). Calcitonin stimulation of urine flow and sodium excretion in the starling. *Amer. J. Physiol.* **238**, R406–R412.

Clarke, W. C., Bern, H. A., Li, C. H., and Cohen, D. C. (1973). Somatotropic and sodium-retaining effects of human growth hormone and placental lactogen in lower vertebrates. *Endocrinology* **93**, 960–964.

Clemons, G. K. and Nicoll, C. S. (1977). Effects of antisera to bullfrog prolactin and growth hormone on metamorphosis of *Rana catesbeiana* tadpoles. *Gen. Comp. Endocr.* **31**, 495–497.

Cockburn, F., Hull, D., and Walton, I. (1968). The effect of lipolytic hormones and theophylline on heat production in brown adipose tissue in vivo. *Brit. J. Pharmacol.* **31**, 568–577.

Cofré, G. and Crabbé, J. (1965). Stimulation by aldosterone of active sodium transport by the isolated colon of the toad *Bufo marinus*. *Nature, Lond.* **207**, 1299–1300.

Cohen, S. and Carpenter, G. (1975). Human epidermal growth factor: isolation and chemical and biological properties. *Proc. Natl. Acad. Sci., USA* **72**, 1317–1321.

Cohen, S., Carpenter, G., and King, L. (1980). Epidermal growth-factor-receptor-protein kinase reactions. *J. Biol. Chem.* **255**, 4834–4842.

Cohen, S. and Savage, C. (1974). Recent studies on the chemistry and biology of epidermal growth factor. *Rec. Prog. Hormone Res.* **30**, 551–572.

Cohen, S. and Taylor, J. M. (1974). Epidermal growth factor: chemical and biological characterization. *Rec. Prog. Hormone Res.* **30**, 533–550.

Cohn, D. V., Smardo, F. L., and Morrissey, J. J. (1979). Evidence for internal homology in bovine preproparathyroid hormone. *Proc. Natl. Acad. Sci., USA* **76**, 1469–1471.

Collins, K. J. and Weiner, J. S. (1968). Endocrinological aspects of exposure to high environmental temperatures. *Physiol. Rev.* **48**, 785–839.

Colombo, L. and Yaron, Z. (1976). Steroid 21-hydroxylase activity in the ovary of the snake *Storeria dekayi* during pregnancy. *Gen. Comp. Endocr.* **28**, 403–412.

Cook, A. F., Stacey, N. E., and Peter, R. E. (1980). Periovulatory changes in serum cortisol levels in the goldfish. *Gen. Comp. Endocr.* **40**, 507–510.

Cooper, W. E. and Ferguson, G. W. (1972). Steroids and color change during gravidity in the lizard *Crotaphytus collaris*. *Gen. Comp. Endocr.* **18**, 69–72.

Coote, J. H., Johns, E. J., Macleod, V. H., and Singer, B. (1972). Effect of renal nerve stimulation, renal blood flow and adrenergic blockade on plasma renin activity in the cat. *J. Physiol., Lond.* **226**, 15–36.

Copp, D. H. (1969). The ultimobranchial glands and calcium regulation. In *Fish Physiology* (edited by W. S. Hoar and D. J. Randall), vol. II *The Endocrine System*, pp. 377–398. New York: Academic Press.

(1972). Calcium regulation in birds. *Gen. Comp. Endocr. Suppl.* **3**, 441–447.

(1976). Comparative endocrinology of calcitonin. In *Handbook of Physiology*, sect. 7 *Endocrinology*, vol. 7, *Parathyroid Gland*, pp. 431–442. Washington, D.C.: American Physiological Society.

Copp, D. H., Cameron, E. C., Cheney, B. A., Davidson, A.G.F., and Henze, K. G. (1962). Evidence for calcitonin — a new hormone from the parathyroid that lowers blood calcium. *Endocrinology* **70**, 638–649.

Copp, D. H., Cockcroft, D. W., and Keuk, Y. (1967*a*). Calcitonin from ultimobranchial glands of dogfish and chickens. *Science* **158**, 924–926.

(1967*b*). Ultimobranchial origin of calcitonin, hypocalcemic effect of extracts from chicken glands. *Can. J. Physiol. Pharmacol.* **45**, 1095–1099.

Cortelyou, J. R. (1967). The effect of commercially prepared parathyroid extracts on plasma and urine calcium levels in *Rana pipiens. Gen. Comp. Endocr.* **9**, 234–240.

Costanzo, L. S., Sheehe, P. R., and Weiner, I. M. (1974). Renal actions of vitamin D in D-deficient rats. *Amer. J. Physiol.* **226**, 1490–1495.

Cote, T., Munemura, M., Eskay, R. L., and Kebabian, J. W. (1980). Biochemical identification of the β-adrenoreceptor and evidence for the involvement of an adenosine 3',5'-monophosphate system in the β-adrenergically induced release of β-melanocyte-stimulating hormone in the intermediate lobe of the rat pituitary gland. *Endocrinology* **107**, 108–116.

Coupland, R. E. (1968). Corticosterone and methylation of noradrenaline by extra-adrenal chromaffin tissue. *J. Endocr.* **41**, 487–490.

Courty, Y. and Dufaure, J. P. (1980). Levels of testosterone, dihydrotestosterone, and androstenedione in the plasma and testis of a lizard (*Lacerta vivipara* Jacquin) during the annual cycle. *Gen. Comp. Endocr.* **42**, 325–333.

Cowie, A. T. (1972). Lactation and its hormonal control. In *Hormones in Reproduction* (edited by C. R. Austin and R. V. Short), pp. 106–143. Cambridge University Press.

Cowie, A. T., Forsyth, I. A., and Hart, I. C. (1980). *Hormonal Control of Lactation.* Berlin: Springer-Verlag.

Crabbé, J. and De Weer, P. (1964). Action of aldosterone on the bladder and skin of the toad. *Nature, Lond.* **202**, 278–279.

Craik, J.C.A. (1978). Effects of hypophysectomy on vitellogenesis in the elasmo-branch *Scyliorhinus canicula* L. *Gen. Comp. Endocr.* **36**, 63–67.

Crews, D. and Licht, P. (1974). Inhibition by corpora atretica of ovarian sensitivity and hormonal stimulation in the lizard, *Anolis carolinensis. Endocrinology* **95**, 102–106.

Crews, D. and Morgentaler, A. (1979). Effects of intracranial implantation of oestradiol and dihydrotestosterone on the sexual behaviour of the lizard *Anolis carolinensis. J. Endocr.* **82**, 373–381.

Crighton, D. B., Hartley, B. M., and Lamming, G. E. (1973). Changes in the luteinizing hormone releasing activity of the hypothalamus, and in the pituitary gland and plasma luteinizing hormone during the oestrous cycle of the sheep. *J. Endocr.* **58**, 377–385.

Csaba, G. (1980). Phylogeny and ontogeny of hormone receptors: the selection theory of receptor formation and hormonal imprinting. *Biol. Rev.* **55**, 47–63.

Currie, W. B. and Thorburn, G. D. (1977*a*). The fetal role in timing the initiation of parturition in the goat. In *The Fetus and Birth*, Ciba Foundation Symposium 47, pp. 49–66. Amsterdam: Elsevier Excerpta Medica, North-Holland.

(1977*b*). Parturition in goats: studies on the interaction between the foetus, placenta, prostaglandin F and progesterone before parturition induced prematurely by corticotrophin infusion of the foetus. *J. Endocr.* **73**, 263–278.

Dacke, C. G. (1979). *Calcium Regulation in Sub-mammalian Vertebrates.* New York: Academic Press.

Daniels-McQueen, S., McWilliams, D., Birken, S., Canfield, R., Landefeld, T., and Boime, I. (1978). Identification of mRNAs encoding the α and β subunits of human chorionic gonadotropin. *J. Biol. Chem.* **253**, 7109–7114.

Davies, N. T., Munday, K. A., and Parsons, B. J. (1970). The effect of angiotensin on rat intestinal fluid transfer. *J. Endocr.* **48**, 39–46.

Davis, P. J., Gregerman, R. I., and Poole, W. E. (1969). Thyroxine-binding proteins in the serum of the grey kangaroo. *J. Endocr.* **45**, 477–478.

Davis, P. J. and Jurgelski, W. (1973). Thyroid hormone-binding in opossum serum: evidence for polymorphism and relationship to haptoglobin polymorphism. *Endocrinology* **92**, 822–832.

Daughaday, W. H. (1971). Sulfation factor regulation of skeletal growth. *Amer. J. Med.* **50**, 277–280.

Daughaday, W. H., Hall, K., Raben, M. S., Salmon, W. D., Van den Brande, J. L., and Van Wyke, J. J. (1972). Somatomedin: proposed designation for sulphation factor. *Nature, Lond.* **235**, 107.

Deavers, D. R. and Musacchia, X. J. (1979). The function of glucocorticoids in the thermogenesis. *Fedn. Proc.* **38**, 2177–2181.

Deftos, L. J., Burton, D. W., Watkins, W. B., and Catherwood, B. D. (1980). Immunohistological studies of artiodactyl and teleost pituitaries with antisera to calcitonin. *Gen. Comp. Endocr.* **42**, 9–18.

De Häen, C. (1976). The non-stoichiometric floating receptor model for hormone sensitive adenyl cyclase. *J. Theor. Biol.* **58**, 383–400.

Della-Fera, M. A., Baile, C. A., Schneider, B. S., and Grinker, J. A. (1981). Cholecystokinin antibody injected in cerebral ventricles stimulates feeding in sheep. *Science* **212**, 687–689.

DeLuca, H. F. (1971). The role of vitamin D and its relationship to parathyroid hormone and calcitonin. *Rec. Prog. Hormone Res.* **27**, 479–510.

(1974). Vitamin D: the vitamin and the hormone. *Fedn. Proc.* **33**, 2211–2219.

DeLuca, H. F., Morii, H., and Melancon, M. J. (1968). The interaction of vitamin D, parathyroid hormone and thyrocalcitonin. In *Parathyroid Hormone and Thyrocalcitonin (Calcitonin)* (edited by R. V. Talmage and F. F. Belanger), pp. 448–454. Amsterdam: Excerpta Medica Foundation.

DeLuise, M., Martin, T. J., Greenberg, P. B., and Michelangeli, V. (1972). Metabolism of porcine, human and salmon calcitonin in the rat. *J. Endocr.* **53**, 475–482.

Demoulin, A., Hudson, B., Franchimont, P., and Legros, J. J. (1977). Arginine-vasotocin does not affect gonadotrophin secretion *in vitro*. *J. Endocr.* **73**, 105–106.

Dennis, V. W., Stead, W. W., and Myers, J. L. (1979). Renal handling of phosphate and calcium. *Ann. Rev. Physiol.* **41**, 257–271.

Dent, J. N. (1968). Survey of amphibian metamorphosis. In *Metamorphosis, a Problem in Developmental Biology* (edited by W. Etkin and L. I. Gilbert), pp. 271–311. New York: Appleton-Century-Crofts.

(1975). Integumentary effects of prolactin in lower vertebrates. *Amer. Zool.* **15**, 923–935.

DeRoos, R. and DeRoos, C. C. (1972). Comparative effects of the pituitary-adrenocortical axis and catecholamines on carbohydrate metabolism in elasmobranch fish. *Gen. Comp. Endocr. Suppl.* **3**, 192–197.

DeRoos, R. and DeRoos, C. C. (1979). Severe insulin-induced hypoglycemia in the spiny dogfish shark (*Squalus acanthias*). *Gen. Comp. Endocr.* **37**, 186–191.

Desranleau, R., Gilardeau, C., and Chrétien, M. (1972). Radioimmunoassay of ovine beta-lipotropic hormone. *Endocrinology* **91**, 1004–1010.

de Wied, D. (1977). Peptides and behavior. *Life Sciences* **20**, 195–204.

Dharmamba, M., Mayer-Gostan, N., Maetz, J., and Bern, H. A. (1973). Effect of prolactin on sodium movement in *Tilapia mossambica* adapted to sea water. *Gen. Comp. Endocr.* **21**, 179–187.

Dicker, S. E. and Elliot, A. B. (1973). Neurohypophysial hormones and homeostasis in the crab-eating frog, *Rana cancrivora*. *Hormone Res.* **4**, 224–260.

Dickhoff, W. W. and Nicoll, C. S. (1979). Studies on the melanocyte-stimulating hormones of the neurointermediate lobe of the American bullfrog *Rana catesbeiana*. *Gen. Comp. Endocr.* **39**, 313–321.

Dobson, S. and Dodd, J. M. (1977a). Endocrine control of the testis in the dogfish *Scyliorhinus canicula* L. I. Effects of partial hypophysectomy on gravimetric, hormonal and biochemical aspects of testis function. *Gen. Comp. Endocr.* **32**, 41–52.

(1977b). Endocrine control of the testis in the dogfish *Scyliorhinus canicula* L. II. Histological and ultrastructural changes in the testis after partial hypophysectomy (ventral lobectomy). *Gen. Comp. Endocr.* **32**, 53–71.

Dockray, C. J. (1975). Comparative studies on secretin. *Gen. Comp. Endocr.* **25**, 203–210.

Dockray, C. J. (1978). Evolution of secretin-like hormones. In *Gut Hormones* (edited by S. R. Bloom), pp. 64–67. London: Churchill Livingstone.

Dockray, C. J. (1979). Comparative biochemistry and physiology of gut hormones. *Ann. Rev. Physiol.* **41**, 83–95.

Dockray, C. J. and Gregory, R. A. (1980). Relations between neuropeptides and gut hormones. *Proc. Roy. Soc.* **B210**, 151–164.

Dodd, J. M. (1960). Gonadal and gonadotrophic hormones in lower vertebrates. In *Marshall's Physiology of Reproduction* (edited by A. S. Parkes), vol. 1 (Pt. 2), pp. 417–582. London: Longmans.

(1972a). Ovarian control in cyclostomes and elasmobranchs. *Amer. Zool.* **12**, 325–339.

(1972b). The endocrine regulation of gametogenesis and gonad maturation in fishes. *Gen. Comp. Endocr. Suppl.* **3**, 675–687.

(1975). The hormones of sex and reproduction and their effects in fish and lower chordates: twenty years on. *Amer. Zool.* **15**, Suppl. 137–171.

Dodd, J. M. and Dodd, M.H.I. (1969). Phylogenetic specificity of thyroid stimulating hormone with special reference to the Amphibia. *Colloq. Int. C.N.R.S.* **177**, 277–285.

Dolman, D. and Edmonds, C. J. (1975). The effect of aldosterone and the renin-angiotensin system on sodium, potassium and chloride transport by proximal and distal rat colon *in vivo*. *J. Physiol., Lond.* **250**, 597–611.

Donaldson, E. M., Yamzaki, F., Dye, H. M., and Philleo, W. W. (1972). Preparation of gonadotrophin from salmon (*Oncorhynchus tshawytsha*) pituitary glands. *Gen. Comp. Endocr.* **18**, 469–481.

Donoso, A. O. and Segura, E. T. (1965). Seasonal variations of plasma adrenaline and noradrenaline in toads. *Gen. Comp. Endocr.* **5**, 440–443.

Dorner, G. (1979). Hormones and sexual differentiation of the brain. *In Sex, Hormones and Behaviour*, Ciba Foundation symposium 62, pp. 81–101. Amsterdam: Excerpta Medica.

Doty, S. B., Robinson, R. A., and Schofield, B. (1976). Morphology of bone and histochemical staining characteristics of bone cells. In *Handbook of Physiology*, Sect. 7 *Endocrinology*, vol. 7, *The Parathyroid Gland*, pp. 3–23. Washington D.C.: American Physiological Society.

Douglas, W. W. (1972). Secretomotor control of adrenal medullary secretion: synaptic, membrane and ionic events in stimulus-secretion coupling. In *Handbook of Physiology*, Sect. 7 *Endocrinology*, vol. 6, *Adrenal Gland*, pp. 367–388. Washington D.C.: American Physiological Society.

    (1974). Mechanisms of release of neurohypophysial hormones: stimulation secretion coupling. In *Handbook of Physiology*, Sect. 7 *Endocrinology*, vol. 4, *The Pituitary Gland and its Neuroendocrine Control* (P. 1) pp. 191–224. Washington D.C.: American Physiological Society.

Dousa, T., Hecter, O., Schwartz, I. L., and Walter, R. (1971). Neurohypophyseal hormone-responsive adenylate cyclase from mammalian kidney. *Proc. Natl. Acad. Sci., USA* **68**, 1693–1697.

Dousa, T., Walter, R., Schwartz, I. L., Sands, H., and Hechter, O. (1972). Role of cyclic AMP in the action of neurohypophyseal hormones on kidney. *Advances in Cyclic Nucleotide Research* (edited by P. Greengard and G. A. Robison), I, pp. 121–135. New York: Raven Press.

Duggan, R. T. (1981). Plasma corticosteroids in marine, terrestrial and freshwater snakes. *Comp. Biochem. Physiol.* **68A**, 115–118.

Dunn, A. D. (1980). Studies on iodoproteins and thyroid hormones in Ascidians. *Gen. Comp. Endocr.* **40**, 473–483.

Duve, H. and Thorpe, A. (1981). Gastrin/cholecystokinin(CCK)-like immunoreactive neurons in the brain of the blowfly, *Calliphora erythrocephala* (Diptera). *Gen. Comp. Endocr.* **43**, 381–391.

Ebling, F. J. (1974). Hormonal control and methods measuring sebaceous gland activity. *J. Invest. Dermatol.* **62**, 161–171.

Ebling, F. J., Ebling, E., Randall, V., and Skinner, J. (1975). The synergistic action of α-melanocyte-stimulating hormone and testosterone on the sebaceous, prostate, preputial, Harderian and lachrymal glands, seminal vesicles and brown adipose tissue in the hypophysectomized-castrated rat. *J. Endocr.* **66**, 407–412.

Ebling, F. J. and Hale, P. A. (1970). The control of the mammalian molt. *Mem. Soc. Endocr.* **18**, 215–235.

Eddy, J.M.P. and Strahan, R. (1968). The role of the pineal complex in the pigmentary effector system of lampreys, *Mordacia mordax* (Richardson) and *Geotria australis* Gray. *Gen. Comp. Endocr.* **11**, 528–534.

Eddy, L. and Lipner, H. (1976). Amphibian metamorphosis: the role of thyrotropin-like hormone. *Gen. Comp. Endocr.* **29**, 333–336.

Edelman, I. S. (1976). Transition from the poikilotherm to the homeotherm: possible role of sodium transport and thyroid hormone. *Fedn. Proc.* **35**, 2180–2184.

Egami, N. and Ishii, S. (1962). Hypophysial control of reproductive functions in teleost fishes. *Gen. Comp. Endocr. Suppl.* **1**, 248–253.

Ehrhardt, A. A. and Meyer-Bahlburg, H.F.L. (1981). Effects of prenatal sex hormones on gender-related behaviour. *Science* **211**, 1312–1318.

Eipper, B. A. and Mains, R. E. (1980). Structure and biosynthesis of pro-adrenocorticotropin/endorphin and related peptides. *Endocr. Rev.* **1**, 1–27.

Elgar, W., Neumann, F., and von Berswordt-Wallrabe, R. (1971). The influence of androgen antagonists and progestogens on the sex differentiation of different mammalian species. In *Hormones in Development* (edited by M. Hamburg, and E.J.W. Barrington), pp. 651–657. New York: Appleton-Century-Crofts.

Elliot, A. B. (1968). Effects of adrenaline on water uptake in *Bufo melanostictus*. *J. Physiol., Lond.* **197**, 87–88P.

Else, P. L. and Hulbert, A. J. (1981). Comparison of the "mammalian machine" and the "reptile machine": energy production. *Amer. J. Physiol.* **240**, R3–R9.

Emdin, S. O., Gammeltoft, S., and Gliemann, J. (1977). Degradation, receptor binding affinity, and potency of insulin from the Atlantic hagfish (*Myxine glutinosa*) determined in isolated rat fat cells. *J. Biol. Chem.* **253**, 602–608.

Ensor, D. M. (1978). *Comparative Endocrinology of Prolactin.* London: Chapman and Hall.

Ensor, D. M. and Ball, J.M. (1972). Prolactin and osmoregulation in fishes. *Fedn. Proc.* **31**, 1615–1623.

Ensor, D. M., Edmondson, M. R., and Phillips, J. G. (1972). Prolactin and dehydration in rats. *J. Endocr.* **52**, lix–lx.

Ensor, D. M., Simons, I. M., and Phillips, J. G. (1973). The effect of hypophysectomy and prolactin replacement therapy on salt and water metabolism in *Anas platyrhynchos. J. Endocr.* **57**, xi.

Epple, A. (1969). The endocrine pancreas. In *Fish Physiology* (edited by W. S. Hoar and D. J. Randall), vol. II, *The Endocrine System,* pp. 275–319. New York: Academic Press.

Epstein, F. H., Cynamon, M., and McKay, W. (1971). Endocrine control of Na-K-ATPase and seawater adaptation in *Anguilla rostrata. Gen. Comp. Endocr.* **16**, 323–328.

Epstein, F. H., Katz, A. I., and Pickford, G. E. (1967). Sodium- and potassium-activated adenosine triphosphatase of gills: role in adaptation of teleosts to salt water. *Science* **156**, 1245–1247.

Epstein, F. H., Silva, P., and Kormanik, G. (1980). Role of Na-K-ATPase in chloride cell function. *Amer. J. Physiol.* **238**, R246–250.

Estler, C. J. and Ammon, H.P.T. (1969). The importance of the adrenergic beta-receptors for thermogenesis and survival of acutely cold-exposed mice. *Can. J. Physiol. Pharmacol.* **47**, 427–434.

Etkin, W. (1970). The endocrine mechanism of amphibian metamorphosis, an evolutionary achievement. *Mem. Soc. Endocr.* **18**, 137–153.

Etkin, W. and Gona, A. G. (1967). Antagonism between prolactin and thyroid hormone in amphibian development. *J. Exp. Zool.* **165**, 249–258.

Fagerlund, U.H.M. (1967). Plasma cortisol concentration in relation to stress in adult sockeye salmon during the freshwater stage in their life cycle. *Gen. Comp. Endocr.* **8**, 197–207.

Falkmer, S., Elde, R. P., Hellerström, C., Petersson, B., Efendíc, S., Fohlman, J., and Siljevall, J-B. (1977). Some phylogenetical aspects on the occurrence of somatostatin in the gastro-entero-pancreatic endocrine system. A histological and immunocytochemical study, combined with quantitative radioimmunological assays of tissue extracts. *Arch. Histol. Jap.* **40**, Suppl. 99–117.

Falkmer, S., and Östberg, Y. (1977). Comparative morphology of pancreatic islets in animals. In *The Diabetic Pancreas* (edited by B. W. Volk and K. F. Wellmann), pp. 15–59. New York: Plenum.

Falkmer, S., Östberg, Y., and Van Noorden, S. V. (1978). Entero-insular endocrine systems of cyclostomes: a clue to hormone evolution. In *Gut Hormones* (edited by S. R. Bloom), pp. 57–63. London: Churchill Livingston.

Falkmer, S. and Patent, G. J. (1972). Comparative and embryological aspects of the pancreatic islets. In *Handbook of Physiology,* Sect. 7 *Endocrinology,* vol. 1, *Endocrine pancreas,* pp. 1–23. Washington, D.C.: American Physiological Society.

Farer, L. S., Robbins, J., Blumberg, B. S., and Rall, J. E. (1962). Thyroxine-serum protein complexes in various animals. *Endocrinology* **70**, 686–696.

Farfel, Z., Salomon, M. R., and Bourne, H. R. (1981). Genetic investigation of adenylate cyclase: mutations in mouse and man. *Ann. Rev. Pharmacol. Toxicol.* **21**, 251–264.

Farmer, S. W., Hayashida, T., Papkoff, H., and Polenov, A. L. (1981). Characteristics of growth hormone isolated from sturgeon (*Acipenser güldenstadti*) pituitaries. *Endocrinology* **108**, 377–381.

Farmer, S. W., Papkoff, H., Bewley, T. A., Hayashida, T., Nishioka, R. S., Bern, H. A., and Li, C. H. (1977). Isolation and properties of teleost prolactin. *Gen. Comp. Endocr.* **31**, 60–71.

Farmer, S. W., Papkoff, H., and Hayashida, T. (1974). Purification and properties of avian growth hormones. *Endocrinology* **95**, 1560–1565.

(1976). Purification and properties of reptilian and amphibian growth hormones. *Endocrinology* **99**, 692–700.

Farmer, S. W., Papkoff, H., Hayashida, T., Bewley, T. A., Bern, H. A., and Li, C. H. (1976). Purification and properties of teleost growth hormone. *Gen. Comp. Endocr.* **30**, 91–100.

Farrar, E. S. and Frye, B. E. (1977). Seasonal variation in the effects of adrenalin and glucagon in *Rana pipiens*. *Gen. Comp. Endocr.* **33**, 76–81.

Feher, J. J. and Wasserman, R. H. (1978). Evidence for a membrane-bound fraction of chick intestinal calcium-binding protein. *Biochim. Biophys. Acta* **540**, 134–143.

(1979). Intestinal calcium-binding protein and calcium absorption in cortisol-treated chicks: effects of vitamin D_3. *Endocrinology* **104**, 547–551.

Fenwick, J. C. (1970). The pineal organ: photoperiod and reproductive cycles in the goldfish *Carassius auratus* L. *J. Endocrin.* **46**, 101–111.

(1976). Effects of stanniectomy on calcium activated adenosinetriphosphatase activity in the gills of fresh water adapted North American eels, *Anguilla rostrata* LeSueur. *Gen. Comp. Endocr.* **29**, 383–387.

Ferguson, G. W. and Chen, C. L. (1973). Steroid hormones, color change and ovarian cycling in free-living female collared lizard, *Crotaphytus collaris*. *Amer. Zool.* **13**, 1277.

Ferguson, D. R. and Heller, H. (1965). Distribution of neurohypophysial hormones in mammals. *J. Physiol., Lond.* **180**, 846–863.

Fernholm, B. (1972). Neurohypophysial-adenohypophysial relations in hagfish (Myxinoidea, Cyclostomata). *Gen. Comp. Endocr. Suppl.* **3**, 1–10.

Feyrter, F. (1938). *Über diffuse endokrine Epitheliale Organe*. Leipzig: Barth.

Fiddes, J. C. and Goodman, H. M. (1980). The cDNA for the β-subunit of human chorionic gonadotropin suggests evolution of a gene by readthrough into the 3'-untranslated region. *Nature, Lond.* **286**, 684–687.

Firth, B. T., Kennaway, D. J., and Rozenbilds, M.A.M. (1979). Plasma melatonin in the scincid lizard, *Trachydosaurus rugosus*: diel rhythm, seasonality, and the effect of constant light and constant darkness. *Gen. Comp. Endocr.* **37**, 493–500.

Fitzsimons, J. T. (1972). Thirst. *Physiol. Rev.* **52**, 468–561.

(1979). *The Physiology of Thirst and Sodium Appetite*. Monographs of the Physiological Society no. 35. Cambridge University Press.

Fitzsimons, J. T. and Kaufman, S. (1977). Cellular and extracellular dehydration, and angiotensin as stimuli to drinking in the common iguana, *Iguana iguana*. *J. Physiol., Lond.* **265**, 443–463.

Fleming, W. R., Brehe, J., and Hanson, R. (1973). Some complicating factors in the study of calcium metabolism in teleosts. *Amer. Zool.* **13**, 793–797.

Fleming, W. R., Stanley, J. G., and Meier, A. H. (1964). Seasonal effects of external calcium, estradiol, and ACTH on the serum calcium and sodium levels of *Fundulus kansae*. *Gen. Comp. Endocr.* **4**, 61–67.

Foà, P. P. (1972). The secretion of glucagon. In *Handbook of Physiology*, Sect. 7 *Endocrinology*, vol. I, *Endocrine pancreas*, pp. 261–277. Washington, D.C.: American Physiological Society.

Follénius, E. and Dubois, M. P. (1978). Immunocytological detection and localization of a peptide reacting with an α-endorphin antiserum in the corticotropic and melanotropic cells of the trout pituitary (*Salmo irideus* Gibb). *Cell Tiss. Res.* **188**, 273–283.

Follett, B. K. (1963). Mole ratios of the neurohypophysial hormones in the vertebrate neural lobe. *Nature, Lond.* **198**, 693–694.

— (1976). Plasma follicle-stimulating hormone during photoperiodically induced sexual maturation in male Japanese quail. *J. Endocr.* **69**, 117–126.

Follett, B. K. and Davies, D. T. (1979). The endocrine control of ovulation in birds. In *Animal Reproduction* (Barc. Symposium 3 edited by H. W. Hawk), pp. 323–344. New York: Halsted Press.

Follett, B. K. and Maung, S. L. (1978). Rate of testicular maturation, in relation to gonadotrophin and testosterone levels, in quail exposed to various artificial photoperiods and to natural daylengths. *J. Endocr.* **78**, 267–280.

Follett, B. K. and Redshaw, M. R. (1974). The physiology of vitellogenesis. In *Physiology of the Amphibia* (edited by B. Lofts), vol. II, pp. 219–298. New York: Academic Press.

Follett, B. K. and Riley, J. (1967). Effect of the length of the daily photoperiod on thyroid activity in the female Japanese quail (*Coturnix coturnix japonica*). *J. Endocr.* **39**, 615–616.

Follett, B. K. and Robinson, J. E. (1980). Photoperiod and gonadotrophin secretion in birds. *Prog. Reprod. Biol.* **5**, 39–51.

Follett, B. K. and Sharp, P. J. (1969). Circadian rhythmicity in photoperiodically induced gonadotrophin release and gonadal growth in the quail. *Nature, Lond.* **223**, 968–971.

Fontaine, J., Le Lièvre, C., and Le Douarin, N. M. (1977). What is the developmental fate of the neural crest cells which migrate into the pancreas in the avian embryo?. *Gen. Comp. Endocr.* **33**, 394–404.

Fontaine, M. (1954). Du déterminisme physiologique des migrations. *Biol. Rev.* **29**, 390–418.

— (1964). Corpuscules de Stannius et régulation ionique (Ca, K, NA) du milieu interiéur de l'Anguille (*Anguilla anguilla* L.). *C.R. Acad. Sci., Paris* **259**, 875–878.

Fontaine, M., Callamand, O., and Olivereau, M. (1949). Hypophyse et euryhalinité chez l'anguille. *C.R. Acad. Sci., Paris* **228**, 513–514.

Fontaine, Y-A. (1969a). La spécificité zoologique des protéines hypophysaires capables de stimuler la thyroide. *Acta Endocrin.*, Kobn, *Suppl.* **136**, 1–154.

— (1969b). La spécificité zoologique d'action des hormones thyréotropes. *Colloq. Int. C.N.R.S.* **177**, 267–275.

Fontaine, Y-A. and Burzawa-Gerard, E. (1977). Esquisse de l'evolution des hormones gonadotropes et thyreotropes des Vertébrés. *Gen. Comp. Endocr.* **32**, 341–347.

Forrest, J. N., Cohen, A. D., Schon, D. A., and Epstein, F. H. (1973*a*). Na transport and Na-K-ATPase in gills during adaptation to seawater: effects of cortisol. *Amer. J. Physiol.* **224**, 709–713.

Forrest, J. N., MacKay, W. C., Gallagher, B., and Epstein, F. H. (1973*b*). Plasma cortisol response to saltwater adaptation in the American eel *Anguilla rostrata*. *Amer. J. Physiol.* **224**, 714–717.

Fouchereau-Peron, M., Laburthe, M., Besson, J., Rosselin, G., and Le Gal, Y. (1980). Characterization of the vasoactive intestinal polypeptide (VIP) in the gut of fishes. *Comp. Biochem. Physiol.* **65A**, 489–492.

Fouchereau-Peron, M., Moukhtar, M. S., Le Gal, Y., and Milhaud, G. (1981). Demonstration of specific receptors for calcitonin in isolated trout gill cells. *Comp. Biochem. Physiol.* **68A**, 417–421.

Frantz, A. G., Kleinberg, D. L., and Noel, G. L. (1972). Studies on prolactin in man. *Rec. Prog. Hormone Res.* **28**, 527–573.

Fraps, R. M. (1955). Egg production and fertility in poultry. In *Progress in the Physiology of Farm Animals* (edited by J. Hammond), vol. II, pp. 661–740. London: Butterworths.

Fraser, D. R. (1980). Regulation of the metabolism of vitamin D. *Physiol. Rev.* **60**, 551–613.

Freeman, H. C. and Idler, D. R. (1973). Effects of corticosteroids on liver transaminases in two salmonids, the rainbow trout (*Salmo gairdnerii*) and the brook trout (*Salvelinus fontinalis*). *Gen. Comp. Endocr.* **20**, 69–75.

Fregley, M. J., Field, F. P., Katovich, M. J., and Barney, C. C. (1979). Catecholaminethyroid hormone interaction in cold acclimated rats. *Fedn. Proc.* **38**, 2162–2169.

Fridberg, G. and Bern, H. A. (1968). The urophysis and the caudal neurosecretory system of fishes. *Biol. Rev.* **43**, 175–199.

Fritsch, H.A.R. and Sprang, R. (1977). On the ultrastructure of polypeptide hormone-producing cells in the gut of the Ascidian, *Ciona intestinalis* L. and the Bivalve, *Mytilus edulis* L. *Cell Tiss. Res.* **177**, 407–413.

Fritz, I. B. (1972). Insulin actions on carbohydrate and lipid metabolism. In *Biochemical Actions of Hormones* (edited by G. Litwack), vol. II, pp. 165–214. New York: Academic Press.

Fritz, I. B. and Lee, L.P.K. (1972). Fat mobilization and ketogenesis. In *Handbook of Physiology*, Sect. 7 *Endocrinology*, vol. I, *Endocrine pancreas*, pp. 579–596. Washington, D.C.: American Physiological Society.

Froesch, E. R., Bürgi, H., Ramseier, E. B., Bally, P., and Labhart, A. (1963). Antibody-supressible and non-supressible insulin-like activities in human serum and their physiologic significance. An insulin assay with adipose tissue of increased precision and specificity. *J. Clin. Invest.* **42**, 1816–1834.

Frye, B. E. (1967). *Hormonal Control in Vertebrates*, p. 104. New York: Macmillan.

Frye, B. E., Brown, P. S., and Snyder, B. W. (1972). Effects of prolactin and somatotropin on growth and metamorphosis of amphibians. *Gen. Comp. Endocr. Suppl.* **3**, 209–220.

Funder, J. W., Feldman, D., and Edelman, I. S. (1973). The roles of plasma binding and receptor specificity in the mineralocorticoid action of aldosterone. *Endocrinology* **92**, 994–1004.

Furr, B.J.A., Bonney, R. C., England, R. J., and Cunningham, F. J. (1973). Luteinizing hormone and progesterone in peripheral blood during the ovulatory cycle in the hen, *Gallus domesticus*. *J. Endocr.* **57**, 159–169.

Furr, B.J.A. and Smith, G. K. (1975). Effects of antisera against gonadal steroids on ovulation in the hen *Gallus domesticus*. *J. Endocr.* **66**, 303–304.

Galli-Gallardo, S. M., Pang, P.K.T., and Oguro, C. (1979). Renal responses of the Chilean toad, *Calyptocephalella caudiverbera*, and the mud puppy *Necturus maculosus*, to mesotocin. *Gen. Comp. Endocr.* **37**, 134–136.

Gelbard, H. A., Stern, P. H., and U'Prichard, D. C. (1980). 1,25-dihydroxyvitamin D_3 nuclear receptors in the pituitary. *Science* **209**, 1247–1249.

Gern, W. A., Owens, D. W., and Ralph, C. L. (1978). Plasma melatonin in the trout: day-night change demonstrated by radioimmunoassay. *Gen. Comp. Endocr.* **34**, 453–458.

Gern, W. A. and Ralph. C. L. (1979). Melatonin synthesis by the retina. *Science* **204**, 183–184.

Geschwind, I. I. (1967). Growth hormone activity in the lungfish pituitary. *Gen. Comp. Endocr.* **8**, 82–83.

(1969). The main lines of evolution of the pituitary hormones. *Colloq. Int. C.N.R.S.* **177**, 385–400.

Ghijsen, W.E.J.M., De Jong, M. D., and Van Os, C. H. (1980). Dissociation between Ca^{2+}-ATPase and alkaline photophatase activities in plasma membranes of rat duodenum. *Biochem. Biophys. Acta* **599**, 538–551.

Gibson, R. G., Mihas, A. A., Colvin, H. W., and Hirschowitz, B. I. (1976). The search for submammalian gastrins: the identification of amphibian gastrin. *Proc. Soc. Exp. Biol. Med., New York* **53**, 284–288.

Gill. V. E., Burford, G. D., Lederis, K., and Zimmerman, E. A. (1977). An immunocytochemical investigation for arginine vasotocin and neurophysin in the pituitary gland and the caudal neurosecretory system of *Catostomous commersoni*. *Gen. Comp. Endocr.* **32**, 505–511.

Gillies, G. and Lowry, P. (1979). Corticotrophin releasing factor may be modulated vasopressin. *Nature, Lond.* **278**, 463–464.

Ginsburg, M. (1968). Production, release, transportation and elimination of the neurohypophysial hormone. In *Neurohypophysial Hormones and Similar Polypeptides* (edited by B. Berde), pp. 286–371. Berlin: Springer-Verlag.

Girgis, S. I., Galan, F. G., Arnett, T. R., Rogers, R. M., Bone, Q., Ravazzola, M., and MacIntyre, I. (1980). Immunoreactive human calcitonin-like molecule in the nervous systems of protochordates and a cyclostome, *Myxine*. *J. Endocr.* **87**, 375–382.

Glass, J. D. and Lynch, G. R. (1981). Melatonin: identification of sites of antigonadal action in the mouse brain. *Science* **214**, 821–823.

Godet, M. (1961). Le problème hydrique et son controle hypophysaire chez le Protoptère. *Ann. Faculty Sciences de l'Universite Dakar* **6**, 183–201.

Godine, J. E., Chin, W. W., and Habener, J. F. (1980). Luteinizing and follicle-stimulating hormones. *J. Biol. Chem.* **255**, 8780–8783.

Goldman, J. M. and Hadley, M. E. (1969). The beta receptor and cyclic 3',–5'-adenosine monophosphate: possible roles in the regulation of melanophore responses of the spadefoot toad, *Scaphiopus couchi*. *Gen. Comp. Endocr.* **13**, 151–163.

Goldsmith, A. R. and Williams, D. M. (1980). Incubation in mallards (*Anas platyrhynchos*): changes in plasma levels of prolactin and luteinizing hormone. *J. Endocr.* **86**, 371–379.

Goldstein, A. (1976). Opioid peptides (endorphins) in pituitary and brain. *Science* **193**, 1081–1086.

Goldstein, A. L., Hooper, J. A., Schulof, R. S., Cohen, G. H., Thurman, G. B., McDaniel, M. C., White, A., and Dardenne, M. (1974). Thymosin and the immunopathology of aging. *Fedn. Proc.* **33**, 2053–2056.

Gona, A. G. and Etkin, W. (1970). Inhibition of metamorphosis in *Ambystoma tigrinum* by prolactin. *Gen. Comp. Endocr.* **14**, 589–591.

Gona, A. G., Pearlman, T., and Etkin, W. (1970). Prolactin-thyroid interaction in the newt, *Diemictlylus viridescens*. *J. Endocr.* **48**, 585–590.

Gona, O. and Gona, A. G. (1973). Action of human placental lactogen on second metamorphosis in the newt *Notophthalmus viridescens*. *Gen. Comp. Endocr.* **21**, 377–380.

Goodridge, A. G. (1964). The effect of insulin, glucagon and prolactin on lipid synthesis and related metabolic activity in migratory and non-migratory finches. *Comp. Biochem. Physiol.* **13**, 1–26.

Gorbman, A. (1940). Suitability of the common goldfish for assay of thyrotropic hormone. *Proc. Soc. Exp. Biol. Med., New York* **45**, 772–773.

Gorbman, A. and Bern, H. A. (1962). *A Textbook of Comparative Endocrinology*, p. 220. New York: Wiley.

Gorbman, A. and Hyder, M. (1973). Failure of mammalian TRH to stimulate thyroid function in the lungfish. *Gen. Comp. Endocr.* **20**, 588–589.

Gotshall, R. W., Davis, J. O., Shade, R. E., Spielman, W., Johnson, J. A., and Braverman, B. (1973). Effects of renal denervation on renin release in sodium-depleted dogs. *Amer. J. Physiol.* **225**, 344–349.

Graham, J. M. and Desjardins, C. (1980). Classical conditioning: induction of luteinizing hormone and testosterone secretion in anticipation of sexual activity. *Science* **210**, 1039–1041.

Greenwood, A. W. and Blyth, J.S.S. (1935). Variation in plumage response of brown leghorn breast feather and its reaction to oestrone. *Proc. Zool. Soc. Lond. Ser. A* **109**, 247–288.

Greer, M. A. and Haibach, H. (1974). Thyroid secretion. In *Handbook of Physiology*, Sect. 7 *Endocrinology*, vol. III, *Thyroid*, pp. 135–146. Washington, D.C.: American Physiological Society.

Gregory, H. (1975). Isolation and structure of urogastrone and its relationship to epidermal growth factor. *Nature, Lond.* **257**, 325–327.

Gregory, R. A. (1962). *Secretory Mechanisms of the Gastrointestinal Tract*, p. 153. London: Edward Arnold.

Greiner, A. C. and Chan, S. C. (1978). Melatonin content of the human pineal gland. *Science* **199**, 83–84.

Griffiths, M. (1978). *The Biology of the Monotremes*, pp. 209–254. New York: Academic Press.

Guernsey, D. L. and Stevens, E. D. (1977). The cell membrane sodium pump as a mechanism for increasing thermogenesis during cold acclimation in rats. *Science* **196**, 908–910.

Guillemin, R. (1978). Peptides in the brain: the new endocrinology of the neuron. *Science* **202**, 390–402.

Guillemin, R. and Burgus, R. (1972). The hormones of the hypothalamus. *Scientific American* **227** (November), 24–33.

Gurdon, J. B. (1977). Egg cytoplasm and gene control in development. *Proc. Roy. Soc.* **B198**, 211–247.

Habener, J. F., Singh, F. R., Deftos, L. J., Neer, R. M., and Potts, J. T. (1971). Explanation for unusual potency of salmon calcitonin. *Nature New Biol.* **232**, 91–92.

Hadley, M. E. (1972). Functional significance of vertebrate integumental pigmentation. *Amer. Zool.* **12**, 63–76.

(1980). Control of release and mechanism of action of melanocyte stimulating hormone. *Gen. Comp. Endocr.* **40**, 311.

Hall, P. F. (1969). Hormonal control of melanin synthesis in birds. *Gen. Comp. Endocr. Suppl.* **2**, 451–458.

Handler, J. S., Bensinger, R., and Orloff, J. (1968). Effects of adrenergic agents on toad bladder response to ADH, 3′, 5′-AMP, and theophylline. *Amer. J. Physiol.* **215**, 1024–1031.

Hansel, W. and Echternkamp, S. E. (1972). Control of ovarian function in domestic animals. *Amer. Zool.* **12**, 225–243.

Hanstrom, B. (1966). Gross anatomy of the hypophysis in mammals. In *The Pituitary Gland* (edited by G. W. Harris and B. T. Donovan), vol. I, pp. 1–57. Berkeley: University of California Press.

Harder, J. D. and Fleming, M. W. (1981). Estradiol and progesterone profiles indicate a lack of endocrine recognition of pregnancy in the opossum. *Science* **212**, 1400–1402.

Harmeyer, J. and DeLuca, H. F. (1969). Calcium-binding protein and calcium absorption after vitamin D administration. *Arch. Biochem. Biophys.* **133**, 247–254.

Harper, A. A. and Raper, H. S. (1943). Pancreozymin, a stimulant of the secretion of pancreatic enzymes in extracts of the small intestine. *J. Physiol., Lond.* **102**, 115–125.

Harper, C. and Toverud, S. U. (1973). Ability of thyrocalcitonin to protect against hypercalcemia in adult rats. *Endocrinology* **93**, 1354–1359.

Harri, M.N.E. (1972). Effect of season and temperature acclimation on the tissue catecholamine level and utilization in the frog *Rana temporaria*. *Comp. Gen. Pharmacol.* **3**, 101–112.

Harri, M. and Hedenstam, R. (1972). Calorigenic effect of adrenaline and noradrenaline in the frog, *Rana temporaria*. *Comp. Biochem. Physiol.* **41A**, 409–419.

Harrington, R. W. (1968). Delimitation of the thermolabile phenocritical period of sex determination and differentiation in the ontogeny of the normally hermaphroditic fish *Rivulus marmoratus*, Poey. *Physiol. Zool.* **41**, 447–459.

Hartman, F. A. and Brownell, K. A. (1949). *The Adrenal Gland.* London: Henry Kimpton.

Haseltine, F. P. and Ohno, S. (1981). Mechanisms of gonadal differentiation. *Science* **211**, 1272–1278.

Hasan, S. H. and Heller, H. (1968). The clearance of neurohypophysial hormones from the circulation of non-mammalian vertebrates. *Brit. J. Pharmacol.* **33**, 523–530.

Hayashida, T. (1970). Immunological studies with rat pituitary growth hormone (RGH). II. Comparative immunochemical investigation of GH from representatives of various vertebrate classes with monkey antiserum to RGH. *Gen. Comp. Endocr.* **15**, 432–452.

(1971). Biological and immunochemical studies with growth hormone in pituitary extracts of holostean and chondrostean fishes. *Gen. Comp. Endocr.* **17**, 278–280.

(1973). Biological and immunochemical studies with growth hormone in pituitary extracts of elasmobranchs. *Gen. Comp. Endocr.* **20**, 377–385.

Hayashida, T. and Lagios, M. D. (1969). Fish growth hormone: a biological, immunochemical, and ultrastructural study of sturgeon and paddlefish pituitaries. *Gen. Comp. Endocr.* **13**, 403–411.

Hayashida, T. and Lewis, U. J. (1978). Immunochemical and biological studies with antiserum to shark growth hormone. *Gen. Comp. Endocr.* **36**, 530–542.

Hayashida, T., Licht, P. and Nicoll, C. S. (1973). Amphibian pituitary growth hormone and prolactin: immunochemical relatedness to rat growth hormone. *Science* **182**, 169–171.

Hayslett, J. P., Epstein, M., Spector, D., Myers, J. D., and Murdaugh, H. V. (1972). Lack of effect of calcitonin on renal function in the elasmobranch *Squalus acanthias*. *Comp. Biochem. Physiol.* **43A**, 223–226.

Hazelwood, R. L. (1973). The avian endocrine pancreas. *Amer. Zool.* **13**, 699–709.

Heap, R. B. (1972). Role of hormones in pregnancy. In *Hormones in Reproduction* (edited by C. R. Austin and R. V. Short), pp. 73–105. Cambridge University Press.

Heap, R. B, Flint, A. P. and Gadsby, J. E. (1979). Role of embryonic signals in the establishment of pregnancy. *Brit. Med. Bull.* 35, No. 2, 129–135.

Heap, R. B., Perry, J. S., and Challis, J.R.G. (1973). Hormonal maintenance of pregnancy. In *Handbook of Physiology*, Sect. 7 *Endocrinology*, vol. II, *Female reproductive system* (Pt. 2), pp. 217–260. Washington, D.C.: American Physiological Society.

Heap, R. B., Renfree, M. B., and Burton, R. D. (1980). Steroid metabolism in the yolk sac placenta and endometrium of the tammar wallaby, *Macropus eugenii*. *J. Endocr.* **87**, 339–349.

Heding, L. G. (1971). Radioimmunological determination of pancreatic and gut glucagon in plasma. *Diabetologia* 7, 10–19.

Heins, J. N., Garland, J. T., and Daughaday, W. H. (1970). Incorporation of ^{35}S-sulfate into rat cartilage explants *in vitro*: effects of aging on responsiveness to stimulation by sulfation factor. *Endocrinology* 87, 688–692.

Heller, H. (1941). Differentiation of an (amphibian) water balance principle from the antidiuretic principle of the posterior pituitary gland. *J. Physiol., Lond.* **100**, 125–141.

(1966). The hormone content of the vertebrate hypothalamo-neurohypophysial system. *Brit. Med. Bull.* **22**, 227–231.

(1972). The effect of neurohypophysial hormones on the female reproductive tract of lower vertebrates. *Gen. Comp. Endocr. Suppl.* **3**, 703–714.

(1973). The effects of oxytocin and vasopressin during the oestrous cycle and pregnancy on the uterus of a marsupial species, the quokka (*Setonix brachyurus*). *J. Endocr.* **58**, 657–671.

(1974). Molecular aspects of comparative endocrinology. *Gen. Comp. Endocr.* **22**, 315–332.

Heller, H., Ferreri, E., and Leathers, D.H.G. (1970). The effect of neurohypophysial hormones on the amphibian oviduct *in vitro*, with some remarks on the histology of this organ. *J. Endocr.* **47**, 495–509.

Heller, J. (1961). The physiology of the antidiuretic hormone. VIII. The antidiuretic activity in the plasma of the mouse, guinea pig, cat, rabbit and dog. *Physiol. Bohemoslov.* **10**, 167–172

Heller, J. and Štulc, J. (1960). The physiology of the antidiuretic hormone. III. The antidiuretic activity of plasma in normal dehydrated rats. *Physiol. Bohemoslov.* **9**, 93–98.

Henderson, I. W. and Chester Jones, I. (1967). Endocrine influences on the net extrarenal fluxes of sodium and potassium in the European eel (*Anguilla anguilla* L.). *J. Endocr.* **37**, 319–325.

Henderson, I. W., Jotisankasa, V., Mosley, W., and Oguri, M. (1976). Endocrine and environmental influences upon plasma cortisol concentrations and plasma renin activity of the eel, *Anguilla anguilla* L. *J. Endocr.* **70**, 81–95.

Henderson, I. W. and Wales, N.A.M. (1974). Renal diuresis and antidiuresis after injections of arginine vasotocin in the freshwater eel (*Anguilla anguilla* L.). *J. Endocr.* **41**, 487–500.

Henderson, J. R. (1969). Why are the Islets of Langerhans? *Lancet* **ii**, 469–470.

Henderson, K. M. (1979). Gonadotrophic regulation of ovarian activity. *Brit. Med. Bull.* **35**, 161–166.

Henry, H. and Norman, A. W. (1975). Presence of renal 25-hydroxyvitamin-D-1-hydroxylase in species of all vertebrate classes. *Comp. Biochem. Physiol.* **50B**, 431–434.

Henry, H. L. and Norman, A. W. (1978). Vitamin D: two dihydroxylated metabolites are required for normal chicken egg hatchability. *Science* **201**, 835–837.

Herbert, J. (1972). Behavioural patterns. In *Reproduction in Mammals. 4. Reproductive Patterns* (edited by C. R. Austin and R. V. Short), pp. 34–68. Cambridge University Press.

Higgs, D. A. and Eales, J. G. (1973). Measurement of circulating thyroxine in several freshwater teleosts by competitive binding analysis. *Canad. J. Zool.* **51**, 49–53.

Highfill, D. R. and Mead, R. A. (1975*a*). Function of corpora lutea of pregnancy in the viviparous snake, *Thamnophis elegans. Gen. Comp. Endocr.* **27**, 401–407.

(1975*b*). Sources and levels of progesterone during pregnancy in the garter snake, *Thamnophis elegans. Gen. Comp. Endocr.* **27**, 389–400.

Hill, C. W. and Fromm, P. O. (1968). Response of the interrenal gland of rainbow trout (*Salmo gairdneri*) to stress. *Gen. Comp. Endocr.* **11**, 69–77.

Hinde, R. A., Steel, E., and Follett, B. K. (1974). Effect of photoperiod on oestrogen-induced nest-building in ovariectomized or refractory female canaries (*Serinus canarius*). *J. Reprod. Fert.* **40**, 383–399.

Hirano, T. (1969). Effects of hypophysectomy and salinity change on plasma cortisol concentration in the Japanese eel, *Anguilla japonica. Endocr. Japon.* **16**, 557–560.

(1979). Effects of carp urophysial extract on renal function in the eel, *Anguilla japonica. Gunm. Symp. Endocr.* **16**, 59–67.

Hirano, T., Hasegawa, S., Yamauchi, H., and Orimo, H. (1981). Further studies on the absence of hypocalcemic effects of eel calcitonin in the eel, *Anguilla japonica. Gen. Comp. Endocr.* **43**, 42–50.

Hirsch, P. F. and Munson, P. L. (1969). Thyrocalcitonin. *Physiol. Rev.* **49**, 548–622.

Hisaw, F. L. (1926). Experimental relation of the pubic ligament of the guinea pig. *Proc. Soc. Exp. Biol. Med., New York* **23**, 661–663.

(1959). The corpora lutea of elasmobranch fishes. *Anat. Rec.* **133**, 289.

Hobart, P., Crawford, R., Shen, L., Pictet, R., and Rutter, W. J. (1980). Cloning and sequence analysis of cDNAs encoding two distinct somatostatin precursors found in the pancreas of the anglerfish. *Nature, Lond.* **288**, 137–141.

Hobden, A. N., Harding, M., and Lawson, D.E.M. (1980). 1,25-Dihydroxycholecalciferol stimulation of a mitochondrial protein in chick intestinal cells. *Nature, Lond.* **288**, 718–720.

Hoffman, C. W. and Dent, J. N. (1977). Hormonal regulation of cellular proliferation in the epidermis of the red-spotted newt. *Gen. Comp. Endocr.* **32**, 522–530.

Hogben, L. T. (1924). *The Pigmentary Effector System.* p. 67. Edinburgh: Oliver & Boyd.

(1942). Chromatic behaviour. *Proc. Roy. Soc. Lond.* **B131**, 111–136.

Holder, F. C., Schroeder, M. D., Guerne, J. M., and Vivien-Roels, B. (1979). A preliminary comparative immunohistochemical, radioimmunological, and biological study of arginine vasotocin (AVT) in the pineal gland and urophysis of some Teleosteii. *Gen. Comp. Endocr.* **37**, 15–25.

Hollenberg, M. D. (1979). Epidermal growth-factor-urogastrone, a polypeptide acquiring hormonal status. *Vit. Hormone* **37**, 69–110.

Holmes, R. L. and Ball, J. N. (1974). *The Pituitary Gland. A Comparative Account.* Cambridge University Press.

Holmes, W. N. (1978). Endocrine responses in osmoregulation: hormonal adaptation in aquatic birds. In *Environmental Endocrinology* (edited by I. Assenmacher and D. S. Farner), pp. 230–239. Berlin: Springer-Verlag.

Holmes, W. N., Butler, D. G., and Phillips, J. G. (1961). Observations on the effects of maintaining glaucous-winged gulls (*Larus glaucescens*) on fresh water and sea water for long periods. *J. Endocr.* **23**, 53–61.

Holmquist. A. L., Dockray, C. J., Rosenquist, G. L., and Walsh, J. H. (1979). Immunochemical characterization of cholecystokinin-like peptides in lamprey gut and brain. *Gen. Comp. Endocr.* **37**, 474–481.

Holst, J. J. (1978). Physiology of enteric glucagon-like immunoreactivity. In *Gut Hormones* (edited by S. R. Bloom), pp. 383–386. London: Churchill Livingstone.

Holt, W. F. and Idler, D. R. (1975). Influence of the interrenal gland on the rectal gland of the skate. *Comp. Biochem. Physiol.* **50C**, 111–119.

Hontela, A. and Peter, R. E. (1980). Effects of pinealectomy, blinding, and sexual condition on serum gonadotropin levels in the goldfish. *Gen. Comp. Endocr.* **40**, 168–179.

Hornsey, D. J. (1977). Triiodothyronine and thyroxine levels in the thyroid and serum of the sea lamprey *Petromyzon marinus* L. *Gen. Comp. Endocr.* **31**, 381–383.

Horuk, R., Blundell, T. R., Lazarus, N. R., Neville, R.W.J., Stone, D., and Wollmer, A. (1980). A monomeric insulin from the porcupine (*Hystrix cristata*), an Old World hystricomorph. *Nature, Lond.* **286**, 822–824.

Horuk, R., Goodwin, P., O'Connor, K., Neville, R.W.J., Lazarus, N. R., and Stone, D. (1979). Evolutionary change in the insulin receptors of hystricomorph rodents. *Nature, Lond.* **279**, 439–440.

Horwitz, B. A. (1979). Cellular events underlying catecholamine-induced thermogenesis: cation transport in brown adipocytes. *Fedn. Proc.* **38**, 2170–2176.

Howe, A. (1973). The mammalian pars intermedia: a review of its structure and function. *J. Endocr.* **59**, 385–409.

Hughes, J., Smith, T. W., Kosterlitz, H. W., Fothergill, L. A., Morgan, B. A., and Morris, H. R. (1975). Identification of two related pentapeptides from the brain with potent opiate agonist activity. *Nature, Lond.* **258**, 557–579.

Hughes, M. R. and Haussler, M. R. (1978). 1,25-dihydroxyvitamin D_3 receptors on parathyroid glands. *J. Biol. Chem.* **253**, 1065–1073.

Hulbert, A. J. and Hudson, J. W. (1976). Thyroid function in a hibernator, *Spermophilus tridecemlineatus. Amer. J. Physiol.* **230**, 1211–1216.

Humbel, R. E., Bosshard, H. R., and Zahn, H. (1972). Chemistry of insulin. In *Handbook of Physiology*, Sect. 7 *Endocrinology*, vol. I, *Endocrine pancreas*, pp. 111–132. Washington, D.C.: American Physiological Society.

Hunter, C. and Baker, B. I. (1979). The distribution of opiate activity in the trout pituitary gland. *Gen. Comp. Endocr.* **37**, 111–114.

Hutchison, J. B. (1971). Effects of hypothalamic implants of gonadal steroids on courtship behaviour in barbary doves (*Streptopelia risoria*). *J. Endocr.* **50**, 97–113.

Hutchison, R. E. (1975). Effects of ovarian steroids and prolactin on the sequential development of nesting behaviour in female budgerigars. *J. Endocr.* **67**, 29–39.

Hutchison, R. E., Hinde, R. A., and Steel, E. (1967). The effects of oestrogen, progesterone and prolactin on brood patch formation in ovariectomized canaries. *J. Endocr.* **39**, 379–385.

Idelman, S. (1979). The structure of the mammalian adrenal cortex. In *General, Comparative and Clinical Endocrinology of the Adrenal Cortex* (edited by I. Chester Jones and I. W. Henderson), vol. 2, pp. 1–199. London: Academic Press.

Idler, D. R. (editor). (1972). *Steroids in Nonmammalian Vertebrates*. New York: Academic Press.

Idler, D. R. and Burton, M.P.M. (1976). The pronephroi as the site of presumptive interrenal cells in the hagfish *Myxine glutinosa* L. *Comp. Biochem. Physiol.* **53A**, 73–77.

Idler, D. R. and Kane, K. M. (1976). Interrenalectomy and Na-K ATPase activity in the rectal gland of the skate *Raja ocellata. Gen. Comp. Endocr.* **28**, 100–102.

Idler, D. R. and Ng, T. B. (1979). Studies on two types of gonadotropins from both salmon and carp pituitaries. *Gen. Comp. Endocr.* **38**, 421–440.

Idler, D. R., Sangalang, G. B., and Truscott, B. (1972). Corticosteroids in the South American lungfish. *Gen. Comp. Endocr. Suppl.* **3**, 238–244.

Idler, D. R., Shamsuzzaman, K. M., and Burton, M. P. (1978). Isolation of prolactin from salmon pituitary. *Gen. Comp. Endocr.* **35**, 409–418.

Idler, D. R. and Truscott, B. (1972). Corticosteroids in fish. In *Steroids in Nonmammalian Vertebrates* (edited by D. R. Idler), pp. 126–252. New York: Academic Press.

Ireland, M. P. (1969). Effect of urophysectomy in *Gasterosteus aculeatus* on survival in fresh water and sea-water. *J. Endocr.* **43**, 133–134.

    (1973). Effects of arginine vasotocin on sodium and potassium metabolism in *Xenopus laevis* after skin gland stimulation and sympathetic blockade. *Comp. Biochem. Physiol.* **44A**, 487–493.

Isaia, J., Maetz, J., and Haywood, G. P. (1978). Effects of epinephrine on branchial non-electrolyte permeability in rainbow trout. *J. Exp. Biol.* **74**, 227–237.

Iturriza, F. C. (1969). Further evidences for the blocking effect of catecholamines on the secretion of melanocyte-stimulating hormone in toads. *Gen. Comp. Endocr.* **12**, 417–426.

Ivy, A. C. and Oldberg, E. (1928). A hormone mechanism for gall bladder contraction and evacuation. *Amer. J. Physiol.* **86**, 599–613.

Jackson, I.M.D. (1981*a*). Abundance of immunoreactive thyrotropin-releasing hormone-like material in the alfalfa plant. *Endocrinology* **108**, 344–346.

    (1981*b*). Evolutionary significance of the phylogenetic distribution of the mammalian hypothalamic releasing hormones. *Fedn. Proc.* **40**, 2545–2552.

Jackson, I.M.D. and Reichlin, S. (1974). Thyrotropin-releasing hormone distribution in hypothalamic and extrahypothalamic brain tissues of mammalian and submammalian chordates. *Endocrinology* **95**, 854–862.

    (1977). Thyrotropin-releasing hormone: abundance in the skin of the frog, *Rana pipiens. Science* **198**, 414–415.

Jackson, R. G. and Sage, M. (1973). A comparison of the effects of mammalian TSH on the thyroid glands of the teleost *Galeichthys felis* and the elasmobranch *Dasyatis sabina. Comp. Biochem. Physiol.* **44A**, 867–870.

Jacobsen, H., Demandt, A., Moody, A. J., and Sundby, F. (1979). Sequence analysis of porcine gut CLI-1. *Biochem. Biophys. Acta* **493**, 452–459.

Jande, S. S., Maler, L., and Lawson, D.E.M. (1981). Immunohistochemical mapping of vitamin D-dependent calcium-binding protein in brain. *Nature, Lond.* **294**, 765–767.

Janský, L. (1973). Non-shivering thermogenesis and its thermoregulatory significance. *Biol. Rev.* **48**, 85–132.

Janssens, P. A. (1964). The metabolism of the aestivating African lungfish. *Comp. Biochem. Physiol.* **11**, 105–117.

——— (1967). Interference of metyrapone with the actions of cortisol in *Xenopus laevis* Daudin and the laboratory rat. *Gen. Comp. Endocr.* **8**, 94–100.

Janssens, P. A., Vinson, G. P., Chester Jones, I., and Mosley, W. (1965). Amphibian characteristics of the adrenal cortex of the African lungfish (*Protopterus* sp.) *J. Endocr.* **32**, 373–382.

Jensen, E. V. and DeSombre, E. R. (1972). Estrogens and progestins. In *Biochemical Actions of Hormones* (edited by G. Litwack), vol. II, pp. 215–255. New York: Academic Press.

——— (1973). Estrogen-receptor interaction. *Science* **182**, 126–134.

Joel, C. D. (1965). The physiological role of brown adipose tissue. In *Handbook of Physiology*, Sect. 5 *Adipose Tissue*, pp. 59–85. Washington, D.C.: American Physiological Society.

John, M. J., Borjesson, B. W., Walsh, J. R., and Niall, H. D. (1981). Limited sequence homology between porcine and rat relaxins: implications for physiological studies. *Endocrinology* **108**, 726–729.

John, T. M. and George, J. C. (1973). Influence of glucagon and neurohypophysial hormones on plasma free fatty acid levels in the pigeon. *Comp. Biochem. Physiol.* **45A**, 541–547.

Johnson, E. (1977). Environmental effects on the hair follicle. In *The Physiology and Pathophysiology of the Skin*, vol. 4 *The hair follicle* (edited by A. Jarrett), pp. 1389–1416. London: Academic Press.

Johnston, C. I., Davis, J. O., Wright, F. S., and Howards, S. S. (1967). Effects of renin and ACTH on adrenal steroid secretion in the American bullfrog. *Amer. J. Physiol.* **213**, 393–399.

Jollès, J., Burzawa-Gerard, E., Fontaine, Y-A., and Jollès, P. (1977). The evolution of gonadotropins: some molecular data concerning a non-mammalian pituitary gonadotropin, the hormone from a teleost fish (*Cyprinus carpio* L.). *Biochimie* **59**, 893–898.

Jorgenson, C. B., Hede, K-E., and Larsen, L. O. (1978). Environmental control of annual ovarian cycle in the toad *Bufo bufo* L.: role of temperature. In *Environmental Physiology* (edited by I. Assenmacher and D. S. Farner), pp. 28–36. Berlin: Springer-Verlag.

Jorgenson, C. B., Larsen, L. O., and Rosenkilde, P. (1965). Hormonal dependency of molting in amphibians: effect of radiothyroidectomy in the toad *Bufo bufo* L. *Gen. Comp. Endocr.* **5**, 248–251.

Jorgenson, C. B. and Vijayakumar, S. (1970). Annual oviduct cycle and its control in the toad *Bufo bufo* L. *Gen. Comp. Endocr.* **14**, 404–411.

Jörnvall, H., Carlström, A., Pettersson, T., Jacobssen, B., Persson, M., and Mutt, V. (1981). Structural homologies between prealbumin, gastrointestinal prohormones and other proteins. *Nature, Lond.* **291**, 261–263.

Jorpes, E. and Mutt, V. (1966). Cholecystokinin and pancreozymin, one single molecule? *Acta Physiol. Scand.* **66**, 196–202.

Joss, J.M.P. (1973). Pineal-gonad relationships in the lamprey *Lampetra fluviatilis*. *Gen. Comp. Endocr.* **21**, 118–122.

Josso, N., Picard, J-Y., and Tran, D. (1977). The antimullerian hormone. *Rec. Prog. Hormone Res.* **33**, 117–163.

Jost, A. (1971). Hormones in development; past and present prospects. In *Hormones and Development* (edited by M. Hamburgh and E.J.W. Barrington), pp. 1–18. New York: Apple-Century-Crofts.

    (1979). Basic sexual trends in the development of vertebrates. In *Sex, Hormones and Behaviour*, Ciba Foundation Symposium 62, pp. 5–18. Amsterdam: Excerpta Medica.

Kamiya, M. (1972). Sodium-potassium-activated adenosinetriphosphatase in isolated chloride cells from eel gills. *Comp. Biochem. Physiol.* **43B**, 611–617.

Kanis, J. A., Cundy, T., Bartlett, M., Smith, R., Heynen, G., Warner, G. T., and Russell, R.G.G. (1978). Is 24,25-dihydroxycholecalciferol a calcium-regulating hormone in man? *Brit. Med. J.* **1**, 1382–1386.

Karlson, P. (1963). New concepts on the mode of action of hormones. *Perspectives in Biology and Medicine* **6**, 203–214.

Karsch, F. J. (1980). Seasonal reproduction: a saga of reversible fertility. *The Physiologist* **23**, 29–38.

Kastin, A. J., Schally, A. V., and Kostrzewa, R. M. (1980). Possible aminergic mediation of MSH release and of the CNS effects of MSH and MIF-I. *Fedn. Proc.* **39**, 2931–2936.

Keller, N., Richardson, I. U., and Yates, F. E. (1969). Protein binding and the biological activity of corticosteroids: *in vivo* induction of hepatic and pancreatic alanine amino-transferases by corticosteroids in normal and estrogen-treated rats. *Endocrinology* **84**, 49–62.

Kelley, D. B. (1978). Neuroanatomical correlates of hormone sensitive behaviors in frogs and birds. *Amer. Zool.* **18**, 477–488.

    (1981). Social signals — an overview. *Amer. Zool.* **21**, 111–116.

Kenny, A. D. (1971). Determination of calcitonin in plasma by bioassay. *Endocrinology* **89**, 1005–1013.

    (1976). Vitamin D metabolism: physiological regulation in egg-laying Japanese quail. *Amer. J. Physiol.* **230**, 1609–1615.

Kenny, A. D. and Dacke, C. G. (1974). The hypercalcaemic response to parathyroid hormone in Japanese quail. *J. Endocr.* **62**, 15–23.

Kerkof, P. R., Boschwitz, D., and Gorbman, A. (1973). The response of hagfish thyroid tissue to thyroid inhibitors and to mammalian thyroid-stimulating hormone. *Gen. Comp. Endocr.* **21**, 231–240.

Kesner, J. S., Convey, E. M., and Anderson, C. R. (1981). Evidence that estradiol induces the preovulatory LH surge in cattle by increasing pituitary sensitivity to LHRH and then increasing LHRH release. *Endocrinology* **108**, 1386–1391.

Keutmann, H. T., Sauer, M. M., Hendy, G. N., O'Riordan, J.L.H., and Potts, J. T. (1978). Complete amino acid sequence of human parathyroid hormone. *Biochemistry* **17**, 5723–5729.

Kikuyama, S., Yamamoto, K., and Mayumi, M. (1980). Growth-promoting and antimetamorphic hormone in pituitary glands of bullfrogs. *Gen. Comp. Endocr.* **41**, 212–216.

Kimmel, J. R. and Pollock, H. G. (1981). Target organs for avian pancreatic polypeptide. *Endocrinology* **109**, 1693–1699.

Kimmel, J. R., Pollock, H. G., and Hayden, L. J. (1978). Biological activity of the avian PP in the chicken. In *Gut Hormones* (edited by S. R. Bloom), pp. 234–241. London: Churchill Livingstone.

King, J. A. and Millar, R. P. (1979*a*). Phylogenetic and anatomical distribution of somatostatin in vertebrates. *Endocrinology* **105**, 1322–1329.

(1979*b*). Hypothalamic luteinizing hormone-releasing hormone content in relation to the seasonal reproductive cycle of *Xenopus laevis*. *Gen. Comp. Endocr.* **39**, 309–312.

(1980*a*). Comparative aspects of luteinizing hormone-releasing hormone structure and function in vertebrate phylogeny. *Endocrinology* **106**, 707–717.

(1980*b*). Radioimmunoassay of methionine[5]-enkephalin sulphoxide: phylogenetic and anatomical distribution. *Peptides* **1**, 211–216.

(1981). TRH, GH-RIH, and LH-RH in metamorphosing *Xenopus laevis*. *Gen. Comp. Endocr.* **44**, 20–27.

King, J. R. and Farner, D. S. (1965). Studies of fat deposition in migratory birds. *Ann. N.Y. Acad. Sci.* **131**, 422–440.

Kirschner, L. B. (1980). Comparison of vertebrate salt-excreting organs. *Amer. J. Physiol.* **238**, R219–R223.

Kleiber, M. (1961). *The Fire of Life. An Introduction to Animal Energetics*, p. 312. New York: Wiley.

Klicka, J. and Mahmoud, I. Y. (1972). Conversion of pregenolone-4[14]C to progesterone-4[14]C by turtle corpus luteum. *Gen. Comp. Endocr.* **19**, 367–369.

(1977). The effects of hormones on the reproductive physiology of the painted turtle, *Chrysemys picta*. *Gen. Comp. Endocr.* **31**, 407–413.

Knobil, E. (1980). The neuroendocrine control of the menstrual cycle. *Rec. Prog. Hormone Res.* **36**, 53–88.

Kobayashi, H. (1981). Angiotensin-induced drinking in parrots. *Gen. Comp. Endocr.* **43**, 399–401.

Kobayashi, H., Uemura, H., Wada, M., and Takei, Y. (1979). Ecological adaptation of angiotensin-induced thirst mechanism in tetrapods. *Gen. Comp. Endocr.* **38**, 93–104.

Kodicek, E. (1974). The story of vitamin D from vitamin to hormone. *Lancet* **i**, 325–329.

Koike, T. I., Pryor, L. R., Neldon, H. L., and Venable, R. S. (1977). Effect of water deprivation of plasma radioimmunoassayable arginine vasotocin in conscious chickens (*Gallus domesticus*). *Gen. Comp. Endocr.* **33**, 359–364.

Kono, T., Robinson, F. W., and Sarver, J. A. (1975). Insulin sensitive phosphodiesterase. Its localization, hormone stimulation and oxidative stabilization. *J. Biol. Chem.* **250**, 7826–7835.

Korc, M., Owenbach, D., Quinto, C., and Rutter, W. J. (1981). Pancreatic islet-acinar cell interaction: amylase messenger RNA levels are determined by insulin. *Science* **213**, 351–363.

Kourides, I. A. and Weintraub, B. D. (1979). mRNA-directed biosynthesis of α-subunit of thyrotropin: translation in cell-free and whole-cell systems. *Proc. Natl. Acad. Sci., USA* **76**, 298–302.

Kouvonen, I. and Gräsbeck, R. (1979). A simplified technique to isolate the porcine and human ileal intrinsic factor receptors and studies on their subunit structures. *Biochem. Biophys. Res. Commun.* **86**, 358–364.

Krebs, E. G. (1972). Protein kinases. *Curr. Top. Cell Regul.* **5**, 99–133.

Krieger, D. T. (1971). The hypothalamus and neuroendocrinology. *Hospital Practice* September, 87–99.

(1972). Circadian corticosteroid periodicity: critical period for abolition by neonatal injection of corticosteroid. *Science* **178**, 1205–1207.

Krieger, D. T. and Liotta, A. S. (1979). Pituitary hormones in brain: where, how and why? *Science* **205**, 366–372.

Krishna, G., Hynie, S., and Brodie, B. B. (1968). Effects of thyroid hormones on adenyl cyclase in adipose tissue and on free fatty acid mobilization. *Proc. Natl. Acad. Sci., USA* **59**, 884–889.

Krishnamurthy, V. G. and Bern, H. A. (1969). Correlative histological study of the corpuscles of Stannius and the juxtaglomerular cells of teleost fishes. *Gen. Comp. Endocr.* **13**, 313–335.

Kumar, R. (1980). The metabolism of 1,25-dihydroxyvitamin D_3. *Endocrine Rev.* **1**, 258–267.

Kumar, M. A. and Sturtridge, W. C. (1973). The physiological role of calcitonin assessed through chronic calcitonin deficiency in rats. *J. Physiol., Lond.* **233**, 33–43.

Kwok, S. and Bryant-Greenwood, G. (1977). Primary structure of porcine relaxin: homology with insulin and related growth factors. *Nature, Lond.* **267**, 544–546.

Lagios, M. D. and Stasko-Concannon, S. (1979). Presumptive interrenal tissue (adrenocortical homolog) in the coelacanth *Latimeria chalumnae*. *Gen. Comp. Endocr.* **37**, 404–406.

Lam, T. J. (1972). Prolactin and hydromineral metabolism in fishes. *Gen. Comp. Endocr. Suppl.* **3**, 328–338.

Landas, S., Phillips, M. I., Stamler, J. F., and Raizada, M. K. (1980). Visualization of specific angiotensin II binding sites in the brain by fluorescent microscopy. *Science* **210**, 791–793.

Langslow, D. R., Kimmel, J. R., and Pollock, H. G. (1973). Studies of the distribution of a new avian pancreatic polypeptide and insulin among birds, reptiles, amphibians and mammals. *Endocrinology* **93**, 558–565.

LaPointe, J. L. and Jacobson, E. R. (1974). Hyperglycemic effect of neurohypophysial hormones in the lizard, *Klauberina riversiana*. *Gen. Comp. Endocr.* **22**, 135–136.

Larsen, L. O. (1965). Effects of hypophysectomy in the cyclostome, *Lampetra fluviatilis* (L) Gray. *Gen. Comp. Endocr.* **5**, 16–30.

(1969). Effects of gonadectomy in the cyclostome, *Lampetra fluviatilis*. *Gen. Comp. Endocr.* **13**, 516–517.

(1973). Development in adult, freshwater river lampreys and its hormonal control. Thesis: University of Copenhagen.

(1978). Hormonal control of sexual maturation in lampreys. In *Comparative Endocrinology* (edited by F. J. Gaillard and H. H. Boer), pp. 105–108. Amsterdam: Elsevier North-Holland.

(1980). Physiology of adult lampreys, with special regard to natural starvation, reproduction, and death after spawning. *Can. J. Fish. Aquat. Sci.* **37**, 1762–1779.

Larsen, L. O. and Rosenkilde, P. (1971). Iodine metabolism in normal, hypophysectomized, and thyrotropin-treated river lampreys, *Lampetra fluviatilis* (Gray) L. (Cyclostomata). *Gen. Comp. Endocr.* **17**, 94–104.

Larsson, A. and Lewander, K. (1972). Effects of glucagon administration to eels (*Anguilla anguilla* L.). *Comp. Biochem. Physiol.* **43A**, 831–836.

Larsson, A. L. (1973). Metabolic effects of epinephrine and norepinephrine in the eel *Anguilla anguilla* L. *Gen. Comp. Endocr.* **20**, 155–167.

Larsson, L-I. (1980). Peptide secretory pathways in GI tract: cytochemical contributions to regulatory physiology of the gut. *Amer. J. Physiol.* **239**, G237–G246.

Larsson, L-I. and Rehfeld, J. F. (1977). Evidence for a common evolutionary origin of gastrin and cholecystokinin. *Nature, Lond.* **269**, 335–338.

(1978). Evolution of CCK-like hormones. In *Gut Hormones* (edited by S. R. Bloom), pp. 68–73. London: Churchill Livingstone.

Lawson, D.E.M., Fraser, D. R., Kodicek, E., Morris, H. R., and Williams, D. H. (1971). Identification of 1,25-dihydroxycholecalciferol, a new kidney hormone controlling calcium metabolism. *Nature, Lond.* **230**, 228–230.

LeBrie, S. J. (1972). Endocrines and water and electrolyte balance in reptiles. *Fedn. Proc.* **31**, 1599–1608.

Lederis, K. (1973). Current studies on urotensin. *Amer. Zool.* **13**, 771–773.

Lee, A. K. and Mercer, E. H. (1967). Cocoon surrounding desert-dwelling frogs. *Science* **157**, 87–88.

Lee, T. H., Lee, M. S., and Lu, M-Y. (1972). Effects of α-MSH on melanogenesis and tyrosinase in B-16 melanoma. *Endocrinology* **91**, 1180–1188.

LeFevre, M. D. (1973). Effects of aldosterone on the isolated substrate-depleted turtle bladder. *Amer. J. Physiol.* **225**, 1252–1256.

Legan, S. J., Karsch, F. J., and Foster, D. L. (1977). The endocrine control of seasonal reproductive function in the ewe: a marked change in response to negative feedback action of estradiol on luteinizing hormone secretion. *Endocrinology* **101**, 818–823.

Leibson, L. and Plisetskaya, E. M. (1968). Effect of insulin on blood sugar level and glycogen content in organs of some cyclostomes and fish. *Gen. Comp. Endocr.* **11**, 381–392.

Leloup, J. and Fontaine, M. (1960). Iodine metabolism in lower vertebrates. *Ann N.Y. Acad. Sci.* **86**, 316–353.

Lemon, M. (1972). Peripheral plasma progesterone during pregnancy and the oestrous cycle in the tammar wallaby, *Macropus eugenii. J. Endocr.* **55**, 63–71.

Leopold, A. S., Erwin, M., Oh, J. C., and Browning, B. (1976). Phytoestrogens: adverse effects on reproduction of California quail. *Science* **191**, 98–100.

Levitin, H. P. (1980). Monoaminergic control of MSH release in the lizard *Anolis carolinensis. Gen. Comp. Endocr.* **41**, 279–286.

Lewander, K., Dave, G., Johansson-Sjöbeck, M. L., Larsson, A., and Lidman, U. (1976). Metabolic effects of insulin in the European eel *Anguilla anguilla* L. *Gen. Comp. Endocr.* **29**, 455–467.

Lewis, S. A. and Diamond, J. M. (1976). Na^+ transport by rabbit urinary bladder, a tight epithelium. *J. Membrane Biol.* **28**, 1–40.

Li, C. H. (1969). Recent studies on the chemistry of human growth hormone. *Colloq. Int. C.N.R.S.* **177**, 175–179.

(1972). Recent knowledge of the chemistry of lactogenic hormones. In *Lactogenic Hormones* (edited by G.E.W. Wolstenholme and J. Knight), pp. 7–22. London: Churchill.

Li, C. H., Barnafi, L., Chrétien, M., and Chung, D. (1965). Isolation and amino-acid sequence of β-LPH from sheep pituitary glands. *Nature, Lond.* **208**, 1093–1094.

Li, C. H. and Chung, D. (1976). Primary structure of human β-lipotrophin *Nature, Lond.* **260**, 622–624.

Li, C. H., Chung, D., and Doneen, B. A. (1976). Isolation, characterization and opiate activity of β-endorphin from human pituitary glands. *Biochem. Biophys. Res. Commun.* **72**, 1542–1547.

Licht, P. (1970). Effects of mammalian gonadotropins (ovine FSH and LH) in female lizards. *Gen. Comp. Endocr.* **14**, 98–106.

(1972). Environmental physiology of reptilian breeding cycles: role of temperature. *Gen. Comp. Endocr. Suppl.* **3**, 477–487.

(1979). Reproductive endocrinology of reptiles and amphibians: gonadotropins. *Ann. Rev. Physiol.* **41**, 337–351.

Licht, P. and Bona Gallo, A. B. (1978). Immunochemical relatedness among pituitary follicle-stimulating hormones of tetrapod vertebrates. *Gen. Comp. Endocr.* **36**, 575–584.

Licht, P., Farmer, S. W., Gallo, A. B., and Papkoff, H. (1979). Pituitary gonadotropins in snakes. *Gen. Comp. Endocr.* **39**, 34–52.

Licht, P., Farmer, S. W., and Papkoff, H. (1978). Biological activity of hybrid combinations of ovine and sea turtle subunits. *Gen. Comp. Endocr.* **35**, 289–294.

Licht, P. and Hoyer, H. (1968). Somatotropic effects of exogenous prolactin and growth hormone in juvenile lizards (*Lacerta s. sicula*). *Gen. Comp. Endocr.* **11**, 338–346.

Licht, P. and Papkoff, H. (1974). Separation of two distinct gonadotropins from the pituitary gland of the snapping turtle (*Chelydra serpentina*). *Gen. Comp. Endocr.* **22**, 218–237.

Licht, P., Papkoff, H., Farmer, S. W., Muller, C. H., Tsui, H. W., and Crews, D. (1977). Evolution of gonadotropin structure and function. *Rec. Progr. Hormone Res.* **33**, 169–243.

Licht, P. and Stockell Hartree, A. (1971). Actions of mammalian, avian and piscine gonadotrophins in the lizard. *J. Endocr.* **49**, 113–124.

Lieberburg, I., Macluskey, N. J., Roy, E. J., and McEwen, B. S. (1978). Sex steroid receptors in the perinatal rat brain. *Amer. Zool.* **18**, 539–544.

Liggins, G. C. (1979). Initiation of parturition. *Brit. Med. Bull.* **35**, 145–150.

Liggins, G. C., Fairclough, R. J., Grieves, S. A., Forster, C. S., and Knox, B. S. (1977). Parturition in the sheep. In *The Fetus and Birth*, Ciba Foundation Symposium 47, pp. 5–25. Amsterdam: Elsevier, North-Holland.

Lillie, F. R. (1917). The free-martin; a study of the actions of sex hormones in the foetal life of cattle. *J. Exp. Biol.* **23**, 371–452.

Lin, T-M., Evans, D. C., Chance, R. E., and Spray, G. F. (1977). Bovine pancreatic peptide: action on gastric and pancreatic secretion in dogs. *Amer. J. Physiol.* **232**, E311–E315.

Lincoln, D. W. and Renfree, M. B. (1981). Milk ejection in a marsupial, *Macropus agilis*. *Nature, Lond.* **289**, 504–506.

Lincoln, G. A. (1979). Use of a pulsed infusion of luteinizing hormone releasing hormone to mimic seasonally induced endocrine changes in the ram. *J. Endocr.* **83**, 251–260.

Lincoln, G. A. and Short, R. V. (1980). Seasonal breeding: nature's contraceptive. *Rec. Prog. Hormone Res.* **36**, 1–52.

Ling, J. K. (1972). Adaptive functions of vertebrate molting cycles. *Amer. Zool.* **12**, 77–93.

Lofts, B. (1964). Seasonal changes in the functional activity of the interstitial and spermatogenetic tissues of the green frog *Rana esculenta*. *Gen. Comp. Endocr.* **4**, 550–562.

(1968). Patterns of testicular activity. In *Perspectives in Endocrinology. Hormones in the Lives of Lower Vertebrates* (edited by E.J.W. Barrington and C. B. Jorgenson), pp. 239–304. New York: Academic Press.

(1969). Seasonal cycles in reptilian testes. *Gen. Comp. Endocr. Suppl.* **2**, 147–155.

Lofts, B. and Bern, H. A. (1972). The functional morphology of steroidogenic tissues. In *Steroids in Nonmammalian Vertebrates* (edited by D. R. Idler), pp. 37–125. New York: Academic Press.

Lofts, B., Follett, B. K., and Murton, R. K. (1970). Temporal changes in the pituitary gonadal axis. *Mem. Soc. Endocr.* **18**, 545–575.

Lofts, B., Murton, R. K., and Thearle, R.J.P. (1973). The effects of testosterone propionate and gonadotropins on the bill pigmentation and testes of the house sparrow (*Passer domesticus*). *Gen. Comp. Endocr.* **21**, 202–209.

Lofts, B., Phillips, J. G., and Tam, W. H. (1971). Seasonal changes in the histology of the adrenal gland of the cobra, *Naja naja. Gen. Comp. Endocr.* **16**, 121–131.

Lomedico, P. T., Chan, S. J., Steiner, D. F., and Saunders, G. F. (1977). Immunological and chemical characteristics of bovine prepro-insulin. *J. Biol. Chem.* **252**, 7971–7978.

Lonovics, J., Guzman, S., Devitt, P. G., Hejtmancik, K. E., Suddith, R. L., Rayford, P. L., and Thompson, J. C. (1981). Action of pancreatic polypeptide on exocrine pancreas and on release of cholecystokinin and secretin. *Endocrinology* **108**, 1925–1930.

Lopez, E., Chartier-Baraduc, M-M., and Deville, J. (1971). Mise en evidence de l'action de la calcitonine porcine sur l'os de la Truite *Salmo gairdnerii* soumise à un traitement déminéralisant. *C. R. Acad. Sci. Paris* **272**, 2600–2603.

Lord, J.A.H., Waterfield, A. A., Hughes, J., and Kosterlitz, H. W. (1977). Endogenous opioid peptides: multiple agonists and receptors. *Nature, Lond.* **267**, 495–499.

Loretz, C. A. and Bern, H. A. (1981). Stimulation of sodium transport across the teleost urinary bladder by urotensin II. *Gen. Comp. Endocr.* **43**, 325–330.

Lowry, P. J. and Scott, A. P. (1975). The evolution of vertebrate corticotrophin and melanocyte stimulating hormone. *Gen. Comp. Endocr.* **26**, 16–23.

Luckey, T. D. (1973). Perspective of thymic hormones. In *Thymic Hormones* (edited by T. D. Luckey), pp. 275–314. Baltimore: University Park Press.

Lutherer, L. O., Fregly, M. J., and Anton, A. H. (1969). An interrelationship between theophylline and catecholamines in the hypothyroid rat acutely exposed to cold. *Fedn. Proc.* **28**, 1238–1242.

Lyons, W. R. (1958). Hormonal synergism in mammary growth. *Proc. Roy. Soc. Lond.* **B149**, 303–325.

Ma, S.W.Y., Shami, Y., Messer, H. H., and Copp, D. H. (1974). Properties of Ca^{2+}-ATPase from the gill of rainbow trout (*Salmo gairdneri*). *Biochim. Biophys. Acta* **345**, 243–251.

MacIntyre, I., Colston, K. W., Evans, I.M.A., Lopez, E., MacAuley, S. J., Piegnoux-Deville, J., Spanos, E., and Szelke, M. (1976). Regulation of vitamin D: an evolutionary view. *Clin. Endocr.* **5**, Suppl. 85s–95s.

MacKenzie, D. S., Licht, P., and Papkoff, H. (1978). Thyrotropin from amphibian (*Rana catesbeiana*) pituitaries and evidence for heterothyrotropic activity of bullfrog luteinizing hormone in reptiles. *Gen. Comp. Endocr.* **36**, 566–574.

MacLeod, R. M. and Lehmeyer, J. E. (1974). Studies on the mechanism of the dopamine-mediated inhibition of prolactin secretion. *Endocrinology* **94**, 1077–1085.

Maclusky, N. J. and Naftolin, F. (1981). Sexual differentiation of the central nervous system. *Science* **211**, 1294–1303.

Maderson, P.F.A., Chiu, K. W., and Phillips, J. G. (1970). Endocrine-epidermal relationships in squamate reptiles. *Mem. Soc. Endocr.* **18**, 259–284.

Maderson, P.F.A. and Licht, P. (1967). Epidermal morphology and sloughing frequency in normal and prolactin treated *Anolis carolinensis* (Iguanidae, Lacertilia). *J. Morphol.* **123**, 157–172.

Maetz, J. (1969). Observations on the role of the pituitary-interrenal axis in the ion regulation of the eel and other teleosts. *Gen. Comp. Endocr. Suppl.* **2**, 299–316.

(1971). Fish gills: mechanisms of salt transfer in fresh water and sea water. *Phil. Trans. Roy. Soc. Lond.* **B262**, 209–249.

Maetz, J., Mayer, N., and Chartier-Baraduc, M. M. (1967). La balance minérale du sodium chez *Anguilla anguilla* en eau de mer, en eau douce et au cours de transfert d'un milieu à l'autre: effets de l'hypophysectomie et de la prolactine. *Gen. comp. Endocr.* **8**, 177–188.

Maetz, J. and Rankin, J. C. (1969). Quelques aspects du rôle biologique des hormones neurohypophysaires chez les poissons. *Colloq. Int. C.N.R.S.* **177**, 45–54.

Maher, M. J. (1965). The role of the thyroid gland in the oxygen consumption of lizards. *Gen. Comp. Endocr.* **5**, 320–325.

Manning, M. and Sawyer, W. H. (1970). 4-Threonine-oxytocin: a more active and specific oxytocic agent than oxytocin. *Nature, Lond.* **227**, 715–716.

Manns, J. G., Boda, J. M., and Willes, R. F. (1967). Probable role of propionate and butyrate in control of insulin secretion in sheep. *Amer. J. Physiol.* **212**, 756–764.

Marshall, F.H.A. (1956). The breeding season. In *Marshall's Physiology of Reproduction* (edited by A. S. Parkes), vol. I (Pt. 1), pp. 1–42. London: Longmans.

Marshall, W. S. and Bern, H. A. (1979). Teleostean urophysis: urotensin II and ion transport across the isolated skin of a marine teleost. *Science* **204**, 519–521.

(1980). Ion transport across the isolated skin of the teleost *Gillichthys mirabilis*. In *Epithelial Transport in Lower Vertebrates* (edited by B. Lahlou), pp. 337–350. Cambridge University Press.

Martin, B. (1975). Steroid-protein interactions in nonmammalian vertebrates. *Gen. Comp. Endocr.* **25**, 42–51.

Martin, R. and Voigt, K. H. (1981). Enkephalins co-exist with oxytocin and vasopressin in nerve terminals of rat neurohypophysis. *Nature, Lond.* **289**, 502–504.

Marx, S. J. and Aurbach, G. D. (1975). Renal receptors for calcitonin: coordinate occurrence with calcitonin-activated adenylate cyclase. *Endocrinology* **97**, 448–453.

Marx, S. J., Woodward, C. J., and Aurbach, G. D. (1972). Calcitonin receptors in kidney and bone. *Science* **178**, 999–1001.

Matsuoka, H., Mulrow, P. J., Franco-Saenz, R., and Li, C. H. (1981). Stimulation of aldosterone production by β-melanotropin. *Nature, Lond.* **291**, 155–156.

Matsuura, S., Ohashi, M., Chen, H. C., Shownkeen, R. C., Stockell Hartree, A., Reichert, L. E., Stevens, V. C., and Powell, J. E. (1980). Physicochemical and immunological characterization of an HCG-like substance from human pituitary glands. *Nature, Lond.* **286**, 740–741.

Mautalen, C. A. (1972). Mechanism of action of *Solanum malacoxylon* upon calcium and phosphate metabolism in the rabbit. *Endocrinology* **90**, 563–567.

Mayer, N., Maetz, J., Chan, D.K.O., Forster, M., and Chester Jones, I. (1967). Cortisol, a sodium excreting factor in the eel (*Anguilla anguilla* L.) adapted to sea water. *Nature, Lond.* **214**, 1118–1120.

McClanahan, L. (1967). Adaptations of the spadefoot toad *Scaphiopus couchi* to desert environments. *Comp. Biochem. Physiol.* **20**, 73–99.

McDonald, I. R. (1980). Physiology of the adrenal cortex in mammals. In *Comparative Physiology: Primitive Mammals* (edited by K. Schmidt-Nielsen, L. Bolis, and C. R. Taylor), pp. 316–323. Cambridge University Press.

McEwen, B. S. (1981). Neural gonadal steroid actions. *Science* **211**, 1303–1311.

McLean, C. and Lowry, P. J. (1981). Natural occurrence but lack of melanotropic activity of γ-MSH in fish. *Nature, Lond.* **290**, 314–343.

McMillan, J. E. and Wilkinson, R. F. (1972). The effect of pancreatic hormones on blood glucose in *Ambystoma annulatum*. *Copeia* 1972 (no. 4), 664–668.

McNabb, R. A. (1969). The effects of thyroxine on glycogen stores and oxygen consumption in the leopard frog, *Rana pipiens*. *Gen. Comp. Endocr.* **12**, 276–281.

McNatty, K. P., Cashmore, M., and Young, A. (1972). Diurnal variation in plasma cortisol levels in sheep. *J. Endocr.* **54**, 361–362.

Means, A. R. and Dedman, J. R. (1980). Calmodulin – an intracellular calcium receptor. *Nature,`Lond.* **285**, 73–77.

Medica, P. A., Turner, F. B., and Smith, D. D. (1973). Hormonal induction of color change in female leopard lizards *Crotaphytus wislizenii*. *Copeia* 1973 (no. 4), 658–661.

Meier, A. H. (1970). Thyroxine phases the circadian fattening response to prolactin. *Proc. Soc. Exp. Biol. Med., New York* **133**, 1113–1116.

Meier, A. H. and Farner, D. S. (1964). A possible endocrine basis for premigratory fattening in the white-crowned sparrow, *Zonotrichia leucophrys gambelli* (Nuttall). *Gen. Comp. Endocr.* **4**, 584–595.

Meier, A. H., Trobec, T. N., Joseph, M. M., and John, T. M. (1971). Temporal synergism of prolactin and adrenal steroids in the regulation of fat stores. *Proc. Soc. Exp. Biol. Med., New York* **137**, 408–415.

Meier, S. and Solursh, M. (1972). The comparative effects of several mammalian growth hormones on sulfate incorporation into acid mucopolysaccharides by cultured chick embryo chondrocytes. *Endocrinology* **90**, 1447–1451.

Milet, C., Peignoux-Deville, J., and Martelly, E. (1979). Gill calcium fluxes in the eel, *Anguilla anguilla* (L.). Effects of Stannius corpuscles and ultimobranchial body. *Comp. Biochem. Physiol.* **63A**, 63–70.

Milhaud, G., Rankin, J. C., Bolis, L., and Benson, A. A. (1977). Calcitonin: its hormonal action on the gill. *Proc. Natl. Acad. Sci., USA* **74**, 4693–4696.

Minick, M. C. and Chavin, W. (1973). Effects of catecholamines upon serum FFA levels in normal and diabetic goldfish, *Carassius auratus* L. *Comp. Biochem. Physiol.* **44A**, 1003–1008.

Mirsky, I. A. (1965). Effect of biologically active peptides on adipose tissue. In *Handbook of Physiology*, Sect. 5 *Adipose Tissue*, pp. 407–415. Washington, D.C.: American Physiological Society.

Mise, T. and Bahl, Om. P. (1980). Assignment of disulfide bonds in the α subunit of human chorionic gonadotropin. *J. Biol. Chem.* **255**, 8516–8522.

Moore, W. T. and Ward, D. N. (1980). Pregnant mare serum gonadotropin. *J. Biol. Chem.* **255**, 6930–6936.

Moreau, R., Raoelison, C., and Sutter, B. Ch. J. (1981). An intestinal insulin-like molecule in *Apis mellifica* L. (Hymeoptera). *Comp. Biochem. Physiol.* **69A**, 79–83.

Morgan, F. J., Birken, S., and Canfield, R. E. (1975). The amino acid sequence of human chorionic gonadotropin. *J. Biol. Chem.* **250**, 5247–5258.

Morley, M., Scanes, C. G., and Chadwick, A. (1981). The effect of ovine prolactin of sodium and water transport across the intestine of the fowl (*Gallus domesticus*). *Comp. Biochem. Physiol.* **68A**, 61–66.

Morris, T. R. (1979). The influence of light on ovulation in domestic birds. In *Animal Reproduction* (edited by H. W. Hawk), Beltsville Symposia in Agricultural Research, 3, pp. 307–322. Montclair, N.J.: Allenheld Osmun.

Mueller, W. J., Brubaker, R. L., Gay, C. V., and Boelkins, J. N. (1973a). Mechanisms of bone resorption in laying hens. *Fedn. Proc.* **32**, 1951–1954.

Mueller, W. J., Hall, K. L., Maurer, C. A., and Joshua, I. G. (1973b). Plasma calcium and inorganic phosphate response of laying hens to parathyroid hormone. *Endocrinology* **92**, 853–856.

Muggeo, M., Ginsberg, B. H., Roth, J., Neville, D. M., De Meyts, P., and Kahn, C. R. (1979). The insulin receptor in vertebrates is functionally more conserved during evolution than insulin itself. *Endocrinology* **104**, 1393–1402.

Munday, K. A., Parsons, B. J., and Poat, J. A. (1971). The effect of angiotensin on cation transport by rat kidney cortex slices. *J. Physiol., Lond.* **215**, 269–282.

Munemura, M., Cote, T. E., Tsuruta, K., and Kebabian, J. W. (1980). The dopamine receptor in the intermediate lobe of the rat pituitary gland: pharmacological characterization. *Endocrinology* **107**, 1676–1683.

Munro, C. J., McNatty, K. P., and Renshaw, L. (1980). Circaannual rhythms of prolactin secretion in ewes and the effect of pinealectomy. *J. Endocr.* **84**, 83–89.

Muramatsu, I. and Bern, H. A. (1979). Effects of urotensin I on the isolated aorta of the chukar, *Alectoris graeca*. *Gen. Comp. Endocr.* **37**, 150–155.

Murat, J. C., Plisetskaya, E. M., and Woo, N.Y.S. (1981). Endocrine control of nutrition in cyclostomes and fish. *Comp. Biochem. Physiol.* **68A**, 149–158.

Myant, N. B. (1971). The role of thyroid hormone in the fetal and postnatal development of mammals. In *Hormones in Development* (edited by M. Hamburgh and E. J. Barrington), pp. 465–471. New York: Appleton-Century-Crofts.

Nakanishi, S., Inoue, A., Kita, T., Nakamura, M., Chang, A.C.Y., Cohen, S. N., and Numa, S. (1979). Nucleotide sequence of cloned cDNA for bovine corticotropin-β-lipotropin precursor. *Nature, Lond.* **278**, 423–427.

Nalbandov, A. V. (1969). Specificity of action of gonadotrophic hormones. *Colloq. Int. C.N.R.S.* **177**, 335–342.

(1973). Control of luteal function in mammals. In *Handbook of Physiology*, Sect. 7 *Endocrinology*, vol. II, *Female Reproductive system* (Pt. 1), pp. 153–167. Washington, D.C.: American Physiological Society.

Nath, P. and Sundararaj, B. I. (1981). Induction of vitellogenesis in the hypophysectomized catfish, *Heteropneustes fossilis* (Bloch): effects of piscine and mammalian hormones. *Gen. Comp. Endocr.* **43**, 191–200.

Nature (1973). Reference to newt in Siberia. **242**, 369.

Nelson, R. A. (1980). Protein and fat metabolism in hibernating bears. *Fedn. Proc.* **39**, 2955–2958.

Ng, T. B. and Idler, D. R. (1980). Gonadotropic regulation of androgen production in flounder and salmonids. *Gen. Comp. Endocr.* **42**, 25–38.

Ng, T. B., Idler, D. R., and Burton, M. (1980a). A comparison of prolactins from a marine, an estuarine, and a freshwater teleost. *Gen. Comp. Endocr.* **42**, 141–146.

Ng, T. B., Idler, D. R., and Burton, M. P. (1980b). Effects of teleost gonadotropins and their antibodies on gonadal histology in winter flounder. *Gen. Comp. Endocr.* **42**, 355–364.

Niall, H. D., Hogan, M. L., Sauer, R., Rosenblum, I. Y., and Greenwood, F. (1971). Sequences of pituitary and placental lactogenic and growth hormones: evolution from a primordial peptide by gene reduplication. *Proc. Natl. Acad. Sci., USA* **68**, 866–869.

Nicoll, C. S. (1980). Ontogeny and evolution of prolactin's function. *Fedn. Proc.* **39**, 2563–2566.

Nicoll, C. S., Bern, H. A., Dunlop, D., and Strohman, R. C. (1965). Prolactin, growth hormone, thyroxine and growth in tadpoles of *Rana catesbeiana*. *Amer. Zool.* **5**, 738–739.

Nicoll, C. S. and Licht, P. (1971). Evolutionary biology of prolactins and somatotropins: ii. Electrophoretic comparison of tetrapod somatotropins. *Gen. Comp. Endocr.* **17**, 490–507.

Nicoll, C. S., Wilson, S. W., Nishioka, R., and Bern, H. (1981). Blood and pituitary prolactin levels in tilapia (*Sarotherodon mossambicus*; Teleostei) from different salinities as measured by a homologous radioimmunoassay. *Gen. Comp. Endocr.* **44**, 365–373.

Nicholls, T. J., Scanes, C. J., and Follett, B. K. (1973). Plasma pituitary luteinizing hormone in Japanese quail during photoperiodically induced gonadal growth and regression. *Gen. Comp. Endocr.* **21**, 84–98.

Nichols, C. W. (1973). Somatotropic effects of prolactin and growth hormone in juvenile snapping turtles (*Chelydra serpentina*). *Gen. Comp. Endocr.* **21**, 219–224.

Nilsson, A. (1970). Gastrointestinal hormones in the holocephalian fish *Chimaera monstrosa* (L.). *Comp. Biochem. Physiol.* **32**, 387–390.

Nishimura, H. (1978). Physiological evolution of the renin-angiotensin system. *Jap. Heart J.* **19**, 806–822.

Nishimura, H., Ogawa, M., and Sawyer, W. H. (1973). Renin-angiotensin system in primitive bony fishes and a holocephalian. *Amer. J. Physiol.* **224**, 950–956.

Nishimura, H., Oguri, M., Ogawa, M., Sokabe, H., and Imai, M. (1970). Absence of renin in kidneys of elasmobranchs and cyclostomes. *Amer. J. Physiol.* **218**, 911–915.

Nishimura, H. and Sawyer, W. H. (1976). Vasopressor, diuretic, and natriuretic responses to angiotensins by the American eel. *Anguilla rostrata. Gen. Comp. Endocr.* **29**, 337–348.

Nishimura, H., Sawyer, W. H., and Nigrelli, R. F. (1976). Renin, cortisol and plasma volume in marine teleost fishes adapted to dilute media. *J. Endocr.* **70**, 47–59.

Nolly, H. L. and Fasciola, J. C. (1973). The specificity of the renin-angiotensinogen reaction through the phylogenetic scale. *Comp. Biochem. Physiol.* **44A**, 639–645.

Norman, A. W. and Henry, H. (1974). 1,25-dihydroxycholecalciferol-A hormonally active form of vitamin D_3. *Rec. Prog. Hormone Res.* **30**, 431–473.

Norris, D. O. and Gern, W. A. (1976). Thyroxine-induced activation of hypothalamo-hypophysial axis in neotenic salamander larvae. *Science* **194**, 525–527.

Novales, R. R. (1972). Recent studies of the melanin-dispersing effect of MSH on melanophores. *Gen. Comp. Endocr. Suppl.* **3**, 125–135.

(1973). Discussion of "Endocrine regulation of pigmentation" by Frank S. Abbott. *Amer. Zool.* **13**, 895–897.

O'Connor, J. M. (1972). Pituitary gonadotropin release patterns in pre-spawning brook trout, *Salvelinus fontinalis*, rainbow trout, *Salmo gairdneri* and leopard frogs, *Rana pipiens. Comp. Biochem. Physiol.* **43A**, 739–746.

Odum, E. P. (1965). Adipose tissue in migratory birds. In *Handbook of Physiology*, Sect. 5 *Adipose Tissue*, pp. 37–43. Washington, D.C.: American Physiological Society.

Ogawa, M., Oguri, M., Sokabe, H., and Nishimura, H. (1972). Juxtaglomerular apparatus in vertebrates. *Gen. Comp. Endocr. Suppl.* **3**, 374–380.

Ogawa, M., Yagasaki, M., and Yamazaki, J. (1973). The effect of prolactin on water influx in isolated gills of the goldfish *Carassius auratus* L. *Comp. Biochem. Physiol.* **44A**, 1177–1183.

Oguro, C. (1973). Parathyroid gland and serum calcium concentration in the giant salamander, *Megalobatrachus davidianus. Gen. Comp. Endocr.* **21**, 565–568.

Oguro, C. and Sasayama, Y. (1976). Morphology and function of the parathyroid gland of the caiman, *Caiman crocodilus. Gen. Comp. Endocr.* **29**, 161–169.

——— (1978). Function of the parathyroid gland in serum calcium regulation in the newt, *Tylotriton andersoni* Boulenger. *Gen. Comp. Endocr.* **35**, 10–15.

Oguro, C. and Tomisawa, A. (1972). Effects of parathyroidectomy on serum calcium concentration of the turtle *Geoclemys reevesii. Gen. Comp. Endocr.* **19**, 587–588.

Oguro, C., Tomisawa, A., and Matuoka, N. (1974). Effect of parathyroidectomy on the serum calcium and phosphorus concentrations in the tortoise, *Testudo graeca. Zool. Magazine* **83**, 201–202.

Oishi, T. and Lauber, J. K. (1974). Pineal control of photoendocrine responses in growing Japanese quail. *Endocrinology* **94**, 1731–1734.

Oksche, A. (1965). Survey of the development and comparative morphology of the pineal organ. *Prog. Brain Res.* **10**, 3–28.

Oliver, C. and Porter, J. C. (1978). Distribution and characterization of α-melanocyte-stimulating hormone in the rat brain. *Endocrinology* **102**, 697–705.

Olivereau, M. (1967). Observations sur l'hypophyse de l'anguille femelle en particulier lors de la maturation sexuelle. *Z. Zellforsch. mikrosk. Anat.* **80**, 286–306.

Olivereau, M. (1978). Serotonin and MSH secretion: effect of parachlorophenylalanine on the pituitary cytology of the eel. *Cell Tiss. Res.* **191**, 83–92.

Olivereau, M. and Olivereau, J. (1978). Prolactin, hypercalcemia and corpuscles of Stannius in seawater eels. *Cell Tiss. Res.* **186**, 81–96.

O'Malley, B. W. and Means, A. R. (1974). Female steroid hormones and target cells nuclei. *Science* **183**, 610–620.

O'Malley, B. W., Roop, D. R., Lai, E. C., Nordstrom, J. L., Catterall, J. F., Swanek, G. E., Colbert, D. A., Tsai, M.-J., Dugaiczyk, A., and Woo, S.L.C. (1979). The ovalbumin gene: organization, structure, transcription and regulation. *Rec. Prog. Hormone Res.* **35**, 1–42.

Omdahl, J. L. and DeLuca, H. F. (1973). Regulation of vitamin D metabolism and function. *Physiol. Rev.* **53**, 327–372.

Oppenheimer, J. H. (1979). Thyroid hormone at the cellular level. *Science* **203**, 971–979.

Orci, L. and Unger, R. H. (1975). Functional subdivision of the islets of Langerhans and possible role of D cells. *Lancet* **ii**, 1243–1244.

Ornoy, A., Goodwin, D., Noff, D., and Edlestein, S. (1978). 24,25-dihydroxyvitamin D is a metabolite of vitamin D essential for bone formation. *Nature, Lond.* **276**, 517–519.

Owens, D. W., Gern, W. A., Ralph, C. L., and Boardman, T. J. (1978). Nonrelationship between plasma melatonin and background adaptation in the rainbow trout (*Salmo gairdneri*). *Gen. Comp. Endocr.* **34**, 459–467.

Oyer, P. E., Cho, S., Peterson, J. D., and Steiner, D. F. (1971). Studies on human proinsulin. *J. Biol. Chem.* **246**, 1375–1386.

Ozon, R. (1972). Androgens in fishes, amphibians, reptiles and birds. In *Steroids in Nonmammalian Vertebrates* (edited by D. R. Idler), pp. 329–389. New York: Academic Press.

Packard, G. C., Packard, M. J., and Gorbman, A. (1976). Serum thyroxine concentrations in the pacific hagfish and lamprey and in the leopard frog. *Gen. Comp. Endocr.* **28**, 365–367.

Pahuja, D. N. and DeLuca, H. F. (1981). Stimulation of intestinal calcium transport and bone calcium mobilization by prolactin in vitamin D-deficient rats. *Science* **214**, 1038–1039.

Pang, P.K.T. (1973). Endocrine control of calcium metabolism in teleosts. *Amer. Zool.* **13**, 775–792.

Pang, P.K.T., Galli-Gallardo, S. M., Collie, N., and Sawyer, W. H. (1980). Renal and peripheral vascular responsiveness to arginine vasotocin in the bullfrog, *Rana catesbeiana. Amer. J. Physiol.* **239**, R156–R160.

Pang, P.K.T., Pang, R. K., and Griffith, R. W. (1975). Corpuscle of Stannius: lack of direct involvement in regulation of serum sodium, potassium, and chloride in the Teleost, *Fundulus heteroclitus. Gen. Comp. Endocr.* **26**, 179–185.

Pang, P.K.T., Pang, R. K., and Sawyer, W. H. (1973). Effects of environmental calcium and replacement therapy on the killifish, *Fundulus heteroclitus*, after the surgical removal of the corpuscles of Stannius. *Endocrinology* **93**, 705–710.

— (1974). Environmental calcium and the sensitivity of the killifish (*Fundulus heteroclitus*) in bioassays for the hypocalcemic response to Stannius corpuscles from killifish and cod (*Gadus morhua*). *Endocrinology* **94**, 548–555.

Pang, P.K.T. and Sawyer, W. H. (1974). Effects of prolactin on hypophysectomized mud puppies *Necturus maculosus. Amer. J. Physiol.* **226**, 458–462.

— (1975). Parathyroid hormone preparations, salmon calcitonin, and urine flow in the South American lungfish, *Lepidosiren paradoxa. J. Exp. Zool.* **193**, 407–412.

— (1978). Renal and vascular responses of the bullfrog (*Rana catesbeiana*) to mesotocin. *Amer. J. Physiol.* **235**, F151–F155.

Pang, P.K.T., Schreibman, M. P., Balbontin, F., and Pang, R. K. (1978). Prolactin and pituitary control of calcium regulation in the killifish, *Fundulus heteroclitus. Gen. Comp. Endocr.* **36**, 306–316.

Papkoff, H. (1972). Subunit interrelationships among the pituitary glycoprotein hormones. *Gen. Comp. Endocr. Suppl.* **3**, 609–616.

Parkes, A. S. and Marshall, A. J. (1960). The reproductive hormones in birds. In *Marshall's Physiology of Reproduction* (edited by A. S. Parkes), vol. I (Pt. 2), pp. 583–706. London: Longmans.

Patent, C. J. (1970). Comparison of some hormonal effects on carbohydrate metabolism in an elasmobranch (*Squalus acanthias*) and a holocephalan (*Hydrolagus collei*). *Gen. Comp. Endocr.* **14**, 215–242.

Patzelt, C., Tager, H. S., Carroll, R. J., and Steiner, D. F. (1979). Identification and processing of proglucagon in pancreatic islets. *Nature, Lond.* **282**, 260–266.

Patzner, R. A. and Ichikawa, T. (1977). Effects of hypophysectomy on the testis of the hagfish, *Eptatretus burgeri* Girard (Cyclostomata). *Zool. Anz.*, Jena **199**, 5/6, pp. 371–380.

Pavel, S., Dorcescu, M., Petrescu-Holban, R., and Ghinea, E. (1973). Biosynthesis of a vasotocin-like peptide in cell cultures from pineal glands of human fetuses. *Science* **181**, 1252–1253.

Payan, P. and Maetz, J. (1971). Balance hydrique chez les elasmobranches: arguments en faveur d'un contrôle endocrinien. *Gen. Comp. Endocr.* **16**, 535–554.

Peach, M. J. (1977). Renin-angiotensin system: biochemistry and mechanism of action. *Physiol. Rev.* **57**, 313–369.

Peaker, M. (1971). Avian salt glands. *Phil. Trans. Roy. Soc. Lond.* **B262**, 289–300.

Peaker, M. and Linzell, J. L. (1975). *Salt Glands in Birds and Reptiles.* Cambridge University Press.

Pearse, A.G.E. (1968). Common cytochemical and ultrastructural characteristics of cells producing polypeptide hormones (the APUD series) and their relevance to thyroid and ultimobranchial C cells and calcitonin. *Proc. Roy. Soc.* **B170**, 71–80.

(1976). Morphology and cytochemistry of thyroid and ultimobranchial C cells. In *Handbook of Physiology*, Sect. 7 *Endocrinology*, vol. 7, *Parathyroid Gland*, pp. 411–421. Washington, D.C.: American Physiological Society.

Pearse, A.G.E. and Takor, T. T. (1979). Embryology of the diffuse neuroendocrine system and its relationship to the common peptides. *Fedn. Proc.* **38**, 2288–2294.

Pearson, D., Shively, J. E., Clark, B. R., Geschwind, I. I., Barkley, M., Nishioka, R. S., and Bern, H. A. (1980). Urotensin II: a somatostatin-like peptide in the caudal neurosecretory system of fishes. *Proc. Natl. Acad. Sci., USA* **77**, 5021–5024.

Peart, W. S. (1977). The kidney as an endocrine organ. *Lancet* **2**, 543–548.

Penhos, J. C. and Ramey, E. (1973). Studies on the endocrine pancreas of amphibians and reptiles. *Amer. Zool.* **12**, 667–698.

Penny, R. J., Tilders, F.J.H., and Thody, A. J. (1979). The effect of hypothalamic lesions on immuno-reactive α-melanocyte stimulating hormone secretion in the rat. *J. Physiol. Lond.* **292**, 59–67.

Peter, R. E. (1971). Feedback effects of thyroxine on the hypothalamus and pituitary of the goldfish, *Carassius auratus*. *J. Endocr.* **51**, 31–39.

Peter, R. E. and Crim, L. W. (1979). Reproductive endocrinology of fishes: gonadal cycles and gonadotropin in teleosts. *Ann. Rev. Physiol.* **41**, 323–335.

Peterson, J. D., Steiner, D. F., Emdin, S. O., and Falkmer, S. (1975). The amino acid sequence of the insulin from a primitive vertebrate, the Atlantic hagfish (*Myxine glutinosa*). *J. Biol. Chem.* **250**, 5183–5191.

Phillips, J. G. and Ensor, D. M. (1972). The significance of environmental factors in the hormone mediated changes of nasal (salt) gland activity in birds. *Gen. Comp. Endocr. Suppl.* **3**, 393–404.

Phillips, L. S. and Vassilopoulou-Sellin, R. (1980). Somatomedins (parts 1 and 2). *New Engl. J. Med.* **302**, 371–380; 438–446.

Pic, P., Mayer-Gostan, N., and Maetz, J. (1973). Sea-water teleosts: Presence of α- and β-adrenergic receptors in the gill regulating salt extrusion and water permeability. In *Comparative Physiology* (edited by L. Bolis, K. Schmidt-Nielsen, and S.H.P. Maddrell), pp. 292–322. Amsterdam: Elsevier North-Holland.

Pickering, A. D. (1972). Effects of hypophysectomy on the activity of the endostyle and thyroid gland in the larval and adult river lamprey, *Lampetra fluviatilus* L. *Gen. Comp. Endocr.* **18**, 335–343.

(1976). Effects of gonadectomy, oestradiol and testosterone on the migrating river lamprey *Lampetra fluviatilis* L. *Gen. Comp. Endocr.* **28**, 473–480.

Pickford, G. E. and Kosto, B. (1957). Hormonal induction of melanogenesis in hypophysectomized killifish (*Fundulus heteroclitus*). *Endocrinology* **61**, 177–196.

Pickford, G. E., Pang, P.K.T., Weinstein, E., Torretti, J., Hendler, E., and Epstein, F. H. (1970). The response of the hypophysectomized Cyprinodont, *Fundulus heteroclitus*, to replacement therapy with cortisol: Effects on blood serum and sodium-potassium activated adenosine triphosphatase in the gills, kidney, and intestinal mucosa. *Gen. Comp. Endocr.* **14**, 524–534.

Pickford, G. E. and Phillips, J. G. (1959). Prolactin, a factor in promoting survival of hypophysectomized killifish in fresh water. *Science* **130**, 454–455.

Pickford, G. E. and Strecker, E. L. (1977). The spawning reflex response of the killifish, *Fundulus heteroclitus*: isotocin is relatively inactive in comparison with arginine-vasotocin. *Gen. Comp. Endocr.* **32**, 132–137.

Pictet, R. and Rutter, W. J. (1972). Development of the embryonic endocrine pancreas. In *Handbook of Physiology*, Sect. 7 *Endocrinology*, vol. 1, *Endocrine pancreas*, pp. 25–66. Washington, D.C.: American Physiological Society.

Pike, J. W., Spanos, E., Colston, K. W., MacIntyre, I., and Haussler, M. R. (1978). Influence of estrogen on renal vitamin D hydroxylase and serum 1α, $25\text{-}(OH)_2D_3$ in chicks. *Amer. J. Physiol.* **235**, E338-E343.

Plisetskaya, E., Kazakov, V. K., Soltitskaya, L., and Leibson, L. G. (1978). Insulin-producing cells in the gut of freshwater bivalve molluscs *Anodonta cygnea* and *Unio picotrum* and the role of insulin in the regulation of their carbohydrate metabolism. *Gen. Comp. Endocr.* **35**, 133–145.

Poffenbarger, P. L., Burns, R., and Bennett-Novak, A. (1976). A phylogenetic study of serum nonsuppressible insulin-like activity (NSILA). *Comp. Biochem. Physiol.* **52A**, 223–226.

Pohorecky, L. A. and Wurtman, R. J. (1971). Adrenocortical control of epinephrine synthesis. *Pharmacol. Rev.* **23**, 1–35

Porthé-Nibelle, J. and Lahlou, B. (1978). Uptake and binding of glucocorticoids in fish tissues. *J. Endocr.* **78**, 407–416.

Potts, J. T., Keutmann, H. T., Niall, H. D., Habener, J. F., and Tregear, G. W. (1972). Comparative biochemistry of parathyroid hormone. *Gen. Comp. Endocr. Suppl.* **3**, 405–410.

Price D., Zaaijer, J.J.P., Ortiz, E., and Brinkmann, A. C. (1975). Current views on the embryonic sex differentiation in reptiles, birds and mammals. *Amer. Zool. Suppl. 1975*, 173–195.

Prigge, W. F. and Grande, F. (1971). Effects of glucagon, epinephrine and insulin on *in vitro* lipolysis of adipose tissue from mammals and birds. *Comp. Biochem. Physiol.* **39B**, 69–82.

Proux, J. and Rougon-Rapuzzi, G. (1980). Evidence for a vasopressin-like molecule in migratory locust. Radioimmunological measurements in different tissues: correlation with various states of hydration. *Gen. Comp. Endocr.* **42**, 378–383.

Puzas, J. E. and Brand, J. S. (1979). Parathyroid hormone stimulation of collagenase secretion by isolated bone cells. *Endocrinology* **104**, 559–562.

Quamme, G. A. (1980). Effect of calcitonin on calcium and magnesium transport in rat nephron. *Amer. J. Physiol.* **238**, E573–E578.

Quay, W. B. (1970). Endocrine effects on the mammalian pineal. *Amer. Zool.* **10**, 237–246.

———— (1972). Integument and the environment: glandular composition, function and evolution. *Amer. Zool.* **12**, 95–108.

Quevedo, W. C. (1972). Epidermal melanin units: melanocyte-keratinocyte interactions. *Amer. Zool.* **12**, 35–41.

Raisz, L. G. (1976). Mechanisms of bone resorption. In *Handbook of Physiology*: Sect. 7, *Endocrinology*, vol. 7, *The Parathyroid Gland*, pp. 117–136. Washington, D.C.: American Physiological Society.

Rall, J. E., Robbins, J., and Lewallen, C. G. (1964). The thyroid. In *The Hormones* (edited by G. Pincus, K. V. Thimann, and E. B. Astwood), vol. V, pp. 159–439. New York: Academic Press.

Ralph, C. L. (1970). Structure and alleged functions of avian pineals. *Amer. Zool.* **10**, 217–235.

Ralph, C. L. (1975). The pineal complex: a retrospective view. *Amer. Zool. Suppl.*, pp. 105–116.

Ramsey, D. H. and Bern, H. A. (1972). Stimulation by ovine prolactin of fluid transfer in everted sacs of rat small intestine. *J. Endocr.* **53**, 453–459.

Rance, T. and Baker, B. I. (1979). The teleost melanin-concentrating hormone—a pituitary hormone of hypothalamic origin. *Gen. Comp. Endocr.* **37**, 64–73.

Rankin, J. C. and Maetz, J. (1971). A perfused teleostean gill preparation: vascular actions of neurohypophysial hormones and catecholamines. *J. Endocr.* **51**, 621–635.

Rasmussen H. and Goodman, D.B.P. (1977). Relationships between calcium and cyclic nucleotides in cell activation. *Physiol. Rev.* **57**, 421–509.

Rasquin, P. and Rosenbloom, L. (1954). Endocrine imbalance and tissue hyperplasia in teleosts maintained in darkness. *Bull. Amer. Mus. Nat. Hist.* **104**, 359–420.

Rayford, P. L., Miller, T. A., and Thompson, J. C. (1976). Secretin, cholecystokinin and newer gastrointestinal hormones. *New Engl. J. Med.* **294**, 1157–1164.

Redshaw, M. R. (1972). The hormonal control of the amphibian ovary. *Amer. Zool.* **12**, 289–306.

Regard, E., Taurog, A., and Nakashima, T. (1978). Plasma thyroxine and triiodothyronine levels in spontaneously metamorphosing *Rana catesbeiana* tadpoles and in adult anuran amphibia. *Endocrinology* **102**, 674–684.

Rehfeld, J. F. (1981). Four basic characteristics of the gastrin-cholecystokinin system. *Amer. J. Physiol.* **240**, G255–G266.

Reinbloth, R. (1970). Intersexuality in fishes. *Mem. Soc. Endocr.* **18**, 515–541.

(1972). Hormonal control of the teleost ovary. *Amer. Zool.* **12**, 307–324.

Reinig, J. W., Daniel, L. N., Schwabe, C., Gowan, L. K., Steinetz, B. G., and O'Bryne, E. M. (1981). Isolation and characterization of relaxin from the sand tiger shark (*Odontaspis taurus*). *Endocrinology* **109**, 537–543.

Reiter, R. J. (1980). The pineal and its hormones in the control of reproduction in mammals. *Endocrine Rev.* **1**, 109–131.

Reiter, R. J. and Sorrentino, S. (1970). Reproductive effects of the mammalian pineal. *Amer. Zool.* **10**, 247–258.

Renfree, M. B. (1979). Initiation of development of diapausing embryo by mammary denervation during lactation in a marsupial. *Nature, Lond.* **278**, 549–551.

Renfree, M. B. (1980). Placental function and embryonic development in marsupials. In *Comparative Physiology: Primitive Mammals* (edited by K. Schmidt-Nielsen, L. Bolis, and C. R. Taylor), pp. 269–284. Cambridge University Press.

Renfree, M. B., Lincoln, D. W., Almeida, O.F.X., and Short, R. V. (1981). Abolition of seasonal embryonic diapause in a wallaby by pineal denervation. *Nature, Lond.* **293**, 138–139.

Renoir, J-M, Mercier-Bodard, C., and Baulieu, E. M. (1980). Hormonal and immunological aspects of the phylogeny of sex steroid binding plasma protein. *Proc. Natl. Acad. Sci., USA* **77**, 4578–4582.

Rillema, J. A. (1980). Mechanism of prolactin action. *Fedn. Proc.* **39**, 2593–2598.

Rinderknecht, E. and Humbel, R. E. (1976*a*). Polypeptides with nonsuppressible insulin-like and cell-growth promoting activities in human serum: isolation, chemical characterization, and some biological properties of forms I and II. *Proc. Natl. Acad. Sci., USA* **73**, 2365–2369.

(1976*b*). Amino-terminal sequences of two polypeptides from human serum with nonsuppressible insulin-like and cell-growth promoting activities: evidence for structural homology with insulin B chain. *Proc. Natl. Acad. Sci., USA* **73**, 4379–4381.

(1978). The amino acid sequence of human insulin-like growth factor I and its structural homology with proinsulin. *J. Biol. Chem.* **253**, 2769–2776.

Ritter, R. C., Slusser, P. G., and Stone, S. (1981). Glucoreceptors controlling feeding and blood glucose: location in the hindbrain. *Science* **213**, 451–453.

Rizzo, A. J. and Goltzman, D. (1981). Calcitonin receptors in the central nervous system of the rat. *Endocrinology* **108**, 1672–1677.

Robertshaw, D., Taylor, C. R., and Mazzia, L. M. (1973). Sweating in primates: role of secretion of the adrenal medulla during exercise. *Amer. J. Physiol.* **224**, 678–681.

Robertson, D. R. (1968). The ultimobranchial gland in *Rana pipiens*. IV. Hypercalcemia and glandular hypertrophy. *Z. Zellforsch. mikroskop. Anat.* **85**, 441–542.

(1969*a*). The ultimobranchial body of *Rana pipiens*. VIII. Effects of extirpation upon calcium distribution and bone cell types. *Gen. Comp. Endocr.* **12**, 479–490.

(1969*b*). The ultimobranchial body in *Rana pipiens*. IX. Effects of extirpation and transplantation on urinary calcium excretion. *Endocrinology* **84**, 1174–1178.

(1971). Cytological and physiological activity of ultimobranchial gland in the premetamorphic anuran *Rana catesbeiana*. *Gen. Comp. Physiol.* **16**, 329–341.

Robertson, O. H., Krupp, M. A., Thomas, S. F., Favour, C. B., Hane, S., and Wexler, B. C. (1961). Hyperadrenocorticoidism in spawning migratory and non-migratory rainbow trout (*Salmo gairdnerii*): comparison with Pacific salmon (Genus *Oncorhynchus*). *Gen. Comp. Endocr.* **1**, 473–484.

Robertson, O. H. and Wexler, B. C. (1959). Histological changes in the organs and tissues of migrating and spawning Pacific salmon (Genus *Oncorhynchus*). *Endocrinology* **66**, 222–239.

Robinson, K. W. and MacFarlane, W. V. (1957). Plasma antidiuretic activity of marsupials during exposure to heat. *Endocrinology* **60**, 679–680.

Robison, G. A., Butcher, R. W., and Sutherland, E. W. (1971). *Cyclic AMP*. New York: Academic Press.

Rodbell, M., Lin, M. C., Salomon, Y., Londos, C., Harwood, J. P., Martin, B. R., Rendell, M., and Berman, M. (1975). Role of adenine and guanine nucleotides in the activity and responses of adenylate cyclase systems to hormones: evidence for multisite transition sites. In *Advances in Nucleotide Research* (edited by G. I. Drummond, P. Greengard, and G. A. Robison), vol. 5, pp. 3–29. New York: Raven Press.

Roos, B. A., Yoon, M., Cutshaw, S. V., and Kalu, D. N. (1980). Calcium regulatory action of endogenous calcitonin demonstrated by passive immunization with calcitonin antibodies. *Endocrinology* **107**, 1320–1326.

Rosen, J. M., Matusik, R. J., Richards, D. A., Gupta, P., and Rodgers, J. R. (1980). Multihormonal regulation of casein gene expression at the transcriptional and posttranscriptional levels in the mammary gland. *Rec. Prog. Hormone. Res.* **36**, 157–187.

Rossier, B. C., Rossier, M., and Lo, C. S. (1979). Thyroxine and Na^+ transport in toad: role in transition from poikilo- to homeothermy. *Amer. J. Physiol.* **236**, C117–C124.

Roth, J. J., Jones, R. E., and Gerrard, A. M. (1973). Corpora lutea and oviposition in the lizard *Sceloporus undulatus*. *Gen. Comp. Endocr.* **21**, 569–572.

Roth, S. I. and Schiller, A. L. (1976). Comparative anatomy of the parathyroid glands. In *Handbook of Physiology*, Sect. 7 *Endocrinology*, vol. 7, *Parathyroid Gland*, pp. 281–311. Washington, D.C.: American Physiological Society.

Rowan, W. (1925). Relation of light to bird migration and developmental changes. *Nature, Lond.* **115**, 494–495.

Rubin, R. T., Reinisch, J. M., and Haskett, R. F., (1981). Postnatal gonadal steroid effects on human behavior. *Science* **211**, 1318–1324.

Rubinstein, M., Stein, S., and Udenfriend, S. (1978). Characterization of pro-opiocortin, a precursor to opioid peptides and corticotropin. *Proc. Natl. Acad. Sci., USA* **75**, 669–671.

Rudinger, J. (1968). Synthetic analogues of oxytocin: an approach to problems of hormone action. *Proc. Roy. Soc., Lond.* **B170**, 17–26.

Russell, J. T., Brownstein, M. J., and Gainer, H. (1980). Biosynthesis of vasopressin, oxytocin and neurophysins: isolation and characterization of two common precursors (prepressophysin and prooxyphysin). *Endocrinology* **107**, 1880–1891.

Rust, C. C. and Meyer, R. K. (1968). Effects of pituitary autografts on hair color in the short-tailed weasel. *Gen. Comp. Endocr.* **11**, 548–551.

(1969). Hair color, molt, and testis size in male, short-tailed weasels treated with melatonin. *Science* **165**, 921–922.

Ryan, G. B., Coghlan, J. P., and Scoggins, B. A. (1979). The granulated peripolar epithelial cell: a potential secretory component of the renal juxtaglomerular complex. *Nature, Lond.* **277**, 655–656.

Sage, M. (1973). The evolution of thyroidal function in fishes. *Amer. Zool.* **13**, 899–905.

Sairam, M. R., Papkoff, H., and Li, C. H. (1972). Human pituitary interstitial stimulating hormone: primary structure of the α subunit. *Biochem. Biophys. Res. Commun.* **48**, 530–537.

Sakamoto, S., Sakamoto, M., Goldhaber, P., and Glimcher, M. (1975). Collagenase and bone resorption: isolation of collagenase from culture medium containing serum after stimulation of bone resorption by addition of parathyroid hormone extract. *Biochem. Biophys. Res. Commun.* **63**, 172–178.

Salmon, W. D. and Daughaday, W. H. (1957). A hormonally controlled serum factor which stimulates sulfate incorporation by cartilage *in vitro*. *J. Lab. Clin. Med.* **49**, 825–836.

Samols, E., Tyler, J., Megyesi, C., and Marks, V. (1966). Immunochemical glucagon in human pancreas, gut, and plasma. *Lancet* **ii**, 727–729.

Sandor, T. (1969). A comparative survey of steroids and steroidogenic pathways throughout the vertebrates. *Gen. Comp. Endocr. Suppl.* **2**, 284–298.

Sandor, T. and Idler, D. R. (1972). Steroid methodology. In *Steroids in Non-mammalian Vertebrates* (edited by D. R. Idler), pp. 6–36. New York: Academic Press.

Sandor, T., Mehdi, A. Z., and Fazekas, A. G. (1977). Corticosteroid-binding macromolecules in the salt-activated nasal gland of the domestic duck (*Anas platyrhynchos*). *Gen. Comp. Endocr.* **32**, 348–359.

Sasayama, Y. and Oguro, C. (1975). Effects of parathyroidectomy on calcium and sodium concentrations of serum and coelomic fluid in bullfrog tadpoles. *J. Exp. Zool.* **192**, 293–298.

(1976). Effects of ultimobranchialectomy on calcium and sodium concentrations of serum and coelomic fluid in bullfrog tadpoles under high calcium and high sodium environment. *Comp. Biochem. Physiol.* **55A**, 35–37.

Sassin, J. F., Frantz, A. G., Weiztman, E. D., and Kapen, S. (1972). Human prolactin: 24-hour pattern with increased release during sleep. *Science* **177**, 1205–1207.

Sawin, C. T., Bolaffi, J. L., Callard, I. P., Bacharach, P., and Jackson, I.M.D. (1978). Induced metamorphosis in *Ambystoma mexicanum*: lack of effect of triiodothyronine on tissue or blood levels of thyrotropin-releasing hormone (TRH). *Gen. Comp. Endocr.* **36**, 427–432.

Sawyer, W. H. (1972*a*). Lungfishes and amphibians: endocrine adaptation and the transition from aquatic to terrestrial life. *Fedn. Proc.* **31**, 1609–1614.

(1972*b*). Neurohypophysial hormones and water and sodium excretion in African lungfish. *Gen. Comp. Endocr. Suppl.* **3**, 345–349.

Sawyer, W. H., Blair-West, J. R., Simpson, P. A., and Sawyer, M. K. (1976). Renal responses of Australian lungfish to vasotocin, angiotensin II, and NaCl infusion. *Amer. J. Physiol.* **231**, 593–602.

Sawyer, W. H. and Pickford, G. E. (1963). Neurohypophyseal principles of *Fundulus heteroclitus*: characteristics and seasonal changes. *Gen. Comp. Endocr.* **3**, 439–445.

Saxena, B. B. and Rathnam, P. (1976). Amino acid sequence of the β subunit of follicle-stimulating hormone from human pituitary glands. *J. Biol. Chem.* **251**, 993–1005.

Scanes, C. G., Bolton, N. J., and Chadwick, A. (1975). Purification and properties of an avian prolactin. *Gen. Comp. Endocr.* **27**, 371–379.

Scanes, C. G., Dobson, S., Follett, B. K., and Dodd, J. M. (1972). Gonadotrophic activity in the pituitary gland of the dogfish (*Scyliorhinus canicula*). *J. Endocr.* **54**, 343–344.

Scanes, C. G., Follett, B. K., and Goos, H. J. Th. (1972). Cross-reaction in a chicken LH radioimmunoassay with plasma and pituitary extracts from various species. *Gen. Comp. Endocr.* **19**, 596–600.

Schally, A. V., Arimura, A., and Kastin, A. J. (1973). Hypothalamic regulatory hormones. *Science* **179**, 341–350.

Schnoes. H. K. and DeLuca, H. F. (1980). Recent progress in vitamin D metabolism and the chemistry of vitamin D metabolites. *Fedn. Proc.* **39**, 2723–2729.

Schramm, M., Orly, J., Eimerl, S., and Korner, M. (1977). Coupling of hormone receptors to adenylate cyclase of different cells by cell fusion. *Nature, Lond.* **268**, 310–313.

Schreibman, M. P. and Kallman, K. D. (1969). The effect of hypophysectomy on freshwater survival in teleosts of the order Antheriniformes. *Gen. Comp. Endocr.* **13**, 27–38.

Schussler, G. C. and Orlando, J. (1978). Fasting decreases triiodothyronine receptor capacity. *Science* **199**, 686–688.

Schwabe, C. and McDonald, J. K. (1977). Relaxin: a disulfide homolog of insulin. *Science* **197**, 914–915.

Schwartz, N. B. (1973). Mechanisms controlling ovulation in small mammals. In *Handbook of Physiology*, Sect. 7 *Endocrinology*, vol. II, *Female reproductive system* (Pt. 1), pp. 125–141. Washington, D.C.: American Physiological Society.

Schwartz, R. J., Kuhn, R. W., Buller, R. E., Schrader, W. T., and O'Malley, B. W. (1976). Progesterone-binding components of chick oviduct. *In vitro* effects of purified hormone receptor complexes on the initiation of RNA synthesis in chromatin. *J. Biol. Chem.* **251**, 5166–5177.

Schwerdtfeger, W. K. (1979). Qualitative and quantitative data on the fine structure of the guppy (*Poecilia reticulata* Peters) epidermis following treatment with thyroxine and testosterone. *Gen. Comp. Endocr.* **38**, 484–490.

Seal, U. S. and Doe, R. P. (1963). Corticosteroid-binding globulin: species distribution and small-scale purification. *Endocrinology* **73**, 371–376.

Sealey, J. E., Atlas, S. A., and Laragh, J. H. (1980). Prorenin and other large molecular weight forms of renin. *Endocrine Rev.* **1**, 365–391.

Seino, Y., Porte, D., and Smith, P. H. (1979*a*). Immunohistochemical localization of somatostatin-containing cells in the intestinal tract: a comparative study. *Gen. Comp. Endocr.* **38**, 229–233.

Seino, Y., Porte, D., Yanaihara, N., and Smith P. H. (1979*b*). Immunocytochemical localization of motilin-containing cells in the intestines of several vertebrate species and a comparison of antisera against natural and synthetic motilin. *Gen. Comp. Endocr.* **38**, 234–37.

Seligman, P. A. and Allen, R. H. (1978). Characterization of the receptor for transcolbalamin isolated from human placenta. *J. Biol. Chem.* **253**, 1766–1772.

Sellers, E. A., Flattery, K. V., and Steiner, G. (1974). Cold acclimation in hypothyroid rats. *Amer. J. Physiol.* **226**, 290–294.

Senior, B. E. and Cunningham, F. J. (1974). Oestradiol and luteinizing hormone during the ovulatory cycle of the hen. *J. Endocr.* **60**, 201–202.

Sernia, C. (1980). Physiology of the adrenal cortex in monotremes. In *Comparative physiology: primitive mammals* (edited by K. Schmidt-Nielsen, L. Bolis, and C. R. Taylor), pp. 308–315. Cambridge University Press.

Sernia, C., Bradley, A. J., and McDonald, I. R. (1979). High affinity binding of adrenocortical and gonadal steroids by plasma proteins of Australian marsupials. *Gen. Comp. Endocr.* **38**, 496–503.

Sernia, C. and McDonald, I. R. (1977). Metabolic effects of cortisol and adrenocorticotrophin in a prototherian mammal *Tachyglossus aculeatus* (SHAW). *J. Endocr.* **75**, 261–269.

Sernia, C. and Tyndale-Biscoe, C. H. (1979). Prolactin receptors in the mammary gland, corpus luteum and other tissues of the tammar wallaby, *Macropus eugenii*. *J. Endocr.* **83**, 79–89.

Shafrir, E. and Wertheimer, E. (1965). Comparative physiology of adipose tissue in different sites and in different species. In *Handbook of Physiology*, Sect. 5 *Adipose Tissue*, pp. 417–429. Washington, D.C.: American Physiological Society.

Shapiro, B. and Pimstone, B. L. (1977). A phylogenetic study of sulphation factor activity in 26 species. *J. Endocr.* **74**, 129–135.

Shapiro, M., Nicholson, W. E., Orth, D. N., Mitchel, W. M., Island, D. P., and Liddle, G. W. (1972). Preliminary characterization of the pituitary melanocyte stimulating hormones of several vertebrate species. *Endocrinology* **90**, 249–256.

Shareghi, G. R. and Stoner, L. C. (1978). Calcium transport across segments of the rabbit distal nephron *in vitro*. *Amer. J. Physiol.* **235**, F367–F375.

Sharman, G. B. (1970). Reproductive physiology of marsupials. *Science* **167**, 1221–1228.
(1976). Evolution of viviparity in mammals. In *Reproduction in Mammals*, 6. *The Evolution of Reproduction* (edited by C. R. Austin and R. V. Short), pp. 32–70. Cambridge University Press.

Shire, J.G.M. (1970). Genetic variation in adrenal structure: quantitative measurements on the cortex and medulla in hybrid mice. *J. Endocr.* **48**, 419–431.

Shoemaker, V. H., Nagy, K. A., and Bradshaw, S. D. (1972). Studies on the control of electrolyte excretion by the nasal gland of the lizard *Dipsosaurus dorsalis*. *Comp. Biochem. Physiol.* **43A**, 749–757.

Short, R. V. (1979). Sex determination and differentiation. *Brit. Med. Bull.* **35**, 121–127.

Shuster, S., Burton, J. L., Thody, A. J., Plummer, N., Goolamali, S. K., and Bates, D. (1973). Melanocyte-stimulating hormone and Parkinsonism. *Lancet* i, 463–465.

Shuster, S. and Thody, A. J. (1974). The control and measurement of sebum secretion. *J. Invest. Dermatol.* **62** 172–190.

Siberian correspondent (1973). Centenarian triton. *Nature (Lond.)* **242**, 369.

Silver, R., Goldsmith, A. R., and Follett, B. K. (1980). Plasma luteinizing hormone in male ring doves during the breeding cycle. *Gen. Comp. Endocr.* **42**, 19–24.

Simantov, R., Goodman, R., Aposhian, D., and Snyder, S. H. (1976). Phylogenetic distribution of a morphine-like peptide 'enkephalin'. *Brain Res.* **111**, 204–211.

Simmons D. J. (1971). Calcium and skeletal tissue physiology in teleost fishes. *Clin. Orthopoedics* **76**, 244–280.

Simpson, P. A. and Blair-West, J. R. (1972). Estimation of marsupial renin using marsupial renin-substrate. *J. Endocr.* **53**, 125–130.

Siris, E. S., Nisula, B. C., Catt, K. J., Horner, K., Birkin, S., Canfield, R. A., and Ross, G. T. (1978). New evidence for intrinsic follicle-stimulating hormone-like activity in human gonadotropin and luteinizing hormone. *Endocrinology* **102**, 1356–1361.

Skadhauge, E. (1969). Activities biologique des hormones neurohypophysaires chez les oiseaux et les reptiles. *Colloq. Int. C.N.R.S.* **177**, 63–68.

Skeggs, L. T., Levine, M., Lentz, K. E., Kahn, J. R., and Dorer, F. E. (1977). New developments in our knowledge of the chemistry of renin. *Fedn. Proc.* **36**, 1755–1759.

Smith, G. P., Gibbs, J., and Young, R. C. (1974). Cholecystokinin and intestinal satiety in the rat. *Fedn. Proc.* **33**, 1146–1149.

Smith, G. P., Jerome, C., Cushin, B. J., Eterno, R., and Simansky, K. J. (1981). Abdominal vagotomy blocks the satiety effect of cholecystokinin in the rat. *Science* **213**, 1036–1037.

Smith, H. W. (1930). Metabolism of the lungfish, *Protopterus aethiopicus*. *J. Biol. Chem.* **88**, 97–130.

Smith, L. F. (1966). Species variation in the amino acid sequence of insulin. *Amer. J. Med.* **40**, 662–666.

Smith, P. H. and Porte, D. (1976). Neuropharmacology of the pancreatic islets. *Ann. Rev. Pharmacol. Toxicol.* **16**, 269–285.

Smith, T. J. and Edelman, I. S. (1979). The role of sodium transport in thyroid thermogenesis. *Fedn. Proc.* **38**, 2150–2153.

Snyder, S. H. (1980). Brain peptides as neurotransmitters. *Science* **209**, 976–983.

So, Y. P. and Fenwick, J. C. (1979). *In vivo* and *in vitro* effects of Stannius corpuscle extract on the branchial uptake of ^{45}Ca in stanniectomized North American eels (*Anguilla rostrata*). *Gen. Comp. Endocr.* **37**, 143–149.

Sokabe, H. and Nakajima, T. (1972). Chemical structure and role of angiotensins in the vertebrates. *Gen. Comp. Endocr. Suppl.* **3**, 382–392.

Sokabe, H., Nishimura, H., Kawabe, K., Tenmoku, S., and Arai, T. (1972). Plasma renin activity in varying hydrated states in the bullfrog. *Amer. J. Physiol.* **222**, 142–146.

Sokabe, H., Nishimura, H., Ogawa, M., and Oguri, M. (1970). Determination of renin in the corpuscles of Stannius of the teleost. *Gen. Comp. Endocr.* **14**, 510–516.

Sokabe, H., Ogawa, M., Oguri, M., and Nishimura, H. (1969). Evolution of the juxtaglomerular apparatus in the vertebrate kidneys. *Texas Reports Biol. Med.* **27**, 867–885.

Solcia, E. et al. (1978). Lausanne 1977 classification of gastroenteropancreatic endocrine cells. In *Gut Hormones* (edited by S. R. Bloom), pp. 40–48. London: Churchill Livingstone.

Spallanzani (1784). *Dissertations Relative to the Natural History of Animals and Vegetables 2*. Trans. from the Italian, London. Quoted by F.H.A. Marshall, 1965.

Spanos, E., Colston, K. W., Evans, I.M.S., Galante, L.S., MacAuley, S.J., and MacIntyre, I. (1976a). Effect of prolactin on vitamin D metabolism. *Molecular Cell Endocr.* **5**, 163–167.

Spanos, E., Pike, J. W., Haussler, M. R., Colston, K. W., Evans, I.M.A., Goldner, A. M., McCain, T. A., and MacIntyre, I. (1976*b*). Circulating 1α, 25-dihydroxyvitamin D in the chicken: enhancement by injection of prolactin and during egg laying. *Life Sciences* **19**, 1751–1756.

Specker, J. L. and Moore, F. L. (1980). Annual cycle of plasma androgens and testicular composition in the rough-skinned newt, *Taricha granulosa*. *Gen. Comp. Endocr.* **42**, 297–303.

Speers, G. M., Perey, D.Y.E., and Brown, D. M. (1970). Effect of ultimobranchialectomy in the laying hen. *Endocrinology* **87**, 1292–1297.

Speight, A., Popkin, R., Watts, A. G., and Fink, G. (1980). Oestradiol-17β increases pituitary responsiveness by a mechanism that involves the release and the priming effect of luteinizing hormone releasing factor. *J. Endocr.* **88**, 301–308.

Spencer, E. M. and Tobiassen, O. (1981). The mechanism of the action of growth hormone on vitamin D metabolism in the rat. *Endocrinology* **108**, 1064–1070.

Srivastava, A. K. and Meier, A. H. (1972). Daily variation in concentration of cortisol in plasma in intact and hypophysectomized gulf killifish. *Science* **177**, 185–187.

Stacey, N. E., Cook, A. F., and Peter, R. E. (1979). Ovulatory surge of gonadotropin in the goldfish, *Carassius auratus*. *Gen. Comp. Endocr.* **37**, 246–249.

Stannius, H. (1839). Die Nebennieren bei Knochenfischen. *Arch. Anat. Physiol.* **97**, 97–101.

Steiner, D. F., Kemmler, W., Clark, J. L., Oyer, P. E., and Rubinstein, A. H. (1972). The biosynthesis of insulin. In *Handbook of Physiology*, Sect. 7 *Endocrinology* vol. 1, *Endocrine pancreas*, pp. 175–198. Washington, D.C.: American Physiological Society.

Steiner, D. F., Kemmler, W., Tager, H. S., and Peterson, J. D. (1974). Proteolytic processing in the biosynthesis of insulin and other proteins. *Fedn. Proc.* **33**, 2105–2115.

Steiner, D. F., Quin, P. S., Chan, S. J., Marsh, J., and Tager, H. S. (1980). Processing mechanisms in the biosynthesis of proteins. *Ann. N.Y. Acad. Sci.* **343**, 1–16.

Sterling, K. (1977). The mitochondrial route of thyroid hormone action. *Bull. N.Y. Acad. Med.* **53**, 260–276.

(1979). Thyroid hormone action at the cell level. Parts 1 and 2, *New England J. Med.* **300**, 117–123, 173–177.

Sterling, K., Brenner, M. A., and Sakurada, T. (1980). Rapid effect of triiodothyronine on the mitochondrial pathway in rat liver *in vivo*. *Science* **210**, 340–342.

Sterling, K., Brenner, M. A., and Saldanha, V. F. (1973). Conversion of thyroxine to triiodothyronine by cultured human cells. *Science* **179**, 1000–1001.

Stern, J. M. and Lehrman, D. S. (1969). Role of testosterone in progesterone-induced incubation behaviour in male ring doves (*Streptopelia risoria*). *J. Endocr.* **44**, 13–22.

Stetson, M. H. and Erickson, J. E. (1972). Hormonal control of photoperiodically induced fat deposition in white-crowned sparrows. *Gen. Comp. Endocr.* **19**, 355–362.

Stewart, A. D. (1968). Genetic variation in the neurohypophysial hormones of the mouse. *J. Endocr.* **41**, xix-xx.

(1972). Genetic determination of the storage of vasopressin and oxytocin in neural lobes of mice. *J. Physiol., Lond.* **222**, 157P

(1973). Sensitivity of mice to (8-arginine)- and (8-lysine)-vasopressins as antidiuretic hormones. *J. Endocr.* **59**, 195–196.

Stewart, J., Fraser, R., Papaioannou, V., and Tait, A. (1972). Aldosterone production and the zona glomerulosa: genetic study. *Endocrinology* **90**, 968–972.

Stoff, J. S., Rosa, R., Hallac, R., Silva, P., and Epstein, F. H. (1979). Hormonal regulation of active chloride transport in the dogfish rectal gland. *Amer. J. Physiol.* **237**, F138–F144.

Strauss, J. S. and Ebling, F. J. (1970). Control and function of skin glands in mammals. *Mem. Soc. Endocr.* **18**, 341–368.

Suki, W. N. (1979). Calcium transport in the nephron. *Amer. J. Physiol.* **237**, F1–F6.

Sumpter, J. P., Follett, B. K., Jenkins, N., and Dodd, J. M. (1978). Studies on the purification and properties of gonadotrophin from ventral lobes of the pituitary gland of the dogfish (*Scyliorhinus canicula* L.). *Gen. Comp. Endocr.* **36**, 264–274.

Sutherland, E. W. (1972). Studies on the mechanism of hormone action. *Science* **177**, 401–408.

Suzuki, S., Gorbman, A., Rolland, M., Montfort, M-F., and Lissitzky, S. (1975). Thyroglobulins of cyclostomes and elasmobranch. *Gen. Comp. Endocr.* **26**, 59–69.

Suzuki, S. and Kondo, Y. (1973). Thyroidal morphogenesis and biosynthesis of thyroglobulin before and after metamorphosis in the lamprey, *Lampetra reissneri*. *Gen. Comp. Endocr.* **21**, 451–460.

Swaminathan, R., Bates, R.F.L., and Care, A. R. (1972). Fresh evidence for a physiological role for calcitonin in calcium homeostasis. *J. Endocr.* **54**, 525–526.

Swaminathan, R., Ker, J., and Care, A. D. (1974). Calcitonin and intestinal calcium reabsorption. *J. Endocr.* **61**, 83–94.

Swift, D. R. and Pickford, G. E. (1965). Seasonal variations in the hormone content of the pituitary gland of the perch *Perca fluviatilis* L. *Gen. Comp. Endocr.* **5**, 354–365.

Tager, H. S. and Markese, J. (1979). Intestinal and pancreatic glucagon-like peptides. *J. Biol. Chem.* **254**, 2229–2233.

Tait, J. F. and Tait, S.A.S. (1979). Recent perspectives on the history of the adrenal cortex. *J. Endocr.* **83**, 3p–24p

Takasugi, N. and Bern, H. A. (1962). Experimental studies on the caudal neurosecretory system in *Tilapia mossambica*. *Comp. Biochem. Physiol.* **6**, 289–303.

Takei, Y. (1977). Angiotensin and water intake in the Japanese quail (*Coturnix coturnix japonica*). *Gen. Comp. Endocr.* **31**, 364–372.

Takei, Y., Hirano, T., and Kobayashi, H. (1979). Angiotensin and water intake in the Japanese eel, *Anguilla japonica*. *Gen. Comp. Endocr.* **38**, 466–475.

Talamantes, F. (1975). Comparative study of the occurrence of placental prolactin in mammals. *Gen. Comp. Endocr.* **27**, 115–121.

Talamantes, F., Ogren, I., Markoff, E., Woodard, S., and Madrid, J. (1980). Phylogenetic distribution, regulation of secretion, and prolactin-like effects of placental lactogens. *Fedn. Proc.* **39**, 2582–2587.

Talmage, R. V. and Meyer, R. A. (1976). Physiological role of parathyroid hormone. In *Handbook of Physiology*: Sect., 7, *Endocrinology*, vol. 7, *The Parathyroid Gland*, pp. 343–351. Washington, D.C.: American Physiological Society.

Tamarkin, L., Westrom, W. K., Hamill, A. I., and Goldman, B. D. (1976). Effect of melatonin on the reproductive systems of male and female syrian hamsters: a diurnal rhythm in sensitivity to melatonin. *Endocrinology* **99**, 1534–1541.

Tanabe, Y., Ishii, T., and Tamaki, Y. (1969). Comparison of thyroxine-binding plasma proteins of various vertebrates and their evolutionary aspects. *Gen. Comp. Endocr.* **13**, 14–21.

Tanaka, Y., Frank, H., and DeLuca, H. F. (1973). Intestinal calcium transport: stimulation by low phosphorus diets. *Science* **181**, 564–566.

Tashjian, A. H., Levine, L., and Wilhelmi, A. E. (1965). Immunochemical relatedness of porcine, bovine, ovine and primate growth hormones. *Endocrinology* **77**, 563–573.

Tata, J. R. (1975). How specific are nuclear "receptors" for thyroid hormones? *Nature, Lond.* **257**, 18–23.

——— (1980). The action of growth and developmental hormones. *Biol. Rev.* **55**, 285–319.

Taylor, A. A. (1977). Comparative physiology of the renin-angiotensin system. *Fedn. Proc.* **36**, 1776–1780.

Taylor, J. D. and Bagnara, J. T. (1972). Dermal chromatophores. *Amer. Zool.* **12**, 43–62.

Taylor, R. E., Tu, T., and Barker, S. B. (1967). Thyroxine-like actions of 3′-*iso*propyl-3′,5′′-dibromo-L-thyronine, a potent iodine-free analog. *Endocrinology* **80**, 1143–1147.

Temple, S. A. (1974). Plasma testosterone titers during the annual reproductive cycle of starlings (*Sturnus vulgaris*). *Gen. Comp. Endocr.* **22**, 470–479.

Tepperman, J. and Tepperman, H. M. (1970). Gluconeogenesis, lipogenesis and the Sherringtonian metaphor. *Fedn. Proc.* **29**, 1284–1293.

Tewary, P. D. and Farner, D. S. (1973). Effect of castration and estrogen administration on the plumage pigment of the male house finch (*Carpodacus mexicanus*). *Amer. Zool.* **13**, 1278.

Thapliyal, J. P. and Misra nee Haldar, C. (1979). Effect of pinealectomy on the photoperiodic gonadal response of the Indian garden lizard, *Calotes versicolor.* *Gen. Comp. Endocr.* **39**, 79–86.

Thody, A. J., Cooper, M. F., Bowden, P. E., Meddis, D., and Shuster, S. (1976). Effect of α-melanocyte-stimulating hormone and testosterone on cutaneous and modified sebaceous glands in the rat. *J. Endocr.* **71**, 279–288.

Thomas, D. H. and Phillips, J. G. (1975). Studies in avian adrenal steroid function. 1. Survival and mineral balance following adrenalectomy in domestic ducks (*Anas platyrhynchos* L.). *Gen. Comp. Endocr.* **26**, 394–403.

Thomas, D. H. and Skadhauge, E. (1979). Chronic aldosterone therapy and the control of transepithelial transport of ions and water by the colon and coprodeum of the domestic fowl (*Gallus domesticus*) *in vivo. J. Endocr.* **83**, 239–250.

Thorndyke, M. C. (1978). Evidence for a 'mammalian' thyroglobulin in endostyle of the ascidian *Stylea clava. Nature, Lond.* **271**, 61–62.

Thornton, V. F. (1972). A progesterone-like factor detected by bioassay in the blood of the toad (*Bufo bufo*) shortly before induced ovulation. *Gen. Comp. Endocr.* **18**, 133–139.

Toran-Allerand, C. D. (1978). Gonadal hormones and brain development: cellular aspects of sexual differentiation. *Amer. Zool.* **18**, 553–565.

Torresani, J., Gorbman, A., Lachiver, F., and Lissitzky, S. (1973). Immunological cross-reactivity between thyroglobulins of mammals and reptiles. *Gen. Comp. Endocr.* **21**, 530–535.

Torrey, T. W. (1971). *Morphogenesis of the Vertebrates* (3rd edition), pp. 44–45. New York: Wiley.

Tregear, G. W., Rietschoten, J. V., Greene, E., Keutmann, H. T., Niall, H. D., Reit, B., Parsons, J. A., and Potts, J. T. (1973). Bovine parathyroid hormone: minimum chain length of synthetic peptide required for biological activity. *Endocrinology* **93**, 1349–1353.

Turek, F. W. and Wolfson, A (1978). Lack of an effect of melatonin treatment via silastic capsules on photic-induced gonadal growth and the photorefractory condition in white-throated sparrows. *Gen. Comp. Endocr.* **34**, 471–474.

Turkington, R. W. (1972). Multiple hormonal interactions. The mammary gland. In *Biochemical Actions of Hormones* (edited by G. Litwack), vol II, pp. 55–80. New York: Academic Press.

Tyndale-Biscoe, C. H. and Evans, S. M. (1980). Pituitary-ovarian interactions in marsupials. In *Comparative Physiology: Primitive Mammals* (edited by K. Schmidt-Nielsen, L. Bolis, and C. R. Taylor), pp. 259–268. Cambridge University Press.

Tyndale-Biscoe, H. (1973). *Life of Marsupials.* New York: Elsevier North-Holland.

Urasaki, H. (1972). Effects of restricted photoperiod and melatonin administration on gonadal weight in the Japanese killifish. *J. Endocr.* **55**, 619–620.

Urist, M. R. (1962). The bone-body fluid continuum: calcium and phosphorus in the skeleton and blood of extinct and living vertebrates. *Perspectus Biol. Med.* **6**, 75–115.

—— (1963). The regulation of calcium and other ions in the serums of hagfish and lampreys. *Proc. N.Y. Acad. Sci.* **109**, 294–311.

—— (1973). Testosterone-induced development of limb gills of the lungfish, *Lepidosiren paradoxa. Comp. Biochem. Physiol.* **44A**, 131–135.

—— (1976). Biogenesis of bone: calcium and phosphorus in the skeleton and blood in vertebrate evolution. In *Handbook of Physiology*, Sect. 7, *Endocrinology,* vol. 7, *The Parathyroid Gland*, pp. 183–213. Washington, D.C.: American Physiological Society.

Urist, M. R. and Scheide, A. O. (1961). Partition of calcium and proteins in the blood of oviparous vertebrates during estrus. *J. Gen. Physiol.* **44**, 743–756.

Urist, M. R., Uyeno, S., King, E., Okada, M., and Applegate, S. (1972). Calcium and phosphorus in the skeleton and blood of the lungfish, *Lepidosiren paradoxa*, with comment on humoral factors in calcium homeostasis in the Osteichthyes. *Comp. Biochem. Physiol.* **42A**, 393–408.

Utida, S., Hirano, T., Ando, M., Johnson, D. W., and Bern, H. A. (1972). Hormonal control of the intestine and urinary bladder in teleost osmoregulation. *Gen. Comp. Endocr. Suppl.* **3**, 317–327.

Vaes, G. (1968). On the mechanisms of bone resorption. *J. Cell. Biol.* **39**, 676–697.

Vale, W., Rivier, C., and Brown, M. (1977). Regulatory peptides of the hypothalamus. *Ann. Rev. Physiol.* **39**, 473–527.

Vale, W., Speiss, J., Rivier, C., and Rivier, J. (1981). Characterization of a 41-residue ovine hypothalamic peptide that stimulates secretion of corticotropin and β-endorphin. *Science* **213**, 1394–1397.

Van Herle, A. J., Vassart, G., and Dumont, J. E. (1979). Control of thyroglobulin synthesis and secretion. *New Engl. J. Med.* **301**, 239–249.

van Tienhoven, A. and Planck, R. J. (1973). The effect of light on avian reproductive activity. In *Handbook of Physiology*, Sect. 7 *Endocrinology*, vol. II, *Female reproductive tract* (Pt. 1), pp. 79–107. Washington, D.C.: American Physiological Society.

Varma, M. M. (1977). Ultrastructural evidence for aldosterone and corticosterone-secreting cells in the adrenocortical tissue of the American bullfrog (*Rana catesbeiana*). *Gen. Comp. Endocr.* **33**, 61–75.

Vatlin, H., Sawyer, W. H., and Sokol, H. W. (1965). Neurohypophysial principles in rats homozygous and heterozygous for hypothalamic diabetes insipidus (Brattleboro strain). *Endocrinology* **77**, 701–706.

Vigier, B., Picard, J-Y., and Josso, N. (1982). Monoclonal antibody against bovine anti-Müllerian hormone. *Endocrinology,* **110**, 131–137.

Vigna, S. R. (1979). Distinction between cholecystokinin-like and gastrin-like biological activities extracted from gastrointestinal tissues of some lower vertebrates. *Gen. Comp. Endocr.* **39**, 512–520.

Vinson, G. P. and Kenyon, C. J. (1979). Steroidogenesis in the zones of the mammalian adrenal. In *General, Comparative and Clinical Endocrinology of the Adrenal Cortex* (edited by I. Chester Jones and I. W. Henderson), vol. 2, pp. 201–264. New York: Academic Press.

Vizsolyi, E. and Perks, A. M. (1969). New neurohypophysial principle in foetal mammals. *Nature, Lond.* **223**, 1169–1171.

Vorherr, H., Vorherr, U. F., and Solomon, S. (1978). Contamination of prolactin preparations by antidiuretic hormone and oxytocin. *Amer. J. Physiol.* **234**, F318–F324.

Wachtel, S. S., Koo, G. C., and Boyse, E. A. (1975). Evolutionary conservation of H-Y ('male') antigen. *Nature, Lond.* **254**, 270–272.

Wachtel, S. S., Ohno, S., Koo, G. C., and Boyse, E. A. (1975). Possible role for H-Y antigen in the primary determination of sex. *Nature, Lond.* **257**, 235–236.

Wada, M., Kobayashi, H., and Farner, D. S. (1975). Induction of drinking in the white-crowned sparrow, *Zonotrichia leucophrys gambelii*, by intracranial injection of angiotensin II. *Gen. Comp. Endocr.* **26**, 192–197.

Wade-Smith, J., Richmond, M. E., Mead, R. A., and Taylor, H. (1980). Hormonal and gestational evidence for delayed implantation in the striped skink, *Mephitis mephitis*. *Gen. Comp. Endocr.* **42**, 509–515.

Wahli, W., Dawid, I. B., Ryffel, G. U., and Weber, R. (1981). Vitellogenesis and the vitellogenin gene family. *Science* **212**, 298–304.

Waldo, C. M. and Wislocki, G. B. (1951). Observations on the shedding of the antlers of the virginia deer (*Odocoileus virginianus borealis*). *Amer. J. Anat.* **88**, 351–395.

Walker, J. M., Akil, H., and Watson, S. J. (1980). Evidence for homologous actions of pro-opiocortin products. *Science* **210**, 1247–1249.

Wallis, M. (1975). The molecular evolution of pituitary hormones. *Biol. Rev.* **50**, 35–98.

Waring, H. (1936). Colour changes in the dogfish (*Scyllium canicula*). *Proc. Liverpool Biol. Soc.* **49**, 17–64.

  (1938). Chromatic behaviour of elasmobranchs. *Proc. Roy. Soc. Lond.* **B125**, 264–282.

  (1942). The co-ordination of vertebrate melanophore responses. *Biol. Rev.* **17**, 120–150.

  (1963). *Color Change Mechanisms in Cold-blooded Vertebrates.* London: Academic Press.

Wasserman, R. H., Henion, J. D., Haussler, M. R., and McCain, T. A. (1976). Calcinogenic factor in *Solanum malacoxylon*: evidence that it is 1,25-dihydroxyvitamin D_3-glycoside. *Science* **194**, 853–855.

Wasserman, R. H. and Taylor, A. N. (1976). Gastrointestinal absorption of calcium and phosphorus. In *Handbook of Physiology*, Sect. 7 *Endocrinology*, vol. 7, *The Parathyroid Gland*, pp. 137–155. Washington, D.C.: American Physiological Society.

Watts, E. G., Copp, D. H., and Deftos, L. J. (1975). Changes in plasma calcitonin and calcium during the migration of salmon. *Endocrinology* **96**, 214–218.

Weinstein, B. (1968). On the relationship between glucagon and secretin. *Experientia* **24**, 406–408.

(1972). A generalized homology correlation for various hormones and proteins. *Experientia* **28**, 1517–1522.

Weisinger, R. S., Coghlan, J. P., Denton, D. A., Fan, J.S.K., Hatzikostas, S., McKinley, M. J., Nelson, J. F., and Scoggins, B. A. (1980). ACTH-elicited sodium appetite in sheep. *Amer. J. Physiol.* **239**, E45–E50.

Weiss, G., O'Byrne, E. M., and Steinetz, B. G. (1976). Relaxin: a product of the human corpus luteum of pregnancy. *Science* **194**, 948–949.

Weiss, M. (1980). Adrenocorticosteroids in prototherian, metatherian and eutherian mammals. In *Comparative physiology: primitive mammals* (edited by K. Schmidt-Nielsen, L. Bolis, and C. R. Taylor), pp. 285–96. Cambridge University Press.

Weiss, M. and McDonald, I. R. (1965). Corticosteroid secretion in the monotreme *Tachyglossus aculeatus*. *J. Endocr.* **33**, 203–210.

Wenberg, G. M. and Holland, J. C. (1973). The circannual variations of some of the hormones of the woodchuck (*Marmota monax*). *Comp. Biochem. Physiol.* **46A**, 523–535.

Wendelaar Bonga, S. E. (1978). The effects of changes in external sodium, calcium, and magnesium concentrations on prolactin cells, skin, and plasma electrolytes of *Gasterosteus aculeatus*. *Gen. Comp. Endocr.* **34**, 265–275.

Wendelaar Bonga, S. E. and Van der Meij, J.C.A. (1980). The effect of ambient calcium on prolactin cell activity and plasma electrolytes in *Sarotherodin mossambicus* (*Tilapia mossambica*). *Gen. Comp. Endocr.* **40**, 391–401.

West, G. B. (1955). The comparative pharmacology of the suprarenal medulla. *Quart. Rev. Biol.* **30**, 116–137.

Westerfield, D. B., Pang, P.K.T., and Burns, J. M. (1980). Some characteristics of melanophore-concentrating hormone (MCH) from teleost pituitary glands. *Gen. Comp. Endocr.* **42**, 494–499.

Wildt, L., Hausler, A., Hutchison, J. S., Marshall, G., and Knobil, E. (1981). Estradiol as a gonadotropin releasing hormone in the rhesus monkey. *Endocrinology* **108**, 2011–2013.

Williams, L. T., Lefkowitz, R. J., Watanabe, A. M., Hathaway, D. R., and Besch, H. R. (1977). Thyroid hormone regulation of β-adrenergic receptor number. *J. Biol. Chem.* **252**, 2787–2789.

Wilson, J. D., George, F. W., and Griffin, J. E. (1981). The hormonal control of sexual development. *Science* **211**, 1278–1284.

Wilson, J. F. and Dodd, J. M. (1973*a*). The role of the pineal complex and lateral eyes in the colour change response of the dogfish, *Scyliorhinus canicula* L. *J. Endocr.* **58**, 591–598.

(1973*b*). The role of melonophore-stimulating hormone in melanogenesis in the dogfish, *Scyliorhinus canicula* L. *J. Endocr.* **58**, 685–686.

Wilson, P. W. and Lawson, D.E.M. (1981). Vitamin D-dependent phosphorylation of an intestinal protein. *Nature, Lond.* **289**, 600–602.

Wilson, S. C. and Cunningham, F. J. (1980). Effects of increasing day length and intermittent lighting schedules in the domestic hen on plasma concentrations of luteinizing hormone (LH) and the LH response to exogenous progesterone. *Gen. Comp. Endocr.* **41**, 546–553.

Wingstrand, K. G. (1951). *The Structure and Development of the Avian Pituitary*. Lund: C.W.K. Gleerup.

(1966). Comparative anatomy and evolution of the hypophysis. In *The Pituitary Gland* (edited by G. W. Harris and B. T. Donovan), vol. 1, pp. 58–146. Berkeley: University of California Press.

Wittle, L. W. and Dent, J. N. (1979). Effects of parathyroidectomy and of parathyroid extract on levels of calcium and phosphate in the blood and urine of the red-spotted newt. *Gen. Comp. Endocr.* **37**, 428–439.

Wong, E. and Flux, D. S. (1962). The oestrogenic activity of red clover isoflavones and some of their degradation products. *J. Endocr.* **24**, 341–348.

Wong, G. L., Luben, R. A., and Cohn, D. V. (1977). 1,25-dihydroxycholecalciferol and parathormone: effects on isolated osteoclast-like and osteoblast-like cells. *Science* **197**, 663–665.

Woodhead, P.M.J. (1969). Effect of oestradiol and thyroxine upon the plasma calcium content of a shark, *Scyliorhinus canicula*. *Gen. Comp. Endocr.* **13**, 310–312.

Woolley, P. (1957). Colour change in a chelonian. *Nature, Lond.* **179**, 1255–1256.

Wright, A., Chester Jones, I., and Phillips, J. G. (1957). The histology of the adrenal gland of prototheria. *J. Endocr.* **15**, 100–107.

Wurtman, R. J., Axelrod, J., and Kelly, D. E. (1968). *The Pineal.* New York: Academic Press.

Xavier, F. (1974). La pseudogestation chez *Nectophyrnoïdes occidentalis* ANGEL. *Gen. Comp. Endocr.* **22**, 98–115.

Xavier, F. and Ozon, R. (1971). Recherches sur l'activité endocrine de l'ovaire de *Nectophrynoïdes occidentalis* ANGEL (amphibien anoure vivipare). ii. Synthèse *in vitro* de stéroids. *Gen. Comp. Endocr.* **16**, 30–40.

Yagil, R., Etzion, Z., and Berlyne, G. M. (1973). The effect of *d*-aldosterone and spironolactone on the concentration of sodium and potassium in the milk of rats. *J. Endocr.* **59**, 633–636.

Yamashita, K., Mieno, M., Shimizu, T., and Yamashita, Er. (1978). Inhibition by melatonin of the pituitary response to luteinizing hormone releasing hormone *in vivo*. *J. Endocr.* **76**, 487–491.

Yamauchi, H., Matsuo, M., Yoshida, A., and Orimo, H. (1978). Effect of eel calcitonin on serum electrolytes in the eel *Anguilla japonica*. *Gen. Comp. Endocr.* **34**, 343–346.

Yaron, Z. (1972). Endocrine aspects of gestation in viviparous snakes. *Gen. Comp. Endocr. Suppl.* **3**, 663–673.

Yorio, T. and Bentley, P. J. (1977). Asymmetrical permeability of the integument of tree frogs (HYLIDAE). *J. Exp. Biol.* **76**, 197–204.

Young J. Z. (1935). The photoreceptors of lampreys. ii. The function of the pineal complex. *J. Exp. Biol.* **12**, 254–270.

Yu, J.Y.L., Dickoff, W. W., Swanson, P., and Gorbman, A. (1981). Vitellogenesis and its hormonal regulation in the Pacific hagfish, *Eptatretus stouti* L. *Gen. Comp. Endocr.* **43**, 492–502.

Zadunaisky, J. A. and Degnan, K. J. (1980). Chloride active transport and osmoregulation. In *Epithelial Transport in Lower Vertebrates* (edited by B. Lahlou), pp. 185–196. Cambridge University Press.

Zelnik, P. R., Hornsey, D. J., and Hardisty, M. W. (1977). Insulin and glucagon-like immunoreactivity in the river lamprey (*Lampetra fluviatilis*). *Gen. Comp. Endocr.* **33**, 55–60.

Zelnik, P. R. and Lederis, K. (1973). Chromatographic separation of urotensins. *Gen. Comp. Endocr.* **20**, 392–400.

Zimmerman, E. A., Carmel, P. W., Husain, M. K., Ferin, M., Tannenbaum, M., Frantz, A. G., and Robison, A. G. (1973). Vasopressin and neurophysin: high concentrations in monkey hypophyseal portal blood. *Science* **182**, 925–927.

Zinder, O., Hamosh, M., Fleck, T.R.C., and Scow, R. O. (1974). Effect of prolactin on lipoprotein lipase in mammary gland and adipose tissue of rats. *Amer. J. Physiol.* **226**, 744–748.

Zipser, R. D., Licht, P., and Bern, H. A. (1969). Comparative effects of mammalian prolactin and growth hormone on growth in the toads *Bufo boreas* and *Bufo marinus*. *Gen. Comp. Endocr.* **13**, 382–391.

Zuber-Vogeli, M. and Xavier, F. (1973). Les modifications cytologique de l'hypophyse distale des femelles de *Nectophrynoïdes occidentalis* Angel après ovariectomie. *Gen. Comp. Endocr.* **20**, 199–213.

Index

[474]